SI PREFIXES

Fraction	Prefix	Symbol	Multiple	Prefix	Symbol
10^{-1}	deci	d	1		
10^{-2}	centi	c	1		
10^{-3}	milli	m	1		
10^{-6}	micro	μ	1		
10^{-9}	nano	n	10	giga	
10^{-12}	pico	p	10^{12}	tera	T
10^{-15}	femto	f	10^{15}	peta	P
10^{-18}	atto	a	10^{18}	exa	E

GREEK ALPHABET

Alpha	A	α	Iota	I	ι	Rho	P	ρ
Beta	B	β	Kappa	K	κ	Sigma	Σ	σ
Gamma	Γ	γ	Lambda	Λ	λ	Tau	T	τ
Delta	Δ	δ	Mu	M	μ	Upsilon	Υ	υ
Epsilon	E	ϵ, ε	Nu	N	ν	Phi	Φ	ϕ
Zeta	Z	ζ	Xi	Ξ	ξ	Chi	X	χ
Eta	H	η	Omicron	O	o	Psi	Ψ	ψ
Theta	Θ	θ	Pi	$\prod$	π	Omega	Ω	ω

SOME COMMONLY USED NON-SI UNITS

Unit	Quantity	Symbol	SI value
Angstrom	length	Å	10^{-10} m = 100 pm
Calorie	energy	cal	4.184 J (defined)
Debye	dipole moment	D	3.3356×10^{-30} C·m
			0.39345 au
Gauss	magnetic field strength	G	10^{-4} T

E_h	cm^{-1}	Hz
$2.293\,710 \times 10^{17}$	$5.034\,11 \times 10^{22}$	$1.509\,189 \times 10^{33}$
$3.808\,798 \times 10^{-4}$	83.5935	$2.506\,069 \times 10^{12}$
$3.674\,931 \times 10^{-2}$	8065.54	$2.417\,988 \times 10^{14}$
1	$2.194\,7463 \times 10^{5}$	$6.579\,684 \times 10^{15}$
$4.556\,355 \times 10^{-6}$	1	$2.997\,925 \times 10^{10}$
$1.519\,830 \times 10^{-16}$	$3.335\,64 \times 10^{-11}$	1

QUANTUM CHEMISTRY

SECOND EDITION

QUANTUM CHEMISTRY

SECOND EDITION

Donald A. McQuarrie

DEPARTMENT OF CHEMISTRY

UNIVERSITY OF CALIFORNIA, DAVIS

UNIVERSITY SCIENCE BOOKS

Sausalito, California

University Science Books
www.uscibooks.com

Production Manager: *Jennifer Uhlich at Wilsted and Taylor*
Manuscript Editor: *John Murdzek*
Proofreader: *Jennifer McClain*
Design: *Yvonne Tsang at Wilsted and Taylor*
Illustrator: *Mervin Hanson*
Compositor: *Windfall Software, using ZzT$_E$X*
Printer & Binder: *Edwards Brothers, Inc.*

This book is printed on acid-free paper.

Library of Congress Cataloging-in-Publication Data

McQuarrie, Donald A. (Donald Allan)
   Quantum chemistry / Donald A. McQuarrie.—2nd ed.
     p. cm.
   Includes index.
   ISBN 978-1-891389-50-4 (alk. paper)
Quantum chemistry.   I. Title.
   QD462.M4 2007
   541′.28—dc22

                                 2007023879

Printed in the United States of America
10 9 8 7 6 5 4 3 2 1

Contents

Preface to the Second Edition

The first edition of this book was written in the early 1980s. At that time molecular calculations were pretty much in the province of professional quantum chemists. An enormous change has occurred since that time. The explosive growth and availability of computer power has placed in the hands of undergraduate students the ability to carry out molecular calculations routinely that were unimaginable twenty years ago. This new edition incorporates this ability by discussing and encouraging the use of quantum chemistry programs such as Gaussian and WebMO, which most chemistry departments have access to. Not only can undergraduates do quantum chemical calculations nowadays, there is even a program in North Carolina, North Carolina High School Computational Chemistry Server (*http://chemistry.ncssm.edu*), that encourages high school students to do so.

In addition to these quantum chemistry programs, there are a number of general mathematical programs such as *MathCad* or *Mathematica* that make it easy to do calculations routinely that were formerly a drudgery. These programs not only perform numerical calculations, but they can also perform algebraic manipulations as well. They are relatively easy to learn and use and every serious scientific student should know how to use one of them. They allow you to focus on the underlying physical ideas and free you from getting bogged down in algebra. They also allow you to explore the properties of equations by varying parameters and plotting the results. There are a number of problems in this edition that require the use of one of these programs.

Another product of the computer revolution is the availability of so much material on-line. We refer to a number of websites throughout the chapters, but one that is particularly useful is the Computational Chemistry Comparison and Benchmark Data Base (*http://srdata.nist.gov/cccbdb*) maintained by the National Institute of Science and Technology (NIST). This website lists numerical results of quantum chemical calculations for hundreds of molecules using a great variety of computational methods. It also has an excellent tutorial that discusses a number of topics that are not treated in this book. I have utilized this website a great deal in Chapter 12 , which treats ab initio molecular orbital theory. If a student can navigate around this website and understand, or at least appreciate, most of the material presented in it, then I will consider this book to have been successful. Websites have the distressing property of disappearing, and so

I have usually included only websites that are government sponsored, but even these websites change their addresses every so often. I checked every website that I refer to just before the book went to press, but if you have difficulty finding one of them, putting the topic into Google seems to work.

The early chapters of this revision do not differ significantly from the first edition. They have been well received and constitute a rather timeless introduction to basic quantum mechanics. One small addition, however, is the introduction of the Dirac bracket notation for state functions and integrals, which is used freely throughout the remainder of the book. Rather than devote a single chapter to molecular spectroscopy, I have included it in Chapter 5 (The Harmonic Oscillator and Vibrational Spectroscopy) and Chapter 6 (The Rigid Rotator and Rotational Spectroscopy). Chapter 7 (The Hydrogen Atom) discusses the hydrogen atomic orbitals as the solutions to the Schrödinger equation for this system, and also uses the results of the Stern–Gerlach experiment and the fine structure of the spectrum of atomic hydrogen to motivate the introduction of electron spin. Chapters 8 and 9 (Approximation Methods and Many-Electron Atoms, respectively) are not too different from the earlier edition, except that a little more emphasis is placed on the Hartree–Fock method. Chapter 9 has an appendix that actually carries out a Hartree–Fock calculation for a helium atom step by step. Chapter 10 (The Chemical Bond: One- and Two-Electron Molecules) is a fairly detailed discussion of the bonding in H_2^+ and H_2, and we utilize these simple systems to introduce many of the techniques that are used in modern molecular calculations. The last section of the chapter carries out a minimal basis set Hartree–Fock–Roothaan calculation for H_2 step by step. Once a student carries through such a calculation for a two-electron system, calculations on larger molecules should pose no conceptual difficulties. Chapter 11 is a standard discussion of qualitative molecular orbital theory, molecular term symbols, and π-electron molecular orbital theory. The final chapter (The Hartree–Fock–Roothaan Method) introduces the use of basis sets consisting of Gaussian functions in modern molecular calculations and the use of computational chemistry programs such as Gaussian and WebMO. One goal of the chapter, and the book itself for that matter, is for a student to be comfortable in carrying out a Hartree–Fock calculation for a given basis set. Much of Chapter 12 is built around the NIST Computational Chemistry Comparison and Benchmark Data Base website that I mentioned previously.

As with the first edition, the mathematical background required of the students is one year of calculus, with no knowledge of differential equations. All the necessary mathematical techniques are developed in the text through a number of short units called MathChapters. These units are self-contained and present just enough material to give a student the ability and the confidence to use the techniques in subsequent chapters. The point of these units is to present the mathematics before it is required so that a student can focus more on the physical principles involved rather than on the mathematics. There are MathChapters on complex numbers, probability and statistics, vectors, series and limits, spherical coordinates, determinants, and matrices. Most of current computational chemistry is formulated in terms of matrices, and I have used matrix notation in a number of places, particularly toward the end of the book.

No one can learn this material (nor any thing else in the physical sciences for that matter) without doing lots of problems. For this reason, I have provided about

50 problems at the end of each chapter. These problems range from filling in gaps to extending the material presented in the chapter, but most illustrate applications of the material. All told, there are over 600 problems in the book. I have provided answers to many of them at the back of the book. In addition, Helen Leung and Mark Marshall of Amherst College have written a Solutions Manual in which the complete solution to every problem is given.

A singular feature of the book is the inclusion of biographies at the beginning of each chapter. I wish to thank my publisher for encouraging me to include them and my wife, Carole, for researching the material for them and writing every one of them. Each one could easily have been several pages long and it was difficult to cut them down to one page.

You read in many prefaces that "this book could not have been written and produced without the help of many people," and it is definitely true. I am particularly grateful to my reviewers, Bill Fink of UC Davis, Scott Feller of Wabash College, Atilla Szabo of NIH, Will Polik of Hope College, Helen Leung and Mark Marshall of Amherst College, and Mervin Hansen of Humboldt State University, who slogged through numerous drafts of chapters and who made many great suggestions. I also wish to give special thanks to Gaussian, Inc., who gave me a copy of Gaussian 03 to use in the preparation of the manuscript and to Will Polik, who set me up to use WebMO. I also wish to thank Christine Taylor and her crew at Wilsted & Taylor Publishing Services and particularly Jennifer Uhlich for transforming a pile of manuscript pages into a beautiful-looking and inviting book without a hitch, Jennifer McClain for doing a superb job of proofreading, Jane Ellis for dealing with many of the production details and procuring all the photographs for the biographies, Mervin Hanson for rendering hundreds of figures in Mathematica and keeping them all straight in spite of countless alterations, John Murdzek for a helpful copyediting, Paul Anagnostopoulos for composing the entire book, and my publisher Bruce Armbruster and his wife and associate Kathy for being the best publishers around and good friends in addition. Finally, I wish to thank my wife, Carole, for preparing the manuscript in LaTeX, for reading the entire manuscript, and for being my best critic in general (in all things).

There are bound to be both typographical and conceptual errors in the book and I would appreciate your letting me know about them so that they can be corrected in subsequent printings. I also would welcome general comments, questions, and suggestions at *mquarrie@mcn.org*, or through the University Science Books website *www.uscibooks.com*, where any ancillary material or notices will be posted.

Max Planck was born in Kiel, Germany (then Prussia) on April 23, 1858, and died in 1948. He showed early talent in both music and science. He received his Ph.D. in theoretical physics in 1879 at the University of Munich for his dissertation on the second law of thermodynamics. He joined the faculty of the University of Kiel in 1885, and in 1888 he was appointed director of the Institute of Theoretical Physics, which was formed for him at the University of Berlin, where he remained until 1926. His application of thermodynamics to physical chemistry won him an early international reputation. Planck was president of the Kaiser Wilhelm Society, later renamed the Max Planck Society, from 1930 until 1937, when he was forced to retire by the Nazi government. Planck is known as the father of the quantum theory because of his theoretical work on blackbody radiation at the end of the 1890s, during which time he introduced a quantum hypothesis to achieve agreement between his theoretical equations, which were based solely on the second law of thermodynamics, contrary to most popular accounts, and experimental data. He maintained his interest in thermodynamics throughout his long career in physics. Planck was awarded the Nobel Prize in Physics in 1918 "in recognition of services he rendered to the advancement of physics by his discovery of energy quanta." Planck's personal life was clouded by tragedy. His two daughters died in childbirth, one son died in World War I, and another son was executed in World War II for his part in an assassination attempt on Hitler in 1944.

CHAPTER 1

The Dawn of the Quantum Theory

Toward the end of the nineteenth century, many scientists believed that all the fundamental discoveries of science had been made and little remained but to clear up a few minor problems and to improve experimental methods to measure physical results to a greater number of decimal places. This attitude was somewhat justified by the great advances that had been made up to that time. Chemists had finally solved the seemingly insurmountable problem of assigning a self-consistent set of atomic masses to the elements. Stanislao Cannizzaro's concept of the molecule, while initially controversial, was finally widely accepted. The great work of Dmitri Mendeleev had resulted in a periodic table of the elements, although the underlying reasons that such periodic behavior occurred in nature were not understood. Friedrich Kekulé had solved the controversy concerning the structure of benzene. The fundamentals of chemical reactions had been elucidated by Svante Arrhenius, and the remaining work seemed to consist primarily of cataloging the various types of chemical reactions.

In the related field of physics, Newtonian mechanics had been extended by Joseph-Louis Lagrange and Sir William Hamilton. The resulting theory was applied to planetary motion and could also explain other complicated natural phenomena such as elasticity and hydrodynamics. Count Rumford and James Joule had demonstrated the equivalence of heat and work, and investigations by Sadi Carnot resulted in the formulation of what is now entropy and the second law of thermodynamics. This work was followed by Josiah Gibbs's complete development of the field of thermodynamics. In fact, Gibbs's treatment of thermodynamics is so relevant to chemistry that it is taught in a form that is essentially unchanged from Gibbs's original formulation. Shortly, scientists would discover that the laws of physics were also relevant to the understanding of chemical systems. The interface between these two seemingly unrelated disciplines formed the modern field of physical chemistry.

The related fields of optics and electromagnetic theory were undergoing similar maturation. The nineteenth century witnessed a continuing controversy as to whether light was wavelike or particle-like. Many diverse and important observations were unified by James Clerk Maxwell in a series of deceptively simple-looking equations that bear his name. Not only did Maxwell's predictions of the electromagnetic behavior of light unify the fields of optics with electricity and magnetism, but their subsequent experimental demonstration by Heinrich Hertz in 1887 appeared to finally demonstrate that light was wavelike. The implications of these fields to chemistry would not be appreciated for several decades, but are now important aspects of the discipline of physical chemistry, particularly in spectroscopy.

The body of these accomplishments in physics is considered the development of what we now call *classical physics*. Little did scientists realize in that justifiably heady era of success that the fundamental tenets of how the physical world works were to be shortly overturned. Fantastic discoveries not only were about to revolutionize physics, chemistry, biology, and engineering, but would have significant effects on technology and politics as well. The early twentieth century saw the birth of the theory of relativity and quantum mechanics. The first, due to the work of Albert Einstein alone, which completely altered scientists' ideas of space and time, was an extension of the classical ideas to include high velocities and astronomical distances. Quantum mechanics, the extension of classical ideas into the behavior of subatomic, atomic, and molecular species, on the other hand, resulted from the efforts of many creative scientists over several decades. To date, the effect of relativity on chemical systems has been limited. Although it is important in understanding electronic properties of heavy atoms, it does not play much of a role in molecular structure and reactivity and so is not generally taught in physical chemistry. Quantum mechanics, however, forms the foundation upon which all of chemistry is built. Our current understanding of atomic structure and molecular bonding is cast in terms of the fundamental principles of quantum mechanics, and no understanding of chemical systems is possible without knowing the basics of this current theory of matter.

Great changes in science are spurred by observations and new creative ideas. Let's go back to the complacent final years of the nineteenth century to see just what were the events that so shook the world of science.

1.1 Blackbody Radiation Could Not Be Explained by Classical Physics

The series of experiments that revolutionized the concepts of physics was concerned with the radiation given off by material bodies when they are heated. We all know, for instance, that when the burner of an electric stove is heated, it first turns a dull red and progressively becomes redder as the temperature increases. We also know that as a body is heated even further, the radiation becomes white and then blue as the temperature continues to increase. Thus, we see that there is a continual shift of the color of a heated body from red through white to blue as the body is heated to higher temperatures. In terms of frequency, the radiation emitted goes from a lower frequency to a higher frequency as the temperature increases, because red is in a lower frequency region of the spectrum than is blue. The exact frequency spectrum emitted by the body depends on the particular body itself, but an *ideal body*, which absorbs and emits all frequencies, is called a *blackbody* and serves as an idealization for any radiating material. The radiation emitted by a blackbody is called *blackbody radiation*.

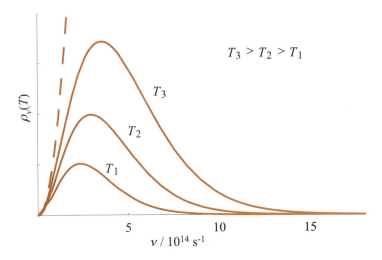

FIGURE 1.1
Spectral distribution of the intensity of blackbody radiation as a function of frequency for several temperatures. The intensity is given in arbitrary units. The dashed line is the prediction of classical physics. As the temperature increases, the maximum shifts to higher frequencies and the total radiated energy (the area under each curve) increases sharply. Note that the horizontal axis is labeled as $\nu/10^{14}$ s^{-1}. This notation means that the dimensionless numbers on that axis are frequencies divided by 10^{14} s^{-1}. We shall use this notation to label columns in tables and axes in figures because of its unambiguous nature and algebraic convenience.

A plot of the intensity of blackbody radiation versus frequency for several temperatures is given in Figure 1.1. Many theoretical physicists tried to derive expressions consistent with these experimental curves of intensity versus frequency, but they were all unsuccessful. In fact, the expression that is derived according to the laws of nineteenth century physics is

$$d\rho(\nu, T) = \rho_\nu(T)d\nu = \frac{8\pi k_B T}{c^3}\nu^2 d\nu \tag{1.1}$$

where $\rho_\nu(T)d\nu$ is the radiant energy density between the frequencies ν and $\nu + d\nu$ and has units of joules per cubic meter (J·m^{-3}). In Equation 1.1, T is the kelvin temperature, and c is the speed of light. The quantity k_B is called the *Boltzmann constant* and is equal to the ideal gas constant R divided by the Avogadro constant (formerly called Avogadro's number). The units of k_B are J·K^{-1}·particle^{-1}, but particle^{-1} is usually not expressed. (Another case is the Avogadro constant, 6.022×10^{23} particle·mol^{-1}, which we will write as 6.022×10^{23} mol^{-1}; the unit "particle" is not expressed.) Equation 1.1 came from the work of Lord Rayleigh and J. H. Jeans and is called the *Rayleigh–Jeans law*. The dashed line in Figure 1.1 shows the prediction of the Rayleigh–Jeans law. Note that the Rayleigh–Jeans law reproduces the experimental data at low frequencies. At high frequencies, however, the Rayleigh–Jeans law predicts that the radiant energy density diverges as ν^2. Because the frequency increases as the radiation enters the ultraviolet region, this divergence was termed the *ultraviolet catastrophe*, a phenomenon that classical physics could not reconcile theoretically. This was the first such failure to explain an important naturally occurring phenomenon and therefore is of great historical interest. Rayleigh and Jeans did not simply make a mistake or misapply some of the ideas of physics; many other people reproduced the equation of Rayleigh and Jeans, showing that this equation was correct according to the physics of the time. This result was very disconcerting, and many people struggled to find a theoretical explanation of blackbody radiation.

1.2 Planck Used a Quantum Hypothesis to Derive the Blackbody Radiation Law

The first person to offer a successful explanation of blackbody radiation was the German physicist Max Planck in 1900. Like Rayleigh and Jeans before him, Planck assumed that the radiation emitted by the blackbody was caused by the oscillations of the electrons in the constituent particles of the material body. These electrons were pictured as oscillating in an atom much like electrons oscillate in an antenna to give off radio waves. In these "atomic antennae," however, the oscillations occur at a much higher frequency; hence, we find frequencies in the visible, infrared, and ultraviolet regions rather than in the radio-wave region of the spectrum. Implicit in the derivation of Rayleigh and Jeans is the assumption that the energies of the electronic oscillators responsible for the emission of the radiation could have any value whatsoever. This assumption is one of the basic assumptions of classical physics. In classical physics, the variables that represent observables (such as position, momentum, and energy) can take on a continuum of values. Planck had the great insight to realize that he had to break away from this mode of thinking to derive an expression that would reproduce experimental data such as those shown in Figure 1.1. He made the revolutionary assumption that the energies of the oscillators were discrete and had to be proportional to an integral multiple of the frequency or, in equation form, that $E = nh\nu$, where E is the energy of an oscillator, n is an integer, h is a proportionality constant, and ν is the frequency. Using this quantization of energy and some statistical thermodynamic ideas, Planck derived the equation

$$d\rho(\nu, T) = \rho_\nu(T)d\nu = \frac{8\pi h}{c^3} \frac{\nu^3 d\nu}{e^{h\nu/k_B T} - 1} \tag{1.2}$$

All the symbols except h in Equation 1.2 have the same meaning as in Equation 1.1. The only undetermined constant in Equation 1.2 is h. Planck showed that this equation gives excellent agreement with the experimental data for all frequencies and temperatures if h has the value 6.626×10^{-34} joule·seconds (J·s). This constant is now one of the most famous and fundamental constants of physics and is called the *Planck constant*. Equation 1.2 is known as the *Planck distribution law for blackbody radiation*. For small frequencies, Equations 1.1 and 1.2 become identical (Problem 1–4), but the Planck distribution does not diverge at large frequencies and, in fact, looks like the curves in Figure 1.1.

EXAMPLE 1–1

Show that $\rho_\nu(T)d\nu$ in both Equations 1.1 and 1.2 has units of energy per unit volume, $J \cdot m^{-3}$.

SOLUTION: The units of T are K, of k_B are $J \cdot K^{-1}$, of ν and $d\nu$ are s^{-1}, and of c are $m \cdot s^{-1}$. Therefore, for the Rayleigh–Jeans law (Equation 1.1),

$$d\rho(\nu, T) = \rho_\nu(T)d\nu = \frac{8\pi k_B T}{c^3} \nu^2 d\nu$$

$$\sim \frac{(J \cdot K^{-1})(K)}{(m \cdot s^{-1})^3}(s^{-1})^2(s^{-1}) = J \cdot m^{-3}$$

For the Planck distribution (Equation 1.2),

$$d\rho(\nu, T) = \rho_\nu(T)d\nu = \frac{8\pi h}{c^3} \frac{\nu^3 d\nu}{e^{h\nu/k_B T} - 1}$$

$$\sim \frac{(J \cdot s)(s^{-1})^3(s^{-1})}{(m \cdot s^{-1})^3} = J \cdot m^{-3}$$

Thus, we see that $\rho_\nu(T)d\nu$, the radiant energy density, has units of energy per unit volume.

Equation 1.2 expresses Planck's radiation law in terms of frequency. Because wavelength (λ) and frequency (ν) are related by $\lambda\nu = c$, then $d\nu = -cd\lambda/\lambda^2$, and we can express Planck's radiation law in terms of wavelength rather than frequency (Problem 1–12):

$$d\rho(\lambda, T) = \rho_\lambda(T)d\lambda = \frac{8\pi hc}{\lambda^5} \frac{d\lambda}{e^{hc/\lambda k_B T} - 1} \tag{1.3}$$

The quantity $\rho_\lambda(T)d\lambda$ is the radiant energy density between λ and $\lambda + d\lambda$. Equation 1.3 is plotted in Figure 1.2 for several values of T.

We can use Equation 1.3 to justify an empirical relationship known as the *Wien displacement law*. The Wien displacement law says that if λ_{max} is the wavelength at which $\rho_\lambda(T)$ is a maximum, then

$$\lambda_{max} T = 2.90 \times 10^{-3} \, m \cdot K \tag{1.4}$$

By differentiating $\rho_\lambda(T)$ with respect to λ, we can show (Problem 1–5) that

$$\lambda_{max} T = \frac{hc}{4.965 k_B} \tag{1.5}$$

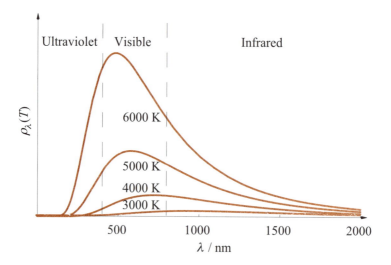

FIGURE 1.2
The distribution of the intensity of the radiation emitted by a blackbody versus wavelength for
various temperatures. As the temperature increases, the total radiation emitted (the area under
the curve) increases.

in accord with the Wien displacement law. Using the modern values of h, c, and k_B given
inside the front cover, we obtain 2.898×10^{-3} m·K for the right side of Equation 1.5,
in excellent agreement with the experimental value given in Equation 1.4.

The theory of blackbody radiation is used regularly in astronomy to estimate the
surface temperatures of stars. Figure 1.3 shows the electromagnetic spectrum of the sun
measured at the earth's upper atmosphere. A comparison of Figure 1.3 with Figure 1.2
suggests that the solar spectrum can be described by a blackbody at approximately
6000 K. If we estimate λ_{max} from Figure 1.3 to be 500 nm, then the Wien displacement
law (Equation 1.4) gives the temperature of the surface of the sun to be

$$T = \frac{2.90 \times 10^{-3}\,\text{m·K}}{500 \times 10^{-9}\,\text{m}} = 5800\text{ K}$$

The star Sirius, which appears blue, has a surface temperature of about 11 000 K (cf.
Problem 1–7).

Equation 1.2 can be used to derive another law that was known at the time. It can
be shown by thermodynamic arguments that the total energy radiated per square meter
per unit time from a blackbody is given by

$$R = \frac{c}{4}E_V = \sigma T^4 \tag{1.6}$$

where E_V is the total radiation energy density. Equation 1.6 is known as the *Stefan–
Boltzmann law* and σ is known as the *Stefan–Boltzmann constant*. The experimental

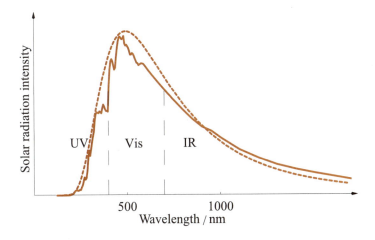

FIGURE 1.3
The electromagnetic spectrum of the sun as measured in the upper atmosphere of the earth. A comparison of this figure with Figure 1.2 shows that the sun's surface radiates as a blackbody at a temperature of about 6000 K (dashed line).

value of σ is 5.6697×10^{-8} J·m^{-2}·K^{-4}·s^{-1}. Note that the units of σ are consistent with Equation 1.6.

EXAMPLE 1–2
Planck's distribution of blackbody radiation gives the energy density between ν and $\nu + d\nu$. Integrate the Planck distribution over all frequencies and compare the result to Equation 1.6.

SOLUTION: The integral of Equation 1.2 over all frequencies is

$$E_V = \int_0^\infty \rho(\nu, T)\, d\nu = \frac{8\pi h}{c^3} \int_0^\infty \frac{\nu^3 d\nu}{e^{h\nu/k_\text{B}T} - 1} \tag{1.7}$$

If we use the fact that

$$\int_0^\infty \frac{x^3 dx}{e^x - 1} = \frac{\pi^4}{15}$$

then we obtain

$$E_V = \frac{8\pi h}{c^3} \left(\frac{k_\text{B}T}{h} \right)^4 \int_0^\infty \frac{x^3 dx}{e^x - 1}$$

$$= \frac{8\pi^5 k_\text{B}^4 T^4}{15 h^3 c^3} \tag{1.8}$$

By comparing this result to the Stefan–Boltzmann law (Equation 1.6), we see that

$$\sigma = \frac{2\pi^5 k_B^4}{15 h^3 c^2} \tag{1.9}$$

Using the values of k_B, h, and c given inside the front cover, the calculated value of σ is 5.670×10^{-8} J·m^{-2}·K^{-4}·s^{-1}, in excellent agreement with the experimental value. Certainly, Planck's derivation of the blackbody distribution law was an impressive feat. Nevertheless, Planck's derivation and, in particular, Planck's assumption that the energies of the oscillators have to be an integral multiple of $h\nu$ was not accepted by most physicists at the time and was considered to be simply an ad hoc derivation. It was felt that in time a satisfactory classical derivation would be found. In a sense, Planck's derivation was little more than a curiosity. Just a few years later, however, in 1905, Einstein used the very same idea to explain the *photoelectric effect*.

1.3 Einstein Explained the Photoelectric Effect with a Quantum Hypothesis

In 1886 and 1887, while carrying out the experiments that supported Maxwell's theory of the electromagnetic nature of light, the German physicist Heinrich Hertz discovered that ultraviolet light causes electrons to be emitted from a metallic surface. The ejection of electrons from the surface of a metal by radiation is called the *photoelectric effect*. Two experimental observations of the photoelectric effect are in stark contrast with the classical wave theory of light. According to classical physics, electromagnetic radiation is an electric field oscillating perpendicular to its direction of propagation, and the intensity of the radiation is proportional to the square of the amplitude of the electric field. As the intensity increases, so does the amplitude of the oscillating electric field. The electrons at the surface of the metal should oscillate along with the field and so, as the intensity (amplitude) increases, the electrons oscillate more violently and eventually break away from the surface with a kinetic energy that depends on the amplitude (intensity) of the field. This classical picture is in complete disagreement with the experimental observations. Experimentally, the kinetic energy of the ejected electrons is independent of the intensity of the incident radiation. Furthermore, the classical picture predicts that the photoelectric effect should occur for any frequency of light as long as the intensity is sufficiently high. The experimental fact, however, is that there is a *threshold frequency*, ν_0, characteristic of the metallic surface, below which no electrons are ejected, regardless of the intensity of the radiation. Above ν_0, the kinetic energy of the ejected electrons varies linearly with the frequency ν. These observations served as an embarrassing contradiction of classical theory.

To explain these results, Albert Einstein used Planck's hypothesis but extended it in an important way. Recall that Planck had applied his energy quantization concept, $E = nh\nu$ or $\Delta E = h\nu$, to the emission and absorption mechanism of the atomic electronic oscillators. Planck believed that once the light energy was emitted, it behaved like a classical wave. Einstein proposed instead that the radiation itself existed as small packets of energy, $E = h\nu$, now known as *photons*. Using a simple conservation-of-energy argument, Einstein showed that the kinetic energy (KE) of an ejected electron is equal to the energy of the incident photon ($h\nu$) minus the minimum energy required to remove an electron from the surface of the particular metal (ϕ). In an equation,

$$\text{KE} = \frac{1}{2}mv^2 = h\nu - \phi \tag{1.10}$$

where ϕ, called the *work function* of the metal, is analogous to an ionization energy of an isolated atom. The left side of Equation 1.10 cannot be negative, so Equation 1.10 predicts that $h\nu \geq \phi$. The minimum frequency that will eject an electron is just the frequency required to overcome the work function of the metal. Thus, there is a threshold frequency, ν_0, given by

$$h\nu_0 = \phi \tag{1.11}$$

Using Equations 1.10 and 1.11, we can write

$$\text{KE} = h\nu - h\nu_0 \qquad \nu \geq \nu_0 \tag{1.12}$$

Equation 1.12 shows that a plot of KE versus ν should be linear and that the slope of the line should be h, in complete agreement with the data in Figure 1.4.

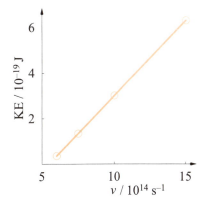

FIGURE 1.4
The kinetic energy of electrons ejected from the surface of sodium metal versus the frequency of the incident ultraviolet radiation. The threshold frequency here is 5.51×10^{14} Hz ($1\,\text{Hz} = 1\,\text{s}^{-1}$).

Before we can discuss Equation 1.12 numerically, we must consider the units involved. The work function ϕ is customarily expressed in units of electron volts (eV). One electron volt is the energy picked up by a particle with the same charge as an electron (or a proton) when it falls through a potential drop of one volt. If you recall that (1 coulomb) $\times$ (1 volt) = 1 joule and use the fact that the charge on a proton is 1.602×10^{-19} C, then

$$1\,\text{eV} = (1.602 \times 10^{-19}\,\text{C})(1\,\text{V}) = 1.602 \times 10^{-19}\,\text{J}$$

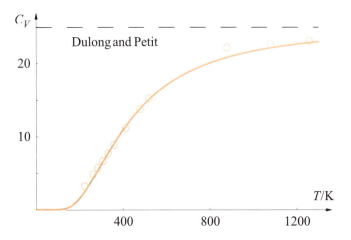

FIGURE 1.5
The molar heat capacity at constant volume of diamond as a function of temperature. The solid curve is the theoretical curve and the circles are experimental data. Classical physics is unable to predict the shape of this curve and predicts that C_V is equal to 25 J·K^{-1}·mol^{-1} at all temperatures. The decrease of C_V with decreasing temperature requires a quantum-theoretical explanation.

1.5 The Hydrogen Atomic Spectrum Consists of Several Series of Lines

For some time, scientists had known that every atom, when subjected to high temperatures or an electrical discharge, emits electromagnetic radiation of characteristic frequencies. In other words, each atom has a characteristic emission spectrum. Because the emission spectra of atoms consist of only certain discrete frequencies, they are called *line spectra*. Hydrogen, the lightest and simplest atom, has the simplest spectrum. Figure 1.6 shows the part of the hydrogen atom emission spectrum that occurs in the visible and near ultraviolet region.

Because atomic spectra are characteristic of the atoms involved, it is reasonable to suspect that the spectrum depends on the electron distribution in the atom. A detailed analysis of the hydrogen atomic spectrum turned out to be a major step in the elucidation of the electronic structure of atoms. For many years, scientists had tried to find a pattern in the wavelengths or frequencies of the lines in the hydrogen atomic spectrum. Finally, in 1885, an amateur Swiss scientist, Johann Balmer, showed that a plot of the frequency of the lines versus $1/n^2$ ($n = 3, 4, 5, \ldots$) is linear, as shown in Figure 1.7.

In particular, Balmer showed that the frequencies of the emission lines in the visible region of the spectrum could be described by the equation

$$\nu = 8.2202 \times 10^{14} \left(1 - \frac{4}{n^2}\right) \text{ Hz}$$

where $n = 3, 4, 5, \ldots$. This equation is now customarily written in terms of the quantity $1/\lambda$ instead of ν. Reciprocal wavelength is denoted by $\tilde{\nu}$. The standard units used

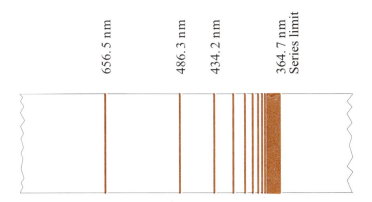

FIGURE 1.6
Emission spectrum of the hydrogen atom in the visible and the near ultraviolet region showing that the emission spectrum of atomic hydrogen is a line spectrum.

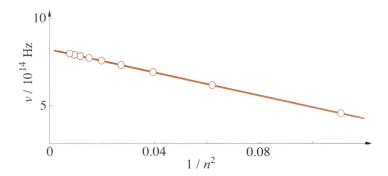

FIGURE 1.7
A plot of frequency versus $1/n^2$ ($n = 3, 4, 5, \ldots$) for the series of lines of the hydrogen atomic spectrum that occurs in the visible and near ultraviolet regions. The actual spectrum is shown in Figure 1.6. The linear nature of this plot leads directly to Equation 1.13.

for $\tilde{v}$ in spectroscopy are cm^{-1}, called *wave numbers*. Although wave number is not an SI unit, its use is so prevalent in spectroscopy that we will use wave numbers in this book. Thus, if we divide the previous equation by c and factor a 4 out of the two terms in parentheses, then we have

$$\tilde{v} = \frac{v}{c} = \frac{1}{\lambda} = 109\,680 \left(\frac{1}{2^2} - \frac{1}{n^2} \right) cm^{-1} \qquad n = 3, 4, \ldots \qquad (1.13)$$

This equation is called *Balmer's formula*.

EXAMPLE 1–5
Using Balmer's formula, calculate the wavelengths of the first few lines of the visible region of the hydrogen atomic spectrum and compare them to the experimental values given in Figure 1.6.

SOLUTION: The first line is obtained by setting $n = 3$, in which case we have

$$\tilde{\nu} = 109\,680 \left(\frac{1}{2^2} - \frac{1}{3^2} \right) \text{cm}^{-1}$$

$$= 1.523 \times 10^4 \text{ cm}^{-1}$$

and

$$\lambda = 6.565 \times 10^{-5} \text{ cm} = 656.5 \text{ nm}$$

The next line is obtained by setting $n = 4$, and so

$$\tilde{\nu} = 109\,680 \left(\frac{1}{2^2} - \frac{1}{4^2} \right) \text{cm}^{-1}$$

$$= 2.056 \times 10^4 \text{ cm}^{-1}$$

and

$$\lambda = 4.863 \times 10^{-5} \text{ cm} = 486.3 \text{ nm}$$

Thus, we see that the agreement with the experimental data (Figure 1.6) is excellent.

Note that Equation 1.13 predicts a series of lines as n takes on the values 3, 4, 5, ... This series of lines, the ones occurring in the visible and near ultraviolet regions of the hydrogen atomic spectrum and predicted by Balmer's formula, is called the *Balmer series*. The Balmer series is shown in Figure 1.6. Note also that Equation 1.9 predicts that the lines in the hydrogen atomic spectrum bunch up as n increases. As n increases, $1/n^2$ decreases, and eventually we can ignore this term compared with the $\frac{1}{4}$ term; and so in the limit $n \to \infty$, we have

$$\tilde{\nu} \longrightarrow 109\,680 \left(\frac{1}{4} \right) \text{cm}^{-1} = 2.742 \times 10^4 \text{ cm}^{-1}$$

or $\lambda = 364.7$ nm, in excellent agreement with the data in Figure 1.6. This value is essentially that for the last line in the Balmer series and is called the *series limit*.

The Balmer series occurs in the visible and near ultraviolet regions. The hydrogen atomic spectrum has lines in other regions; in fact, series of lines similar to the Balmer series appear in the ultraviolet and infrared regions (cf. Figure 1.8).

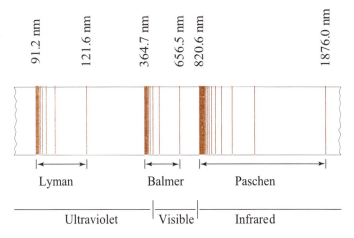

FIGURE 1.8
A schematic representation of the various series in the hydrogen atomic spectrum. The Lyman series lies in the ultraviolet region, the Balmer series lies in the visible region, and the other series lie in the infrared region (see Table 1.1).

TABLE 1.1
The First Four Series of Lines Making Up the Hydrogen Atomic Spectrum

Series name	n_1	n_2	Region of spectrum
Lyman	1	2, 3, 4, . . .	Ultraviolet
Balmer	2	3, 4, 5, . . .	Visible
Paschen	3	4, 5, 6, . . .	Near infrared [a]
Brackett	4	5, 6, 7, . . .	Infrared

a. The term "near infrared" denotes the part of the infrared region of the spectrum that is near to the visible region.

1.6 The Rydberg Formula Accounts for All the Lines in the Hydrogen Atomic Spectrum

The Swiss spectroscopist Johannes Rydberg accounted for all the lines in the hydrogen atomic spectrum by generalizing the Balmer formula to

$$\tilde{v} = \frac{1}{\lambda} = 109\ 680 \left(\frac{1}{n_1^2} - \frac{1}{n_2^2} \right) \text{cm}^{-1} \qquad (n_2 > n_1) \qquad (1.14)$$

where both n_1 and n_2 are integers but n_2 is always greater than n_1. Equation 1.14 is called the *Rydberg formula*. Note that the Balmer series is recovered if we let $n_1 = 2$.

The other series are obtained by letting n_1 be 1, 3, 4, ... The names associated with these various series are given in Figure 1.8 and Table 1.1. The constant in Equation 1.14 is called the *Rydberg constant* and Equation 1.10 is commonly written as

$$\tilde{v} = R_H \left(\frac{1}{n_1^2} - \frac{1}{n_2^2} \right) \tag{1.15}$$

where R_H is the Rydberg constant. The modern value of the Rydberg constant is 109 677.57 cm^{-1}, one of the most accurately known physical constants.

EXAMPLE 1–6

Calculate the wavelength of the second line in the Paschen series, and show that this line lies in the near infrared—that is, in the infrared region near the visible.

SOLUTION: In the Paschen series, $n_1 = 3$ and $n_2 = 4, 5, 6, \ldots$, according to Table 1.1. Thus, the second line in the Paschen series is given by setting $n_1 = 3$ and $n_2 = 5$ in Equation 1.15:

$$\tilde{v} = 109\ 677.57 \left(\frac{1}{3^2} - \frac{1}{5^2} \right) \text{cm}^{-1}$$

$$= 7.799 \times 10^3 \text{ cm}^{-1}$$

and

$$\lambda = 1.282 \times 10^{-4} \text{ cm} = 1282 \text{ nm}$$

The fact that the formula describing the hydrogen spectrum is in a sense controlled by two integers is truly amazing. Why should a hydrogen atom care about our integers? We will see that integers play a special role in quantum theory.

The spectra of other atoms were also observed to consist of series of lines, and in the 1890s Rydberg found approximate empirical laws for many of them. The empirical laws for other atoms were generally more involved than Equation 1.15, but the really interesting feature is that all the observed lines could be expressed as the difference between terms such as those in Equation 1.15. This feature was known as the *Ritz combination rule*, and we will see that it follows immediately from our modern view of atomic structure. At the time, however, it was just an empirical rule waiting for a theoretical explanation.

1.7 Angular Momentum Is a Fundamental Property of Rotating Systems

The theoretical explanation of the atomic spectrum of hydrogen was to come from a young Dane named Niels Bohr. In 1911, the New Zealand physicist Ernest Rutherford, based upon the α-particle scattering experiments of his collegues Hans Geiger and Ernest Marsden, had proposed the nuclear model of the atom. Bohr was working in Rutherford's laboratory at the time and saw how to incorporate this new viewpoint of the atom and the quantization condition of Planck into a successful theory of the hydrogen atom. Before discussing this, however, we must have a digression on classical mechanics because Bohr's model of the hydrogen atom deals with some classical mechanical ideas of circular motion.

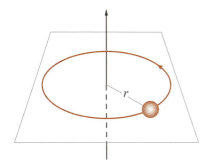

FIGURE 1.9
The rotation of a single particle about a fixed point.

Linear momentum is given by mv and is usually denoted by the symbol p. Now consider a particle rotating in a plane about a fixed center, as in Figure 1.9. Let v_{rot} be the frequency of rotation (cycles/second). The velocity of the particle, then, is $v = 2\pi r v_{\text{rot}} = r\omega_{\text{rot}}$, where $\omega_{\text{rot}} = 2\pi v_{\text{rot}}$ has units of radians/second and is called the *angular velocity*. The kinetic energy of the revolving particle (we'll now use the standard symbol, T, for kinetic energy) is

$$T = \frac{1}{2}mv^2 = \frac{1}{2}mr^2\omega^2 = \frac{1}{2}I\omega^2 \tag{1.16}$$

where $I = mr^2$ is the *moment of inertia*. By comparing the first and the last expressions for the kinetic energy in Equation 1.16, we can make the correspondences $\omega \leftrightarrow v$ and $I \leftrightarrow m$, where ω and I are angular quantities and v and m are linear quantities. According to this correspondence, there should be a quantity $I\omega$ corresponding to the linear momentum mv. In fact, the quantity l, defined by

$$l = I\omega = mr^2\frac{v}{r} = mvr \tag{1.17}$$

is called the *angular momentum* and is a fundamental quantity associated with rotating systems, just as linear momentum is a fundamental quantity in linear systems.

Kinetic energy can be written in terms of momentum. For a linear system, we have

$$T = \frac{mv^2}{2} = \frac{(mv)^2}{2m} = \frac{p^2}{2m} \tag{1.18}$$

TABLE 1.2
The Correspondences Between the Motion of Linear Systems and Rotating Systems

Linear motion	Angular motion
Mass (m)	Moment of inertia (I)
Speed (v)	Angular speed (ω)
Momentum ($p = mv$)	Angular momentum ($l = I\omega$)
Kinetic energy $\left(T = \dfrac{mv^2}{2} = \dfrac{p^2}{2m} \right)$	Rotational kinetic energy $\left(T = \dfrac{I\omega^2}{2} = \dfrac{l^2}{2I} \right)$

and for a rotating system,

$$T = \frac{I\omega^2}{2} = \frac{(I\omega)^2}{2I} = \frac{l^2}{2I} \tag{1.19}$$

The correspondences between the motion of linear systems and rotating systems are given in Table 1.2.

Recall from general physics that a particle revolving around a fixed point as in Figure 1.9 experiences an outward acceleration, and requires an inward force

$$f = \frac{mv^2}{r} \tag{1.20}$$

to keep it moving in a circular orbit. (Equation 1.20 is derived in Problem 1–52.) For a mass tied to the fixed center by a string, this force is supplied by the tension in the string.

1.8 Bohr Assumed That the Angular Momentum of the Electron in a Hydrogen Atom Is Quantized

According to the nuclear model of the atom, the hydrogen atom can be pictured as a central, rather massive nucleus with one electron. Because the nucleus is so much more massive than the electron, we can consider the nucleus to be fixed and the electron to be revolving about it, much like the diagram in Figure 1.9. The force holding the electron in a circular orbit is supplied by the coulombic force of attraction between the proton and the electron. If we equate Coulomb's force law ($e^2/4\pi\epsilon_0 r^2$) with Equation 1.20, then we have

$$\frac{e^2}{4\pi\epsilon_0 r^2} = \frac{m_e v^2}{r} \tag{1.21}$$

where ϵ_0 is the permittivity of free space and is equal to

$$8.854\ 19 \times 10^{-12}\ \text{coulomb}^2/\text{newton}\cdot\text{meter}^2\ (\text{C}^2\cdot\text{N}^{-1}\cdot\text{m}^{-2})$$

The occurrence of the factor $4\pi\epsilon_0$ in Coulomb's law is a result of using SI units.

We are tacitly assuming here that the electron is revolving around a fixed nucleus in a circular orbit of radius r. Classically, however, because the electron is constantly being accelerated according to Equation 1.20, it should emit electromagnetic radiation and lose energy just as electrons accelerated in an antenna. Consequently, classical physics predicts that an electron revolving around a nucleus will lose energy and spiral into the nucleus, and so a stable orbit is classically forbidden. It was Bohr's great contribution to make two nonclassical assumptions. The first of these was to assume the existence of stationary orbits, in denial of classical physics. He then specified these orbits by invoking a quantization condition, and in this case, he assumed that the angular momentum of the electron must be quantized according to

$$l = m_e v r = n\hbar \qquad n = 1, 2, \ldots \tag{1.22}$$

where $\hbar = h/2\pi$ (called h-bar), which occurs often in quantum mechanics. Solving Equation 1.22 for v and substituting into Equation 1.21, we obtain

$$r = \frac{4\pi\epsilon_0\hbar^2 n^2}{m_e e^2} \qquad n = 1, 2, \ldots \tag{1.23}$$

Thus, we see that the radii of the allowed orbits, or *Bohr orbits*, are *quantized*. According to this picture, the electron can move around the nucleus only in circular orbits with radii given by Equation 1.23. The orbit with the smallest radius is the orbit with $n = 1$:

$$r = \frac{4\pi(8.854\ 19 \times 10^{-12}\ \text{C}^2\cdot\text{N}^{-1}\cdot\text{m}^{-2})(1.055 \times 10^{-34}\ \text{J}\cdot\text{s})^2}{(9.110 \times 10^{-31}\ \text{kg})(1.602 \times 10^{-19}\ \text{C})^2}$$

$$= 5.29 \times 10^{-11}\ \text{m} = 52.9\ \text{pm} = 0.529\ \text{Å} \tag{1.24}$$

The radius of the first Bohr orbit is often denoted by a_0.

The total energy of the electron is equal to the sum of its kinetic energy and potential energy. The potential energy of an electron and a proton separated by a distance r is

$$V(r) = -\frac{e^2}{4\pi\epsilon_0 r} \tag{1.25}$$

The negative sign here indicates that the proton and electron attract each other; their energy is less than it is when they are infinitely separated [$V(\infty) = 0$]. The total energy of the electron in a hydrogen atom is

$$E = \frac{1}{2}m_e v^2 - \frac{e^2}{4\pi\epsilon_0 r} \tag{1.26}$$

Using Equation 1.21 to eliminate the $m_e v^2$ in the kinetic energy term, Equation 1.26 becomes

$$E = \frac{1}{2} \left(\frac{e^2}{4\pi \epsilon_0 r} \right) - \frac{e^2}{4\pi \epsilon_0 r}$$

$$= - \frac{e^2}{8\pi \epsilon_0 r}$$

The only allowed values of r are those given by Equation 1.23 and so if we substitute Equation 1.23 into the previous equation, then we find that the only allowed energies are

$$E_n = -\frac{m_e e^4}{8\epsilon_0^2 h^2} \frac{1}{n^2} \qquad n = 1, 2, \ldots \qquad (1.27)$$

The negative sign in Equation 1.27 indicates that the energy states are bound states; the energies given by Equation 1.27 are less than when the proton and electron are infinitely separated. Note that $n = 1$ in Equation 1.27 corresponds to the state of lowest energy, called the *ground-state energy*. At ordinary temperatures, hydrogen atoms, as well as most other atoms and molecules, will be found almost exclusively in their ground electronic state. The states of higher energy are called *excited states* and are generally unstable with respect to the ground state. An atom or molecule in an excited state will usually relax back to the ground state and give off the energy as electromagnetic radiation (see Figure 1.10).

We can display the energies given by Equation 1.27 in an energy-level diagram like that in Figure 1.10. Note that the energy levels merge as $n \to \infty$. Bohr assumed that the observed emission spectrum of the hydrogen atom was due to transitions from one allowed energy state to a lower state, and so

$$\Delta E = \frac{m_e e^4}{8\epsilon_0^2 h^2} \left(\frac{1}{n_1^2} - \frac{1}{n_2^2} \right) = h\nu \qquad (1.28)$$

where $n_2 > n_1$. Setting $\Delta E = h\nu$ is called the *Bohr frequency condition* and is the basic assumption that as the electron falls from one level to another, the energy evolved is given off as a photon of energy $E = h\nu$. Figure 1.10 groups the various transitions that occur according to the final state into which the electron falls. We can see, then, that the various observed spectral series arise in a natural way from the Bohr model. The Lyman series occurs when electrons that are excited to higher levels relax to the $n = 1$ state, the Balmer series occurs when excited electrons fall back into the $n = 2$ state, and so on.

We can write the theoretical formula (Equation 1.28) in the form of the empirical Rydberg formula by writing $h\nu = hc\tilde{\nu}$:

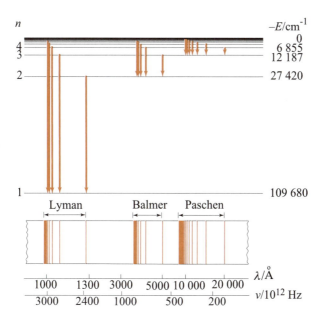

FIGURE 1.10
The energy-level diagram for the hydrogen atom, showing how transitions from higher states into some particular lower state lead to the observed spectral series for hydrogen.

$$\tilde{\nu} = \frac{m_e e^4}{8\epsilon_0^2 ch^3}\left(\frac{1}{n_1^2} - \frac{1}{n_2^2}\right)$$ (1.29)

If we compare Equations 1.16 and 1.29, then we conclude that

$$R_{\mathrm{H}} = \frac{m_e e^4}{8\epsilon_0^2 ch^3}$$ (1.30)

EXAMPLE 1–7
Using the values of the physical constants given inside the front cover of this book, calculate R_{H} and compare the result to its experimental value, 109 677.6 cm^{-1}.

SOLUTION:

$$R_{\mathrm{H}} = \frac{(9.109\ 3897 \times 10^{-31}\ \mathrm{kg})(1.602\ 177 \times 10^{-19}\ \mathrm{C})^4}{(8)(8.854\ 187 \times 10^{-12}\ \mathrm{C}^2\cdot\mathrm{N}^{-1}\cdot\mathrm{m}^{-1})^2(2.997\ 924\ 58 \times 10^8\ \mathrm{m\cdot s}^{-1})(6.626\ 076 \times 10^{-34}\ \mathrm{J\cdot s})^3}$$

$$= 1.097\ 37 \times 10^7\ \mathrm{m}^{-1} = 109\ 737\ \mathrm{cm}^{-1}$$

which is within 0.5% of the experimental value of 109 677.6 cm^{-1}, surely a remarkable agreement.

EXAMPLE 1–8
Calculate the ionization energy of the hydrogen atom.

SOLUTION: The ionization energy IE is the energy required to take the electron from the ground state to the first unbound state, which is obtained by letting $n_2 = \infty$ in Equation 1.29. Thus, we write

$$\text{IE} = R_{\text{H}} \left(\frac{1}{1^2} - \frac{1}{\infty^2} \right)$$

or

$$\text{IE} = R_{\text{H}} = 109\ 677.6\ \text{cm}^{-1}$$

$$= 2.179 \times 10^{-18}\ \text{J} = 13.6\ \text{eV}$$

Note that we have expressed the energy in units of wave numbers (cm^{-1}). This is not strictly a unit of energy, but because of the simple relation between wave number and energy, $\varepsilon = hc\tilde{\nu}$, one often does express energy in this way.

1.9 The Electronic Mass Should Be Replaced by a Reduced Mass in the Bohr Theory

In deriving Equations 1.23 and 1.27, we have assumed that because the proton is so much more massive than the electron, we can regard the proton as being a fixed center around which the electron revolves. It is not really necessary to make this assumption. Consider the general case of two masses rotating about each other, as shown in Figure 1.11. The center of mass of this system is fixed, and each of the masses will be rotating about that point. The center of mass lies along the line joining their centers and is defined through the condition

$$m_1 r_1 = m_2 r_2 \tag{1.31}$$

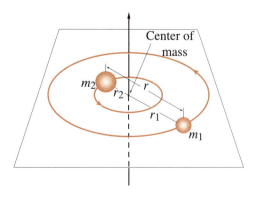

FIGURE 1.11
Two masses rotating about their center of mass. The center of mass lies along the line joining the two masses and is given by the condition $r_1 m_1 = r_2 m_2$, where r_1 and r_2 are the distances of m_1 and m_2, respectively, from the center of mass.

where r_1 and r_2 are defined in Figure 1.11. The distance between the particles is

$$r = r_1 + r_2 \tag{1.32}$$

Using Equations 1.31 and 1.32, simple algebra shows that

$$r_1 = \frac{m_2}{m_1 + m_2} r \quad \text{and} \quad r_2 = \frac{m_1}{m_1 + m_2} r \tag{1.33}$$

The total kinetic energy is

$$T = \frac{1}{2} m_1 v_1^2 + \frac{1}{2} m_2 v_2^2 \tag{1.34}$$

Now if ω is the angular velocity of the two masses about the fixed center of mass, then

$$v_1 = r_1 \omega \quad \text{and} \quad v_2 = r_2 \omega \tag{1.35}$$

If we substitute Equations 1.35 into Equation 1.34, then we obtain

$$
\begin{aligned}
T &= \frac{1}{2} m_1 r_1^2 \omega^2 + \frac{1}{2} m_2 r_2^2 \omega^2 \\
&= \frac{1}{2} (m_1 r_1^2 + m_2 r_2^2) \omega^2 \\
&= \frac{1}{2} I \omega
\end{aligned}
\tag{1.36}
$$

where

$$I = m_1 r_1^2 + m_2 r_2^2 \tag{1.37}$$

is called the *moment of inertia* of the system. Substituting Equations 1.33 into Equation 1.37, we find that I can be written as

$$I = \left(\frac{m_1 m_2}{m_1 + m_2} \right) r^2 \tag{1.38}$$

The factor $m_1 m_2 / (m_1 + m_2)$ has units of mass and occurs often in problems involving two masses interacting along their line of center; this quantity is called the *reduced mass* μ:

$$\mu = \frac{m_1 m_2}{m_1 + m_2} \tag{1.39}$$

Using Equation 1.39, we can write Equation 1.38 in the form

$$I = \mu r^2 \tag{1.40}$$

In Equation 1.34, both masses are rotating about the center of mass of the system, as in Figure 1.11. Recall that in Figure 1.9 we had one mass revolving about a fixed point, and in that case I was given by

$$I = mr^2 \tag{1.41}$$

A comparison of Equations 1.40 and 1.41 allows us to give a useful interpretation of reduced mass, one that we shall use several times throughout the book. Equation 1.40 means that we can treat the two-body system with masses m_1 and m_2 revolving about each other as simply one body of mass μ revolving about the other one fixed in position—that is, with an infinite mass. This is an extremely important and useful result regarding two-body systems. We have reduced the two-body problem to a one-body problem, the effective mass of the one body being the reduced mass $\mu = m_1 m_2 / (m_1 + m_2)$.

We can apply this result to our treatment of the hydrogen atom. We can obtain a rigorous result that is as simple as the one obtained by fixing the proton at the center of revolution if we use not the actual mass of the electron but the reduced mass of the proton–electron system:

$$\mu = \frac{m_p m_e}{m_p + m_e} \tag{1.42}$$

where m_p and m_e are the masses of the proton and electron, respectively. Note that since $m_p \gg m_e$, the denominator is essentially m_p, and so $\mu \approx m_e$. This says that it is indeed a good approximation to consider the proton to be fixed at the center of revolution. The reduced mass is

$$\mu = \frac{(9.1094 \times 10^{-31}\,\text{kg})(1.672\,62 \times 10^{-27}\,\text{kg})}{9.1094 \times 10^{-31}\,\text{kg} + 1.672\,62 \times 10^{-27}\,\text{kg}}$$

$$= 9.1048 \times 10^{-31}\,\text{kg} = 0.9995\,m_e \tag{1.43}$$

Equation 1.30 for the Rydberg constant now reads

$$R_H = \frac{\mu e^4}{8\epsilon_0^2 c h^3} \tag{1.44}$$

and now one calculates R_H to be $109\,676\;\text{cm}^{-1}$ instead of $109\,737\;\text{cm}^{-1}$ (Problem 1–29). Recall that the accepted experimental value is $109\,677.6\;\text{cm}^{-1}$.

In spite of its algebraic simplicity, the Bohr theory gives a very nice picture of the hydrogen atom. It can also be directly applied to any hydrogen-like ion, such as He^+ and Li^{2+}, consisting of one electron around a nucleus. It is a simple matter to extend the above results to these ions. Instead of starting with Equation 1.21, we use the following equation where the charge on the nucleus is Ze instead of just e:

$$\frac{Ze^2}{4\pi\epsilon_0 r^2} = \frac{\mu v^2}{r} \tag{1.45}$$

Everything else now follows directly and eventually we have (Problem 1–32)

$$\tilde{v} = \frac{Z^2 e^4 \mu}{8\epsilon_0^2 c h^3}\left(\frac{1}{n_1^2} - \frac{1}{n_2^2}\right) \tag{1.46}$$

or simply

$$\tilde{v} = Z^2 \tilde{v}_H \tag{1.47}$$

EXAMPLE 1–9

Calculate the radius of the first Bohr orbit for He^+.

SOLUTION: By eliminating v between Equations 1.22 and 1.45, we obtain

$$r = \frac{4\pi\epsilon_0 \hbar^2 n^2}{Z\mu e^2} = \frac{\epsilon_0 h^2 n^2}{Z\pi\mu e^2}$$

Note that this reduces to Equation 1.23 when $Z = 1$. If we let $Z = 2$ and $n = 1$, we find that

$$r = \frac{(8.854 \times 10^{-12}\ \text{C}^2\cdot\text{J}^{-1}\cdot\text{m}^{-1})(6.626 \times 10^{-34}\ \text{J}\cdot\text{s})^2}{2(3.1416)(9.110 \times 10^{-31}\ \text{kg})(1.602 \times 10^{-19}\ \text{C})^2}$$

$$= 2.65 \times 10^{-11}\ \text{m} = 26.5\ \text{pm} = 0.265\ \text{Å}$$

One spectacular success of the Bohr theory was the correct assignment of some solar spectral lines of He^+. These lines were previously thought to be due to atomic hydrogen and to be anomalous because they did not fit the Rydberg formula (Problem 1–33). In spite of a number of successes and the beautiful simplicity of the Bohr theory, it could not be extended successfully even to a two-electron system such as helium. Furthermore, even for simple systems such as hydrogen, it was never able to explain the spectra that arise when a magnetic field is applied to the system, nor was it able to predict the intensities of the spectral lines. In spite of ingenious efforts by Bohr and others, they were never able to extend the theory to explain such phenomena.

1.10 Louis de Broglie Postulated That Matter Has Wavelike Properties

Although we have an intriguing partial insight into the electronic structure of atoms, something is missing. To explore this further, let's go back to a discussion of the nature of light.

Scientists have always had trouble describing the nature of light. In many experiments light shows a definite wavelike character, but in many others light seems to behave as a stream of photons. The dispersion of white light into its spectrum by a prism is an example of the first type of experiment, and the photoelectric effect is an example of the second. Because light appears wavelike in some instances and particle-like in others, this disparity is referred to as the *wave–particle duality of light*. In 1924, a young French scientist, Louis de Broglie, reasoned that if light can display this wave–particle duality, then matter, which certainly appears particle-like, might also display wavelike properties under certain conditions. This proposal is rather strange at first, but it does suggest a nice symmetry in nature. Certainly, if light can be particle-like at times, why should matter not be wavelike at times?

De Broglie was able to put his idea into a quantitative scheme. Einstein had shown from relativity theory that the wavelength, λ, and the momentum, p, of a photon are related by

$$\lambda = \frac{h}{p} \tag{1.48}$$

De Broglie argued that both light *and* matter obey this equation. Because the momentum of a particle is given by mv, this equation predicts that a particle of mass m moving with a velocity v will have a *de Broglie wavelength* given by $\lambda = h/mv$.

EXAMPLE 1–10
Calculate the de Broglie wavelength for a baseball (5.0 oz) traveling at 90 mph.

SOLUTION: Five ounces corresponds to

$$m = (5.0 \text{ oz}) \left(\frac{1 \text{ lb}}{16 \text{ oz}} \right) \left(\frac{0.454 \text{ kg}}{1 \text{ lb}} \right) = 0.14 \text{ kg}$$

and 90 mph corresponds to

$$v = \left(\frac{90 \text{ mi}}{1 \text{ hr}} \right) \left(\frac{1610 \text{ m}}{1 \text{ mi}} \right) \left(\frac{1 \text{ hr}}{3600 \text{ s}} \right) = 40 \text{ m} \cdot \text{s}^{-1}$$

The momentum of the baseball is

$$p = mv = (0.14 \text{ kg})(40 \text{ m} \cdot \text{s}^{-1}) = 5.6 \text{ kg} \cdot \text{m} \cdot \text{s}^{-1}$$

If we substitute de Broglie's relation (Equation 1.48) into Equation 1.49, we obtain the Bohr quantization condition

$$mvr = n\hbar \qquad (1.50)$$

An interesting application of Equation 1.50 is to use it to calculate the velocity of an electron in a Bohr orbit.

EXAMPLE 1–12

Use Equation 1.50 to calculate the velocity of an electron in the first Bohr orbit.

SOLUTION: If we solve Equation 1.50 for v, we find

$$v = \frac{n\hbar}{mr}$$

For the first Bohr orbit, $n = 1$, and the radius is given by Equation 1.24. If we substitute $n = 1$ and Equation 1.24 into the above equation for v, we find that

$$v = \frac{1.055 \times 10^{-34}\,\text{J}\cdot\text{s}}{(9.11 \times 10^{-31}\,\text{kg})(5.29 \times 10^{-11}\,\text{m})}$$

$$= 2.19 \times 10^{6}\,\text{m}\cdot\text{s}^{-1}$$

Note that this is almost 1% of the speed of light, a very large speed.

1.12 De Broglie Waves Are Observed Experimentally

When a beam of X rays is directed at a crystalline substance, the beam is scattered in a definite manner characteristic of the atomic structure of the crystalline substance. This phenomenon is called *X-ray diffraction* and occurs because the interatomic spacings in the crystal are about the same as the wavelength of the X rays. The X-ray diffraction pattern from aluminum foil is shown in Figure 1.13a. The X rays scatter from the foil in rings of different diameters. The distances between the rings are determined by the interatomic spacing in the metal foil. Figure 1.13b shows an electron diffraction pattern from aluminum foil that results when a beam of electrons is similarly directed. The similarity of the two patterns shows that both X rays and electrons do indeed behave analogously in these experiments.

The wavelike property of electrons is used in electron microscopes. The wavelengths of the electrons can be controlled through an applied voltage, and the small de Broglie wavelengths attainable offer a more precise probe than an ordinary light microscope. In addition, in contrast to electromagnetic radiation of similar wavelengths (X rays and ultraviolet), the electron beam can be readily focused by using electric and

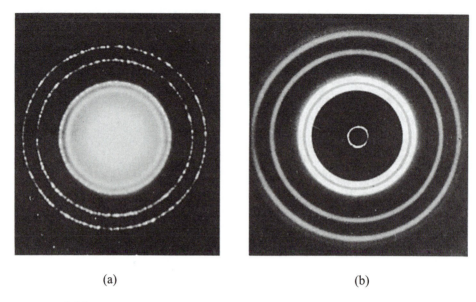

(a)   (b)

FIGURE 1.13
(a) The X-ray diffraction pattern of aluminum foil. (b) The electron diffraction pattern of
aluminum foil. The similarity of these two patterns shows that electrons can behave like X rays
and display wavelike properties. Reproduced courtesy of Education Development Center, Inc.,
Newton, MA, from PSCC physics film, *Matter Waves*.

magnetic fields, generating sharper images. Electron microscopes are used routinely in
chemistry and biology to investigate atomic and molecular structures.

An interesting aside in the concept of the wave–particle duality of matter is that it
was J. J. Thomson who first showed in 1895 that the electron was a subatomic particle
and it was G. P. Thomson who was one of the first to show experimentally in 1926 that
the electron could act as a wave. These two Thomsons are father and son. The father
won a Nobel Prize in 1906 for showing that the electron is a particle, and the son won
a Nobel Prize in 1937 for showing that it is a wave.

1.13 Certain Two-Slit Experiments Exemplify Wave–Particle Duality

One of the principal experiments that displays the wave properties of light is the
interference of two beams of light. The first experiment showing interference of light
was performed by the British scientist Thomas Young around 1800.

Consider a light wave impinging on an opaque screen with two very narrow slits, as
shown in Figure 1.14. As the light passes through the slits, each slit acts as a new source
of light (this is called Huygen's principle). The light reaching a point P on a second
screen a distance l away from the first travels different path lengths, $S_1 P$ and $S_2 P$, as
shown in Figure 1.15. For simplicity, we assume that l is large enough compared to the
separation of the two slits that the two rays of light follow essentially parallel paths.

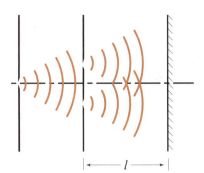

FIGURE 1.14
An illustration of a waveform from a single source impinging on an opaque screen with two narrow slits. Huygen's principle from optics says that each slit acts as a source of a new wave.

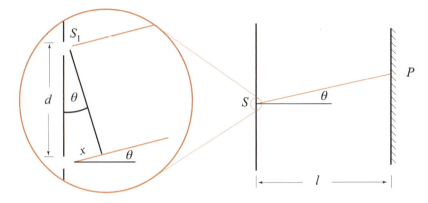

FIGURE 1.15
A schematic diagram of a two-slit interference experiment (right). The distance l is large enough compared to the distance between the slits, d, so that the two light waves are essentially parallel. The angle θ is the angle that the beams make with the perpendicular line between the two screens. The diagram on the left shows a magnification of the circled region on the right.

The difference in the distance each light wave follows is $x = d\cos(90° - \theta) = d\sin\theta$, where θ is shown in Figure 1.15. If this difference is equal to an integral number of wavelengths, then $d\sin\theta = n\lambda$ and the two light waves will be in phase and constructively interfere. If the difference is equal to a half odd integral number of wavelengths, then the two light waves will be exactly out of phase and destructively interfere. Thus, as θ varies, the condition $d\sin\theta = n\lambda$ produces an interference pattern consisting of alternating bright and dark areas on the second screen, like that shown in Figure 1.16.

After the discussion of the previous two sections, it shouldn't be surprising that a beam of electrons or any other atomic or subatomic particle will produce an interference pattern. Interference patterns have been observed experimentally for electrons and neutrons, and Figure 1.17 shows an interference pattern obtained for helium atoms taken from a paper titled "Young's Double Slit Experiment with Atoms" by the German physicists O. Carnal and J. Mlynek.

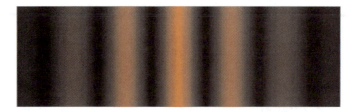

FIGURE 1.16
An illustration of the interference pattern produced by a light wave in a two-slit experiment.

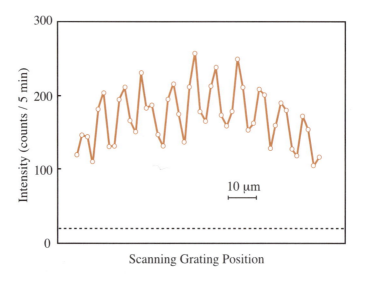

FIGURE 1.17
Data illustrating an interference pattern for helium atoms from an 8 μm grating. The figure shows the number of atoms recorded in five-minute intervals plotted against the distance along the recording screen. The dashed line is the detector background and the solid lines are simply an aid to the eye. Adapted from Carnal, O., and Mlynek, J., Young's Double Slit Experiment with Atoms: A Simple Atom Interferometer. *Phys. Rev. Lett.,* **66**, 2689 (1991).

It is possible to perform a two-slit experiment such that only one particle at a time passes through the slits and to record its arrival at a second screen coated with a fluorescent film, which emits a brief burst of light at the point where it is struck. Using a suitable position-dependent recording device, we can record the arrival of each particle and accumulate the results (Figure 1.18). At low exposures, these tiny bursts of light appear more or less randomly over the film; but for longer exposures, hints of a pattern begin to emerge. The bursts appear to be preferentially occurring in certain regions in the second panel on the left, and a full interference pattern is discernable

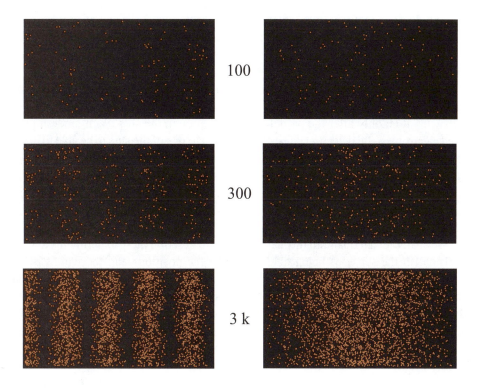

FIGURE 1.18
When a particle passes through the two-slit screen in Figure 1.14, its arrival at the second screen is recorded by a dot. As more and more particles arrive at the screen, an interference pattern slowly builds up. The three panels on the left record the arrival of 100, 300, and 3000 particles. The three panels on the right show the arrival of the same number of particles as those on the left, but one of the slits is closed, yielding no interference pattern. These figures are computer simulations of actual experiments carried out by A. Tomomura et al., *Am. J. Phys.,* **57**, 117 (1980).

in the third panel on the left. Experiments like this have been performed for electrons, neutrons, and even helium and sodium atoms.

In order for interference to occur, the particle must have gone through both slits. But how can one particle go through both slits simultaneously? Or does it? Suppose we do the experiment again, but cover up one of the slits; then the pattern that develops on the far screen looks just like you might expect, as shown in the panels on the right side of Figure 1.18. There is no interference pattern. Let's now uncover both slits and try to detect which slit the particle passes through by setting up some sort of a detector behind one of the slits so that it can detect any particle that passes through that slit before it impinges on the fluorescent screen. We now know which slit the particle passes through on its way to the screen and don't have to assume that it somehow passes through both slits simultaneously. We may know which slit the particle has gone through, but when we look at the fluorescent screen, we see that there is no interference pattern! The pattern is simply the one that would be obtained if one or the other slit had been covered. If we now turn off the detector, we obtain the interference pattern. It's as if each particle behaves as a wave as it confronts the slits, unless we try to observe it, in which case it behaves as a particle.

If this isn't enough to convince you that the subatomic world is weirdly different from our macroscopic world, consider this final twist. Suppose we set up the detector and switch it on only after the particle has gone through the slits. With modern high-speed electronics, the detector can be close enough to one of the slits to be able to detect whether the particle has gone through it, but before it reaches the detector. Surely now it is too late for the particle to decide to behave as a particle that has passed through only one of the slits. Apparently not. Experiments like this, called delayed choice experiments in the literature, have been carried out for photons, and the interference pattern is found to disappear.

We shall see that quantum mechanics is able to describe these experiments to great precision, in the sense of being able to predict the positions and the widths of the interference ridges. The *physical interpretation* of the results, however, is quite another matter. How can a particle go through two slits simultaneously? In the experiments that have been done where one particle at a time goes through the slits, does each particle interfere with itself to produce an interference pattern? Attempts to answer questions like these have raised contentious issues that are not yet settled among those involved in the philosophical interpretations of quantum mechanics. As a computational tool, however, quantum mechanics is probably the most successful scientific theory ever formulated.

1.14 The Heisenberg Uncertainty Principle States That the Position and the Momentum of a Particle Cannot Be Specified Simultaneously with Unlimited Precision

We now know that we must consider light and matter as having the characteristics of both waves and particles. Let's consider a measurement of the position of an electron. If we wish to locate the electron within a distance Δx, then we must use a measuring device that has a spatial resolution less than Δx. One way to achieve this resolution is to use light with a wavelength on the order of $\lambda \approx \Delta x$. For the electron to be "seen," a photon must interact or collide in some way with the electron, for otherwise the photon will just pass right by and the electron will appear transparent. The photon has a momentum $p = h/\lambda$, and during the collision, some of this momentum will be transferred to the electron. The very act of locating the electron leads to a change in its momentum. If we wish to locate the electron more accurately, we must use light with a smaller wavelength. Consequently, the photons in the light beam will have greater momentum because of the relation $p = h/\lambda$. Because some of the photon's momentum must be transferred to the electron in the process of locating it, the momentum change of the electron becomes greater. A careful analysis of this process was carried out in the mid-1920s by the German physicist Werner Heisenberg, who showed that it is not possible to determine exactly how much momentum is transferred to the electron. This difficulty means that if we wish to locate an electron to within a region Δx, there will be an uncertainty in the momentum of the electron. Heisenberg was able to show that if Δp is the uncertainty in the momentum of the electron, then

$$\Delta x \Delta p \geq h \qquad (1.51)$$

Equation 1.51 is called *Heisenberg's uncertainty principle* and is a fundamental principle of nature. The uncertainty principle states that if we wish to locate any particle to within a distance Δx, then we automatically introduce an uncertainty in the momentum of the particle and that the uncertainty is given by Equation 1.51. Note that this uncertainty does not stem from poor measurement or experimental technique but is a fundamental property of the act of measurement itself. The following two examples demonstrate the numerical consequences of the uncertainty principle.

EXAMPLE 1–13
Calculate the uncertainty in the position of a baseball thrown at 90 mph if we measure its speed to a millionth of 1.0%.

SOLUTION: According to Example 1–10, a baseball traveling at 90 mph has a momentum of $5.6 \text{ kg} \cdot \text{m} \cdot \text{s}^{-1}$. A millionth of 1.0% of this value is $5.6 \times 10^{-8} \text{ kg} \cdot \text{m} \cdot \text{s}^{-1}$, so

$$\Delta p = 5.6 \times 10^{-8} \text{ kg} \cdot \text{m} \cdot \text{s}^{-1}$$

The minimum uncertainty in the position of the baseball is

$$\Delta x = \frac{h}{\Delta p} = \frac{6.626 \times 10^{-34} \text{ J} \cdot \text{s}}{5.6 \times 10^{-8} \text{ kg} \cdot \text{m} \cdot \text{s}^{-1}}$$

$$= 1.2 \times 10^{-26} \text{ m}$$

a completely inconsequential distance.

EXAMPLE 1–14
What is the uncertainty in momentum if we wish to locate an electron within an atom, say, so that Δx is approximately 50 pm?

SOLUTION:

$$\Delta p = \frac{h}{\Delta x} = \frac{6.626 \times 10^{-34} \text{ J} \cdot \text{s}}{50 \times 10^{-12} \text{ m}}$$

$$= 1.3 \times 10^{-23} \text{ kg} \cdot \text{m} \cdot \text{s}^{-1}$$

Because $p = mv$ and the mass of an electron is 9.11×10^{-31} kg, this value of Δp corresponds to

$$\Delta v = \frac{\Delta p}{m_e} = \frac{1.3 \times 10^{-23} \text{ kg·m·s}^{-1}}{9.11 \times 10^{-31} \text{ kg}}$$

$$= 1.4 \times 10^7 \text{ m·s}^{-1}$$

which is a very large uncertainty in the speed.

These two examples show that although the Heisenberg uncertainty principle is of no consequence for everyday, macroscopic bodies, it has very important consequences in dealing with atomic and subatomic particles. This conclusion is similar to the one that we drew for the application of the de Broglie relation between wavelength and momentum. The uncertainty principle led to an awkward result. It turns out that the Bohr theory is inconsistent with the uncertainty principle. Fortunately, a new, more general quantum theory was soon presented that is consistent with the uncertainty principle. We will see that this theory is applicable to all atoms and molecules and forms the basis for our understanding of atomic and molecular structure. This theory was formulated by the Austrian physicist Erwin Schrödinger and will be discussed in Chapter 3. In preparation, in the next chapter, we will discuss the classical wave equation, which serves as a useful and informative background to the Schrödinger equation.

Problems

1–1. Radiation in the ultraviolet region of the electromagnetic spectrum is usually described in terms of wavelength, λ, and is given in nanometers (10^{-9} m). Calculate the values of v, $\tilde{v}$, and E for ultraviolet radiation with $\lambda = 200$ nm and compare your results with those in Figure 1.19.

1–2. Radiation in the infrared region is often expressed in terms of wave numbers, $\tilde{v} = 1/\lambda$. A typical value of $\tilde{v}$ in this region is 10^3 cm^{-1}. Calculate the values of v, λ, and E for radiation with $\tilde{v} = 10^3$ cm^{-1} and compare your results with those in Figure 1.19.

1–3. Past the infrared region, in the direction of lower energies, is the microwave region. In this region, radiation is usually characterized by its frequency, v, expressed in units of megahertz (MHz), where the unit hertz (Hz) is a cycle per second. A typical microwave frequency is 2.0×10^4 MHz. Calculate the values of $\tilde{v}$, λ, and E for this radiation and compare your results with those in Figure 1.19.

1–4. Planck's principal assumption was that the energies of the electronic oscillators can have only the values $E = nh v$ and that $\Delta E = h v$. As $v \to 0$, then $\Delta E \to 0$ and E is essentially continuous. Thus, we should expect the nonclassical Planck distribution to go over to the classical Rayleigh–Jeans distribution at low frequencies, where $\Delta E \to 0$. Show that Equation 1.2 reduces to Equation 1.1 as $v \to 0$. (Recall that $e^x = 1 + x + (x^2/2!) + \cdots$, or, in other words, that $e^x \approx 1 + x$ when x is small.)

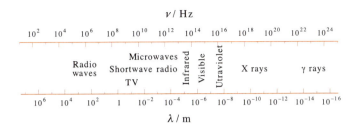

ν / Hz

FIGURE 1.19
The regions of electromagnetic radiation.

1–5. Before Planck's theoretical work on blackbody radiation, Wien showed empirically that (Equation 1.4)

$$\lambda_{max}T = 2.90 \times 10^{-3}\,\text{m}\cdot\text{K}$$

where λ_{max} is the wavelength at which the blackbody spectrum has its maximum value at a temperature T. This expression is called the Wien displacement law; derive it from Planck's theoretical expression for the blackbody distribution by differentiating Equation 1.3 with respect to λ. *Hint:* Set $hc/\lambda_{max}k_BT = x$ and derive the intermediate result $e^{-x} + (x/5) = 1$. This equation cannot be solved analytically but must be solved numerically. Solve it by iteration on a hand calculator, and show that $x = 4.965$ is the solution.

1–6. At what wavelength does the maximum in the energy-density distribution function for a blackbody occur if (a) $T = 300$ K, (b) $T = 3000$ K, and (c) $T = 10\,000$ K?

1–7. Sirius, one of the hottest known stars, has approximately a blackbody spectrum with $\lambda_{max} = 260$ nm. Estimate the surface temperature of Sirius.

1–8. The temperature of the fireball in a thermonuclear explosion can reach temperatures of approximately 10^7 K. What value of λ_{max} does this correspond to? In what region of the spectrum is this wavelength found (cf. Figure 1.19)?

1–9. We can use the Planck distribution to derive the *Stefan–Boltzmann law*, which gives the total energy density emitted by a blackbody as a function of temperature. Derive the Stefan–Boltzmann law by integrating the Planck distribution over all frequencies. *Hint:* You'll need to use the integral $\int_0^\infty dx\, x^3/(e^x - 1) = \pi^4/15$.

1–10. Can you derive the temperature dependence of the result in Problem 1–9 without evaluating the integral?

1–11. Calculate the energy of a photon for a wavelength of 100 pm (about one atomic diameter).

1–12. Express Planck's radiation law in terms of λ (and $d\lambda$) by using the relationship $\lambda\nu = c$.

1–13. Calculate the number of photons in a 2.00 mJ light pulse at (a) 1.06 μm, (b) 537 nm, and (c) 266 nm.

theory of relativity. The mass of a particle varies with its velocity according to

$$m = \frac{m_0}{[1 - (v^2/c^2)]^{1/2}} \tag{1.56}$$

where m_0, the mass as $v \to 0$, is called the *rest mass* of the particle. If we assume that the motion takes place in the x direction, the momentum is mv_x, and so

$$p_x = \frac{m_0 v_x}{[1 - (v_x^2/c^2)]^{1/2}} \tag{1.57}$$

Note that even though $m_0 = 0$ for a photon, p_x is not necessarily zero because $v_x = c$ and so $p_x \approx 0/0$, an indeterminate form. Recall that force F is equal to the rate of change of momentum or that

$$F = \frac{dp}{dt} \tag{1.58}$$

This is just Newton's second law. Kinetic energy can be defined as the work that is required to accelerate a particle from rest to some final velocity v. Because work is the integral of force times distance, the kinetic energy can be expressed as

$$T = \int_{v=0}^{v=v} F \, dx \tag{1.59}$$

where we have dropped the x subscripts for convenience. Equation 1.59 can be manipulated as

$$T = \int_{v=0}^{v=v} F \, dx = \int_{v=0}^{v=v} \frac{dp}{dt} \frac{dx}{dt} \, dt = \int_{v=0}^{v=v} v \frac{dp}{dt} \, dt$$

$$= \int_{v=0}^{v=v} v \frac{d(mv)}{dt} \, dt = \int_{v=0}^{v=v} v \, d(mv) \tag{1.60}$$

Remember now that m in these last two integrals is *not* constant but is a function of v through Equation 1.53. Note that if m were a constant, as it is in nonrelativistic (or classical) mechanics, then $T = \frac{1}{2}mv^2$, the classical result. Now substitute Equation 1.56 into Equation 1.60 to obtain

$$T = m_0 c^2 \left\{ \frac{1}{[1 - (v^2/c^2)]^{1/2}} - 1 \right\} \tag{1.61}$$

To obtain this result, you need the standard integral

$$\int \frac{x \, dx}{(ax^2 + b)^{3/2}} = -\frac{1}{a(ax^2 + b)^{1/2}}$$

Show that Equation 1.61 reduces to the classical result as $v/c \to 0$. By combining Equations 1.56 and 1.61, show that T can be written as

$$T = (m - m_0)c^2 \tag{1.62}$$

This equation is interpreted by considering mc^2 to be the total energy E of the particle and m_0c^2 to be the rest energy of the particle, so that

$$E = T + m_0c^2$$

Lastly now, eliminate v in favor of p by using Equations 1.57 and 1.61, and show that

$$(T + m_0c^2)^2 = (pc)^2 + (m_0c^2)^2$$

and using Equation 1.62, write this as

$$E = [(pc)^2 + (m_0c^2)^2]^{1/2} \tag{1.63}$$

which is our desired equation and a fundamental equation of the special theory of relativity. In the case of a photon, $m_0 = 0$ and $E = pc$. But E also equals $h\nu$ according to the quantum theory, and so we have $pc = h\nu$, which yields $p = h/\lambda$ because $c = \lambda\nu$.

1–54. In this problem we shall derive an expression for the interference pattern for a two-slit experiment. First show that $y(z) = A \cos[2\pi z/\lambda]$ represents a wave with amplitude A and wavelength λ. Now argue that $y(z, t) = A \cos[2\pi(z - vt)/\lambda]$ represents a similar wave that has been moved (translated) to the right by a distance vt. We say that $y(z, t) = A \cos[2\pi(z - vt)/\lambda]$ represents a wave form that is traveling to the right (a traveling wave) with a velocity v. If, for example, the wave is an electromagnetic wave, then $y(z, t)$ represents the electric field at a point (z, t) and we write $E(z, t) = E_0 \cos[2\pi(z - vt)/\lambda]$. If we let z_0 be the distance S_1P in Figure 1.15, then we can write the electric field at the point P as a superposition of the waves coming from the two slits, or

$$E(\theta) = E_0 \cos\left[\frac{2\pi}{\lambda}(z_0 - vt)\right] + E_0 \cos\left[\frac{2\pi}{\lambda}(z_0 + d \sin\theta - vt)\right]$$

where we write $E(\theta)$ to emphasize its dependence on the angle θ. Now use the trigonometric identity

$$\cos\alpha + \cos\beta = 2\cos\left(\frac{\alpha + \beta}{2}\right)\cos\left(\frac{\alpha - \beta}{2}\right)$$

to write $E(\theta)$ as

$$E(\theta) = 2E_0 \cos\left(\frac{\pi d \sin\theta}{\lambda}\right)\cos\left[\frac{2\pi}{\lambda}\left(z_0 - vt + \frac{d \sin\theta}{2}\right)\right]$$

The intensity of a wave is given by the square of its amplitude (this is proven in Problem 2–18), and so

$$I(\theta) = 4E_0^2 \cos^2 \left(\frac{\pi d \sin \theta}{\lambda} \right) \cos^2 \left[\frac{2\pi}{\lambda} \left(z_0 - vt + \frac{d \sin \theta}{2} \right) \right]$$

The recording on the screen in Figure 1.15 is an average of $I(\theta)$ over a period of time that amounts to many cycles of the wave. Using the relation $\cos^2 \alpha = \frac{1}{2}(1 + \cos 2\alpha)$, show that the average of the term $\cos^2[2\pi(z_0 - vt + d \sin \theta/2)/\lambda]$ in $I(\theta)$ is equal to 1/2, giving

$$I(\theta) = 2E_0^2 \cos^2 \left(\frac{\pi d \sin \theta}{\lambda} \right)$$

as our desired result. (We will derive this same result more easily in Section 2.6 using complex numbers.) Plot $I(\theta)/E_0^2$ against θ for typical values of a two-slit interference experiment, $d = 0.010$ mm and $\lambda = 6000$ Å and compare your result to Figure 2.10.

References

Resnick, R., Halliday, D. *Basic Concepts in Relativity and Early Quantum Theory*, 2nd ed. Macmillan Publishing: New York, 1992.

Al-Khalili, J. *Quantum: A Guide for the Perplexed*. Weidenfeld & Nicolson: London, 2004.

Greenstein, G., Zajonic, A. *The Quantum Challenge*, 2nd ed. Jones and Bartlett Publishers: Sudbury MA, 2005.

Halliday, D., Resnick, R., Walker, J. *Fundamentals of Physics*, 7th ed. Wiley & Sons: New York, 2004.

Segré, E. *From X-rays to Quarks: Modern Physicists and Their Discoveries*. W. H. Freeman. New York, 1980.

Gamow, G. *Thirty Years That Shook Physics: The Story of Quantum Theory*. Dover Publications: Mineola, NY, 1985.

Complex Numbers

Throughout chemistry, we frequently use complex numbers. In this MathChapter, we review some of the properties of complex numbers. Recall that complex numbers involve the imaginary unit, i, which is defined to be the square root of -1:

$$i = \sqrt{-1} \tag{A.1}$$

or

$$i^2 = -1 \tag{A.2}$$

Complex numbers arise naturally when solving certain quadratic equations. For example, the two solutions to

$$z^2 - 2z + 5 = 0$$

are given by

$$z = 1 \pm \sqrt{-4}$$

or

$$z = 1 \pm 2i$$

where 1 is said to be the real part and ± 2 the imaginary part of the complex number z. Generally, we write a complex number as

$$z = x + iy \tag{A.3}$$

with

$$x = \text{Re}(z) \qquad y = \text{Im}(z) \tag{A.4}$$

45

We add or subtract complex numbers by adding or subtracting their real and imaginary parts separately. For example, if $z_1 = 2 + 3i$ and $z_2 = 1 - 4i$, then

$$z_1 - z_2 = (2 - 1) + [3 - (-4)]i = 1 + 7i$$

Furthermore, we can write

$$2z_1 + 3z_2 = 2(2 + 3i) + 3(1 - 4i) = 4 + 6i + 3 - 12i = 7 - 6i$$

To multiply complex numbers together, we simply multiply the two quantities as binomials and use the fact that $i^2 = -1$. For example,

$$(2 - i)(-3 + 2i) = -6 + 3i + 4i - 2i^2$$
$$= -4 + 7i$$

To divide complex numbers, it is convenient to introduce the complex conjugate of z, which we denote by z^* and form by replacing i by $-i$. For example, if $z = x + iy$, then $z^* = x - iy$. Note that a complex number multiplied by its complex conjugate is a real quantity:

$$zz^* = (x + iy)(x - iy) = x^2 - i^2y^2 = x^2 + y^2 \tag{A.5}$$

The square root of zz^* is called the magnitude or the absolute value of z, and is denoted by $|z|$.

Consider now the quotient of two complex numbers:

$$z = \frac{2 + i}{1 + 2i}$$

This ratio can be written in the form $x + iy$ if we multiply both the numerator and the denominator by $1 - 2i$, the complex conjugate of the denominator:

$$z = \frac{2 + i}{1 + 2i}\left(\frac{1 - 2i}{1 - 2i}\right) = \frac{4 - 3i}{5} = \frac{4}{5} - \frac{3}{5}i$$

EXAMPLE A–1
Show that

$$z^{-1} = \frac{x}{x^2 + y^2} - \frac{iy}{x^2 + y^2}$$

SOLUTION:

$$z^{-1} = \frac{1}{z} = \frac{1}{x+iy} = \frac{1}{x+iy}\left(\frac{x-iy}{x-iy}\right) = \frac{x-iy}{x^2+y^2}$$

$$= \frac{x}{x^2+y^2} - \frac{iy}{x^2+y^2}$$

Because complex numbers consist of two parts, a real part and an imaginary part, we can represent a complex number by a point in a two-dimensional coordinate system where the real part is plotted along the horizontal (x) axis and the imaginary part is plotted along the vertical (y) axis, as in Figure A.1. The plane of such a figure is called the complex plane. If we draw a vector **r** from the origin of this figure to the point $z = (x, y)$, then the length of the vector, $r = (x^2 + y^2)^{\frac{1}{2}}$, is the magnitude or the absolute value of z. The angle θ that the vector **r** makes with the x axis is the phase angle of z.

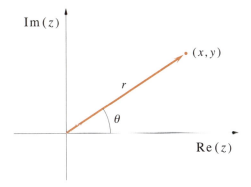

FIGURE A.1
Representation of a complex number $z = x + iy$ as a point in a two-dimensional coordinate system. The plane of this figure is called the complex plane.

EXAMPLE A–2
Given $z = 1 + i$, determine the magnitude, $|z|$, and the phase angle θ of z.

SOLUTION: The magnitude of z is given by the square root of

$$zz^* = (1+i)(1-i) = 2$$

or $|z| = 2^{\frac{1}{2}}$. Figure A.1 shows that the tangent of the phase angle is given by

$$\tan\theta = \frac{y}{x} = 1$$

or $\theta = 45°$, or $\pi/4$ radians. (Recall that 1 radian $= 180°/\pi$, or $1° = \pi/180$ radian.)

We can always express $z = x + iy$ in terms of r and θ by using Euler's formula,

$$e^{i\theta} = \cos\theta + i\sin\theta \tag{A.6}$$

which is derived in Problem A–10. Referring to Figure A.1, we see that

$$x = r\cos\theta \qquad \text{and} \qquad y = r\sin\theta$$

and so

$$z = x + iy = r\cos\theta + ir\sin\theta$$

$$= r(\cos\theta + i\sin\theta) = re^{i\theta} \tag{A.7}$$

where

$$r = \left(x^2 + y^2\right)^{\frac{1}{2}} \tag{A.8}$$

and

$$\tan\theta = \frac{y}{x} \tag{A.9}$$

Equation A.7, the polar representation of z, is often more convenient to use than Equation A.3, the cartesian representation of z.

Note that

$$z^* = re^{-i\theta} \tag{A.10}$$

and that

$$zz^* = \left(re^{-i\theta}\right)\left(re^{-i\theta}\right) = r^2 \tag{A.11}$$

or $r = (zz^*)^{\frac{1}{2}}$. Also note that $z = e^{i\theta}$ is a unit vector in the complex plane because $r^2 = (e^{i\theta})(e^{-i\theta}) = 1$. The following example proves this result in another way.

EXAMPLE A–3
Show that $e^{-i\theta} = \cos\theta - i\sin\theta$ and use this result and the polar representation of z to show that $|e^{i\theta}| = 1$.

SOLUTION: To prove that $e^{-i\theta} = \cos\theta - i\sin\theta$, we use Equation A.6 and the fact that $\cos\theta$ is an even function of θ [$\cos(-\theta) = \cos\theta$] and that $\sin\theta$ is an odd function of θ [$\sin(-\theta) = -\sin\theta$]. Therefore,

$$e^{-i\theta} = \cos\theta + i\sin(-\theta) = \cos\theta - i\sin\theta$$

Furthermore,

$$|e^{i\theta}| = [(\cos\theta + i\sin\theta)(\cos\theta - i\sin\theta)]^{1/2}$$

$$= (\cos^2\theta + \sin^2\theta)^{1/2} = 1$$

Problems

A–1. Find the real and imaginary parts of the following quantities:

(a) $(2 - i)^3$ (b) $e^{\pi i/2}$

(c) $e^{-2+i\pi/2}$ (d) $(\sqrt{2} + 2i)e^{-i\pi/2}$

A–2. If $z = x + 2iy$, then find

(a) $\text{Re}(z^*)$ (b) $\text{Re}(z^2)$

(c) $\text{Im}(z^2)$ (d) $\text{Re}(zz^*)$

(e) $\text{Im}(zz^*)$

A–3. Express the following complex numbers in the form $re^{i\theta}$:

(a) $6i$ (b) $4 - \sqrt{2}i$

(c) $-1 - 2i$ (d) $\pi + ei$

A–4. Express the following complex numbers in the form $x + iy$:

(a) $e^{\pi/4i}$ (b) $6e^{2\pi i/3}$

(c) $e^{-(\pi/4)i+\ln 2}$ (d) $e^{-2\pi i} + e^{4\pi i}$

A–5. Prove that $e^{i\pi} = -1$. Comment on the nature of the numbers in this relation.

A–6. Show that

$$\cos\theta = \frac{e^{i\theta} + e^{-i\theta}}{2}$$

and that

$$\sin\theta = \frac{e^{i\theta} - e^{-i\theta}}{2i}$$

A–7. Use Equation A.6 to derive

$$z^n = r^n(\cos\theta + i\sin\theta)^n = r^n(\cos n\theta + i\sin n\theta)$$

and from this, the formula of de Moivre:

$$(\cos\theta + i\sin\theta)^n = \cos n\theta + i\sin n\theta$$

A–8. Use the formula of de Moivre, which is given in Problem A–7, to derive the trigonometric identities

$$\cos 2\theta = \cos^2\theta - \sin^2\theta$$

$$\sin 2\theta = 2\sin\theta\cos\theta$$

$$\cos 3\theta = \cos^3\theta - 3\cos\theta\sin^2\theta$$

$$= 4\cos^3\theta - 3\cos\theta$$

$$\sin 3\theta = 3\cos^2\theta\sin\theta - \sin^3\theta$$

$$- 3\sin\theta - 4\sin^3\theta$$

A–9. Consider the set of functions

$$\Phi_m(\phi) = \frac{1}{\sqrt{2\pi}}e^{im\phi} \qquad \begin{cases} m = 0, \pm 1, \pm 2, \ldots \\ 0 \le \phi \le 2\pi \end{cases}$$

First show that

$$\int_0^{2\pi} d\phi\, \Phi_m(\phi) = \begin{cases} 0 & \text{for all values of } m \ne 0 \\ \sqrt{2\pi} & m = 0 \end{cases}$$

Now show that

$$\int_0^{2\pi} d\phi\, \Phi_m^*(\phi)\Phi_n(\phi) = \begin{cases} 0 & m \ne n \\ 1 & m = n \end{cases}$$

A–10. This problem offers a derivation of Euler's formula. Start with

$$f(\theta) = \ln(\cos\theta + i\sin\theta) \tag{1}$$

Show that

$$\frac{df}{d\theta} = i \tag{2}$$

Now integrate both sides of equation 2 to obtain

$$f(\theta) = \ln(\cos\theta + i\sin\theta) = i\theta + c \tag{3}$$

where c is a constant of integration. Show that $c = 0$ and then exponentiate equation 3 to obtain Euler's formula.

A–11. Using Euler's formula and assuming that x represents a real number, show that $\cos ix$ and $-i\sin ix$ are equivalent to real functions of the real variable x. These functions are defined as the hyperbolic cosine and hyperbolic sine functions, $\cosh x$ and $\sinh x$, respectively. Sketch these functions. Do they oscillate like $\sin x$ and $\cos x$?

A–12. Show that $\sinh ix = i \sin x$ and that $\cosh ix = \cos x$. (See the previous problem.)

A–13. Evaluate i^i.

A–14. The equation $x^2 = 1$ has two distinct roots, $x = \pm 1$. The equation $x^N = 1$ has N distinct roots, called the N roots of unity. This problem shows how to find the N roots of unity. We shall see that some of the roots turn out to be complex, so let's write the equation as $z^N = 1$. Now let $z = e^{i\theta}$ and obtain $e^{iN\theta} = 1$. Show that this must be equivalent to $e^{iN\theta} = 1$, or

$$\cos N\theta + i \sin N\theta = 1$$

Now argue that $N\theta = 2\pi n$, where n has the N distinct values $0, 1, 2, \ldots, N-1$ or that the N roots of unity are given by

$$z = e^{2\pi in/N} \qquad n = 0, 1, 2, \ldots, N-1$$

Show that we obtain $z = 1$ and $z = \pm 1$, for $N = 1$ and $N = 2$, respectively. Now show that

$$z = 1, \quad -\frac{1}{2} + i\frac{\sqrt{3}}{2}, \quad \text{and} \quad -\frac{1}{2} - i\frac{\sqrt{3}}{2}$$

for $N = 3$. Show that each of these roots is of unit magnitude. Plot these three roots in the complex plane. Now show that $z = 1, i, -1$, and $-i$ for $N = 4$ and that

$$z = 1, -1, \quad \frac{1}{2} \pm i\frac{\sqrt{3}}{2}, \quad \text{and} \quad -\frac{1}{2} \pm i\frac{\sqrt{3}}{2}$$

for $N = 6$. Plot the four roots for $N = 4$ and the six roots for $N = 6$ in the complex plane. Compare the plots for $N = 3$, $N = 4$, and $N = 6$. Do you see a pattern?

A–15. Using the results of Problem A–14, find the three distinct roots of $x^3 = 8$.

A–16. The *Schwartz inequality* says that if $z_1 = x_1 + iy_1$ and $z_2 = x_2 + iy_2$, then $x_1x_2 + y_1y_2 \leq |z_1| \cdot |z_2|$. To prove this inequality, start with its square

$$(x_1x_2 + y_1y_2)^2 \leq |z_1|^2 |z_2|^2 = (x_1^2 + y_1^2)(x_2^2 + y_2^2)$$

Now use the fact that $(x_1y_2 - x_2y_1)^2 \geq 0$ to prove the inequality.

A–17. The *triangle inequality* says that if z_1 and z_2 are complex numbers, then $|z_1 + z_2| \leq |z_1| + |z_2|$. To prove this inequality, start with

$$|z_1 + z_2|^2 = (x_1 + x_2)^2 + (y_1 + y_2)^2 = x_1^2 + y_1^2 + x_2^2 + y_2^2 + 2x_1x_2 + 2y_1y_2$$

Now use the Schwartz inequality (previous problem) to prove the inequality. Why do you think this is called the triangle inequality?

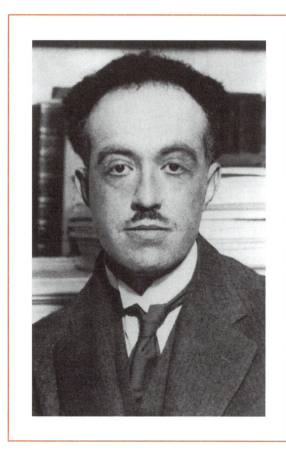

Louis de Broglie was born on August 15, 1892 in Dieppe, France, into an aristocratic family and died in 1987. He studied history as an undergraduate, but his interest turned to science as a result of his working with his older brother, Maurice, who had built his own private laboratory for X-ray research. De Broglie took up his formal studies in physics after World War I, receiving his D.Sc. from the University of Paris in 1924. His dissertation was on the wavelike properties of matter, a highly controversial and original proposal at that time. Using the special theory of relativity, de Broglie postulated that material particles should exhibit wavelike properties under certain conditions, just as radiation was known to exhibit particle-like properties. After receiving his D.Sc., he remained as a free lecturer at the Sorbonne and later was appointed professor of theoretical physics at the new Henri Poincaré Institute. He was professor of theoretical physics at the University of Paris from 1937 until his retirement in 1962. The wavelike properties he postulated were later demonstrated experimentally and are now exploited as a basis of the electron microscope. De Broglie spent the latter part of his career trying to obtain a causal interpretation of the wave mechanics to replace the probabilistic theories. He was awarded the Nobel Prize in Physics in 1929 "for his discovery of the wave nature of electrons."

CHAPTER 2

The Classical Wave Equation

In 1925, Erwin Schrödinger and Werner Heisenberg independently formulated a general quantum theory. At first sight, the two methods appeared to be different because Heisenberg's method is formulated in terms of matrices, whereas Schrödinger's method is formulated in terms of partial differential equations. Just a year later, however, Schrödinger showed that the two formulations are mathematically equivalent. Because most students of physical chemistry are not familiar with matrix algebra, quantum theory is customarily presented according to Schrödinger's formulation, the central feature of which is a partial differential equation now known as the *Schrödinger equation*. Partial differential equations may sound no more comforting than matrix algebra, but fortunarely we require only elementary calculus to treat the problems in this book. We learned in Chapter 1 that matter can behave as a wave, so it's not surprising that the Schrödinger equation (sometimes called the Schrödinger wave equation) describes wavelike behavior. The wave equation of classical physics describes various wave phenomena such as a vibrating string, a vibrating drum head, ocean waves, and acoustic waves. Not only does the classical wave equation provide a physical background to the Schrödinger equation, but, in addition, the mathematics involved in solving the classical wave equation are central to any discussion of quantum mechanics. Because most students of physical chemistry have little experience with classical wave equations, this chapter discusses this topic. In particular, we will solve the standard problem of a vibrating string because not only is the method of solving this problem similar to the method we will use to solve the Schrödinger equation, but it also gives us an excellent opportunity to relate the mathematical solution of a problem to the physical nature of the problem. Many of the problems at the end of the chapter illustrate the connection between physical problems and the mathematics developed in the chapter.

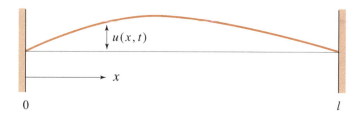

FIGURE 2.1
A vibrating string whose ends are fixed at 0 and l. The displacement of the vibration at position
x and time t is $u(x, t)$.

2.1 The One-Dimensional Wave Equation Describes the Motion of a Vibrating String

Consider a uniform string stretched between two fixed points, as shown in Figure 2.1.
The maximum displacement of the string from its equilibrium horizontal position is
called its *amplitude*. If we let $u(x, t)$ be the displacement of the string, then $u(x, t)$
satisfies the equation

$$\frac{\partial^2 u}{\partial x^2} = \frac{1}{v^2}\frac{\partial^2 u}{\partial t^2} \tag{2.1}$$

where v is the speed with which a disturbance moves along the string. Equation 2.1 is
the *classical wave equation*. Equation 2.1 is a *partial differential equation* because the
unknown, $u(x, t)$ in this case, occurs in partial derivatives. The variables x and t are
said to be the *independent variables* and $u(x, t)$, which depends upon x and t, is said to
be the *dependent variable*. Equation 2.1 is a *linear partial differential equation* because
$u(x, t)$ and its derivatives appear only to the first power and there are no cross terms.

In addition to having to satisfy Equation 2.1, the displacement $u(x, t)$ must satisfy
certain physical conditions as well. Because the ends of the string are held fixed, the
displacement at these two points is always zero, and so we have the requirement that

$$u(0, t) = 0 \quad \text{and} \quad u(l, t) = 0 \quad \text{(for all } t) \tag{2.2}$$

These two conditions are called *boundary conditions* because they specify the behavior
of $u(x, t)$ at the boundaries. Generally, a partial differential equation must be solved
subject to certain boundary conditions, the nature of which will be apparent on physical
grounds.

2.2 The Wave Equation Can Be Solved by the Method of Separation of Variables

The classical wave equation, as well as the Schrödinger equation and many other partial
differential equations that arise in physical chemistry, can be solved readily by a method

called *separation of variables*. We shall use the problem of a vibrating string to illustrate this method.

The key step in the method of separation of variables is to assume that $u(x, t)$ factors into a function of x, $X(x)$, times a function of t, $T(t)$, or that

$$u(x, t) = X(x)T(t) \tag{2.3}$$

If we substitute Equation 2.3 into Equation 2.1, we obtain

$$T(t)\frac{d^2X(x)}{dx^2} = \frac{1}{v^2}X(x)\frac{d^2T(t)}{dt^2} \tag{2.4}$$

Now we divide both sides of Equation 2.4 by $u(x, t) = X(x)T(t)$ and obtain

$$\frac{1}{X(x)}\frac{d^2X(x)}{dx^2} = \frac{1}{v^2T(t)}\frac{d^2T(t)}{dt^2} \tag{2.5}$$

The left side of Equation 2.5 is a function of x only and the right side is a function of t only. Because x and t are independent variables, each side of Equation 2.5 can be varied independently. The only way for the equality of the two sides to be preserved under any variation of x and t is for each side to be equal to a constant. If we let this constant be K, we can write

$$\frac{1}{X(x)}\frac{d^2X(x)}{dx^2} = K \tag{2.6}$$

and

$$\frac{1}{v^2T(t)}\frac{d^2T(t)}{dt^2} = K \tag{2.7}$$

where K is called the *separation constant* and will be determined later. Equations 2.6 and 2.7 can be written as

$$\frac{d^2X(x)}{dx^2} - KX(x) = 0 \tag{2.8}$$

and

$$\frac{d^2T(t)}{dt^2} - Kv^2T(t) = 0 \tag{2.9}$$

Equations 2.8 and 2.9 are called *ordinary differential equations* (as opposed to partial differential equations) because the unknowns, $X(x)$ and $T(t)$ in this case, occur as ordinary derivatives. Both of these differential equations are linear because the unknowns and their derivatives appear only to the first power and there are no cross terms. Furthermore, the coefficients of every term involving the unknowns in these equations are constants—that is, 1 and $-K$ in Equation 2.8 and 1 and $-Kv^2$ in Equation 2.9. These equations are called *linear differential equations with constant coefficients* and are quite easy to solve, as we shall see.

The value of K in Equations 2.8 and 2.9 is yet to be determined. We do not know right now whether K is positive, negative, or even zero. Let's first assume that $K = 0$. In this case, Equations 2.8 and 2.9 can be integrated immediately to find

$$X(x) = a_1 x + b_1 \tag{2.10}$$

and

$$T(t) = a_2 t + b_2 \tag{2.11}$$

where the a's and b's are just integration constants, which can be determined by using the boundary conditions given in Equations 2.2. In terms of $X(x)$ and $T(t)$, the boundary conditions are

$$u(0, t) = X(0)T(t) = 0 \quad \text{and} \quad u(l, t) = X(l)T(t) = 0$$

Because $T(t)$ certainly does not vanish for all t, we must have that

$$X(0) = 0 \quad \text{and} \quad X(l) = 0 \tag{2.12}$$

which is how the boundary conditions affect $X(x)$. Going back to Equation 2.10, we conclude that the only way to satisfy Equations 2.12 is for $a_1 = b_1 = 0$, which means that $X(x) = 0$ and that $u(x, t) = 0$ for all x. This is called a *trivial solution* to Equation 2.1 and is of no physical interest. (Throwing away solutions to mathematical equations should not disturb you. What we know from physics is that every physically acceptable solution $u(x, t)$ must satisfy Equation 2.1, *not* that every solution to the equation is physically acceptable.)

Now let's assume that $K > 0$ in Equation 2.8. To this end, write K as k^2, where k is real. This assures that K is positive because it is the square of a real number. In this case, Equation 2.8 becomes

$$\frac{d^2 X(x)}{dx^2} - k^2 X(x) = 0 \tag{2.13}$$

Experience shows that solutions to a linear differential equation with constant coefficients whose right side is equal to zero are of the form $X(x) = e^{\alpha x}$, where α is a constant to be determined.

EXAMPLE 2–1
Solve the equation

$$\frac{d^2 y}{dx^2} - 3\frac{dy}{dx} + 2y = 0$$

SOLUTION: If we substitute $y(x) = e^{\alpha x}$ into this differential equation, we obtain

$$\alpha^2 y - 3\alpha y + 2y = 0$$

$$\alpha^2 - 3\alpha + 2 = 0$$

$$(\alpha - 2)(\alpha - 1) = 0$$

or that $\alpha = 1$ and 2. The two solutions are $y(x) = e^x$ and $y(x) = e^{2x}$ and

$$y(x) = c_1 e^x + c_2 e^{2x}$$

is also a solution. Prove this by substituting this solution back into the original equation.

We now look for a solution to Equation 2.13 by letting $X(x) = e^{\alpha x}$ and get

$$(\alpha^2 - k^2) X(x) = 0$$

Therefore, either $(\alpha^2 - k^2)$ or $X(x)$ must equal zero. The case $X(x) = 0$ is a trivial solution, and so $\alpha^2 - k^2$ must equal zero. Therefore,

$$\alpha = \pm k$$

Thus, there are two solutions: $X(x) = e^{kx}$ and e^{-kx}. We can easily prove that

$$X(x) = c_1 e^{kx} + c_2 e^{-kx} \tag{2.14}$$

(where c_1 and c_2 are constants) is also a solution. This is the general solution to all differential equations with the form of Equation 2.13. The fact that a sum of the two solutions, e^{kx} and e^{-kx}, is also a solution is a direct consequence of Equation 2.13 being a *linear* differential equation (Problem 2–9). Note that the highest derivative in Equation 2.13 is a second derivative, which implies that in some sense we are performing two integrations when we find its solution. When we do two integrations, we always obtain two constants of integration. The solution we have found has two constants, c_1 and c_2, which suggests that it is the most general solution.

Applying the boundary conditions given by Equations 2.12 to Equation 2.14 gives

$$c_1 + c_2 = 0 \quad \text{and} \quad c_1 e^{kl} + c_2 e^{-kl} = 0$$

The only way to satisfy these conditions is with $c_1 = c_2 = 0$, and so once again, we find only a trivial solution.

So far, we have found only a trivial solution to Equation 2.1 if $K = 0$ or $K > 0$.

2.3 Some Differential Equations Have Oscillatory Solutions

Let's hope that assuming K to be negative gives us something interesting. If we set $K = -\beta^2$, then K is negative if β is real. In this case Equation 2.8 is

$$\frac{d^2 X(x)}{dx^2} + \beta^2 X(x) = 0 \tag{2.15}$$

Let $X(x) = e^{\alpha x}$ to obtain

$$(\alpha^2 + \beta^2) X(x) = 0$$

or that

$$\alpha = \pm i\beta$$

(MathChapter A). The general solution to Equation 2.15 is

$$X(x) = c_1 e^{i\beta x} + c_2 e^{-i\beta x} \tag{2.16}$$

We can easily verify that this is a solution by substituting Equation 2.16 directly into Equation 2.15.

It is sometimes more convenient to rewrite expressions such as $e^{i\beta x}$ or $e^{-i\beta x}$ in Equation 2.16 using Euler's formula (Equation A.6):

$$e^{\pm i\theta} = \cos\theta \pm i\sin\theta$$

If we substitute Euler's formula into Equation 2.16, we find

$$X(x) = c_1(\cos\beta x + i\sin\beta x) + c_2(\cos\beta x - i\sin\beta x)$$
$$= (c_1 + c_2)\cos\beta x + (ic_1 - ic_2)\sin\beta x$$

But $c_1 + c_2$ and $ic_1 - ic_2$ are also just constants, and if we call them c_3 and c_4, respectively, we can write

$$X(x) = c_3 \cos\beta x + c_4 \sin\beta x$$

instead of

$$X(x) = c_1 e^{i\beta x} + c_2 e^{-i\beta x}$$

These two forms for $X(x)$ are equivalent.

EXAMPLE 2–2

Prove that

$$y(x) = A \cos \beta x + B \sin \beta x$$

(where A and B are constants) is a solution to the differential equation

$$\frac{d^2 y}{dx^2} + \beta^2 y(x) = 0$$

SOLUTION: The first derivative of $y(x)$ is

$$\frac{dy}{dx} = -A\beta \sin \beta x + B\beta \cos \beta x$$

and the second derivative is

$$\frac{d^2 y}{dx^2} = -A\beta^2 \cos \beta x - B\beta^2 \sin \beta x$$

Therefore, we see that

$$\frac{d^2 y}{dx^2} + \beta^2 y(x) = 0$$

or that $y(x) = A \cos \beta x + B \sin \beta x$ is a solution of the differential equation

$$\frac{d^2 y}{dx^2} + \beta^2 y(x) = 0$$

The next example is important and one whose general solution should be learned.

EXAMPLE 2–3

Solve the equation

$$\frac{d^2 x}{dt^2} + \omega^2 x(t) = 0$$

subject to the initial conditions $x(0) = A$ and $dx/dt = 0$ at $t = 0$.

SOLUTION: In this case, we find $\alpha = \pm i\omega$ and

$$x(t) = c_1 e^{i\omega t} + c_2 e^{-i\omega t}$$

or

$$x(t) = c_3 \cos \omega t + c_4 \sin \omega t$$

Now

$$x(0) = c_3 = A$$

and

$$\left(\frac{dx}{dt} \right)_{t=0} = \omega c_4 = 0$$

implying that $c_4 = 0$ and that the particular solution we are seeking is

$$x(t) = A \cos \omega t$$

This solution is plotted in Figure 2.2. Note that it oscillates cosinusoidally in time, with an amplitude A and a frequency v, given by (see Problem 2–3)

$$v = \frac{\omega}{2\pi}$$

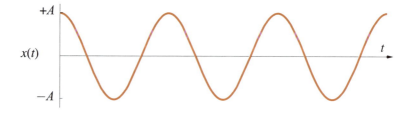

FIGURE 2.2
A plot of $x(t) = A \cos \omega t$, the solution to the problem in Example 2–3. The amplitude is A, the wavelength is $2\pi v/\omega$, and the frequency is $\omega/2\pi$.

Referring to Example 2–2, we see that the general solution to Equation 2.15 can be written as

$$X(x) = A \cos \beta x + B \sin \beta x \tag{2.17}$$

The boundary condition that $X(0) = 0$ implies that $A = 0$. The condition at the boundary $x = l$ says that

$$X(l) = B \sin \beta l = 0 \tag{2.18}$$

Equation 2.18 can be satisfied in two ways. One is that $B = 0$, but this along with the fact that $A = 0$ yields a trivial solution. The other way is to require that $\sin \beta l = 0$. Because $\sin \theta = 0$ when $\theta = 0, \pi, 2\pi, 3\pi, \ldots$, Equation 2.18 implies that

$$\beta l = n\pi \qquad n = 1, 2, 3, \ldots \qquad (2.19)$$

where we have omitted the $n = 0$ case because it leads to $\beta = 0$, and a trivial solution. Equation 2.19 determines the parameter β and hence the separation constant $K = -\beta^2$. So far, then, we have that

$$X(x) = B \sin \frac{n\pi x}{l} \qquad (2.20)$$

2.4 The General Solution to the Wave Equation Is a Superposition of Normal Modes

Remember that we have Equation 2.9 to solve also. Because $K = -\beta^2$, Equation 2.9 can be written as

$$\frac{d^2 T(t)}{dt^2} + \beta^2 v^2 T(t) = 0 \qquad (2.21)$$

where Equation 2.19 says that $\beta = n\pi/l$. Referring to the result obtained in Example 2–2 again, the general solution to Equation 2.21 is

$$T(t) = D \cos \omega_n t + E \sin \omega_n t \qquad (2.22)$$

where $\omega_n = \beta v = n\pi v/l$. We have no conditions to specify D and E, so the amplitude $u(x, t)$ is (cf. Equation 2.3)

$$u(x, t) = X(x)T(t)$$

$$= \left(B \sin \frac{n\pi x}{l} \right) \left(D \cos \omega_n t + E \sin \omega_n t \right)$$

$$= \left(F \cos \omega_n t + G \sin \omega_n t \right) \sin \frac{n\pi x}{l} \qquad n = 1, 2, \ldots$$

where we have let $F = DB$ and $G = EB$. Because there is a $u(x, t)$ for each integer n and because the values of F and G may depend on n, we should write $u(x, t)$ as

$$u_n(x, t) = \left(F_n \cos \omega_n t + G_n \sin \omega_n t \right) \sin \frac{n\pi x}{l} \qquad n = 1, 2, \ldots \qquad (2.23)$$

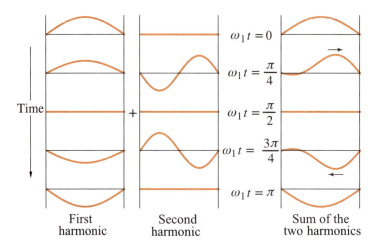

FIGURE 2.4
An illustration of how two standing waves can combine to give a traveling wave. In both
parts, time increases downward. The left portion shows the independent motion of the first two
harmonics. Both harmonics are standing waves; the first harmonic goes through half a cycle
and the second harmonic goes through one complete cycle in the time shown. The right side
shows the sum of the two harmonics. The sum is not a standing wave. As shown, the sum is a
traveling wave that travels back and forth between the fixed ends. The traveling wave has gone
through one-half of a cycle in the time shown.

2.5 A Vibrating Membrane Is Described by a Two-Dimensional Wave Equation

The generalization of Equation 2.1 to two dimensions is

$$\frac{\partial^2 u}{\partial x^2} + \frac{\partial^2 u}{\partial y^2} = \frac{1}{v^2}\frac{\partial^2 u}{\partial t^2} \tag{2.28}$$

where $u = u(x, y, t)$ and x, y, and t are the independent variables. We will apply this
equation to a rectangular membrane whose entire perimeter is clamped. By referring
to the geometry in Figure 2.5, we see that the boundary conditions that $u(x, y, t)$ must
satisfy (because its four edges are clamped) are

$$\left.\begin{array}{l} u(0, y) = u(a, y) = 0 \\ u(x, 0) = u(x, b) = 0 \end{array}\right\} \quad \text{(for all } t) \tag{2.29}$$

By applying the method of separation of variables to Equation 2.28, we assume
that $u(x, y, t)$ can be written as the product of a spatial part and a temporal part or that

$$u(x, y, t) = F(x, y)T(t) \tag{2.30}$$

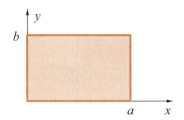

FIGURE 2.5
A rectangular membrane clamped along its perimeter.

We substitute Equation 2.30 into Equation 2.28 and divide both sides by $F(x, y)T(t)$ to find

$$\frac{1}{v^2 T(t)} \frac{d^2 T}{dt^2} = \frac{1}{F(x, y)} \left(\frac{\partial^2 F}{\partial x^2} + \frac{\partial^2 F}{\partial y^2} \right) \tag{2.31}$$

The right side of Equation 2.31 is a function of x and y only and the left side is a function of t only. The equality can be true for all t, x, and y only if both sides are equal to a constant. Anticipating that the separation constant will be negative, as it was in the previous sections, we write it as $-\beta^2$ and obtain the two separate equations

$$\frac{d^2 T}{dt^2} + v^2 \beta^2 T(t) = 0 \tag{2.32}$$

and

$$\frac{\partial^2 F}{\partial x^2} + \frac{\partial^2 F}{\partial y^2} + \beta^2 F(x, y) = 0 \tag{2.33}$$

Equation 2.33 is still a partial differential equation. To solve it, we once again use separation of variables. Substitute $F(x, y) = X(x)Y(y)$ into Equation 2.33 and divide both sides by $X(x)Y(y)$ to obtain

$$\frac{1}{X(x)} \frac{d^2 X}{dx^2} + \frac{1}{Y(y)} \frac{d^2 Y}{dy^2} + \beta^2 = 0 \tag{2.34}$$

Again we argue that because x and y are independent variables, the only way this equation can be valid is that

$$\frac{1}{X(x)} \frac{d^2 X}{dx^2} = -p^2 \tag{2.35}$$

and

$$\frac{1}{Y(y)} \frac{d^2 Y}{dy^2} = -q^2 \tag{2.36}$$

where p^2 and q^2 are separation constants, which according to Equation 2.34 must satisfy

$$p^2 + q^2 = \beta^2 \tag{2.37}$$

Equations 2.35 and 2.36 can be rewritten as

$$\frac{d^2 X}{dx^2} + p^2 X(x) = 0 \tag{2.38}$$

and

$$\frac{d^2 Y}{dy^2} + q^2 Y(y) = 0 \tag{2.39}$$

Equation 2.28, a partial differential equation in three variables, has been reduced to three ordinary differential equations (Equations 2.32, 2.38, and 2.39), each of which is exactly of the form discussed in Example 2–2. The solutions to Equations 2.38 and 2.39 are

$$X(x) = A \cos px + B \sin px \tag{2.40}$$

and

$$Y(y) = C \cos qy + D \sin qy \tag{2.41}$$

The boundary conditions, Equations 2.29, in terms of the functions $X(x)$ and $Y(y)$ are

$$X(0)Y(y) = X(a)Y(y) = 0$$

and

$$X(x)Y(0) = X(x)Y(b) = 0$$

which imply that

$$X(0) = X(a) = 0$$
$$Y(0) = Y(b) = 0 \tag{2.42}$$

Applying the first of Equations 2.42 to Equation 2.40 shows that $A = 0$ and $pa = n\pi$, so that

$$X_n(x) = B \sin \frac{n\pi x}{a} \qquad n = 1, 2, \ldots \tag{2.43}$$

In exactly the same manner, we find that $C = 0$ and $qb = m\pi$, where $m = 1, 2, \ldots$, and so

$$Y_m(y) = D \sin \frac{m\pi y}{b} \qquad m = 1, 2, \ldots \tag{2.44}$$

Recalling that $p^2 + q^2 = \beta^2$, we see that

$$\beta_{nm} = \pi \left(\frac{n^2}{a^2} + \frac{m^2}{b^2} \right)^{1/2} \quad \begin{array}{l} n = 1, 2, \ldots \\ m = 1, 2, \ldots \end{array} \tag{2.45}$$

where we have subscripted β to emphasize that it depends on the two integers n and m. Finally, now we solve Equation 2.32 for the time dependence:

$$T_{nm}(t) = E_{nm} \cos \omega_{nm} t + F_{nm} \sin \omega_{nm} t \tag{2.46}$$

where

$$\omega_{nm} = v \beta_{nm}$$

$$= v \pi \left(\frac{n^2}{a^2} + \frac{m^2}{b^2} \right)^{1/2} \tag{2.47}$$

According to Problem 2–5, Equation 2.46 can be written as

$$T_{nm}(t) = G_{nm} \cos(\omega_{nm} t + \phi_{nm}) \tag{2.48}$$

One solution to Equation 2.28 is given by the product $u_{nm}(x, y, t) = X_n(x) Y_m(y) T_{nm}(t)$, and the general solution is given by

$$u(x, y, t) = \sum_{n-1}^{\infty} \sum_{m=1}^{\infty} u_{nm}(x, y, t)$$

$$= \sum_{n=1}^{\infty} \sum_{m=1}^{\infty} A_{nm} \cos(\omega_{nm} t + \phi_{nm}) \sin \frac{n\pi x}{a} \sin \frac{m\pi y}{b} \tag{2.49}$$

As in the one-dimensional case of a vibrating string, we see that the general vibrational motion of a rectangular drum can be expressed as a superposition of normal modes, $u_{nm}(x, y, t)$. Some of these modes are shown in Figure 2.6. Note that in this two-dimensional problem we obtain *nodal lines*. In two-dimensional problems, the nodes are lines, as compared with points in one-dimensional problems. Figure 2.6 shows the normal modes for a case in which $a \neq b$. The case in which $a = b$ is an interesting

u_{11} u_{21} u_{31}

FIGURE 2.6
The first few normal modes of a rectangular membrane. The positive values of u are in orange and the negative values are in grey.

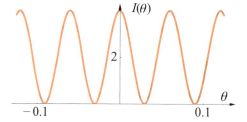

FIGURE 2.10
$I(\theta)$ given by Equation 2.58 plotted against θ for typical values of a two-slit experiment, $d = 0.010$ mm and $\lambda = 6000$ Å.

EXAMPLE 2–6
Suppose the second screen is placed $l = 1.0$ m beyond the slits. What is the separation of adjacent maxima along the second screen? Take $d = 0.010$ mm and $\lambda = 6000$ Å.

SOLUTION: The distance z along the second screen from the perpendicular line from halfway between the slits to the second screen is given by $z = l \sin \theta$ (see Figure 2.9). Therefore, the distance between successive maxima is given by

$$\Delta z = l \left[(n+1)\frac{\lambda}{d} - n\frac{\lambda}{d} \right] = l\frac{\lambda}{d}$$

$$= \frac{(1.0 \text{ m})(6.0 \times 10^{-7} \text{ m})}{1.0 \times 10^{-5} \text{ m}} = 60 \text{ mm}$$

Problem 2–19 has you generalize Equation 2.58 to the case of N slits. This problem nicely illustrates the convenience of using Equation 2.54 instead of Equation 2.53.

This chapter has presented a discussion of the wave equation and its solutions. In Chapter 3, we will use the mathematical methods developed here, and so we recommend doing many of the problems at the end of this chapter before going on. Several problems involve physical systems and serve as refreshers or introductions to classical mechanics.

Problems

2–1. Find the general solutions to the following differential equations.

(a) $\dfrac{d^2 y}{dx^2} - 4\dfrac{dy}{dx} + 3y = 0$

(b) $\dfrac{d^2 y}{dx^2} + 6\dfrac{dy}{dx} = 0$

(c) $\dfrac{dy}{dx} + 3y = 0$

(d) $\dfrac{d^2 y}{dx^2} + 2\dfrac{dy}{dx} - y = 0$

(e) $\dfrac{d^2 y}{dx^2} - 3\dfrac{dy}{dx} + 2y = 0$

2–2. Solve the following differential equations:

(a) $\dfrac{d^2 y}{dx^2} - 4y = 0 \qquad y(0) = 2 \qquad \dfrac{dy}{dx}(\text{at } x = 0) = 4$

(b) $\dfrac{d^2y}{dx^2} - 5\dfrac{dy}{dx} + 6y = 0$ $y(0) = -1$ $\dfrac{dy}{dx}(\text{at } x = 0) = 0$

(c) $\dfrac{dy}{dx} - 2y = 0$ $y(0) = 2$

2–3. Prove that $x(t) = \cos \omega t$ oscillates with a frequency $\nu = \omega/2\pi$. Prove that $x(t) = A \cos \omega t + B \sin \omega t$ oscillates with the same frequency, $\omega/2\pi$.

2–4. Solve the following differential equations:

(a) $\dfrac{d^2x}{dt^2} + \omega^2 x(t) = 0$ $x(0) = 0$ $\dfrac{dx}{dt}(\text{at } t = 0) = v_0$

(b) $\dfrac{d^2x}{dt^2} + \omega^2 x(t) = 0$ $x(0) = A$ $\dfrac{dx}{dt}(\text{at } t = 0) = v_0$

Prove in both cases that $x(t)$ oscillates with frequency $\omega/2\pi$.

2–5. The general solution to the differential equation

$$\dfrac{d^2x}{dt^2} + \omega^2 x(t) = 0$$

is

$$x(t) = c_1 \cos \omega t + c_2 \sin \omega t$$

For convenience, we often write this solution in the equivalent forms

$$x(t) = A \sin(\omega t + \phi) \qquad \text{or} \qquad x(t) = B \cos(\omega t + \psi)$$

Show that all three of these expressions for $x(t)$ are equivalent. Derive equations for A and ϕ in terms of c_1 and c_2, and for B and ψ in terms of c_1 and c_2. Show that all three forms of $x(t)$ oscillate with frequency $\omega/2\pi$. *Hint:* Use the trigonometric identities

$$\sin(\alpha + \beta) = \sin \alpha \cos \beta + \cos \alpha \sin \beta \qquad \text{and} \qquad \cos(\alpha + \beta) = \cos \alpha \cos \beta - \sin \alpha \sin \beta$$

2–6. In all the differential equations that we have discussed so far, the values of the exponents α that we have found have been either real or purely imaginary. Let us consider a case in which α turns out to be complex. Consider the equation

$$\dfrac{d^2y}{dx^2} + 2\dfrac{dy}{dx} + 10y = 0$$

If we substitute $y(x) = e^{\alpha x}$ into this equation, we find that $\alpha^2 + 2\alpha + 10 = 0$ or that $\alpha = -1 \pm 3i$. The general solution is

$$y(x) = c_1 e^{(-1+3i)x} + c_2 e^{(-1-3i)x}$$

$$= c_1 e^{-x} e^{3ix} + c_2 e^{-x} e^{-3ix}$$

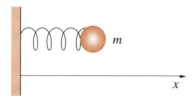

FIGURE 2.11
A body of mass m connected to a wall by a spring.

Show that $y(x)$ can be written in the equivalent form

$$y(x) = e^{-x}(c_3 \cos 3x + c_4 \sin 3x)$$

Thus we see that complex values of the α's lead to trigonometric solutions modulated by an exponential factor. Solve the following equations:

(a) $\dfrac{d^2 y}{dx^2} + 2\dfrac{dy}{dx} + 2y = 0$

(b) $\dfrac{d^2 y}{dx^2} - 6\dfrac{dy}{dx} + 25y = 0$

(c) $\dfrac{d^2 y}{dx^2} + 2\beta\dfrac{dy}{dx} + (\beta^2 + \omega^2)y = 0$

(d) $\dfrac{d^2 y}{dx^2} + 4\dfrac{dy}{dx} + 5y = 0 \qquad y(0) = 1 \qquad \dfrac{dy}{dx}(\text{at } x = 0) = -3$

2–7. This problem develops the idea of a classical harmonic oscillator. Consider a mass m attached to a spring, as shown in Figure 2.11. Suppose there is no gravitational force acting on m so that the only force is from the spring. Let the relaxed or undistorted length of the spring be x_0. Hooke's law says that the force acting on the mass m is $f = -k(x - x_0)$, where k is a constant characteristic of the spring and is called the force constant of the spring. Note that the minus sign indicates the direction of the force: to the left if $x > x_0$ (extended) and to the right if $x < x_0$ (compressed). The momentum of the mass is

$$p = m\frac{dx}{dt} = m\frac{d(x - x_0)}{dt}$$

Newton's second law says that the rate of change of momentum is equal to a force

$$\frac{dp}{dt} = f$$

Replacing $f(x)$ by Hooke's law, show that

$$m\frac{d^2 x}{dt^2} = -k(x - x_0)$$

Upon letting $\xi = x - x_0$ be the displacement of the spring from its undistorted length, then

$$m\frac{d^2 \xi}{dt^2} + k\xi = 0$$

Given that the mass starts at $\xi = 0$ with an intial velocity v_0, show that the displacement is given by

$$\xi(t) = v_0 \left(\frac{m}{k}\right)^{1/2} \sin\left[\left(\frac{k}{m}\right)^{1/2} t\right]$$

Interpret and discuss this solution. What does the motion look like? What is the frequency? What is the amplitude?

2–8. Modify Problem 2–7 to the case where the mass is moving through a viscous medium with a viscous force proportional to but opposite the velocity. Show that the equation of motion is

$$m\frac{d^2\xi}{dt^2} + \gamma\frac{d\xi}{dt} + k\xi^2 = 0$$

where γ is the viscous drag coefficient. Solve this equation and discuss the behavior of $\xi(t)$ for various values of m, γ, and k. This system is called a *damped harmonic oscillator*.

2–9. Consider the linear second-order differential equation

$$\frac{d^2y}{dx^2} + a_1(x)\frac{dy}{dx} + a_0(x)y(x) = 0$$

Note that this equation is linear because $y(x)$ and its derivatives appear only to the first power and there are no cross terms. It does not have constant coefficients, however, and there is no general, simple method for solving it like there is if the coefficients were constants. In fact, each equation of this type must be treated more or less individually. Nevertheless, because it is linear, we must have that, if $y_1(x)$ and $y_2(x)$ are any two solutions, then a linear combination,

$$y(x) = c_1 y_1(x) + c_2 y_2(x)$$

where c_1 and c_2 are constants, is also a solution. Prove that $y(x)$ is a solution.

2–10. We will see in Chapter 3 that the Schrödinger equation for a particle of mass m that is constrained to move freely along a line between 0 and a is

$$\frac{d^2\psi}{dx^2} + \left(\frac{8\pi^2 mE}{h^2}\right)\psi(x) = 0$$

with the boundary condition

$$\psi(0) = \psi(a) = 0$$

In this equation, E is the energy of the particle and $\psi(x)$ is its wave function. Solve this differential equation for $\psi(x)$, apply the boundary conditions, and show that the energy can have only the values

$$E_n = \frac{n^2 h^2}{8ma^2} \qquad n = 1, 2, 3, \ldots$$

or that the energy is quantized.

2–11. Prove that the number of nodes for a vibrating string clamped at both ends is $n - 1$ for the nth harmonic.

2–12. Prove that

$$
y(x, t) = A \sin \left[\frac{2\pi}{\lambda} (x - vt) \right]
$$

is a wave of wavelength λ and frequency $\nu = v/\lambda$ traveling to the right with a velocity v.

2–13. Sketch the normal modes of a vibrating rectangular membrane and show that they look like those shown in Figure 2.6.

2–14. This problem is the extension of Problem 2–10 to two dimensions. In this case, the particle is constrained to move freely over the surface of a rectangle of sides a and b. The Schrödinger equation for this problem is

$$
\frac{\partial^2 \psi}{\partial x^2} + \frac{\partial^2 \psi}{\partial y^2} + \left(\frac{8\pi^2 m E}{h^2} \right) \psi(x, y) = 0
$$

with the boundary conditions

$$
\psi(0, y) = \psi(a, y) = 0 \qquad \text{for all } y, \qquad 0 \le y \le b
$$
$$
\psi(x, 0) = \psi(x, b) = 0 \qquad \text{for all } x, \qquad 0 \le x \le a
$$

Solve this equation for $\psi(x, y)$, apply the boundary conditions, and show that the energy is quantized according to

$$
E_{n_x, n_y} = \frac{n_x^2 h^2}{8ma^2} + \frac{n_y^2 h^2}{8mb^2} \qquad \begin{cases} n_x = 1, 2, 3, \ldots \\ n_y = 1, 2, 3, \ldots \end{cases}
$$

2–15. Extend Problems 2–10 and 2–14 to three dimensions, where a particle is constrained to move freely throughout a rectangular box of sides a, b, and c. The Schrödinger equation for this system is

$$
\frac{\partial^2 \psi}{\partial x^2} + \frac{\partial^2 \psi}{\partial y^2} + \frac{\partial^2 \psi}{\partial z^2} + \left(\frac{8\pi^2 m E}{h^2} \right) \psi(x, y, z) = 0
$$

and the boundary conditions are that $\psi(x, y, z)$ vanishes over all the surfaces of the box.

2–16. Show that Equations 2.46 and 2.48 are equivalent. How are G_{nm} and ϕ_{nm} in Equation 2.48 related to the quantities in Equation 2.46?

2–17. Prove that $u_n(x, t)$, the nth normal mode of a vibrating string (Equation 2.23), can be written as the superposition of two similar traveling waves moving in opposite directions. Let $\phi_n = 0$ in Equation 2.25.

2–18. This problem shows that the intensity of a wave is proportional to the square of its amplitude. Figure 2.12 illustrates the geometry of a vibrating string. Because the velocity

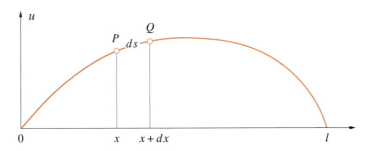

FIGURE 2.12
The geometry of a vibrating string.

at any point of the string is $\partial u/\partial t$, the kinetic energy, T, of the entire string is

$$T = \int_0^l \frac{1}{2}\rho \left(\frac{\partial u}{\partial t}\right)^2 dx$$

where ρ is the linear mass density of the string. The potential energy is found by considering the increase in length of the small arc PQ of length ds in Figure 2.12. The segment of the string along that arc has increased its length from dx to ds. Therefore, the potential energy associated with this increase is

$$V = \int_0^l \tau(ds - dx)$$

where τ is the tension in the string. Using the fact that $(ds)^2 = (dx)^2 + (du)^2$, show that

$$V = \int_0^l \tau \left\{\left[1 + \left(\frac{\partial u}{\partial x}\right)^2\right]^{1/2} - 1\right\} dx$$

Using the fact that $(1+x)^{1/2} \approx 1 + (x/2)$ for small x, show that

$$V = \frac{1}{2}\tau \int_0^l \left(\frac{\partial u}{\partial x}\right)^2 dx$$

for small displacements.

The total energy of the vibrating string is the sum of T and V and so

$$E = \frac{\rho}{2}\int_0^l \left(\frac{\partial u}{\partial t}\right)^2 dx + \frac{\tau}{2}\int_0^l \left(\frac{\partial u}{\partial x}\right)^2 dx$$

Equation 2.25 shows that the nth normal mode can be written in the form

$$u_n(x, l) = D_n \cos(\omega_n t + \phi_n) \sin \frac{n\pi x}{l}$$

where $\omega_n = vn\pi/l$. Using this equation, show that

$$T_n = \frac{\pi^2 v^2 n^2 \rho}{4l} D_n^2 \sin^2(\omega_n t + \phi_n)$$

and

$$V_n = \frac{\pi^2 n^2 \tau}{4l} D_n^2 \cos^2(\omega_n t + \phi_n)$$

Using the fact that $v = (\tau/\rho)^{1/2}$, show that

$$E_n = \frac{\pi^2 v^2 n^2 \rho}{4l} D_n^2$$

Note that the total energy, or intensity, is proportional to the square of the amplitude. Although we have shown this proportionality only for the case of a vibrating string, it is a general result and shows that the intensity of a wave is proportional to the square of the amplitude. If we had carried everything through in complex notation instead of sines and cosines, then we would have found that E_n is proportional to $|D_n|^2$ instead of just D_n^2.

Generally, there are many normal modes present at the same time, and the complete solution is

$$u(x, t) = \sum_{n=1}^{\infty} D_n \cos(\omega_n t + \phi_n) \sin \frac{n\pi x}{l}$$

Using the fact that (see Problem 3–23)

$$\int_0^l \sin \frac{n\pi x}{l} \sin \frac{m\pi x}{l} dx = 0 \qquad \text{if } m \neq n$$

show that

$$E_n = \frac{\pi^2 v^2 \rho}{4l} \sum_{n=1}^{\infty} n^2 D_n^2$$

2–19. In this problem, we'll generalize Equation 2.58 to the case of N slits. Figure 2.13 summarizes the geometrical setup. Realize that all the rays impinge on one point P; they are approximately parallel, as in Figure 2.9. The electric field at the point P is given by

$$E(\theta) = E_0 e^{i(k\rho_1 - \omega t)} + E_0 e^{i(k\rho_2 - \omega t)} + \cdots + E_0 e^{i(k\rho_N - \omega t)}$$

$$= E_0 e^{-i\omega t} \sum_{n=1}^{N} e^{ik\rho_n}$$

which is the generalization of Equation 2.56. Now argue that the path difference between any two adjacent sources is

$$\delta = \rho_{n+1} - \rho_n \qquad n = 1, 2, \ldots, N - 1$$

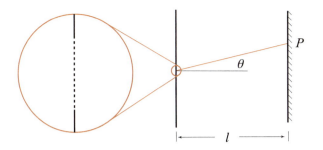

FIGURE 2.13
The geometry used in Problem 2–19 for the derivation of the interference pattern for N slits.

and therefore that $\rho_{n+1} - \rho_1 = n\delta$, where $\delta = d \sin \theta$. Now show that

$$E(\theta) = E_0 e^{i(k\rho_1 - \omega t)} \sum_{n=0}^{N-1} e^{ikn\delta}$$

Now use the fact that

$$\sum_{n=0}^{N-1} x^n = \frac{1 - x^N}{1 - x}$$

(see MathChapter D for a simple proof of this formula) to write $E(\theta)$ as

$$E(\theta) = E_0 e^{i(k\rho_1 - \omega t)} \left(\frac{1 - e^{ikN\delta}}{1 - e^{ik\delta}} \right)$$

Now factor $e^{ikN\delta/2}$ from the numerator and $e^{ik\delta/2}$ from the denominator to get

$$E(\theta) = E_0 e^{i(k\rho_1 - \omega t + kN\delta/2 - k\delta/2)} \left(\frac{e^{ikN\delta/2} + e^{-ikN\delta/2}}{e^{ik\delta/2} - e^{-ik\delta/2}} \right)$$

Finally, use the fact that $(e^{ix} - e^{-ix})/2i = \sin x$ and that $k = 2\pi/\lambda$ to show that the intensity of radiation at the point P is

$$I(\theta) = A^*(\theta)A(\theta) = E_0^2 \left(\frac{\sin[(\pi Nd/\lambda) \sin \theta]}{\sin[(\pi d/\lambda) \sin \theta]} \right)^2 \tag{2.59}$$

Plot $I(\theta)$ against θ for $d = 0.010$ mm and $\lambda = 6000$ Å and show that it is the same result as in Figure 2.10.

2–20. Show that Equation 2.59 from the previous problem reduces to Equation 2.58 when $N = 2$.

2–21. Plot Equation 2.59 for $N = 3$.

2–22. Plot Equation 2.59 for $N = 1$. What's going on here?

2–23. If you plot Equation 2.59 for a number of values of N, you'll see that the number of minima between the large maxima (called the principal maxima) is equal to $N - 1$. Can you prove this?

Problems 2–24 through 2–29 illustrate some other applications of differential equations to classical mechanics.

Many problems in classical mechanics can be reduced to the problem of solving a differential equation with constant coefficients (cf. Problem 2–7). The basic starting point is Newton's second law, which says that the rate of change of momentum is equal to the force acting on a body. Momentum p equals mv, and so if the mass is constant, then in one dimension we have

$$\frac{dp}{dt} = m\frac{dv}{dt} = m\frac{d^2x}{dt^2} = f$$

If we are given the force as a function of x, then this equation is a differential equation for $x(t)$, which is called the trajectory of the particle. Going back to the simple harmonic oscillator discussed in Problem 2–7, if we let x be the displacement of the mass from its equilibrium position, then Hooke's law says that $f(x) = -kx$, and the differential equation corresponding to Newton's second law is

$$\frac{d^2x}{dt^2} + kx(t) = 0$$

a differential equation that we have seen several times.

2–24. Consider a body falling freely from a height x_0 according to Figure 2.14. If we neglect air resistance or viscous drag, the only force acting upon the body is the gravitational force mg. Using the coordinates in Figure 2.14a, mg acts in the same direction as x and

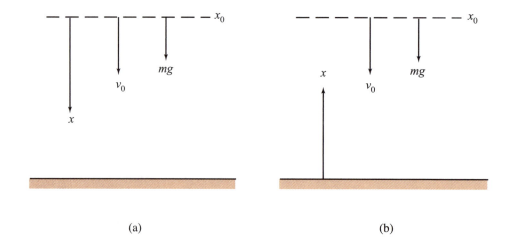

(a)   (b)

FIGURE 2.14
(a) A coordinate system for a body falling from a height x_0, and (b) a different coordinate system for a body falling from a height x_0.

so the differential equation corresponding to Newton's second law is

$$m\frac{d^2x}{dt^2} = mg$$

Show that

$$x(t) = \frac{1}{2}gt^2 + v_0 t + x_0$$

where x_0 and v_0 are the initial values of x and v. According to Figure 2.14a, $x_0 = 0$ and so

$$x(t) = \frac{1}{2}gt^2 + v_0 t$$

If the particle is just dropped, then $v_0 = 0$ and so

$$x(t) = \frac{1}{2}gt^2$$

Discuss this solution.

Now do the same problem using Figure 2.14b as the definition of the various quantities involved, and show that although the equations may look different from those above, they say exactly the same thing because the diagram we draw to define the direction of x, v_0, and mg does not affect the falling body.

2–25. Derive an equation for the maximum height a body will reach if it is shot straight upward with a velocity v_0. Refer to Figure 2.14b but realize that in this case v_0 points upward. How long will it take for the body to return to earth?

2–26. Consider a simple pendulum as shown in Figure 2.15. We let the length of the pendulum be l and assume that all the mass of the pendulum is concentrated at its end, as shown in Figure 2.15. A physical example of this case might be a mass suspended by a string. We assume that the motion of the pendulum is set up such that it oscillates within a plane so that we have a problem in plane polar coordinates. Let the distance along the arc in the figure describe the motion of the pendulum, so that its momentum is $m\,ds/dt = ml\,d\theta/dt$ and its rate of change of momentum is $ml\,d^2\theta/dt^2$. Show that the component of force in the direction of motion is $-mg \sin \theta$, where the minus sign occurs because the direction of this

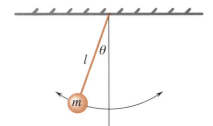

FIGURE 2.15
The coordinate system describing an oscillating pendulum.

force is opposite that of the angle θ. Show that the equation of motion is

$$ml\frac{d^2\theta}{dt^2} = -mg\sin\theta$$

Now assume that the motion takes place only through very small angles and show that the motion becomes that of a simple harmonic oscillator. What is the natural frequency of this harmonic oscillator? *Hint:* Use the fact that $\sin\theta \approx \theta$ for small values of θ.

2–27. Consider the motion of a pendulum like that in Problem 2–26 but swinging in a viscous medium. Suppose that the viscous force is proportional to but oppositely directed to its velocity; that is,

$$f_{\text{viscous}} = -\lambda\frac{ds}{dt} = -\lambda l\frac{d\theta}{dt}$$

where λ is a viscous drag coefficient. Show that for small angles, Newton's equation is

$$ml\frac{d^2\theta}{dt^2} + \lambda l\frac{d\theta}{dt} + mg\theta = 0$$

Show that there is no harmonic motion if

$$\lambda^2 > \frac{4m^2g}{l}$$

Does it make physical sense that the medium can be so viscous that the pendulum undergoes no harmonic motion?

2–28. Consider two pendulums of equal lengths and masses that are connected by a spring that obeys Hooke's law (Problem 2–7). This system is shown in Figure 2.16. Assuming that the motion takes place in a plane and that the angular displacement of each pendulum from the horizontal is small, show that the equations of motion for this system are

$$m\frac{d^2x}{dt^2} = -m\omega_0^2 x - k(x - y)$$

$$m\frac{d^2y}{dt^2} = -m\omega_0^2 y - k(y - x)$$

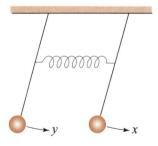

FIGURE 2.16
Two pendulums coupled by a spring that obeys Hooke's law.

where ω_0 is the natural vibrational frequency of each isolated pendulum [i.e., $\omega_0 = (g/l)^{1/2}$] and k is the force constant of the connecting spring. In order to solve these two simultaneous differential equations, assume that the two pendulums swing harmonically and so try

$$x(t) = Ae^{i\omega t} \qquad y(t) = Be^{i\omega t}$$

Substitute these expressions into the two differential equations and obtain

$$\left(\omega^2 - \omega_0^2 - \frac{k}{m}\right) A = -\frac{k}{m} B$$

$$\left(\omega^2 - \omega_0^2 - \frac{k}{m}\right) B = -\frac{k}{m} A$$

Now we have two simultaneous linear homogeneous algebraic equations for the two amplitudes A and B. We shall learn in MathChapter E that the determinant of the coefficients must vanish in order for there to be a nontrivial solution. Show that this condition gives

$$\left(\omega^2 - \omega_0^2 - \frac{k}{m}\right)^2 = \left(\frac{k}{m}\right)^2$$

Now show that there are two natural frequencies for this system—namely,

$$\omega_1^2 = \omega_0^2 \qquad \text{and} \qquad \omega_2^2 = \omega_0^2 + \frac{2k}{m}$$

Interpret the motion associated with these frequencies by substituting ω_1^2 and ω_2^2 back into the two equations for A and B. The motion associated with these values of A and B are called *normal modes*, and any complicated, general motion of this system can be written as a linear combination of these normal modes. Notice that there are two coordinates (x and y) in this problem and two normal modes. We shall see in Chapter 5 that the complicated vibrational motion of molecules can be resolved into a linear combination of natural, or normal, modes.

2–29. Problem 2–28 can be solved by introducing center-of-mass and relative coordinates (cf. Section 5.3). Add and subtract the differential equations for $x(t)$ and $y(t)$ and then introduce the new variables

$$\eta = x + y \qquad \text{and} \qquad \xi = x - y$$

Show that the differential equations for η and ξ are independent. Solve each one and compare your results to those of Problem 2–28.

2–30. Equation 2.51 suggests that $\sin[2\pi(x + vt)/\lambda]$ and $\sin[2\pi(x - vt)/\lambda]$ are solutions to the wave equation. Now prove that

$$u(x, t) = \phi(x + vt) + \psi(x - vt)$$

where ϕ and ψ are suitably well-behaved but otherwise arbitrary functions, is also a solution. This solution is known as d'Alembert's solution of the wave equation. Give a physical interpretation of $\phi(x + vt)$ and $\psi(x - vt)$. D'Alembert's solution can be obtained by transforming the independent variables in the wave equation. Introduce the new variables

$$\eta = x + vt \qquad \text{and} \qquad \xi = x - vt$$

into Equation 2.1 and show that the wave equation becomes

$$\frac{\partial^2 u}{\partial \eta \partial \xi} = 0$$

This form of the wave equation can be solved by two successive integrations. The first integration gives

$$\frac{\partial u}{\partial \eta} = f(\eta)$$

where $f(\eta)$ is an arbitrary function of η only. Show that a second integration gives

$$u(\eta, \xi) = \int f(\eta)d\eta + \psi(\xi) = \phi(\eta) + \psi(\xi)$$

or

$$u(\eta, \xi) = \phi(x + vt) + \psi(x - vt)$$

References

Barrante, J. *Applied Mathematics for Physical Chemistry,* 3rd ed. Prentice Hall: Upper Saddle River, NJ, 2003.

Mortimer, R. G. *Mathematics for Physical Chemistry,* 2nd ed. Academic Press: San Diego, CA, 1999.

Steiner, E. *The Chemistry Maths Book.* Oxford University Press: New York, 1996.

Edwards, C. H., Penney, D. E. *Elementary Differential Equations with Boundary Value Problems,* 5th ed. Prentice Hall: Upper Saddle River, NJ, 2004.

Farlow, S. J. *Partial Differential Equations for Scientists and Engineers.* Dover Publications: Mineola, MN, 1993.

Stephenson, G. *Partial Differential Equations for Scientists and Engineers.* Longman, Inc.: New York, 1985.

McQuarrie, D. A. *Mathematical Methods for Scientists and Engineers.* University Science Books: Sausalito, CA, 2003.

Probability and Statistics

In many of the following chapters, we will deal with probability distributions, average values, and standard deviations. Consequently, we take a few pages here to discuss some basic ideas of probability and show how to calculate average quantities in general.

Consider some experiment, such as the tossing of a coin or the rolling of a die, that has n possible outcomes, each with probability p_j, where $j = 1, 2, \ldots, n$. If the experiment is repeated indefinitely, we intuitively expect that

$$p_j = \lim_{N \to \infty} \frac{N_j}{N} \qquad j = 1, 2, \ldots, n \tag{B.1}$$

where N_j is the number of times that the event j occurs and N is the total number of repetitions of the experiment. Because $0 \leq N_j \leq N$, p_j must satisfy the condition

$$0 \leq p_j \leq 1 \tag{B.2}$$

When $p_j = 1$, we say the event j is a certainty and when $p_j = 0$, we say it is impossible. In addition, because

$$\sum_{j=1}^{n} N_j = N$$

we have the normalization condition,

$$\sum_{j=1}^{n} p_j = 1 \tag{B.3}$$

Equation B.3 means that the probability that some event occurs is a certainty. Suppose now that some number x_j is associated with the outcome j. Then we define the *average* of x or the *mean* of x to be

$$\langle x \rangle = \sum_{j=1}^{n} x_j p_j = \sum_{j=1}^{n} x_j p(x_j) \tag{B.4}$$

where in the last term we have used the expanded notation $p(x_j)$, meaning the probability of realizing the number x_j. We will denote an average of a quantity by enclosing the quantity in angular brackets.

EXAMPLE B–1

Suppose we are given the following data:

x	$p(x)$
1	0.20
3	0.25
4	0.55

Calculate the average value of x.

SOLUTION: Using Equation B.4, we have

$$\langle x \rangle = (1)(0.20) + (3)(0.25) + (4)(0.55) = 3.15$$

It is helpful to interpret a probability distribution like p_j as a distribution of a unit mass along the x axis in a discrete manner such that p_j is the fraction of mass located at the point x_j (Figure B.1). According to this interpretation, the average value of x is the center of mass of this system.

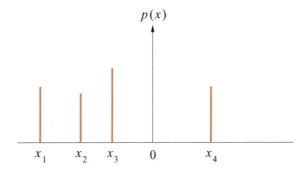

FIGURE B.1
The discrete probability frequency function or probability density, $p(x)$.

Another quantity of importance is

$$\langle x^2 \rangle = \sum_{j=1}^{n} x_j^2 p_j \tag{B.5}$$

The quantity $\langle x^2 \rangle$ is called the *second moment* of the distribution $\{p_j\}$ and is analogous to the moment of inertia.

EXAMPLE B–2

Calculate the second moment of the data given in Example B–1.

SOLUTION: Using Equation B.5, we have

$$\langle x^2 \rangle = (1)^2(0.20) + (3)^2(0.25) + (4)^2(0.55) = 11.25$$

Note from Examples B–1 and B–2 that $\langle x^2 \rangle \neq \langle x \rangle^2$. This nonequality is a general result that we will prove below.

A physically more interesting quantity than $\langle x^2 \rangle$ is the *second central moment*, or the *variance*, defined by

$$\sigma_x^2 = \langle (x - \langle x \rangle)^2 \rangle = \sum_{j=1}^{n}(x_j - \langle x \rangle)^2 p_j \tag{B.6}$$

As the notation suggests, we denote the square root of the quantity in Equation B.6 by σ_x, which is called the *standard deviation*. From the summation in Equation B.6, we can see that σ_x^2 will be large if x_j is likely to differ from $\langle x \rangle$, because in that case $(x_j - \langle x \rangle)$ and so $(x_j - \langle x \rangle)^2$ will be large for the significant values of p_j. On the other hand, σ_x^2 will be small if x_j is not likely to differ from $\langle x \rangle$, or if the x_j cluster around $\langle x \rangle$, because then $(x_j - \langle x \rangle)^2$ will be small for the significant values of p_j. Thus, we see that either the variance or the standard deviation is a measure of the spread of the distribution about its mean.

Equation B.6 shows that σ_x^2 is a sum of positive terms, and so $\sigma_x^2 \geq 0$. Furthermore,

$$\sigma_x^2 = \sum_{j=1}^{n}(x_j - \langle x \rangle)^2 p_j = \sum_{j=1}^{n}(x_j^2 - 2\langle x \rangle x_j + \langle x \rangle^2) p_j$$

$$= \sum_{j=1}^{n} x_j^2 p_j - 2 \sum_{j=1}^{n}\langle x \rangle x_j p_j + \sum_{j=1}^{n}\langle x \rangle^2 p_j \tag{B.7}$$

The first term here is just $\langle x^2 \rangle$ (cf. Equation B.5). To evaluate the second and third terms, we need to realize that $\langle x \rangle$, the average of x_j, is just a number and so can be factored out of the summations, leaving a summation of the form $\sum x_j p_j$ in the second term and

$\sum p_j$ in the third term. The summation $\sum x_j p_j$ is $\langle x \rangle$ by definition and the summation $\sum p_j$ is unity because of normalization (Equation B.3). Putting all this together, we find that

$$\sigma_x^2 = \langle x^2 \rangle - 2\langle x \rangle^2 + \langle x \rangle^2$$

$$= \langle x^2 \rangle - \langle x \rangle^2 \geq 0 \tag{B.8}$$

Because $\sigma_x^2 \geq 0$, we see that $\langle x^2 \rangle \geq \langle x \rangle^2$. A consideration of Equation B.6 shows that $\sigma_x^2 = 0$ or $\langle x \rangle^2 = \langle x^2 \rangle$ only when $x_j = \langle x \rangle$ with a probability of one, a case that is not really probabilistic because the event j occurs on every trial.

So far we have considered only discrete distributions, but continuous distributions are also important in physical chemistry. It is convenient to use the unit mass analogy. Consider a unit mass to be distributed continuously along the x axis, or along some interval on the x axis. We define the linear mass density $\rho(x)$ by

$$dm = \rho(x)dx$$

where dm is the fraction of the mass lying between x and $x + dx$. By analogy, then, we say that the probability that some quantity x, such as the position of a particle in a box, lies between x and $x + dx$ is

$$\text{Prob}\{x, x + dx\} = p(x)dx \tag{B.9}$$

and that

$$\text{Prob}\{a \leq x \leq b\} = \int_a^b p(x)dx \tag{B.10}$$

In the mass analogy, $\text{Prob}\{a \leq x \leq b\}$ is the fraction of mass that lies in the interval $a \leq x \leq b$. The normalization condition is

$$\int_a^b p(x)dx = 1 \tag{B.11}$$

Following Equations B.4 through B.6, we have the definitions

$$\langle x \rangle = \int_a^b x p(x)dx \tag{B.12}$$

$$\langle x^2 \rangle = \int_a^b x^2 p(x)dx \tag{B.13}$$

and

$$\sigma_x^2 = \int_a^b (x - \langle x \rangle)^2 p(x)dx \tag{B.14}$$

EXAMPLE B–3

Perhaps the simplest continuous distribution is the so-called uniform distribution, where

$$p(x) = \begin{cases} \text{constant} = A & a \le x \le b \\ 0 & \text{otherwise} \end{cases}$$

Show that A must equal $1/(b-a)$. Evaluate $\langle x \rangle$, $\langle x^2 \rangle$, σ_x^2, and σ_x for this distribution.

SOLUTION: Because $p(x)$ must be normalized,

$$\int_a^b p(x)dx = 1 = A \int_a^b dx = A(b-a)$$

Therefore, $A = 1/(b-a)$ and

$$p(x) = \begin{cases} \dfrac{1}{b-a} & a \le x \le b \\ 0 & \text{otherwise} \end{cases}$$

The mean of x is given by

$$\langle x \rangle = \int_a^b xp(x)dx = \frac{1}{b-a} \int_a^b x\,dx$$

$$= \frac{b^2 - a^2}{2(b-a)} = \frac{b+a}{2}$$

and the second moment of x by

$$\langle x^2 \rangle = \int_a^b x^2 p(x)dx = \frac{1}{b-a} \int_a^b x^2 dx$$

$$= \frac{b^3 - a^3}{3(b-a)} = \frac{b^2 + ab + a^2}{3}$$

Last, the variance is given by Equation B.6, and so

$$\sigma_x^2 = \langle x^2 \rangle - \langle x \rangle^2 = \frac{(b-a)^2}{12}$$

and the standard deviation is

$$\sigma_x = \frac{(b-a)}{\sqrt{12}}$$

EXAMPLE B–4

The most commonly occurring and most important continuous probability distribution is the *Gaussian distribution*, given by

$$p(x)dx = ce^{-x^2/2a^2}dx \qquad -\infty < x < \infty$$

Find c, $\langle x \rangle$, σ_x^2, and σ_x.

SOLUTION: The constant c is determined by normalization:

$$\int_{-\infty}^{\infty} p(x)dx = 1 = c \int_{-\infty}^{\infty} e^{-x^2/2a^2}dx \tag{B.15}$$

If you look in a table of integrals (e.g., *The CRC Standard Mathematical Tables* or *The CRC Handbook of Chemistry and Physics*; CRC Press: Boca Raton, FL), you won't find the above integral. However, you will find the integral (see also Problem B–6)

$$\int_0^{\infty} e^{-\alpha x^2}dx = \left(\frac{\pi}{4\alpha}\right)^{1/2} \tag{B.16}$$

The reason that you won't find the integral with the limts $(-\infty, \infty)$ is illustrated in Figure B.2a, where $e^{-\alpha x^2}$ is plotted against x. Note that the graph is symmetric about the vertical axis, so that the corresponding areas on the two sides of the axis are equal. Such a function has the mathematical property that $f(x) = f(-x)$ and is called an *even function*. For an even function

$$\int_{-A}^{A} f_{\text{even}}(x)dx = 2 \int_0^{A} f_{\text{even}}(x)dx \tag{B.17}$$

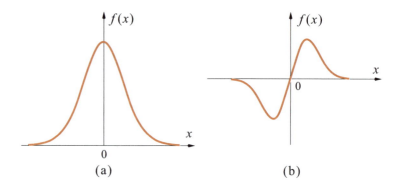

FIGURE B.2
(a) The function $f(x) = e^{-x^2}$ is an even function, $f(x) = f(-x)$. (b) The function $f(x) = xe^{-x^2}$ is an odd function, $f(x) = -f(-x)$.

If we recognize that $p(x) = e^{-x^2/2a^2}$ is an even function and use Equation B.16, then we find that

$$c \int_{-\infty}^{\infty} e^{-x^2/2a^2} dx = 2c \int_0^{\infty} e^{-x^2/2a^2} dx$$

$$= 2c \left(\frac{\pi a^2}{2}\right)^{1/2} = 1$$

or $c = 1/(2\pi a^2)^{1/2}$.

The mean of x is given by

$$\langle x \rangle = \int_{-\infty}^{\infty} x p(x) dx = (2\pi a^2)^{-1/2} \int_{-\infty}^{\infty} x e^{-x^2/2a^2} dx \qquad (B.18)$$

The integrand in Equation B.18 is plotted in Figure B.2b. Notice that this graph is antisymmetric about the vertical axis and that the area on one side of the vertical axis cancels the corresponding area on the other side. This function has the mathematical property that $f(x) = -f(-x)$ and is called an *odd function*. For an odd function,

$$\int_{-A}^{A} f_{odd}(x) dx = 0 \qquad (B.19)$$

The function $xe^{-x^2/2a^2}$ is an odd function, and so

$$\langle x \rangle = \int_{-\infty}^{\infty} x e^{-x^2/2a^2} dx = 0$$

The second moment of x is given by

$$\langle x^2 \rangle = (2\pi a^2)^{-1/2} \int_{-\infty}^{\infty} x^2 e^{-x^2/2a^2} dx$$

The integrand in this case is even because $f(x) = x^2 e^{-x^2/2a^2} = f(-x)$. Therefore,

$$\langle x^2 \rangle = 2(2\pi a^2)^{-1/2} \int_0^{\infty} x^2 e^{-x^2/2a^2} dx$$

The integral

$$\int_0^{\infty} x^2 e^{-\alpha x^2} dx = \frac{1}{4\alpha} \left(\frac{\pi}{\alpha}\right)^{1/2} \qquad (B.20)$$

can be found in integral tables, and so

$$\langle x^2 \rangle = \frac{2}{(2\pi a^2)^{1/2}} \frac{(2\pi a^2)^{1/2} a^2}{2} = a^2$$

Because $\langle x \rangle = 0$, $\sigma_x^2 = \langle x^2 \rangle$, and so σ_x is given by

$$\sigma_x = a$$

The standard deviation of a normal distribution is the parameter that appears in the exponential. The standard notation for a normalized Gaussian distribution function is

$$p(x)dx = (2\pi\sigma_x^2)^{-1/2} e^{-x^2/2\sigma_x^2} dx \tag{B.21}$$

Figure B.3 shows Equation B.21 for various values of σ_x. Note that the curves become narrower and taller for smaller values of σ_x.

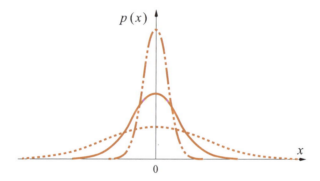

FIGURE B.3
A plot of a Gaussian distribution, $p(x)$, (Equation B.21) for three values of σ_x. The dotted curve corresponds to $\sigma_x = 2$, the solid curve to $\sigma_x = 1$, and the dash–dotted curve to $\sigma_x = 0.5$.

A more general version of a Gaussian distribution is

$$p(x)dx = (2\pi\sigma_x^2)^{-1/2} e^{-(x-\langle x \rangle)^2/2\sigma_x^2} dx \tag{B.22}$$

This expression looks like those in Figure B.3 except that the curves are centered at $x = \langle x \rangle$ rather than $x = 0$. A Gaussian distribution is one of the most important and commonly used probability distributions in all of science.

Problems

B–1. Using the following table

x	$f(x)$
-6	0.05
-2	0.15
0	0.50
1	0.10
3	0.05
4	0.10
5	0.05

calculate $\langle x \rangle$ and $\langle x^2 \rangle$ and show that $\sigma_x^2 > 0$.

B–2. A discrete probability distribution that is commonly used in statistics is the Poisson distribution

$$f_n = \frac{\lambda^n}{n!} e^{-\lambda} \qquad n = 0, 1, 2, \ldots$$

where λ is a positive constant. Prove that f_n is normalized. Evaluate $\langle n \rangle$ and $\langle n^2 \rangle$ and show that $\sigma^2 > 0$. Recall that

$$e^x = \sum_{n=0}^{\infty} \frac{x^n}{n!}$$

B–3. An important continuous distribution is the exponential distribution

$$p(x)dx = ce^{-\lambda x}dx \qquad 0 \le x < \infty$$

Evaluate c, $\langle x \rangle$, and σ^2, and the probability that $x \ge a$.

B–4. Prove explicitly that

$$\int_{-\infty}^{\infty} e^{-\alpha x^2} dx = 2 \int_{0}^{\infty} e^{-\alpha x^2} dx$$

by breaking the integral from $-\infty$ to ∞ into one from $-\infty$ to 0 and another from 0 to ∞. Let $z = -x$ in the first integral and $z = x$ in the second to prove the above relation.

B–5. By using the procedure in Problem B–4, show explicitly that

$$\int_{-\infty}^{\infty} xe^{-\alpha x^2} dx = 0$$

B–6. Integrals of the type

$$I_n(\alpha) = \int_{-\infty}^{\infty} x^{2n} e^{-\alpha x^2} dx \qquad n = 0, 1, 2, \ldots$$

occur frequently in a number of applications. We can simply either look them up in a table of integrals or continue this problem. First, show that

$$I_n(\alpha) = 2 \int_{0}^{\infty} x^{2n} e^{-\alpha x^2} dx$$

The case $n = 0$ can be handled by the following trick. Show that the square of $I_0(\alpha)$ can be written in the form

$$I_0^2(\alpha) = 4 \int_{0}^{\infty} \int_{0}^{\infty} dx dy \, e^{-\alpha(x^2 + y^2)}$$

Now convert to plane polar coordinates, letting

$$r^2 = x^2 + y^2 \qquad \text{and} \qquad dxdy = rdrd\theta$$

Show that the appropriate limits of integration are $0 \le r < \infty$ and $0 \le \theta \le \pi/2$ and that

$$I_0^2(\alpha) = 4 \int_{0}^{\pi/2} d\theta \int_{0}^{\infty} drr e^{-\alpha r^2}$$

which is elementary and gives

$$I_0^2(\alpha) = 4 \cdot \frac{\pi}{2} \cdot \frac{1}{2\alpha} = \frac{\pi}{\alpha}$$

or that

$$I_0(\alpha) = \left(\frac{\pi}{\alpha}\right)^{1/2}$$

Now prove that the $I_n(\alpha)$ may be obtained by repeated differentiation of $I_0(\alpha)$ with respect to α and, in particular, that

$$\frac{d^n I_0(\alpha)}{d\alpha^n} = (-1)^n I_n(\alpha)$$

Use this result and the fact that $I_0(\alpha) = (\pi/\alpha)^{1/2}$ to generate $I_1(\alpha)$, $I_2(\alpha)$, and so forth.

B–7. Without using a table of integrals, show that all of the odd moments of a Gaussian distribution are zero. Using the results derived in Problem B–6, calculate $\langle x^4 \rangle$ for a Gaussian distribution.

B–8. Consider a particle to be constrained to lie along a one-dimensional segment 0 to a. We will learn in the next chapter that the probability that the particle is found to lie between x

and $x + dx$ is given by

$$p(x)dx = \frac{2}{a} \sin^2 \frac{n\pi x}{a} dx$$

where $n = 1, 2, 3, \ldots$. First show that $p(x)$ is normalized. Now show that the average position of the particle along the line segment is $a/2$. Is this result physically reasonable? The integrals that you need are (*The CRC Handbook of Chemistry and Physics* or *The CRC Standard Mathematical Tables*; CRC Press: Boca Raton, FL)

$$\int \sin^2 \alpha x dx = \frac{x}{2} - \frac{\sin 2\alpha x}{4\alpha}$$

and

$$\int x \sin^2 \alpha x dx = \frac{x^2}{4} - \frac{x \sin 2\alpha x}{4\alpha} - \frac{\cos 2\alpha x}{8\alpha^2}$$

B–9. Show that $\langle x \rangle^2 = a^2/4$ and that the variance associated with the probability distribution given in Problem B–8 is given by $\left(\dfrac{a}{2\pi n}\right)^2 \left(\dfrac{\pi^2 n^2}{3} - 2\right)$. The necessary integral is (CRC tables)

$$\int x^2 \sin^2 \alpha x dx = \frac{x^3}{6} - \left(\frac{x^2}{4\alpha} - \frac{1}{8\alpha^3}\right) \sin 2\alpha x - \frac{x \cos 2\alpha x}{4\alpha^2}$$

B–10. Show that

$$\sigma_x = (\langle x^2 \rangle - \langle x \rangle^2)^{1/2}$$

for a particle in a box is less than a, the width of the box, for any value of n. If σ_x is the uncertainty in the position of the particle, could σ_x ever be larger than a?

B–11. All the definite integrals used in Problems B–8 and B–9 can be evaluated from

$$I(\beta) = \int_0^a e^{\beta x} \sin^2 \frac{n\pi x}{a} dx$$

Show that the above integrals are given by $I(0)$, $I'(0)$, and $I''(0)$, respectively, where the primes denote differentiation with respect to β. Using a table of integrals, evaluate $I(\beta)$ and then the above three integrals by differentiation.

B–12. Using the probability distribution given in Problem B–8, calculate the probability that the particle will be found between 0 and $a/2$. The necessary integral is given in Problem B–8.

Erwin Schrödinger was born in Vienna, Austria, on August 12, 1887, and died there in 1961. He received his Ph.D. in theoretical physics in 1910 from the University of Vienna. He then held a number of positions in Germany and in 1927 succeeded Max Planck at the University of Berlin at Planck's request. Schrödinger left Berlin in 1933 because of his opposition to Hitler and Nazi policies and eventually moved to the University of Graz in Austria in 1936. After the invasion of Austria by Germany, he was forcibly removed from his professorship in 1936. He then moved to the Institute of Advanced Studies, which was created for him, at the University College, Dublin, Ireland. He remained there for 17 years and then retired to his native Austria. Schrödinger shared the Nobel Prize in Physics with Paul Dirac in 1933 for the "discovery of new productive forms of atomic theory." Schrödinger rejected the probabilistic interpretation of the wave equation, which led to serious disagreement with Max Born, but they remained warm friends in spite of their scientific disagreement. Schrödinger preferred to work alone, and so no school developed around him, as it did for several other developers of quantum mechanics. His influential book, *What is Life?*, caused a number of physicists to become interested in biology. His personal life, which was rather unconventional, has been engagingly related by Walter Moore in his book *Schrödinger: Life and Thought* (Cambridge University Press: Cambridge, UK, 1989).

CHAPTER 3

The Schrödinger Equation and a Particle in a Box

The Schrödinger equation is our fundamental equation of quantum mechanics. The solutions to the Schrödinger equation are called *wave functions*. We will see that a wave function gives a complete quantum-mechanical description of any system. In this chapter, we present and discuss the version of the Schrödinger equation that does not contain time as a variable. Solutions to the time-independent Schrödinger equation are called *stationary-state wave functions* because they are independent of time. Many problems of interest to chemists can be treated by using only stationary-state wave functions.

In this chapter, we present the time-independent Schrödinger equation and then apply it to a free particle of mass m that is restricted to lie along a one-dimensional interval of length a. This system is called a *particle in a box* and the calculation of its properties is a standard introductory problem in quantum mechanics. The particle-in-a-box problem is simple, yet very instructive. In the course of discussing this problem, we will introduce the probabilistic interpretation of wave functions. We use this interpretation to illustrate the application of the uncertainty principle to a particle in a box.

3.1 The Schrödinger Equation Is the Equation for the Wave Function of a Particle

We cannot derive the Schrödinger equation any more than we can derive Newton's laws, and Newton's second law, $f = ma$, in particular. We shall regard the Schrödinger equation to be a fundamental postulate, or axiom, of quantum mechanics, just as Newton's laws are fundamental postulates of classical mechanics. Even though we cannot derive the Schrödinger equation, we can at least show that it is plausible and perhaps even trace Schrödinger's original line of thought. We finished Chapter 1 with a discussion of matter waves, arguing that matter has wavelike character in addition to its obvious particle-like character. As one story goes, at a meeting at which this new

97

idea of matter waves was being discussed, someone mentioned that if indeed matter does possess wavelike properties, then there must be some sort of wave equation that governs them.

Let's start with the classical one-dimensional wave equation for simplicity:

$$\frac{\partial^2 u}{\partial x^2} = \frac{1}{v^2}\frac{\partial^2 u}{\partial t^2} \tag{3.1}$$

We have seen in Chapter 2 that Equation 3.1 can be solved by the method of separation of variables and that $u(x, t)$ can be written as the product of a function of x and a harmonic or sinusoidal function of time. We will express the temporal part as $\cos \omega t$ (cf. Equation 2.22) and write $u(x, t)$ as

$$u(x, t) = \psi(x)\cos \omega t \tag{3.2}$$

Because $\psi(x)$ is the spatial factor of the amplitude $u(x, t)$, we will call $\psi(x)$ the *spatial amplitude* of the wave. If we substitute Equation 3.2 into Equation 3.1, we obtain an equation for the spatial amplitude $\psi(x)$,

$$\frac{d^2\psi}{dx^2} + \frac{\omega^2}{v^2}\psi(x) = 0 \tag{3.3}$$

Using the fact that $\omega = 2\pi v$ and that $v\lambda = v$, Equation 3.3 becomes

$$\frac{d^2\psi}{dx^2} + \frac{4\pi^2}{\lambda^2}\psi(x) = 0 \tag{3.4}$$

We now introduce the idea of de Broglie matter waves into Equation 3.4. The total energy of a particle is the sum of its kinetic energy and its potential energy,

$$E = \frac{p^2}{2m} + V(x) \tag{3.5}$$

where $p = mv$ is the momentum of the particle and $V(x)$ is its potential energy. If we solve Equation 3.5 for the momentum p, we find

$$p = \{2m[E - V(x)]\}^{1/2} \tag{3.6}$$

According to the de Broglie formula,

$$\lambda = \frac{h}{p} = \frac{h}{\{2m[E - V(x)]\}^{1/2}}$$

Substituting this into Equation 3.4, we find

$$\frac{d^2\psi}{dx^2} + \frac{2m}{\hbar^2}[E - V(x)]\psi(x) = 0 \tag{3.7}$$

where $\hbar$ (called h-bar) $= h/2\pi$.

Equation 3.7 is the *Schrödinger equation*, a differential equation whose solution, $\psi(x)$, describes a particle of mass m moving in a potential field described by $V(x)$. The exact nature of $\psi(x)$ is vague at this point, but in analogy to the classical wave equation, it is a measure of the amplitude of the matter wave and is called the wave function of the particle. Equation 3.7 does not contain time and is called the *time-independent Schrödinger equation*. The wave functions obtained from Equation 3.7 are called *stationary-state wave functions*. Although there is a more general Schrödinger equation that contains a time dependence (Section 4.9), we will see throughout this book that many problems of chemical interest can be described in terms of stationary-state wave functions.

Equation 3.7 can be rewritten in the form

$$-\frac{\hbar^2}{2m}\frac{d^2\psi}{dx^2} + V(x)\psi(x) = E\psi(x) \tag{3.8}$$

Equation 3.8 is a particularly nice way to write the Schrödinger equation when we introduce the idea of an operator in the next section.

3.2 Classical-Mechanical Quantities Are Represented by Linear Operators in Quantum Mechanics

An *operator* is a symbol that tells you to do something to whatever follows the symbol. For example, we can consider dy/dx to be the d/dx operator operating on the function $y(x)$. Some other examples are SQR (square what follows), $\int_0^1 dx$ (integrate what follows from 0 to 1), 3 (multiply by 3), and $\partial/\partial y$. We usually denote an operator by a capital letter with a carat over it (e.g., $\hat{A}$). Thus, we write

$$\hat{A}f(x) = g(x)$$

to indicate that the operator $\hat{A}$ operates on $f(x)$ to give a new function $g(x)$.

EXAMPLE 3–1
Perform the following operations:

(a) $\hat{A}(2x)$, $\hat{A} = \dfrac{d^2}{dx^2}$

(b) $\hat{A}(x^2)$, $\hat{A} = \dfrac{d^2}{dx^2} + 2\dfrac{d}{dx} + 3$

(c) $\hat{A}(xy^3)$, $\hat{A} = \dfrac{\partial}{\partial y}$

(d) $\hat{A}(e^{ikx})$, $\hat{A} = -i\hbar\dfrac{d}{dx}$

SOLUTION:

(a) $\hat{A}(2x) = \dfrac{d^2}{dx^2}(2x) = 0$

(b) $\hat{A}(x^2) = \dfrac{d^2}{dx^2}x^2 + 2\dfrac{d}{dx}x^2 + 3x^2 = 2 + 4x + 3x^2$

(c) $\hat{A}(xy^3) = \dfrac{\partial}{\partial y}xy^3 = 3xy^2$

(d) $\hat{A}(e^{ikx}) = -i\hbar\dfrac{d}{dx}e^{ikx} = k\hbar e^{ikx}$

In quantum mechanics, we deal only with *linear operators*. An operator is said to be linear if

$$\hat{A}\left[c_1 f_1(x) + c_2 f_2(x)\right] = c_1\hat{A}f_1(x) + c_2\hat{A}f_2(x) \tag{3.9}$$

where c_1 and c_2 are (possibly complex) constants. Clearly, the "differentiate" and "integrate" operators are linear because

$$\frac{d}{dx}\left[c_1 f_1(x) + c_2 f_2(x)\right] = c_1\frac{df_1}{dx} + c_2\frac{df_2}{dx}$$

and

$$\int \left[c_1 f_1(x) + c_2 f_2(x)\right] dx = c_1\int f_1(x)dx + c_2\int f_2(x)dx$$

The "square" operator, SQR, on the other hand, is nonlinear because

$$\text{SQR}\left[c_1 f_1(x) + c_2 f_2(x)\right] = c_1^2 f_1^2(x) + c_2^2 f_2^2(x) + 2c_1c_2 f_1(x)f_2(x)$$

$$\neq c_1 f_1^2(x) + c_2 f_2^2(x)$$

and therefore it does not satisfy the definition given by Equation 3.9.

EXAMPLE 3–2

Determine whether the following operators are linear or nonlinear:

(a) $\hat{A}f(x) = \text{SQRT } f(x)$ (take the square root)

(b) $\hat{A}f(x) = x^2 f(x)$

SOLUTION:

(a) $\hat{A}\left[c_1 f_1(x) + c_2 f_2(x)\right] = \text{SQRT}\left[c_1 f_1(x) + c_2 f_2(x)\right]$
$= \left[c_1 f_1(x) + c_2 f_2(x)\right]^{1/2} \neq c_1 f_1^{1/2}(x) + c_2 f_2^{1/2}(x)$
and so SQRT is a nonlinear operator.

(b) $\hat{A}\left[c_1 f_1(x) + c_2 f_2(x)\right] = x^2\left[c_1 f_1(x) + c_2 f_2(x)\right]$
$= c_1 x^2 f_1(x) + c_2 x^2 f_2(x) = c_1\hat{A}f_1(x) + c_1\hat{A}f_1(x)$
and so x^2 (multiply by x^2) is a linear operator.

3.3 The Schrödinger Equation Can Be Formulated as an Eigenvalue Problem

A problem that we frequently encounter in physical chemistry is the following: Given $\hat{A}$, find a function $\phi(x)$ and a constant a such that

$$\hat{A}\phi(x) = a\phi(x) \qquad (3.10)$$

Note that the result of operating on the function $\phi(x)$ by $\hat{A}$ is simply to give $\phi(x)$ back again, only multiplied by a constant factor. Clearly, $\hat{A}$ and $\phi(x)$ have a very special relationship to each other. The function $\phi(x)$ is called an *eigenfunction* of the operator $\hat{A}$, and a is called an *eigenvalue*. The problem of determining $\phi(x)$ and a for a given $\hat{A}$ is called an *eigenvalue problem*.

EXAMPLE 3–3
Show that $e^{\alpha x}$ is an eigenfunction of the operator d^n/dx^n. What is the eigenvalue?

SOLUTION: We differentiate $e^{\alpha x}$ n times and obtain

$$\frac{d^n}{dx^n}e^{\alpha x} = \alpha^n e^{\alpha x}$$

and so the eigenvalue is α^n.

Operators can be imaginary or complex quantities. We will soon learn that the x component of the linear momentum can be represented in quantum mechanics by an operator of the form

$$\hat{P}_x = -i\hbar\frac{\partial}{\partial x} \qquad (3.11)$$

EXAMPLE 3–4
Show that e^{ikx} is an eigenfunction of the operator $\hat{P}_x = -i\hbar\dfrac{\partial}{\partial x}$. What is the eigenvalue?

SOLUTION: We apply $\hat{P}_x$ to e^{ikx} and find

$$\hat{P}_x e^{ikx} = -i\hbar\frac{\partial}{\partial x}e^{ikx} = \hbar k e^{ikx}$$

and so we see that e^{ikx} is an eigenfunction and $\hbar k$ is the eigenvalue of the operator $\hat{P}_x$.

Let's go back to Equation 3.8. We can write the left side of Equation 3.8 in the form

$$\left[-\frac{\hbar^2}{2m}\frac{d^2}{dx^2} + V(x) \right] \psi(x) = E\psi(x) \tag{3.12}$$

If we denote the operator in brackets by $\hat{H}$, then Equation 3.12 can be written as

$$\hat{H}\psi(x) = E\psi(x) \tag{3.13}$$

We have formulated the Schrödinger equation as an eigenvalue problem. The operator $\hat{H}$,

$$\hat{H} = -\frac{\hbar^2}{2m}\frac{d^2}{dx^2} + V(x) \tag{3.14}$$

is called the *Hamiltonian operator*. The wave function is an eigenfunction, and the energy is an eigenvalue of the Hamiltonian operator. This suggests a correspondence between the Hamiltonian operator and the energy. We will see that such correspondences of operators and classical-mechanical variables are fundamental to the formalism of quantum mechanics.

If $V(x) = 0$ in Equation 3.14, the energy is all kinetic energy and so we define a kinetic energy operator according to

$$\hat{T}_x = -\frac{\hbar^2}{2m}\frac{d^2}{dx^2} \tag{3.15}$$

(Strictly speaking, the derivative here should be a partial derivative, but we will consider only one-dimensional systems for the time being.) Furthermore, classically, $T = p^2/2m$, and so we conclude that

$$\hat{P}_x^2 = -\hbar^2\frac{d^2}{dx^2} \tag{3.16}$$

We can interpret the operator $\hat{P}_x^2$ by considering the case of two operators acting sequentially, as in $\hat{A}\hat{B}f(x)$. In cases such as this, we apply each operator in turn, working from right to left. Thus,

$$\hat{A}\hat{B}f(x) = \hat{A}[\hat{B}f(x)] = \hat{A}h(x)$$

where $h(x) = \hat{B}f(x)$. Once again, we require that all the indicated operations be compatible. If $\hat{A} = \hat{B}$, we have $\hat{A}\hat{A}f(x)$ and denote this term as $\hat{A}^2 f(x)$. Note that $\hat{A}^2 f(x) \neq [\hat{A}f(x)]^2$ for arbitrary $f(x)$.

EXAMPLE 3–5

Given $\hat{A} = d/dx$ and $\hat{B} = x^2$ (multiply by x^2), show (a) that $\hat{A}^2 f(x) \neq [\hat{A} f(x)]^2$ and (b) that $\hat{A}\hat{B} f(x) \neq \hat{B}\hat{A} f(x)$ for arbitrary $f(x)$.

SOLUTION:

(a) $$\hat{A}^2 f(x) = \frac{d}{dx}\left(\frac{df}{dx}\right) = \frac{d^2 f}{dx^2}$$

$$[\hat{A} f(x)]^2 = \left(\frac{df}{dx}\right)^2 \neq \frac{d^2 f}{dx^2}$$

for arbitrary $f(x)$.

(b) $$\hat{A}\hat{B} f(x) = \frac{d}{dx}[x^2 f(x)] = 2xf(x) + x^2\frac{df}{dx}$$

$$\hat{B}\hat{A} f(x) = x^2\frac{df}{dx} \neq \hat{A}\hat{B} f(x)$$

for arbitrary $f(x)$. Thus, we see that the order of the application of operators must be specified. If $\hat{A}$ and $\hat{B}$ are such that

$$\hat{A}\hat{B} f(x) = \hat{B}\hat{A} f(x)$$

for any compatible $f(x)$, then the two operators are said to *commute*. The two operators in this example, however, do not commute.

Using the fact that $\hat{P}_x^2$ means two successive applications of $\hat{P}_x$, we see that the operator $\hat{P}_x^2$ in Equation 3.16 can be factored as

$$\hat{P}_x^2 = -\hbar^2\frac{d^2}{dx^2} = \left(-i\hbar\frac{d}{dx}\right)\left(-i\hbar\frac{d}{dx}\right)$$

so that we can say that $-i\hbar d/dx$ is equal to the momentum operator. Note that this definition is consistent with Equation 3.11.

3.4 Wave Functions Have a Probabilistic Interpretation

In this section, we will study the case of a free particle of mass m constrained to lie along the x axis between $x = 0$ and $x = a$. This case is called the *problem of a particle in a one-dimensional box* (cf. Figure 3.1). It is mathematically a fairly simple problem, so we can study the solutions in great detail and extract and discuss their physical

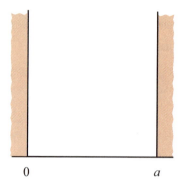

FIGURE 3.1
The geometry of the problem of a particle in a one-dimensional box.

0   a

consequences, which carry over to more complicated problems. In addition, we will see that this simple model has at least a crude application to the π electrons in a linear conjugated hydrocarbon.

The terminology *free particle* means that the particle experiences no potential energy or that $V(x) = 0$. If we set $V(x) = 0$ in Equation 3.7, we see that the Schrödinger equation for a free particle in a one-dimensional box is

$$\frac{d^2\psi}{dx^2} + \frac{2mE}{\hbar^2}\psi(x) = 0 \qquad 0 \leq x \leq a \qquad (3.17)$$

The particle is restricted to the region $0 \leq x \leq a$ and so cannot be found outside this region (see Figure 3.1). To implement the condition that the particle is restricted to the region $0 \leq x \leq a$, we must formulate an interpretation of the wave function $\psi(x)$. We have said that $\psi(x)$ represents the amplitude of the particle in some sense. Because the intensity of a wave is the square of the magnitude of the amplitude (cf. Problem 2–18), we can write that the "intensity of the particle" is proportional to $\psi^*(x)\psi(x)$, where the asterisk here denotes a complex conjugate [recall that $\psi^*(x)\psi(x)$ is a real quantity; see MathChapter A]. The problem lies in just what we mean by intensity. Schrödinger originally interpreted it in the following way. Suppose the particle is an electron. Then Schrödinger considered $e\psi^*(x)\psi(x)$ to be the charge density and $e\psi^*(x)\psi(x)dx$ to be the amount of charge between x and $x + dx$. Thus, he presumably pictured the electron to be spread all over the region. A few years later, however, Max Born, a German physicist working in scattering theory, found that this interpretation led to logical difficulties and replaced Schrödinger's interpretation with $\psi^*(x)\psi(x)dx$ as the *probability that the particle is located between x and $x + dx$*. Born's view is now generally accepted.

Because the particle is restricted to the region $0 \leq x \leq a$, the probability that the particle is found outside this region is zero. Consequently, we shall require that $\psi(x) = 0$ outside the region $0 \leq x \leq a$, which is mathematically how we restrict the particle to this region. Furthermore, because $\psi(x)$ is a measure of the position of the particle, we shall require $\psi(x)$ to be a continuous function. If $\psi(x) = 0$ outside the interval $0 \leq x \leq a$ and is a continuous function, then

$$\psi(0) = \psi(a) = 0$$

These are boundary conditions that we impose on the problem.

3.5 The Energy of a Particle in a Box Is Quantized

The general solution of Equation 3.17 is (see Example 2–1)

$$\psi(x) = A \cos kx + B \sin kx$$

with

$$k = \frac{(2mE)^{1/2}}{\hbar} \tag{3.18}$$

The first boundary condition requires that $\psi(0) = 0$, which implies immediately that $A = 0$ because $\cos(0) = 1$ and $\sin(0) = 0$. The second boundary condition then gives us

$$\psi(a) = B \sin ka = 0 \tag{3.19}$$

We reject the obvious choice that $B = 0$ because it yields a trivial or physically uninteresting solution, $\psi(x) = 0$, for all x. The other choice is that

$$ka = n\pi \qquad n = 1, 2, \ldots \tag{3.20}$$

(cf. Equations 2.18 through 2.20). By using Equation 3.18 for k, we find that

$$E_n = \frac{h^2 n^2}{8ma^2} \qquad n = 1, 2, \ldots \tag{3.21}$$

Thus, the energy turns out to have only the discrete values given by Equation 3.21 and no other values. The energy of the particle is said to be *quantized* and the integer n is called a *quantum number*. Note that the quantization arises naturally from the boundary conditions. We have gone beyond the stage of Planck and Bohr where quantum numbers are introduced in an ad hoc manner. The natural occurrence of quantum numbers was an exciting feature of the Schrödinger equation, and, in the introduction to the first of his now famous series of four papers published in 1926, Schrödinger says:

> In this communication I wish to show that the usual rules of quantization can be replaced by another postulate (the Schrödinger equation) in which there occurs no mention of whole numbers. Instead, the introduction of integers arises in the same natural way as, for example, in a vibrating string, for which the number of nodes is integral. The new conception can be generalized, and I believe that it penetrates deeply into the true nature of the quantum rules. [From *Ann. Phys.*, **79**, 361 (1926).]

The wave function corresponding to E_n is

$$\psi(x) = B \sin kx$$

$$= B \sin \frac{n\pi x}{a} \qquad n = 1, 2, \ldots \qquad (3.22)$$

We will determine the constant B shortly. These wave functions are plotted in Figure 3.2. They look just like the standing waves set up in a vibrating string (cf. Figure 2.3). Note that the energy increases with the number of nodes.

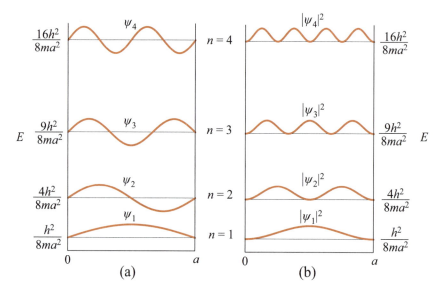

FIGURE 3.2
The energy levels, wave functions (a), and probability densities (b) for the particle in a box.

The model of a particle in a one-dimensional box has been applied to the π electrons in linear conjugated hydrocarbons. Consider butadiene, $H_2C=CHCH=CH_2$, which has four π electrons. Although butadiene, like all polyenes, is not a linear molecule, we will assume for simplicity that the π electrons in butadiene move along a straight line whose length can be estimated as equal to two C=C bond lengths (2×135 pm) plus one C—C bond (154 pm) plus the distance of a carbon atom radius at each end (2×77.0 pm $= 154$ pm), giving a total distance of 578 pm. According to Equation 3.21, the allowed π electronic energies are given by

$$E_n = \frac{h^2 n^2}{8 m_e a^2} \qquad n = 1, 2, \ldots$$

But the Pauli exclusion principle (which we discuss later but is assumed here to be known from general chemistry) says that each of these states can hold only two electrons (with opposite spins), and so the four π electrons fill the first two levels as shown in

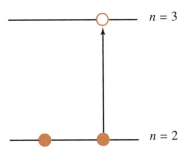

n = 3

n = 2

n = 1

FIGURE 3.3
The free-electron model energy-level scheme for butadiene.

Figure 3.3. The first excited state of this system of four π electrons is that which has one electron elevated from the $n = 2$ state to the $n = 3$ state (cf. Figure 3.3), and the energy to make a transition from the $n = 2$ state to the $n = 3$ state is

$$\Delta E = \frac{h^2}{8m_e a^2}(3^2 - 2^2)$$

The mass m_e is that of an electron (9.109×10^{-31} kg), and the length of the box is taken to be 578 pm, or 578×10^{-12} m. Therefore,

$$\Delta E = \frac{\left(6.626 \times 10^{-34} \text{ J·s}\right)^2 5}{8(9.109 \times 10^{-31} \text{ kg})(578 \times 10^{-12} \text{ m})^2} = 9.02 \times 10^{-19} \text{ J}$$

and

$$\tilde{\nu} = 4.54 \times 10^4 \text{ cm}^{-1}$$

Butadiene has an absorption band at 4.61×10^4 cm^{-1}, and so we see that this very simple model, called the *free-electron model*, can be somewhat successful at explaining the absorption spectrum of butadiene (cf. Problem 3–9).

3.6 Wave Functions Must Be Normalized

According to the Born interpretation,

$$\psi_n^*(x)\psi_n(x)dx = B^*B \sin^2 \frac{n\pi x}{a}dx \tag{3.23}$$

is the probability that the particle is located between x and $x + dx$. Because the particle is restricted to the region $0 \leq x \leq a$, it is certain to be found there and so the probability that the particle lies between 0 and a is unity (Equation B.11), or

$$\int_0^a \psi_n^*(x)\psi_n(x)dx = 1 \tag{3.24}$$

If we substitute Equation 3.23 into Equation 3.24, we find that

$$|B|^2 \int_0^a \sin^2 \frac{n\pi x}{a} dx = 1 \tag{3.25}$$

We let $n\pi x/a$ be z in Equation 3.25 to obtain

$$\int_0^a \sin^2 \frac{n\pi x}{a} dx = \frac{a}{n\pi} \int_0^{n\pi} \sin^2 z \, dz = \frac{a}{n\pi} \left(\frac{n\pi}{2}\right) = \frac{a}{2} \tag{3.26}$$

Therefore, $B^2(a/2) = 1$, $B = (2/a)^{1/2}$, and

$$\psi_n(x) = \left(\frac{2}{a}\right)^{1/2} \sin \frac{n\pi x}{a} \qquad 0 \le x \le a, \quad n = 1, 2, \ldots \tag{3.27}$$

A wave function that satisfies Equation 3.24, and the one given by Equation 3.27 in particular, is said to be *normalized*. When the constant that multiplies a wave function is adjusted to assure that Equation 3.24 is satisfied, the resulting constant is called a *normalization constant*. Because the Hamiltonian operator is a linear operator, if ψ is a solution to $\hat{H}\psi = E\psi$, then any constant, say, A, times ψ is also a solution, and A can always be chosen to produce a normalized solution to the Schrödinger equation, $\hat{H}\psi = E\psi$ (cf. Problem 3–10).

Because $\psi^*(x)\psi(x)dx$ is the probability of finding the particle between x and $x + dx$, the probability of finding the particle within the interval $x_1 \le x \le x_2$ is

$$\text{Prob}(x_1 \le x \le x_2) = \int_{x_1}^{x_2} \psi^*(x)\psi(x)dx \tag{3.28}$$

EXAMPLE 3–6
Calculate the probability that a particle in a one-dimensional box of length a is found between 0 and $a/2$.

SOLUTION: The probability that the particle will be found between 0 and $a/2$ is

$$\text{Prob}(0 \le x \le a/2) = \int_0^{a/2} \psi^*(x)\psi(x)dx = \frac{2}{a} \int_0^{a/2} \sin^2 \frac{n\pi x}{a} dx$$

If we let $n\pi x/a$ be z, then we find

$$\text{Prob}(0 \le x \le a/2) = \frac{2}{n\pi} \int_0^{n\pi/2} \sin^2 z\, dz = \frac{2}{n\pi} \left| \frac{z}{2} - \frac{\sin 2z}{4} \right|_0^{n\pi/2}$$

$$= \frac{2}{n\pi}\left(\frac{n\pi}{4} - \frac{\sin n\pi}{4}\right) = \frac{1}{2} \qquad \text{(for all } n)$$

Thus, the probability that the particle lies in one-half of the interval $0 \le x \le a$ is $\frac{1}{2}$. Does this seem to be physically reasonable to you?

We can use Figure 3.2 and a slight variation of Example 3–6 to illustrate a fundamental principle of quantum mechanics. Figure 3.2 shows that the particle is more likely to be found near the center of the box for the $n = 1$ state but that the probability density becomes more uniformly distributed as n increases. Figure 3.4 shows that the probability density, $\psi_n^*(x)\psi_n(x) = (2/a) \sin^2 n\pi x/a$, for $n = 20$ is fairly uniformly distributed from 0 to a.

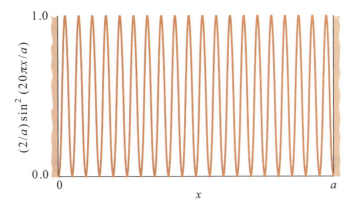

FIGURE 3.4
The probability density, $\psi_n^*(x)\psi_n(x) = (2/a) \sin^2 n\pi x/a$ for $n = 20$, illustrating the correspondence principle, which says that the particle tends to behave classically in the limit of large n.

In fact, a variation of Example 3–6 (Problem 3–12) gives

$$\text{Prob}(0 \le x \le a/4) = \text{Prob}(3a/4 \le x \le a) = \begin{cases} \dfrac{1}{4} & n \text{ even} \\[2mm] \dfrac{1}{4} - \dfrac{(-1)^{\frac{n-1}{2}}}{2\pi n} & n \text{ odd} \end{cases}$$

and

$$\text{Prob}(a/4 \le x \le a/2) = \text{Prob}(a/2 \le x \le 3a/4) = \begin{cases} \dfrac{1}{4} & n \text{ even} \\[2mm] \dfrac{1}{4} + \dfrac{(-1)^{\frac{n-1}{2}}}{2\pi n} & n \text{ odd} \end{cases}$$

In both cases, the probabilities approach 1/4 more and more closely as n grows larger. A similar result is found for any equi-sized intervals. In other words, the probability density becomes uniform as n increases, which is the expected behavior of a classical particle, which has no preferred position between 0 and a.

The results illustrate the *correspondence principle*, according to which quantum-mechanical results and classical-mechanical results tend to agree in the limit of large quantum numbers. The large quantum number limit is often called the classical limit.

3.7 The Average Momentum of a Particle in a Box Is Zero

We can use the probability distribution $\psi_n^*(x)\psi_n(x)$ to calculate averages and standard deviations (MathChapter B) of various physical quantities such as position and momentum. Using the example of a particle in a box, we see that

$$f(x)dx = \begin{cases} \dfrac{2}{a}\sin^2\dfrac{n\pi x}{a}dx & 0 \le x \le a \\ 0 & \text{otherwise} \end{cases} \tag{3.29}$$

is the probability that the particle is found between x and $x + dx$. These probabilities are plotted in Figure 3.2b. The average value of x, or the mean position of the particle, is given by

$$\langle x \rangle = \frac{2}{a}\int_0^a x \sin^2\frac{n\pi x}{a}dx \tag{3.30}$$

The integral in Equation 3.30 equals $a^2/4$ (Problem B–8). Therefore,

$$\langle x \rangle = \frac{2}{a}\cdot\frac{a^2}{4} = \frac{a}{2} \qquad \text{(for all } n\text{)} \tag{3.31}$$

This is the physically expected result because the particle "sees" nothing except the walls at $x = 0$ and $x = a$, and so by symmetry $\langle x \rangle$ must be $a/2$.

We can calculate the spread about $\langle x \rangle$ by calculating the variance, σ_x^2. First we calculate $\langle x^2 \rangle$, which is (Problem B–9)

$$\langle x^2 \rangle = \frac{2}{a}\int_0^a x^2 \sin^2\frac{n\pi x}{a}dx$$

$$= \left(\frac{a}{2\pi n}\right)^2\left(\frac{4\pi^2 n^2}{3} - 2\right) = \frac{a^2}{3} - \frac{a^2}{2n^2\pi^2} \tag{3.32}$$

The variance of x is given by

$$\sigma_x^2 = \langle x^2 \rangle - \langle x \rangle^2 = \frac{a^2}{12} - \frac{a^2}{2n^2\pi^2} = \left(\frac{a}{2\pi n}\right)^2\left(\frac{\pi^2 n^2}{3} - 2\right)$$

and so the standard deviation is

$$\sigma_x = \frac{a}{2\pi n} \left(\frac{\pi^2 n^2}{3} - 2 \right)^{1/2} \tag{3.33}$$

We shall see that σ_x is directly involved in the Heisenberg uncertainty principle. Problem 3–18 has you show that $\langle x \rangle$, $\langle x^2 \rangle$, and σ_x go to the classical limit as $n \to \infty$.

A problem arises if we wish to calculate the average energy or momentum because these quantities are represented by differential operators. Recall that the energy and momentum operators are

$$\hat{H} = -\frac{\hbar^2}{2m} \frac{d^2}{dx^2} + V(x)$$

and

$$\hat{P}_x = -i\hbar \frac{d}{dx}$$

The problem is that we must decide whether the operator works on $\psi^*(x)\psi(x)dx$ or on $\psi(x)$ or on $\psi^*(x)$ alone. To determine this, let's go back to the Schrödinger equation in operator notation:

$$\hat{H}\psi_n(x) = E_n\psi_n(x) \tag{3.34}$$

If we multiply this equation from the left (see Problem 3–25) by $\psi_n^*(x)$ and integrate over all values of x, we obtain

$$\int \psi_n^*(x)\hat{H}\psi_n(x)dx = \int \psi_n^*(x)E_n\psi_n(x)dx = E_n \int \psi_n^*(x)\psi_n(x)dx = E_n \tag{3.35}$$

where the second step follows because E_n is a number and the last step follows because $\psi_n(x)$ is normalized. Equation 3.35 suggests that we sandwich the operator between a wave function $\psi_n(x)$ and its complex conjugate $\psi_n^*(x)$ to calculate the average value of the physical quantity associated with that operator. We will set this up as a formal postulate in Chapter 4, but our assumption is that

$$\langle s \rangle = \int \psi_n^*(x)\hat{S}\psi_n(x)dx \tag{3.36}$$

where $\hat{S}$ is the quantum-mechanical operator associated with the physical quantity s, and $\langle s \rangle$ is the average value of s in the state described by the wave function. For example, the average momentum of a particle in a box in the state described by $\psi_n(x)$ is

$$\langle p \rangle = \int_0^a \left[\left(\frac{2}{a} \right)^{1/2} \sin \frac{n\pi x}{a} \right] \left(-i\hbar \frac{d}{dx} \right) \left[\left(\frac{2}{a} \right)^{1/2} \sin \frac{n\pi x}{a} \right] dx \tag{3.37}$$

In this particular case, $\psi_n(x)$ is real, but generally the operator is sandwiched in between $\psi_n^*(x)$ and $\psi_n(x)$ and so operates only on $\psi_n(x)$ because only $\psi_n(x)$ lies to the right of the operator. We did not have to worry about this when we calculated $\langle x \rangle$ above because the position operator $\hat{X}$ is simply the "multiply by x" operator and its placement in the integrand in Equation 3.36 makes no difference.

If we simplify Equation 3.37, then we find that

$$\langle p \rangle = -i\hbar \frac{2\pi n}{a^2} \int_0^a \sin \frac{n\pi x}{a} \cos \frac{n\pi x}{a} dx$$

By consulting the table of integrals in the inside front cover or Problem 3–19, we find that this integral is equal to zero, and so

$$\langle p \rangle = 0 \tag{3.38}$$

Thus, a particle in a box is equally likely to be moving in either direction.

3.8 The Uncertainty Principle Says That $\sigma_p \sigma_x > \hbar/2$

Now let's calculate the variance of the momentum, $\sigma_p^2 = \langle p^2 \rangle - \langle p \rangle^2$, of a particle in a box. To calculate $\langle p^2 \rangle$, we use

$$\langle p^2 \rangle = \int \psi_n^*(x) \hat{P}_x^2 \psi_n(x) dx \tag{3.39}$$

and remember that $\hat{P}_x^2$ means apply $\hat{P}_x$ twice in succession. Using Equation 3.16,

$$\langle p^2 \rangle = \int_0^a \left[\left(\frac{2}{a} \right)^{1/2} \sin \frac{n\pi x}{a} \right] \left(-\hbar^2 \frac{d^2}{dx^2} \right) \left[\left(\frac{2}{a} \right)^{1/2} \sin \frac{n\pi x}{a} \right] dx$$

$$= \frac{2n^2\pi^2\hbar^2}{a^3} \int_0^a \sin \frac{n\pi x}{a} \sin \frac{n\pi x}{a} dx$$

$$= \frac{2n^2\pi^2\hbar^2}{a^3} \cdot \frac{a}{2} = \frac{n^2\pi^2\hbar^2}{a^2} \tag{3.40}$$

The square root of $\langle p^2 \rangle$ is called the *root-mean-square momentum*. Note how Equation 3.40 is consistent with the equation

$$\langle E \rangle = \left\langle \frac{p^2}{2m} \right\rangle = \frac{\langle p^2 \rangle}{2m} = \frac{n^2 h^2}{8ma^2}$$

Using Equation 3.40 and 3.38, we see that

$$\sigma_p^2 = \frac{n^2 \pi^2 \hbar^2}{a^2}$$

and

$$\sigma_p = \frac{n \pi \hbar}{a} \tag{3.41}$$

Because the variance σ^2, and hence the standard deviation σ, is a measure of the spread of a distribution about its mean value, we can interpret σ as a measure of the uncertainty involved in any measurement. For the case of a particle in a box, we have been able to evaluate σ_x and σ_p explicitly in Equations 3.33 and 3.41. We interpret these quantities as the uncertainty involved when we measure the position or the momentum of the particle, respectively. We expect to obtain a distribution of measured values because the position of the particle is given by the probability distribution, Equation 3.29.

Equation 3.41 shows that the uncertainty in a measurement of p is inversely proportional to a. Thus, the more we try to localize the particle, the greater is the uncertainty in its momentum. The uncertainty in the position of the particle is directly proportional to a (Equation 3.33), which simply means that the larger the region over which the particle can be found, the greater is the uncertainty in its position. A particle that can range over the entire x axis ($-\infty < x < \infty$) is called a *free particle*. In the case of a free particle, $a \to \infty$ in Equation 3.41, and there is no uncertainty in the momentum. The momentum of a free particle has a definite value (see Problem 3–35). The uncertainty in the position, however, is infinite. Thus, we see that there is a reciprocal relation between the uncertainty in momentum and position. If we take the product of σ_x and σ_p, then we have

$$\sigma_x \sigma_p = \frac{\hbar}{2} \left(\frac{\pi^2 n^2}{3} - 2 \right)^{1/2} \tag{3.42}$$

The value of the square-root term here is never less than 1, and so we write

$$\sigma_x \sigma_p > \frac{\hbar}{2} \tag{3.43}$$

Equation 3.43 is one version of the Heisenberg uncertainty principle. We have been able to derive Equation 3.43 explicitly here because the mathematical manipulations for a particle in a box are fairly simple.

Let's try to summarize what we have learned concerning the uncertainty principle. A free particle has a definite momentum, but its position is completely indefinite. When we localize a particle by restricting it to a region of length a, it no longer has a definite momentum, and the spread in its momentum is given by Equation 3.41. If we let the length a of the region go to zero, so that we have localized the particle precisely and there is no uncertainty in its position, then Equation 3.41 shows that there is an infinite uncertainty in the momentum. The uncertainty principle says that the minimum product of the two uncertainties is on the order of the Planck constant.

3.9 The Problem of a Particle in a Three-Dimensional Box Is a Simple Extension of the One-Dimensional Case

The simplest three-dimensional quantum-mechanical system is the three-dimensional version of a particle in a box. In this case, the particle is confined to lie within a rectangular parallelepiped with sides of lengths a, b, and c (Figure 3.5).

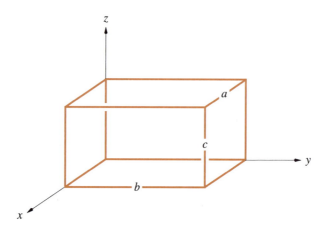

FIGURE 3.5
A rectangular parallelepiped of sides a, b, and c. In the problem of a particle in a three-dimensional box, the particle is restricted to lie within the region shown above.

The Schrödinger equation for this system is the three-dimensional extension of Equation 3.17.

$$-\frac{\hbar^2}{2m}\left(\frac{\partial^2\psi}{\partial x^2} + \frac{\partial^2\psi}{\partial y^2} + \frac{\partial^2\psi}{\partial z^2}\right) = E\psi(x, y, z) \qquad \begin{cases} 0 \leq x \leq a \\ 0 \leq y \leq b \\ 0 \leq z \leq c \end{cases} \qquad (3.44)$$

Equation 3.44 is often written in the form

$$-\frac{\hbar^2}{2m}\nabla^2\psi = E\psi$$

where the operator ("del squared"),

$$\nabla^2 = \frac{\partial^2}{\partial x^2} + \frac{\partial^2}{\partial y^2} + \frac{\partial^2}{\partial z^2} \qquad (3.45)$$

is called the *Laplacian operator*. The Laplacian operator appears in many physical problems.

The wave function $\psi(x, y, z)$ satisfies the boundary conditions that it vanishes at all the walls of the box, and so

$$\psi(0, y, z) = \psi(a, y, z) = 0 \qquad \text{for all } y \text{ and } z$$
$$\psi(x, 0, z) = \psi(x, b, z) = 0 \qquad \text{for all } x \text{ and } z \qquad (3.46)$$
$$\psi(x, y, 0) = \psi(x, y, c) = 0 \qquad \text{for all } x \text{ and } y$$

We will use the method of separation of variables to solve Equation 3.46. We write

$$\psi(x, y, z) = X(x)Y(y)Z(z) \qquad (3.47)$$

Substitute Equation 3.47 into Equation 3.44, and then divide through by $\psi(x, y, z) = X(x)Y(y)Z(z)$ to obtain

$$-\frac{\hbar^2}{2m}\frac{1}{X(x)}\frac{d^2X}{dx^2} - \frac{\hbar^2}{2m}\frac{1}{Y(y)}\frac{d^2Y}{dy^2} - \frac{\hbar^2}{2m}\frac{1}{Z(z)}\frac{d^2Z}{dz^2} = E \qquad (3.48)$$

Each of the three terms on the left side of Equation 3.48 is a function of only x, y, or z, respectively. Because x, y, and z are independent variables, the value of each term can be varied independently, and so each term must equal a constant for Equation 3.48 to be valid for all values of x, y, and z. Thus, we can write Equation 3.48 as

$$-\frac{\hbar^2}{2m}\frac{1}{X(x)}\frac{d^2X}{dx^2} = E_x$$
$$-\frac{\hbar^2}{2m}\frac{1}{Y(y)}\frac{d^2Y}{dy^2} = E_y \qquad (3.49)$$
$$-\frac{\hbar^2}{2m}\frac{1}{Z(z)}\frac{d^2Z}{dz^2} = E_z$$

where E_x, E_y, and E_z are constants and where

$$E_x + E_y + E_z = E \qquad (3.50)$$

From Equation 3.46, the boundary conditions associated with Equation 3.47 are that

$$X(0) = X(a) = 0$$
$$Y(0) = Y(b) = 0 \qquad (3.51)$$
$$Z(0) = Z(c) = 0$$

Thus, we see that Equations 3.50 and 3.51 are the same as for the one-dimensional case of a particle in a box. Following the same development as in Section 3.5, we obtain

$$X(x) = A_x \sin \frac{n_x \pi x}{a} \qquad n_x = 1, 2, 3, \ldots$$

$$Y(y) = A_y \sin \frac{n_y \pi y}{b} \qquad n_y = 1, 2, 3, \ldots \qquad (3.52)$$

$$Z(z) = A_z \sin \frac{n_z \pi z}{c} \qquad n_z = 1, 2, 3, \ldots$$

According to Equation 3.47, the solution to Equation 3.44 is

$$\psi(x, y, z) = A_x A_y A_z \sin \frac{n_x \pi x}{a} \sin \frac{n_y \pi y}{b} \sin \frac{n_z \pi z}{c} \qquad (3.53)$$

with n_x, n_y, and n_z independently assuming the values 1, 2, 3, ... The normalization constant $A_x A_y A_z$ is found from the equation

$$\int_0^a dx \int_0^b dy \int_0^c dz \psi^*(x, y, z) \psi(x, y, z) = 1 \qquad (3.54)$$

Problem 3–30 shows that

$$A_x A_y A_z = \left(\frac{8}{abc} \right)^{1/2} \qquad (3.55)$$

Thus, the normalized wave functions of a particle in a three-dimensional box are

$$\psi_{n_x n_y n_z} = \left(\frac{8}{abc} \right)^{1/2} \sin \frac{n_x \pi x}{a} \sin \frac{n_y \pi y}{b} \sin \frac{n_z \pi z}{c} \qquad \begin{cases} n_x = 1, 2, 3, \ldots \\ n_y = 1, 2, 3, \ldots \\ n_z = 1, 2, 3, \ldots \end{cases} \qquad (3.56)$$

If we substitute Equation 3.56 into Equation 3.44, then we obtain

$$E_{n_x n_y n_z} = \frac{h^2}{8m} \left(\frac{n_x^2}{a^2} + \frac{n_y^2}{b^2} + \frac{n_z^2}{c^2} \right) \qquad \begin{cases} n_x = 1, 2, 3, \ldots \\ n_y = 1, 2, 3, \ldots \\ n_z = 1, 2, 3, \ldots \end{cases} \qquad (3.57)$$

Equation 3.57 is the three-dimensional extension of Equation 3.21.

We should expect by symmetry that the average position of a particle in a three-dimensional box is at the center of the box, but we can show this by direct calculation.

EXAMPLE 3–7
Show that the average position of a particle confined to the region shown in Figure 3.5 is the point $(a/2, b/2, c/2)$.

SOLUTION: The position operator in three dimensions is (see MathChapter C)

$$\hat{\mathbf{R}} = \hat{X}\mathbf{i} + \hat{Y}\mathbf{j} + \hat{Z}\mathbf{k}$$

and the average position is given by

$$\langle \mathbf{r} \rangle = \int_0^a dx \int_0^b dy \int_0^c dz \psi^*(x, y, z)\hat{\mathbf{R}}\psi(x, y, z)$$

$$= \mathbf{i}\langle x \rangle + \mathbf{j}\langle y \rangle + \mathbf{k}\langle z \rangle$$

Let's evaluate $\langle x \rangle$ first. Using Equation 3.56, we have

$$\langle x \rangle = \left[\left(\frac{2}{a}\right) \int_0^a x \sin^2 \frac{n_x \pi x}{a} dx \right] \left[\left(\frac{2}{b}\right) \int_0^b \sin^2 \frac{n_y \pi y}{b} dy \right] \left[\left(\frac{2}{c}\right) \int_0^c \sin^2 \frac{n_z \pi z}{c} dz \right]$$

The second and third integrals here are unity by the normalization condition of a particle in a one-dimensional box (Equation 3.27). The first integral is just $\langle x \rangle$ for a particle in a one-dimensional box. Referring to Equation 3.31, we see that $\langle x \rangle = a/2$. The calculation for $\langle y \rangle$ and $\langle z \rangle$ are similar, and so we see that

$$\langle \mathbf{r} \rangle = \frac{a}{2}\mathbf{i} + \frac{b}{2}\mathbf{j} + \frac{c}{2}\mathbf{k}$$

Thus, the average position of the particle is in the center of the box.

In a similar manner, we should expect from the case of a particle in a one-dimensional box that the average momentum of a particle in a three-dimensional box is zero. The momentum operator in three dimensions is

$$\hat{\mathbf{P}} = -i\hbar \left(\mathbf{i}\frac{\partial}{\partial x} + \mathbf{j}\frac{\partial}{\partial y} + \mathbf{k}\frac{\partial}{\partial z} \right) \tag{3.58}$$

and so

$$\langle \mathbf{p} \rangle = \int_0^a dx \int_0^b dy \int_0^c dz \psi^*(x, y, z)\hat{\mathbf{P}}\psi(x, y, z) \tag{3.59}$$

It is a straightforward exercise to show that $\langle \mathbf{p} \rangle = 0$ (see Problem 3–31).

An interesting feature of a particle in a three-dimensional box occurs when the sides of the box are equal. In this case, $a = b = c$ in Equation 3.57, and so

$$E_{n_x n_y n_z} = \frac{h^2}{8ma^2}(n_x^2 + n_y^2 + n_z^2) \tag{3.60}$$

Only one set of values n_x, n_y, and n_z corresponds to the lowest energy level. This level, E_{111}, is said to be nondegenerate. However, three sets of values of n_x, n_y, and n_z correspond to the second energy level, and we say that this level is three-fold degenerate, or

$$E_{211} = E_{121} = E_{112} = \frac{6h^2}{8ma^2}$$

Figure 3.6 shows the distribution of the first few energy levels of a particle in a cube.

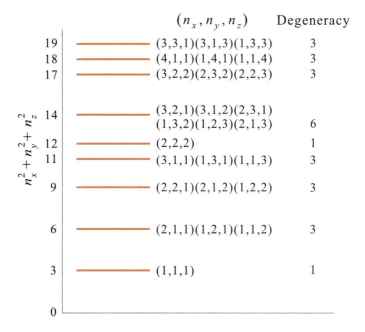

FIGURE 3.6
The energy levels for a particle in a cube, showing degeneracies.

Note that the degeneracy occurs because of the symmetry introduced when the general rectangular box becomes a cube and that the degeneracy is "lifted" when the symmetry is destroyed by making the sides of different lengths. A general principle of quantum mechanics states that degeneracies are the result of underlying symmetry and are lifted when the symmetry is broken.

According to Equation 3.53, the wave functions for a particle in a three-dimensional box factor into products of wave functions for a particle in a one-dimensional box. In addition, Equation 3.57 shows that the energy eigenvalues are sums of terms corresponding to the x, y, and z directions. In other words, the problem of a particle in a three-dimensional box reduces to three one-dimensional problems. This is no accident. It is a direct result of the fact that the Hamiltonian operator for a particle in a three-dimensional box is a sum of three independent terms:

$$\hat{H} = \hat{H}_x + \hat{H}_y + \hat{H}_z$$

where

$$\hat{H}_x = -\frac{\hbar^2}{2m}\frac{\partial^2}{\partial x^2} \qquad \text{etc.}$$

In such a case, we say that the Hamiltonian operator is *separable*.

Thus, we see that if $\hat{H}$ is separable—that is, if $\hat{H}$ can be written as the sum of terms involving independent coordinates, say

$$\hat{H} = \hat{H}_1(s) + \hat{H}_2(w) \tag{3.61}$$

where s and w are the independent coordinates—then the eigenfunctions of $\hat{H}$ are given by the products of the eigenfunctions of $\hat{H}_1$ and $\hat{H}_2$,

$$\psi_{nm}(s,\,w) = \phi_n(s)\varphi_m(w) \tag{3.62}$$

where

$$\hat{H}_1(s)\phi_n(s) = E_n\phi_n(s)$$
$$\hat{H}_2(w)\varphi_m(w) = E_m\varphi_m(w) \tag{3.63}$$

Furthermore, the eigenvalues of $\hat{H}$ are the sums of the eigenvalues of $\hat{H}_1$ and $\hat{H}_2$,

$$E_{nm} = E_n + E_m \tag{3.64}$$

This important result provides a significant simplification because it reduces the original problem to several simpler problems.

We have used the simple case of a particle in a box to illustrate some of the general principles and results of quantum mechanics. In Chapter 4, we present and discuss a set of postulates that we use throughout the remainder of this book.

Problems

3–1. Evaluate $g = \hat{A} f$, where $\hat{A}$ and f are given below:

$\hat{A}$	f
(a) SQRT	x^4
(b) $\dfrac{d^3}{dx^3} + x^3$	e^{-ax}
(c) $\displaystyle\int_0^1 dx$	$x^3 - 2x + 3$
(d) $\dfrac{\partial^2}{\partial x^2} + \dfrac{\partial^2}{\partial y^2} + \dfrac{\partial^2}{\partial z^2}$	$x^3 y^2 z^4$

3–2. Determine whether the following operators are linear or nonlinear:

(a) $\hat{A} f(x) = \text{SQR} f(x)$ [square $f(x)$]

(b) $\hat{A} f(x) = f^*(x)$ [form the complex conjugate of $f(x)$]

(c) $\hat{A} f(x) = 0$ [multiply $f(x)$ by zero]

(d) $\hat{A} f(x) = [f(x)]^{-1}$ [take the reciprocal of $f(x)$]

(e) $\hat{A} f(x) = f(0)$ [evaluate $f(x)$ at $x = 0$]

(f) $\hat{A} f(x) = \ln f(x)$ [take the logarithm of $f(x)$]

3–3. In each case, show that $f(x)$ is an eigenfunction of the operator given. Find the eigenvalue.

$\hat{A}$	$f(x)$
(a) $\dfrac{d^2}{dx^2}$	$\cos \omega x$
(b) $\dfrac{d}{dt}$	$e^{i\omega t}$
(c) $\dfrac{d^2}{dx^2} + 2\dfrac{d}{dx} + 3$	$e^{\alpha x}$
(d) $\dfrac{\partial}{\partial y}$	$x^2 e^{6y}$

3–4. Show that $(\cos ax)(\cos by)(\cos cz)$ is an eigenfunction of the operator,

$$\nabla^2 = \frac{\partial^2}{\partial x^2} + \frac{\partial^2}{\partial y^2} + \frac{\partial^2}{\partial z^2}$$

which is called the Laplacian operator.

3–5. Write out the operator $\hat{A}^2$ for $\hat{A} =$

(a) $\dfrac{d^2}{dx^2}$ (b) $\dfrac{d}{dx} + x$ (c) $\dfrac{d^2}{dx^2} - 2x\dfrac{d}{dx} + 1$

Hint: Be sure to include $f(x)$ before carrying out the operations.

3–6. Determine whether or not the following pairs of operators commute.

	$\hat{A}$	$\hat{B}$
(a)	$\dfrac{d}{dx}$	$\dfrac{d^2}{dx^2} + 2\dfrac{d}{dx}$
(b)	x	$\dfrac{d}{dx}$
(c)	SQR	SQRT
(d)	$\dfrac{\partial}{\partial x}$	$\dfrac{\partial}{\partial y}$

3–7. In ordinary algebra, $(P + Q)(P - Q) = P^2 - Q^2$. Expand $(\hat{P} + \hat{Q})(\hat{P} - \hat{Q})$. Under what conditions do we find the same result as in the case of ordinary algebra?

3–8. If we operate on the particle-in-a-box wave functions (Equations 3.27) with the momentum operator (Equation 3.11), we find

$$\hat{P}B \sin \frac{n\pi x}{a} = -i\hbar B \frac{\partial}{\partial x}\left(\sin \frac{n\pi x}{a}\right)$$

$$= -\frac{i\hbar n\pi}{a} B \cos \frac{n\pi x}{a}$$

Note that this is *not* an eigenvalue equation, and so we say that the momentum of a particle in a box does not have a fixed, definite value. Although the particle does not have a definite momentum, we can use the classical equation $E = p^2/2m$ to define formally some sort of effective momentum. Using Equation 3.21 for E, show that $p = nh/2a$ and that the de Broglie wavelengths associated with these momenta are $\lambda = h/p = 2a/n$. Show that this last equation says that an integral number of half-wavelengths fit into the box or that Figure 3.2 corresponds to standing de Broglie waves or matter waves.

3–9. In Section 3.5, we applied the equations for a particle in a box to the π electrons in butadiene. This simple model is called the free-electron model. Using the same argument, show that the length of hexatriene can be estimated to be 867 pm. Show that the first electronic transition is predicted to occur at 2.8×10^4 cm^{-1}. (Remember that hexatriene has six π electrons.)

3–10. Prove that if $\psi(x)$ is a solution to the Schrödinger equation, then any constant times $\psi(x)$ is also a solution.

3–11. In this problem, we will prove that the form of the Schrödinger equation imposes the condition that the first derivative of a wave function be continuous. The Schrödinger equation is

$$\frac{d^2\psi}{dx^2} + \frac{2m}{\hbar^2}[E - V(x)]\psi(x) = 0$$

If we integrate both sides from $a - \epsilon$ to $a + \epsilon$, where a is an arbitrary value of x and ϵ is infinitesimally small, then we have

$$\left.\frac{d\psi}{dx}\right|_{x=a+\epsilon} - \left.\frac{d\psi}{dx}\right|_{x=a-\epsilon} = \frac{2m}{\hbar^2}\int_{a-\epsilon}^{a+\epsilon}[V(x) - E]\psi(x)dx$$

Now show that $d\psi/dx$ is continuous if $V(x)$ is continuous.

Suppose now that $V(x)$ is *not* continuous at $x = a$, as in

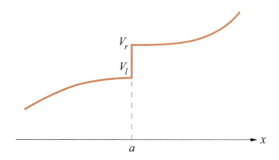

Show that

$$\left.\frac{d\psi}{dx}\right|_{x=a+\epsilon} - \left.\frac{d\psi}{dx}\right|_{x=a-\epsilon} = \frac{2m}{\hbar^2}(V_l + V_r - 2E)\psi(a)\epsilon$$

so that $d\psi/dx$ is continuous even if $V(x)$ has a *finite* discontinuity. What if $V(x)$ has an infinite discontinuity, as in the problem of a particle in a box? Are the first derivatives of the wave functions continuous at the boundaries of the box?

3–12. Show that the probability associated with the state ψ_n for a particle in a one-dimensional box of length a obeys the following relationships:

$$\text{Prob}(0 \leq x \leq a/4) = \text{Prob}(3a/4 \leq x \leq a) = \begin{cases} \dfrac{1}{4} & n \text{ even} \\[2ex] \dfrac{1}{4} - \dfrac{(-1)^{\frac{n-1}{2}}}{2\pi n} & n \text{ odd} \end{cases}$$

and

$$\text{Prob}(a/4 \leq x \leq a/2) = \text{Prob}(a/2 \leq x \leq 3a/4) = \begin{cases} \dfrac{1}{4} & n \text{ even} \\[2ex] \dfrac{1}{4} + \dfrac{(-1)^{\frac{n-1}{2}}}{2\pi n} & n \text{ odd} \end{cases}$$

3–13. What are the units, if any, for the wave function of a particle in a one-dimensional box?

3–14. Using a table of integrals, show that

$$\int_0^a \sin^2\frac{n\pi x}{a}dx = \frac{a}{2}$$

$$\int_0^a x\sin^2\frac{n\pi x}{a}dx = \frac{a^2}{4}$$

and

$$\int_0^a x^2 \sin^2 \frac{n\pi x}{a} dx = \left(\frac{a}{2\pi n}\right)^3 \left(\frac{4\pi^3 n^3}{3} - 2n\pi\right)$$

All of these integrals can be evaluated from

$$I(\beta) = \int_0^a e^{\beta x} \sin^2 \frac{n\pi x}{a} dx$$

Show that the above integrals are given by $I(0)$, $I'(0)$, and $I''(0)$, respectively, where the primes denote differentiation with respect to β. Using a table of integrals, evaluate $I(\beta)$ and then the above three integrals by differentiation.

3–15. Show that

$$\langle x \rangle = \frac{a}{2}$$

for all the states of a particle in a box. Is this result physically reasonable?

3–16. Show that $\langle p \rangle = 0$ for all states of a one-dimensional box of length a.

3–17. Show that

$$\sigma_x = (\langle x^2 \rangle - \langle x \rangle^2)^{1/2}$$

for a particle in a box is less than a, the width of the box, for any value of n. If σ_x is the uncertainty in the position of the particle, could σ_x ever be larger than a?

3–18. A classical particle in a box has an equi-likelihood of being found anywhere within the region $0 \leq x \leq a$. Consequently, its probability distribution is

$$p(x)dx = \frac{dx}{a} \qquad 0 \leq x \leq a$$

Show that $\langle x \rangle = a/2$ and $\langle x^2 \rangle = a^2/3$ for this system. Now show that $\langle x^2 \rangle$ (Equation 3.32) and σ_x (Equation 3.33) for a quantum-mechanical particle in a box take on the classical values as $n \to \infty$. This result is an example of the *correspondence principle*.

3–19. Using the trigonometric identity

$$\sin 2\theta = 2 \sin \theta \cos \theta$$

show that

$$\int_0^a \sin \frac{n\pi x}{a} \cos \frac{n\pi x}{a} dx = 0$$

3–20. Prove that

$$\int_0^a e^{\pm i 2\pi n x/a} dx = 0 \quad n \neq 0$$

3–21. Using the trigonometric identity

$$\sin \alpha \sin \beta = \frac{1}{2} \cos(\alpha - \beta) - \frac{1}{2} \cos(\alpha + \beta)$$

show that the particle-in-a-box wave functions (Equations 3.27) satisfy the relation

$$\int_0^a \psi_n^*(x) \psi_m dx = 0 \qquad m \neq n$$

(The asterisk in this case is superfluous because the functions are real.) If a set of functions satisfies the above integral condition, we say that the set is *orthogonal* and, in particular, that $\psi_m(x)$ is orthogonal to $\psi_n(x)$. If, in addition, the functions are normalized, then we say that the set is *orthonormal*.

3–22. Prove that the set of functions

$$\psi_n(x) = a^{-1/2} e^{i\pi nx/a} \qquad n = 0, \pm 1, \pm 2, \ldots$$

is orthonormal (cf. Problem 3–21) over the interval $-a \leq x \leq a$. A compact way to express orthonormality in the ψ_n is to write

$$\int_0^a \psi_m^*(x) \psi_n dx = \delta_{mn}$$

The symbol δ_{mn} is called a Kronecker delta and is defined by

$$\delta_{mn} = \begin{cases} 1 & \text{if } m = n \\ 0 & \text{if } m \neq n \end{cases}$$

3–23. In problems dealing with a particle in a box, we often need to evaluate integrals of the type

$$\int_0^a \sin \frac{n\pi x}{a} \sin \frac{m\pi x}{a} dx \qquad \text{and} \qquad \int_0^a \cos \frac{n\pi x}{a} \cos \frac{m\pi x}{a} dx$$

Integrals such as these are easy to evaluate if you convert the trigonometric functions to complex exponentials by using the identities (see MathChapter A)

$$\cos \theta = \frac{e^{i\theta} + e^{-i\theta}}{2} \qquad \text{and} \qquad \sin \theta = \frac{e^{i\theta} - e^{-i\theta}}{2i}$$

and then realize that the set of functions

$$\psi_n(x) = a^{-1/2} e^{in\pi x/a} \qquad n = 0, \pm 1, \pm 2, \ldots$$

is orthonormal on the interval $-a \leq x \leq a$ (Problem 3–22). Show that

$$\int_0^a \sin \frac{n\pi x}{a} \sin \frac{m\pi x}{a} dx = \int_0^a \cos \frac{n\pi x}{a} \cos \frac{m\pi x}{a} dx = \frac{a}{2} \delta_{nm}$$

where δ_{nm} is the Kronecker delta (defined in Problem 3–22). Also show that

$$\int_0^a \cos \frac{n\pi x}{a} \sin \frac{m\pi x}{a} dx = 0$$

3–24. Show that the set of functions

$$\phi_n(\theta) = (2\pi)^{-1/2}e^{in\theta} \qquad 0 \le \theta \le 2\pi$$

is orthonormal (Problem 3–21).

3–25. In going from Equation 3.34 to 3.35, we multiplied Equation 3.34 from the left by $\psi^*(x)$ and then integrated over all values of x to obtain Equation 3.35. Does it make any difference whether we multiplied from the left or the right?

3–26. Calculate $\langle x \rangle$ and $\langle x^2 \rangle$ for the $n = 2$ state of a particle in a one-dimensional box of length a. Show that

$$\sigma_x = \frac{a}{4\pi} \left(\frac{4\pi^2}{3} - 2 \right)^{1/2}$$

3–27. Calculate $\langle p \rangle$ and $\langle p^2 \rangle$ for the $n = 2$ state of a particle in a one-dimensional box of length a. Show that

$$\sigma_p = \frac{h}{a}$$

3–28. Consider a particle of mass m in a one-dimensional box of length a. Its average energy is given by

$$\langle E \rangle = \frac{1}{2m} \langle p^2 \rangle$$

Because $\langle p \rangle = 0$, $\langle p^2 \rangle = \sigma_p^2$, where σ_p can be called the uncertainty in p. Using the uncertainty principle, show that the energy must be at least as large as $\hbar^2/8ma^2$ because σ_x, the uncertainty in x, cannot be larger than a.

3–29. Discuss the degeneracies of the first few energy levels of a particle in a three-dimensional box when $a \ne b \ne c$.

3–30. Show that the normalized wave function for a particle in a three-dimensional box with sides of length a, b, and c is

$$\psi(x, y, z) = \left(\frac{8}{abc} \right)^{1/2} \sin \frac{n_x \pi x}{a} \sin \frac{n_y \pi y}{b} \sin \frac{n_z \pi z}{c}$$

3–31. Show that $\langle \mathbf{p} \rangle = 0$ for the ground state of a particle in a three-dimensional box with sides of length a, b, and c.

3–32. What are the degeneracies of the first four energy levels for a particle in a three-dimensional box with $a = b = 1.5c$?

3–33. The Schrödinger equation for a particle of mass m constrained to move on a circle of radius a is

$$-\frac{\hbar^2}{2I} \frac{d^2\psi}{d\theta^2} = E\psi(\theta) \qquad 0 \le \theta \le 2\pi$$

where $I = ma^2$ is the moment of inertia and θ is the angle that describes the position of the particle around the ring. Show by direct substitution that the solutions to this equation are

$$\psi(\theta) = Ae^{in\theta}$$

where $n = \pm(2IE)^{1/2}/\hbar$. Argue that the appropriate boundary condition is $\psi(\theta) = \psi(\theta + 2\pi)$ and use this condition to show that

$$E = \frac{n^2\hbar^2}{2I} \qquad n = 0, \pm1, \pm2, \ldots$$

Show that the normalization constant A is $(2\pi)^{-1/2}$. Discuss how you might use these results for a free-electron model of benzene.

3–34. Set up the problem of a particle in a box with its walls located at $-a$ and $+a$. Show that the energies are equal to those of a box with walls located at 0 and $2a$. (These energies may be obtained from the results that we derived in the chapter simply by replacing a by $2a$.) Show, however, that the wave functions are not the same and are given by

$$\psi_n(x) = \begin{cases} \dfrac{1}{a^{1/2}} \sin \dfrac{n\pi x}{2a} & n \text{ even} \\[2ex] \dfrac{1}{a^{1/2}} \cos \dfrac{n\pi x}{2a} & n \text{ odd} \end{cases}$$

Does it bother you that the wave functions seem to depend upon whether the walls are located at $\pm a$ or 0 and $2a$? Surely the particle "knows" only that it has a region of length $2a$ in which to move and cannot be affected by where you place the origin for the two sets of wave functions. What does this tell you? Do you think that any experimentally observable properties depend upon where you choose to place the origin of the x axis? Show that $\sigma_x\sigma_p > \hbar/2$, exactly as we obtained in Section 3.8.

3–35. The quantized energies of a particle in a box result from the boundary conditions, or from the fact that the particle is restricted to a finite region. In this problem, we investigate the quantum-mechanical problem of a free particle, one that is not restricted to a finite region. The potential energy $V(x)$ is equal to zero and the Schrödinger equation is

$$\frac{d^2\psi}{dx^2} + \frac{2mE}{\hbar^2}\psi(x) = 0 \qquad -\infty < x < \infty$$

Note that the particle can lie anywhere along the x axis in this problem. Show that the two solutions of this Schrödinger equation are

$$\psi_1(x) = A_1 e^{i(2mE)^{1/2}x/\hbar} = A_1 e^{ikx}$$

and

$$\psi_2(x) = A_2 e^{-i(2mE)^{1/2}x/\hbar} = A_2 e^{-ikx}$$

where

$$k = \frac{(2mE)^{1/2}}{\hbar}$$

Show that if E is allowed to take on negative values, then the wave functions become unbounded for large x. Therefore, we will require that the energy, E, be a positive quantity.

To get a physical interpretation of the states that $\psi_1(x)$ and $\psi_2(x)$ describe, operate on $\psi_1(x)$ and $\psi_2(x)$ with the momentum operator $\hat{P}$ (Equation 3.11), and show that

$$\hat{P}\psi_1 = -i\hbar \frac{d\psi_1}{dx} = \hbar k \psi_1$$

and

$$\hat{P}\psi_2 = -i\hbar \frac{d\psi_2}{dx} = -\hbar k \psi_2$$

Notice that these are eigenvalue equations. Our interpretation of these two equations is that ψ_1 describes a free particle with fixed momentum $\hbar k$ and that ψ_2 describes a particle with fixed momentum $-\hbar k$. Thus, ψ_1 describes a particle moving to the right and ψ_2 describes a particle moving to the left, both with a fixed momentum. Notice also that there are no restrictions on k, and so the particle can have any value of momentum. Now show that

$$E - \frac{\hbar^2 k^2}{2m}$$

Notice that the energy is not quantized; the energy of the particle can have any positive value in this case because no boundaries are associated with this problem.

Last, show that $\psi_1^*(x)\psi_1(x) = A_1^* A_1 = |A_1|^2 = $ constant, and that $\psi_2^*(x)\psi_2(x) = A_2^* A_2 = |A_2|^2 = $ constant. Discuss this result in terms of the probabilistic interpretation of $\psi^*\psi$. Also discuss the application of the uncertainty principle to this problem. What are σ_p and σ_x?

3–36. Derive the equation for the allowed energies of a particle in a one-dimensional box by assuming that the particle is described by standing de Broglie waves within the box.

3–37. In Chapter 4, we will encounter the time-dependent Schrödinger equation

$$\hat{H}\Psi(x, t) = i\hbar \frac{\partial \Psi(x, t)}{\partial t}$$

where Ψ is now a function of both position and time. Show that if the Hamiltonian operator does not contain time explicitly $[\hat{H} = \hat{H}(x)]$, then this partial differential equation can be separated into two ordinary differential equations by setting $\Psi(x, t) = \psi(x)f(t)$. What is the separation constant in this problem? Generalize this result to three dimensions. What is the function $f(t)$?

3–38. Using the result of Problem 3–37, what is the time-dependent wave function for the ground state of a particle in a one-dimensional box of length a? Use this wave function to evaluate the average value of x. Are you surprised?

3–39. We can use the wave functions of Problem 3–34 to illustrate some fundamental symmetry properties of wave functions. Show that the wave functions are alternately symmetric and antisymmetric or even and odd with respect to the operation $x \rightarrow -x$, which is a reflection through the $x = 0$ line. This symmetry property of the wave function is a consequence of the symmetry of the Hamiltonian operator, as we shall now show. The Schrödinger equation may be written as

$$\hat{H}(x)\psi_n(x) = E_n\psi_n(x)$$

Reflection through the $x = 0$ line gives $x \rightarrow -x$, and so

$$\hat{H}(-x)\psi_n(-x) = E_n\psi_n(-x)$$

Now show that $\hat{H}(x) = \hat{H}(-x)$ (i.e., that $\hat{H}$ is symmetric), and so show that

$$\hat{H}(x)\psi_n(-x) = E_n\psi_n(-x)$$

Thus, we see that $\psi_n(-x)$ is also an eigenfunction of $\hat{H}$ belonging to the same eigenvalue E_n. Now, if there is only one eigenfunction associated with each eigenvalue (we call this a *nondegenerate case*), then argue that $\psi_n(x)$ and $\psi_n(-x)$ must differ by a multiplicative constant [i.e., that $\psi_n(-x) = c\psi_n(x)$]. By applying the inversion operation again to this equation, show that $c = \pm1$ and that all the wave functions must be either even or odd with respect to reflection through the $x = 0$ line because the Hamiltonian operator is symmetric. Thus, we see that the symmetry of the Hamiltonian operator influences the symmetry of the wave functions.

References

Moore, W. J. *Schrödinger: Life and Thought*. Cambridge University Press: New York, 1992.
Greenspan, N. *The End of the Certain World: The Life and Science of Max Born*. Basic Books: New York, 2005.

Vectors

A vector is a quantity that has both magnitude and direction. Examples of vectors are position, force, velocity, and momentum. We specify the position of something, for example, by giving not only its distance from a certain point but also its direction from that point. We often represent a vector by an arrow, where the length of the arrow is the magnitude of the vector and its direction is the same as the direction of the vector.

Two vectors can be added together to get a new vector. Consider the two vectors $\mathbf{u}$ and $\mathbf{v}$ in Figure C.1. (We denote vectors by boldface symbols.) To find $\mathbf{w} = \mathbf{u} + \mathbf{v}$, we place the tail of $\mathbf{u}$ at the tip of $\mathbf{v}$ and then draw $\mathbf{w}$ from the tail of $\mathbf{v}$ to the tip of $\mathbf{u}$, as shown in the figure. We could also have placed the tail of $\mathbf{u}$ at the origin and then placed the tail of $\mathbf{v}$ at the tip of $\mathbf{u}$ and drawn $\mathbf{w}$ from the tail of $\mathbf{u}$ to the tip of $\mathbf{v}$. As Figure C.1 indicates, we get the same result either way, so we see that

$$\mathbf{w} = \mathbf{u} + \mathbf{v} = \mathbf{v} + \mathbf{u} \tag{C.1}$$

Vector addition is commutative.

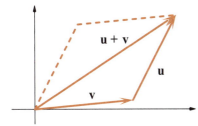

FIGURE C.1
An illustration of the addition of two vectors, $\mathbf{u} + \mathbf{v} = \mathbf{v} + \mathbf{u} = \mathbf{w}$.

To subtract two vectors, we draw one of them in the opposite direction and then add it to the other. Writing a vector in its opposite direction is equivalent to forming the vector $-\mathbf{v}$. Thus, mathematically we have

$$\mathbf{t} = \mathbf{u} - \mathbf{v} = \mathbf{u} + (-\mathbf{v}) \tag{C.2}$$

129

Generally, a number a times a vector is a new vector that is parallel to **u** but whose length is a times the length of **u**. If a is positive, then a**u** lies in the same direction as **u**, but if a is negative, then a**u** lies in the opposite direction.

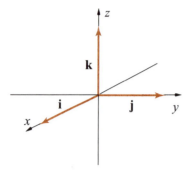

FIGURE C.2
The fundamental unit vectors **i**, **j**, and **k** of a cartesian coordinate system.

A useful set of vectors are the vectors that are of unit length and point along the positive x, y, and z axes of a cartesian coordinate system. These *unit vectors* (unit length), which we designate by **i**, **j**, and **k**, respectively, are shown in Figure C.2. We shall always draw a cartesian coordinate system so that it is right-handed. A *right-handed coordinate system* is such that when you curl the four fingers of your right hand from **i** to **j**, your thumb points along **k** (Figure C.3).

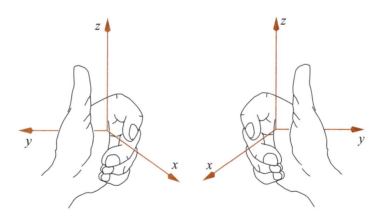

FIGURE C.3
An illustration of a right-handed cartesian coordinate system (right) and a left-handed cartesian coordinate system (left).

Any three-dimensional vector **u** can be described in terms of these unit vectors,

$$\mathbf{u} = u_x\,\mathbf{i} + u_y\,\mathbf{j} + u_z\,\mathbf{k} \tag{C.3}$$

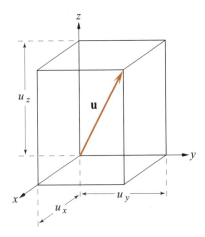

FIGURE C.4
The components of a vector **u** are its projections along the x, y, and z axes, showing that the length of **u** is equal to $(u_x^2 + u_y^2 + u_z^2)^{1/2}$.

where, for example, $u_x \mathbf{i}$ is u_x units long and lies in the direction of $\mathbf{i}$. The quantities u_x, u_y, and u_z in Equation C.3 are the *components* of **u**. They are the projections of **u** along the respective cartesian axes (Figure C.4). In terms of components, the sum or difference of two vectors is given by

$$\mathbf{u} \pm \mathbf{v} = (u_x \pm v_x)\,\mathbf{i} + (u_y \pm v_y)\,\mathbf{j} + (u_z \pm v_z)\,\mathbf{k} \tag{C.4}$$

Figure C.4 shows that the length of **u** is given by

$$u = |\mathbf{u}| = (u_x^2 + u_y^2 + u_z^2)^{1/2} \tag{C.5}$$

EXAMPLE C–1
If $\mathbf{u} = 2\mathbf{i} - \mathbf{j} + 3\mathbf{k}$ and $\mathbf{v} = -\mathbf{i} + 2\mathbf{j} - \mathbf{k}$, then what is the length of $\mathbf{u} + \mathbf{v}$?

SOLUTION: Using Equation C.4, we have

$$\mathbf{u} + \mathbf{v} = (2 - 1)\mathbf{i} + (-1 + 2)\mathbf{j} + (3 - 1)\mathbf{k} = \mathbf{i} + \mathbf{j} + 2\mathbf{k}$$

and using Equation C.5 gives

$$|\mathbf{u} + \mathbf{v}| = (1^2 + 1^2 + 2^2)^{1/2} = \sqrt{6}$$

There are two ways to form the product of two vectors, and both have many applications in physical chemistry. One way yields a scalar quantity (in other words, just a number), and the other yields a vector. Not surprisingly, we call the result of the first method a *scalar product* and the result of the second method a *vector product*.

The scalar product of two vectors **u** and **v** is defined as

$$\mathbf{u} \cdot \mathbf{v} = |\mathbf{u}||\mathbf{v}| \cos \theta \tag{C.6}$$

where θ is the angle between **u** and **v**. Note from the definition that

$$\mathbf{u} \cdot \mathbf{v} = \mathbf{v} \cdot \mathbf{u} \tag{C.7}$$

Taking a scalar product is a *commutative operation*. The dot between **u** and **v** is such a standard notation that $\mathbf{u} \cdot \mathbf{v}$ is often called the *dot product* of **u** and **v**. The dot products of the unit vectors **i**, **j**, and **k** are

$$\mathbf{i} \cdot \mathbf{i} = \mathbf{j} \cdot \mathbf{j} = \mathbf{k} \cdot \mathbf{k} = |1||1| \cos 0° = 1$$
$$\mathbf{i} \cdot \mathbf{j} = \mathbf{j} \cdot \mathbf{i} = \mathbf{i} \cdot \mathbf{k} = \mathbf{k} \cdot \mathbf{i} = \mathbf{j} \cdot \mathbf{k} = \mathbf{k} \cdot \mathbf{j} = |1||1| \cos 90° = 0 \tag{C.8}$$

We can use Equations C.8 to evaluate the dot product of two vectors:

$$\mathbf{u} \cdot \mathbf{v} = (u_x \mathbf{i} + u_y \mathbf{j} + u_z \mathbf{k}) \cdot (v_x \mathbf{i} + v_y \mathbf{j} + v_z \mathbf{k})$$
$$= u_x v_x \mathbf{i} \cdot \mathbf{i} + u_x v_y \mathbf{i} \cdot \mathbf{j} + u_x v_z \mathbf{i} \cdot \mathbf{k}$$
$$+ u_y v_x \mathbf{j} \cdot \mathbf{i} + u_y v_y \mathbf{j} \cdot \mathbf{j} + u_y v_z \mathbf{j} \cdot \mathbf{k}$$
$$+ u_z v_x \mathbf{k} \cdot \mathbf{i} + u_z v_y \mathbf{k} \cdot \mathbf{j} + u_z v_z \mathbf{k} \cdot \mathbf{k}$$

and so

$$\mathbf{u} \cdot \mathbf{v} = u_x v_x + u_y v_y + u_z v_z \tag{C.9}$$

EXAMPLE C–2
Find the length of $\mathbf{u} = 2\mathbf{i} - \mathbf{j} + 3\mathbf{k}$.

SOLUTION: Equation C.9 with $\mathbf{u} = \mathbf{v}$ gives

$$\mathbf{u} \cdot \mathbf{u} = u_x^2 + u_y^2 + u_z^2 = |\mathbf{u}|^2$$

Therefore,

$$|\mathbf{u}| = (\mathbf{u} \cdot \mathbf{u})^{1/2} = (4 + 1 + 9)^{1/2} = \sqrt{14}$$

EXAMPLE C–3

Find the angle between the two vectors $\mathbf{u} = \mathbf{i} + 3\mathbf{j} - \mathbf{k}$ and $\mathbf{v} = \mathbf{j} - \mathbf{k}$.

SOLUTION: We use Equation C.6, but first we must find

$$|\mathbf{u}| = (\mathbf{u} \cdot \mathbf{u})^{1/2} = (1 + 9 + 1)^{1/2} = \sqrt{11}$$

$$|\mathbf{v}| = (\mathbf{v} \cdot \mathbf{v})^{1/2} = (0 + 1 + 1)^{1/2} = \sqrt{2}$$

and

$$\mathbf{u} \cdot \mathbf{v} = 0 + 3 + 1 = 4$$

Therefore,

$$\cos \theta = \frac{\mathbf{u} \cdot \mathbf{v}}{|\mathbf{u}||\mathbf{v}|} = \frac{4}{\sqrt{22}} = 0.8528$$

or $\theta = 31.48°$.

Because $\cos 90° = 0$, the dot product of vectors that are perpendicular to each other is equal to zero. For example, the dot products between the $\mathbf{i}$, $\mathbf{j}$, and $\mathbf{k}$ cartesian unit vectors are equal to zero, as Equation C.8 says.

EXAMPLE C–4

Show that the vectors $\mathbf{v}_1 = \dfrac{1}{\sqrt{3}}\mathbf{i} + \dfrac{1}{\sqrt{3}}\mathbf{j} + \dfrac{1}{\sqrt{3}}\mathbf{k}$, $\mathbf{v}_2 = \dfrac{1}{\sqrt{6}}\mathbf{i} - \dfrac{2}{\sqrt{6}}\mathbf{j} + \dfrac{1}{\sqrt{6}}\mathbf{k}$, and $\mathbf{v}_3 = -\dfrac{1}{\sqrt{2}}\mathbf{i} + \dfrac{1}{\sqrt{2}}\mathbf{k}$ are of unit length and are mutually perpendicular.

SOLUTION: The lengths are given by

$$(\mathbf{v}_1 \cdot \mathbf{v}_1)^{1/2} = \left(\frac{1}{3} + \frac{1}{3} + \frac{1}{3} \right)^{1/2} = 1$$

$$(\mathbf{v}_2 \cdot \mathbf{v}_2)^{1/2} = \left(\frac{1}{6} + \frac{4}{6} + \frac{1}{6} \right)^{1/2} = 1$$

$$(\mathbf{v}_3 \cdot \mathbf{v}_3)^{1/2} = \left(\frac{1}{2} + 0 + \frac{1}{2} \right)^{1/2} = 1$$

The dot products between the different vectors are

$$\mathbf{v}_1 \cdot \mathbf{v}_2 = \frac{1}{\sqrt{18}} - \frac{2}{\sqrt{18}} + \frac{1}{\sqrt{18}} = 0$$

$$\mathbf{v}_1 \cdot \mathbf{v}_3 = -\frac{1}{\sqrt{6}} + 0 + \frac{1}{\sqrt{6}} = 0$$

$$\mathbf{v}_2 \cdot \mathbf{v}_3 = -\frac{1}{\sqrt{12}} + 0 + \frac{1}{\sqrt{12}} = 0$$

None of the vector operations that we have used so far are limited to two or three dimensions. We can easily generalize Equation C.9 to N dimensions by writing

$$\mathbf{u} \cdot \mathbf{v} = \sum_{j=1}^{N} u_j v_j \tag{C.10}$$

The length of an N-dimensional vector is given by

$$l = (\mathbf{u} \cdot \mathbf{u})^{1/2} = \left(\sum_{j=1}^{N} u_j^2 \right)^{1/2} \tag{C.11}$$

If the dot product of two N-dimensional vectors is equal to zero, then we say that the two vectors are *orthogonal*. Thus, the term orthogonal is just a generalization of perpendicular. Furthermore, if the length of a vector is equal to 1, then the vector is said to be *normalized*. A set of mutually orthogonal vectors that are also normalized is said to be *orthonormal*. It is common notation to represent N-dimensional vectors by just listing their components within parentheses. Problem C–7 has you show that the set of vectors $(1/\sqrt{3}, 1/\sqrt{3}, 0, 1/\sqrt{3})$, $(1/\sqrt{3}, -1/\sqrt{3}, 1/\sqrt{3}, 0)$, $(0, 1/\sqrt{3}, 1/\sqrt{3}, -1/\sqrt{3})$, and $(1/\sqrt{3}, 0, -1/\sqrt{3}, -1/\sqrt{3})$ is orthonormal.

One application of a dot product involves the definition of work. Recall that work is defined as force times distance, where "force" means the component of force that lies in the same direction as the displacement. If we let $\mathbf{F}$ be the force and $\mathbf{d}$ be the displacement, then work is defined as

$$\text{work} = \mathbf{F} \cdot \mathbf{d} \tag{C.12}$$

We can write Equation C.12 as $(F \cos \theta)(d)$ to emphasize that $F \cos \theta$ is the component of $\mathbf{F}$ in the direction of $\mathbf{d}$ (Figure C.5).

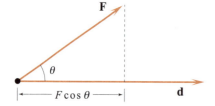

FIGURE C.5
Work is defined as $w = \mathbf{F} \cdot \mathbf{d}$, or $(F \cos \theta)d$, where $F \cos \theta$ is the component of $\mathbf{F}$ along $\mathbf{d}$.

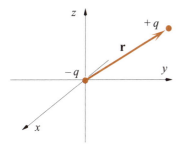

FIGURE C.6
A dipole moment is a vector that points from a negative charge, $-q$, to a positive charge, $+q$, and whose magnitude is qr.

Another important application of a dot product involves the interaction of a dipole moment with an electric field. You may have learned in organic chemistry that the separation of opposite charges in a molecule gives rise to a dipole moment, which is often indicated by an arrow pointing from the negative charge to the positive charge. For example, because a chlorine atom is more electronegative than a hydrogen atom, HCl has a dipole moment, which we indicate by writing $\overleftarrow{\text{HCl}}$. Strictly speaking, a dipole moment is a vector quantity whose magnitude is equal to the product of the positive charge and the distance between the positive and negative charges and whose direction is from the negative charge to the positive charge. Thus, for the two separated charges illustrated in Figure C.6, the dipole moment $\boldsymbol{\mu}$ is equal to

$$\boldsymbol{\mu} = q\,\mathbf{r}$$

We will learn later that if we apply an electric field $\mathbf{E}$ to a dipole moment, then the potential energy of interaction will be

$$V = -\boldsymbol{\mu} \cdot \mathbf{E} \tag{C.13}$$

The vector product of two vectors is a vector defined by

$$\mathbf{u} \times \mathbf{v} = |\mathbf{u}||\mathbf{v}|\,\mathbf{c}\,\sin\theta \tag{C.14}$$

where θ is the angle between $\mathbf{u}$ and $\mathbf{v}$ and $\mathbf{c}$ is a unit vector perpendicular to the plane formed by $\mathbf{u}$ and $\mathbf{v}$. The direction of $\mathbf{c}$ is given by the right-hand rule: If the four fingers of your right hand curl from $\mathbf{u}$ to $\mathbf{v}$, then $\mathbf{c}$ lies along the direction of your thumb. (See Figure C.3 for a similar construction.) The notation given in Equation C.14 is so commonly used that the vector product is usually called the *cross product*. Because the direction of $\mathbf{c}$ is given by the right-hand rule, the cross product operation is not commutative, and, in particular

$$\mathbf{u} \times \mathbf{v} = -\mathbf{v} \times \mathbf{u} \tag{C.15}$$

The cross products of the cartesian unit vectors are

$$
\begin{aligned}
\mathbf{i} \times \mathbf{i} = \mathbf{j} \times \mathbf{j} = \mathbf{k} \times \mathbf{k} = |1||1| \, \mathbf{c} \sin 0° = 0 \\
\mathbf{i} \times \mathbf{j} = -\mathbf{j} \times \mathbf{i} = |1||1| \, \mathbf{k} \sin 90° = \mathbf{k} \\
\mathbf{j} \times \mathbf{k} = -\mathbf{k} \times \mathbf{j} = \mathbf{i} \\
\mathbf{k} \times \mathbf{i} = -\mathbf{i} \times \mathbf{k} = \mathbf{j}
\end{aligned}
\tag{C.16}
$$

In terms of components of $\mathbf{u}$ and $\mathbf{v}$, we have (Problem C–10)

$$
\mathbf{u} \times \mathbf{v} = (u_y v_z - u_z v_y)\,\mathbf{i} + (u_z v_x - u_x v_z)\,\mathbf{j} + (u_x v_y - u_y v_x)\,\mathbf{k} \tag{C.17}
$$

Equation C.17 can be conveniently expressed as a determinant (see MathChapter E):

$$
\mathbf{u} \times \mathbf{v} =
\begin{vmatrix}
\mathbf{i} & \mathbf{j} & \mathbf{k} \\
u_x & u_y & u_z \\
v_x & v_y & v_z
\end{vmatrix}
\tag{C.18}
$$

Equations C.17 and C.18 are equivalent.

EXAMPLE C–5
Given $\mathbf{u} = -2\mathbf{i} + \mathbf{j} + \mathbf{k}$ and $\mathbf{v} = 3\mathbf{i} - \mathbf{j} + \mathbf{k}$, determine $\mathbf{w} = \mathbf{u} \times \mathbf{v}$.

SOLUTION: Using Equation C.17, we have

$$
\mathbf{w} = [(1)(1) - (1)(-1)]\mathbf{i} + [(1)(3) - (-2)(1)]\mathbf{j} + [(-2)(-1) - (1)(3)]\mathbf{k}
$$
$$
= 2\mathbf{i} + 5\mathbf{j} - \mathbf{k}
$$

One physically important application of a cross product involves the definition of angular momentum. If a particle has a momentum $\mathbf{p} = m\mathbf{v}$ at a position $\mathbf{r}$ from a fixed point (as in Figure C.7), then its *angular momentum* is defined by

$$
\mathbf{l} = \mathbf{r} \times \mathbf{p} \tag{C.19}
$$

Note that the angular momentum is a vector perpendicular to the plane formed by $\mathbf{r}$ and $\mathbf{p}$ (Figure C.8). In terms of components, $\mathbf{l}$ is equal to (see Equation C.17)

$$
\mathbf{l} = (yp_x - xp_y)\,\mathbf{i} + (zp_x - xp_z)\,\mathbf{j} + (xp_y - yp_x)\,\mathbf{k} \tag{C.20}
$$

We will see that angular momentum plays an important role in quantum mechanics.

Another example that involves a cross product is the equation that gives the force $\mathbf{F}$ on a particle of charge q moving with velocity $\mathbf{v}$ through a magnetic field $\mathbf{B}$:

$$
\mathbf{F} = q(\mathbf{v} \times \mathbf{B})
$$

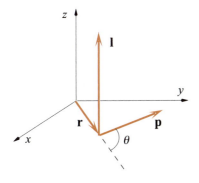

FIGURE C.7
The angular momentum of a particle of momentum **p** and position **r** from a fixed center is a vector perpendicular to the plane formed by **r** and **p** and in the direction of **r** × **p**.

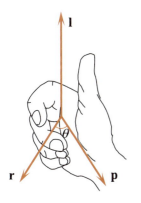

FIGURE C.8
Angular momentum is a vector quantity that lies perpendicular to the plane formed by **r** and **p** and is directed such that the vectors **r**, **p**, and **l** form a right-handed coordinate system.

Note that the force is perpendicular to **v**, and so the effect of **B** is to cause the motion of the particle to curve, not to speed up or slow down.

We can also take derivatives of vectors. Suppose that the components of momentum, **p**, depend upon time. Then

$$\frac{d\mathbf{p}(t)}{dt} = \frac{dp_x(t)}{dt}\mathbf{i} + \frac{dp_y(t)}{dt}\mathbf{j} + \frac{dp_z(t)}{dt}\mathbf{k} \qquad (C.21)$$

(There are no derivatives of **i**, **j**, and **k** because they are fixed in space.) Newton's law of motion is

$$\frac{d\mathbf{p}}{dt} = \mathbf{F} \qquad (C.22)$$

This law is actually three separate equations, one for each component. Because $\mathbf{p} = m\mathbf{v}$, if m is a constant, we can write Newton's equation as

$$m\frac{d\mathbf{v}}{dt} = \mathbf{F}$$

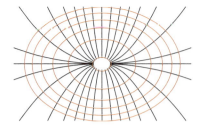

FIGURE C.9
A set of level curves (orange) for the surface $z = f(x, y)$ and the path $\nabla f(x, y)$ (black), which follows the direction of steepest descent.

Furthermore, because $\mathbf{v} = d\mathbf{r}/dt$, we can also express Newton's equations as

$$m\frac{d^2\mathbf{r}}{dt^2} = \mathbf{F} \tag{C.23}$$

Once again, Equation C.23 represents a set of three equations, one for each component.

There are a couple of differential vector operators that occur frequently in chemical and physical problems. One of these is the *gradient*, which is defined by

$$\nabla f(x, y, z) = \operatorname{grad} f(x, y, z) = \mathbf{i}\frac{\partial f}{\partial x} + \mathbf{j}\frac{\partial f}{\partial y} + \mathbf{k}\frac{\partial f}{\partial z} \tag{C.24}$$

Note that the gradient operator, ∇, operates on a scalar function. The vector ∇f is called the *gradient vector* of $f(x, y, z)$. Consider a set of contour lines on a topographical map or a set of isotherms or isobars on a weather map or a set of equipotentials in a potential energy diagram. Those lines are collectively called *level curves*. If a surface is described by $z = f(x, y)$, then the level curves are given by $z = $ constant (Figure C.9). The path traced out by ∇f in Figure C.9 is normal (perpendicular) to each level curve that it crosses and follows the direction of steepest descent. For a set of equipotentials, for example, ∇f represents the corresponding electric field and traces out the path that a charged particle will follow (Figure C.10).

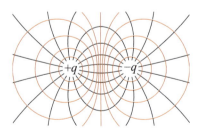

FIGURE C.10
The equipotentials (orange) and the electric field (black) of an electric dipole formed by equal and opposite charges.

Many physical laws are expressed in terms of a gradient vector. For example, Fick's law of diffusion says that the flux of a solute is proportional to the gradient of its concentration, or if $c(x, y, z, t)$ is the concentration of solute at the point (x, y, z) at time t, then

$$\text{flux of solute} = -D\nabla c(x, y, z, t)$$

where D is called the diffusion constant. Similarly, Fourier's law of heat flow says that the flux of heat is described by

$$\text{flux of heat} = -\lambda\nabla T(x, y, z, t)$$

where T is the temperature and λ is the thermal conductivity. If $V(x, y, z)$ is a mechanical potential energy experienced by a body, then the force on the body is given by

$$\mathbf{F} = -\nabla V(x, y, z) \tag{C.25}$$

In addition, if $\phi(x, y, z)$ is an electrostatic potential, then the electric field associated with that potential is given by

$$\mathbf{E} = -\nabla \phi(x, y, z) \tag{C.26}$$

EXAMPLE C–6

Suppose that a particle experiences a potential energy

$$V(x, y, z) = \frac{k_x x^2}{2} + \frac{k_y y^2}{2} + \frac{k_z z^2}{2},$$

where the k's are constant. Derive an expression for the force acting on the particle.

SOLUTION: We use Equation C.25 to write

$$\mathbf{F}(x, y, z) = -\mathbf{i}\frac{\partial V}{\partial x} - \mathbf{j}\frac{\partial V}{\partial y} - \mathbf{k}\frac{\partial V}{\partial z}$$

$$= -\mathbf{i}\,k_x x - \mathbf{j}\,k_y y - \mathbf{k}\,k_z z$$

Problems

C–1. Find the length of the vector $\mathbf{v} = 2\,\mathbf{i} - \mathbf{j} + 3\,\mathbf{k}$.

C–2. Find the length of the vector $\mathbf{r} = x\,\mathbf{i} + y\,\mathbf{j}$ and of the vector $\mathbf{r} = x\,\mathbf{i} + y\,\mathbf{j} + z\,\mathbf{k}$.

C–3. Prove that $\mathbf{u} \cdot \mathbf{v} = 0$ if $\mathbf{u}$ and $\mathbf{v}$ are perpendicular to each other. Two vectors that are perpendicular to each other are said to be orthogonal.

C–4. Show that the vectors $\mathbf{u} = 2\,\mathbf{i} - 4\,\mathbf{j} - 2\,\mathbf{k}$ and $\mathbf{v} = 3\,\mathbf{i} + 4\,\mathbf{j} - 5\,\mathbf{k}$ are orthogonal.

C–5. Show that the vector $\mathbf{r} = 2\,\mathbf{i} - 3\,\mathbf{k}$ lies entirely in a plane perpendicular to the y axis.

C–6. Find the angle between the two vectors $\mathbf{u} = -\mathbf{i} + 2\,\mathbf{j} + \mathbf{k}$ and $\mathbf{v} = 3\,\mathbf{i} - \mathbf{j} + 2\,\mathbf{k}$.

C–7. Show that the set of vectors $(1/\sqrt{3},\ 1/\sqrt{3},\ 0,\ 1/\sqrt{3})$, $(1/\sqrt{3},\ -1/\sqrt{3},\ 1/\sqrt{3},\ 0)$, $(0,\ 1/\sqrt{3},\ 1/\sqrt{3},\ -1/\sqrt{3})$, and $(1/\sqrt{3},\ 0,\ -1/\sqrt{3},\ -1/\sqrt{3})$ is orthonormal.

C–8. Determine $\mathbf{w} = \mathbf{u} \times \mathbf{v}$ given that $\mathbf{u} = -\mathbf{i} + 2\,\mathbf{j} + \mathbf{k}$ and $\mathbf{v} = 3\,\mathbf{i} - \mathbf{j} + 2\,\mathbf{k}$. What is $\mathbf{v} \times \mathbf{u}$ equal to?

C–9. Show that $\mathbf{u} \times \mathbf{u} = 0$.

C–10. Using Equation C.16, prove that $\mathbf{u} \times \mathbf{v}$ is given by Equation C.17.

C–11. Show that $l = |\mathbf{l}| = mvr$ for circular motion.

C–12. Show that

$$\frac{d}{dt}(\mathbf{u} \cdot \mathbf{v}) = \frac{d\mathbf{u}}{dt} \cdot \mathbf{v} + \mathbf{u} \cdot \frac{d\mathbf{v}}{dt}$$

and

$$\frac{d}{dt}(\mathbf{u} \times \mathbf{v}) = \frac{d\mathbf{u}}{dt} \times \mathbf{v} + \mathbf{u} \times \frac{d\mathbf{v}}{dt}$$

C–13. Using the results of Problem C–12, prove that

$$\mathbf{u} \times \frac{d^2\mathbf{u}}{dt^2} = \frac{d}{dt}\left(\mathbf{u} \times \frac{d\mathbf{u}}{dt}\right)$$

C–14. In vector notation, Newton's equations for a single particle are

$$m\frac{d^2\mathbf{r}}{dt^2} = \mathbf{F}(x,\ y,\ z)$$

By operating on this equation from the left by $\mathbf{r} \times$ and using the result of Problem C–13, show that

$$m\frac{d}{dt}\left(\mathbf{r} \times \frac{d\mathbf{r}}{dt}\right) = \mathbf{r} \times \mathbf{F}$$

Because momentum is defined as $\mathbf{p} = m\mathbf{v} = m\dfrac{d\mathbf{r}}{dt}$, the above expression reads

$$\frac{d}{dt}(\mathbf{r} \times \mathbf{p}) = \mathbf{r} \times \mathbf{F}$$

But $\mathbf{r} \times \mathbf{p} = \mathbf{l}$, the angular momentum, and so we have

$$\frac{d\mathbf{l}}{dt} = \mathbf{r} \times \mathbf{F}$$

This is the form of Newton's equation for a rotating system. Notice that $d\mathbf{l}/dt = 0$, or that angular momentum is conserved if $\mathbf{r} \times \mathbf{F} = 0$. Can you identify $\mathbf{r} \times \mathbf{F}$?

C–15. Find the gradient of $f(x, y, z) = x^2 - yz + xz^2$ at the point $(1, 1, 1)$.

C–16. The electrostatic potential produced by a dipole moment μ located at the origin and directed along the x axis is given by

$$\phi(x, y, z) = \frac{\mu x}{(x^2 + y^2 + z^2)^{3/2}} \qquad (x, y, z \neq 0)$$

Derive an expression for the electric field associated with this potential.

C–17. We proved the *Schwartz inequality* for complex numbers in Problem A–16. For vectors, the Schwartz inequality takes the form

$$(\mathbf{u} \cdot \mathbf{v})^2 \leq |\mathbf{u}|^2 |\mathbf{v}|^2$$

Why do you think that this is so? Do you see a parallel between this result for two-dimensional vectors and the complex number version?

C–18. We proved the *triangle inequality* for complex numbers in Problem A–17. For vectors, the triangle inequality takes the form

$$|\mathbf{u} + \mathbf{v}| \leq |\mathbf{u}| + |\mathbf{v}|$$

Prove this inequality by starting with

$$|\mathbf{u} + \mathbf{v}|^2 = |\mathbf{u}|^2 + |\mathbf{v}|^2 + 2\mathbf{u} \cdot \mathbf{v}$$

and then using the Schwartz inequality (previous problem). Why do you think this is called the triangle inequality?

Niels Bohr was born in Copenhagen, Denmark, on October 7, 1885, and died there in 1962. In 1911, Bohr received his Ph.D. in physics from the University of Copenhagen. He then spent a year with J. J. Thompson and Ernest Rutherford in England, where he formulated his theory of the hydrogen atom and its atomic spectrum. In 1913, he returned to the University of Copenhagen, where he remained for the rest of his life. In 1920, he was named director of the Institute of Theoretical Physics, which was supported largely by the Carlsberg Brewery and was an international center for theoretical physics during the 1920s and '30s, when quantum mechanics was being developed. After World War II, Bohr worked energetically for peaceful uses of atomic energy. He organized the first Atoms for Peace Conference in 1955 and received the first Atoms for Peace Prize in 1957. Bohr was awarded the Nobel Prize in Physics in 1922 "for his investigation of the structure of atoms and of the radiation emanating from them."

Werner Heisenberg was born on December 5, 1901, in Duisberg, Germany, grew up in Munich, and died in 1976. In 1923, Heisenberg received his Ph.D. in physics from the University of Munich. He then spent a year working with Max Born at the University of Göttingen and three years with Niels Bohr in Copenhagen. He was chair of theoretical physics at the University of Leipzig from 1927 to 1941, the youngest to have received such an appointment. Because of a deep loyalty to Germany, Heisenberg opted to stay in Germany when the Nazis came to power. After World War II, he was named director of the Max Planck Institute for Physics, where he strove to rebuild German science. Heisenberg developed one of the first formulations of quantum mechanics, which was based on matrix algebra. Heisenberg was awarded the 1932 Nobel Prize in Physics "for the creation of quantum mechanics." His role in Nazi Germany is somewhat clouded, prompting one author (David Cassidy) to title his biography of Heisenberg *Uncertainty* (W. H. Freeman: New York, 1993).

CHAPTER 4

The Postulates and General Principles of Quantum Mechanics

Up to now we have made a number of conjectures concerning the formulation of quantum mechanics. For example, we have been led to suspect that the variables of classical mechanics are represented in quantum mechanics by operators. These operate on wave functions to give the average or expected results of measurements according to Equation 3.13. In this chapter we shall formalize the various conjectures that we have made in Chapter 3 as a set of postulates and then discuss some general theorems that follow from these postulates. This is similar to setting up a set of axioms in geometry and then logically deducing the consequences of these axioms. The ultimate test of whether the axioms or postulates are sensible is to compare the end results to experimental data. As one gains experience and insight in an area, one can propose more abstract and economical postulates, but here we shall present a fairly elementary set that will suffice for all systems that we shall discuss in this book, and for almost all systems of interest in chemistry.

4.1 The State of a System Is Completely Specified by Its Wave Function

Classical mechanics deals with quantities called *dynamical variables*, such as position, momentum, angular momentum, and energy. A measurable dynamical variable is called an *observable*. The classical-mechanical state of a one-body system at any particular time is specified completely by the three position coordinates (x, y, z) and the three momenta or velocities (v_x, v_y, v_z) at that time. The time evolution of the system is governed by Newton's equations,

$$m\frac{d^2x}{dt^2} = F_x \qquad m\frac{d^2y}{dt^2} = F_y \qquad m\frac{d^2z}{dt^2} = F_z \tag{4.1}$$

where F_x, F_y, and F_z are the components of the force $\mathbf{F}(x, y, z)$. Realize that generally each force component depends upon x, y, and z. To emphasize this, we write

$$m\frac{d^2x}{dt^2} = F_x(x, y, z) \qquad m\frac{d^2y}{dt^2} = F_y(x, y, z) \qquad m\frac{d^2z}{dt^2} = F_z(x, y, z) \quad (4.2)$$

Note that each of these equations is a second-order equation, and so there will be two integration constants from each one. We can choose the integration constants to be the initial positions and initial velocities and write them as x_0, y_0, z_0, v_{x0}, v_{y0}, and v_{z0}. The solutions to Equations 4.2 are $x(t)$, $y(t)$, and $z(t)$, which describe the position of the particle as a function of time. The position of the particle depends not only on the time but also on the initial conditions. To emphasize this, we write the solutions to Equations 4.2 as

$$x(t) = x(t; x_0, y_0, z_0, v_{x0}, v_{y0}, v_{z0})$$

$$y(t) = y(t; x_0, y_0, z_0, v_{x0}, v_{y0}, v_{z0})$$

$$z(t) = z(t; x_0, y_0, z_0, v_{x0}, v_{y0}, v_{z0})$$

We can write these three equations in vector notation:

$$\mathbf{r}(t) = \mathbf{r}(t; \mathbf{r}_0, \mathbf{v}_0)$$

The vector $\mathbf{r}(t)$ describes the position of the particle as a function of time; $\mathbf{r}(t)$ is called the *trajectory* of the particle. Classical mechanics provides a method for calculating the trajectory of a particle in terms of the forces acting upon the particle through Newton's equations (Equations 4.2).

If there are N particles in the system, then it takes $3N$ coordinates and $3N$ velocities to specify the state of the system. There are $3N$ second-order differential equations and hence $3N$ initial positions and $3N$ initial velocities. The trajectory of the system is the position of each of the N particles in the system as a function of time and as a function of the initial conditions. An N-body system may be mathematically more complicated than a one-body system, but no new concepts need to be introduced.

Thus we see that in classical mechanics we can specify the state of a system by giving $3N$ positions and $3N$ velocities or momenta. We should suspect immediately that this is not going to be so in quantum mechanics because the uncertainty principle tells us that we cannot specify or determine the position and momentum of a particle simultaneously with infinite precision. The uncertainty principle is of no practical importance for macroscopic bodies, and so classical mechanics is a perfectly adequate prescription for macroscopic bodies; but for small bodies such as electrons, atoms, and molecules, the consequences of the uncertainty principle are far from negligible and so the classical-mechanical picture is not valid. This leads us to our first postulate of quantum mechanics.

Postulate 1

The state of a quantum-mechanical system is completely specified by a function $\Psi(\mathbf{r}, t)$ that depends on the coordinates of the particle and on time. This function, called the wave function or state function, has the important property that $\Psi^*(\mathbf{r}, t)\Psi(\mathbf{r}, t)dxdydz$ is the probability that the particle lies in the volume element $dxdydz$ located at $\mathbf{r}$ at time t.

If there is more than one particle, say two, then we write

$$\Psi^*(\mathbf{r}_1, \mathbf{r}_2, t)\Psi(\mathbf{r}_1, \mathbf{r}_2, t)\, dx_1 dy_1 dz_1 dx_2 dy_2 dz_2$$

for the probability that particle 1 lies in the volume element $dx_1 dy_1 dz_1$ at $\mathbf{r}_1$ and that particle 2 lies in the volume element $dx_2 dy_2 dz_2$ at $\mathbf{r}_2$ at time t. Postulate 1 says that the state of a quantum-mechanical system such as two electrons is completely specified by this function and that nothing else is required.

Because the square of the wave function has a probabilistic interpretation, it must satisfy certain physical requirements. For example, a wave function must be normalized, so that in the case of one particle, for simplicity, we have

$$\iiint_{\text{all space}} dxdydz\, \Psi^*(\mathbf{r}, t)\Psi(\mathbf{r}, t) = 1 \tag{4.3}$$

for all time. The notation "all space" here means that we integrate over all possible values of x, y, and z.

It is convenient to abbreviate Equation 4.3 by letting $dxdydz = d\tau$ and to write

$$\int_{-\infty}^{\infty} d\tau\, \Psi^*(\mathbf{r}, t)\Psi(\mathbf{r}, t) = 1 \tag{4.4}$$

with the understanding that this is really a triple integral over all possible values of x, y, and z. For the specific case of a three-dimensional particle in a box, the limits are $(0, a)$, $(0, b)$, and $(0, c)$.

EXAMPLE 4–1

The wave functions for a particle restricted to lie in a rectangular region of lengths a and b (a particle in a two-dimensional box) are

$$\psi_{n_x n_y}(x, y) = \left(\frac{4}{ab}\right)^{1/2} \sin \frac{n_x \pi x}{a} \sin \frac{n_y \pi y}{b} \qquad n_x, n_y = 1, 2, \ldots \begin{cases} 0 \le x \le a \\ 0 \le y \le b \end{cases}$$

Show that these wave functions are normalized.

SOLUTION: We wish to show that

$$\int_0^a \int_0^b dxdy\, \psi^*(x, y)\psi(x, y) = \left(\frac{4}{ab}\right) \int_0^a \int_0^b dxdy\, \sin^2 \frac{n_x \pi x}{a} \sin^2 \frac{n_y \pi y}{b} = 1$$

This double integral actually factors into a product of two single integrals:

$$\left(\frac{4}{ab}\right) \int_0^a dx \, \sin^2 \frac{n_x \pi x}{a} \int_0^b dy \, \sin^2 \frac{n_y \pi y}{b} \stackrel{?}{=} 1$$

Equation 3.26 shows that the first integral is equal to $a/2$ and that the second is equal to $b/2$, so that we have

$$\left(\frac{4}{ab}\right) \cdot \frac{a}{2} \cdot \frac{b}{2} = 1$$

and thus the above wave functions are normalized.

A less stringent condition than Equation 4.4 is that the wave function must be able to be normalized. This can be so only if the integral in Equation 4.4 is finite. If this is the case, we say that $\Psi(\mathbf{r}, t)$ is *square integrable* (cf. Problem 4–1). If the integral in Equation 4.4 equals a constant c rather than 1, then we can normalize $\Psi(\mathbf{r}, t)$ by dividing it by $c^{1/2}$. In addition to being normalized, or least normalizable, because $\Psi^*(\mathbf{r}, t)\Psi(\mathbf{r}, t) \, d\tau$ is a probability, we require that $\Psi(\mathbf{r}, t)$ and its first spatial derivative be single-valued, continuous, and finite (cf. Problem 4–4). We summarize these requirements by saying that $\Psi(\mathbf{r}, t)$ must be well behaved.

EXAMPLE 4–2
Determine whether each of the following functions is acceptable or not as a state function over the indicated intervals.

(a) e^{-x} $(0, \infty)$

(b) e^{-x} $(-\infty, \infty)$

(c) $\sin^{-1} x$ $(-1, 1)$

(d) $\dfrac{\sin x}{x}$ $(0, \infty)$

(e) $e^{-|x|}$ $(-\infty, \infty)$

SOLUTION:
(a) Acceptable; e^{-x} is single-valued, continuous, finite, and quadratically integrable over the interval $(0, \infty)$.
(b) Not acceptable; e^{-x} cannot be normalized over the interval $(-\infty, \infty)$ because e^{-x} diverges as $x \to -\infty$.
(c) Not acceptable; $\sin^{-1} x$ is a multivalued function. For example,

$$\sin^{-1} 1 = \frac{\pi}{2}, \frac{\pi}{2} + 2\pi, \frac{\pi}{2} + 4\pi, \ldots$$

(d) Acceptable; realize that $\sin x/x$ is finite at $x = 0$.
(e) Not acceptable; the first derivative of $e^{-|x|}$ is not continuous at $x = 0$.

4.2 Quantum-Mechanical Operators Represent Classical-Mechanical Variables

In Chapter 3, we concluded that classical-mechanical quantities are represented by linear operators in quantum mechanics. We now formalize this conclusion by our next postulate.

Postulate 2

To every observable in classical mechanics there corresponds a linear operator in quantum mechanics.

We have seen some examples of the correspondence between observables and operators in Chapter 3. These correspondences are listed in Table 4.1. The only new entry in Table 4.1 is that for the angular momentum, which requires some comment.

We discussed circular motion in some detail in Section 1.7. In that particular case, the angular momentum l is given by $l = mvr$. Just like linear momentum, however, angular momentum is actually a vector quantity. If a particle is at some point $\mathbf{r}$ and has momentum $\mathbf{p}$, then its angular momentum is defined by $\mathbf{l} = \mathbf{r} \times \mathbf{p}$. As we saw in MathChapter C, this vector cross product has a magnitude $|\mathbf{r}| |\mathbf{p}| \sin \theta$, where θ is the angle between $\mathbf{r}$ and $\mathbf{p}$ and its direction is perpendicular to the plane formed by $\mathbf{r}$ and $\mathbf{p}$. The components of $\mathbf{l} = \mathbf{r} \times \mathbf{p}$ are (Equation C.20)

$$l_x = y p_z - z p_y$$

$$l_y = z p_x - x p_z \qquad (4.5)$$

$$l_z = x p_y - y p_x$$

Note that the angular momentum operators given in Table 4.1 can be obtained from Equation 4.5 by letting the linear momenta p_x, p_y, and p_z assume their operator equivalents.

According to Postulate 2, all quantum-mechanical operators are linear. There is an important property of linear operators that we have not discussed yet. Consider an eigenvalue problem with a two-fold degeneracy; that is, consider the two equations

$$\hat{A}\psi_1 = a\psi_1 \qquad \text{and} \qquad \hat{A}\psi_2 = a\psi_2$$

Both ψ_1 and ψ_2 have the same eigenvalue a. If this is the case, then, any linear combination of ψ_1 and ψ_2, say $c_1\psi_1 + c_2\psi_2$, is also an eigenfunction of $\hat{A}$ with the eigenvalue a. The proof relies on the linear property of $\hat{A}$ (Section 3.2):

$$\hat{A}(c_1\psi_1 + c_2\psi_2) = c_1\hat{A}\psi_1 + c_2\hat{A}\psi_2$$

$$= c_1 a\psi_1 + c_2 a\psi_2 = a(c_1\psi_1 + c_2\psi_2)$$

TABLE 4.1
Classical-Mechanical Observables and Their Corresponding Quantum-Mechanical Operators

Observable		Operator	
Name	Symbol	Symbol	Operation
Position	x	$\hat{X}$	Multiply by x
	$\mathbf{r}$	$\hat{\mathbf{R}}$	Multiply by $\mathbf{r}$
Momentum	p_x	$\hat{P}_x$	$-i\hbar\dfrac{\partial}{\partial x}$
	$\mathbf{p}$	$\hat{\mathbf{P}}$	$-i\hbar(\mathbf{i}\dfrac{\partial}{\partial x}+\mathbf{j}\dfrac{\partial}{\partial y}+\mathbf{k}\dfrac{\partial}{\partial z})$
Kinetic energy	T_x	$\hat{T}_x$	$-\dfrac{\hbar^2}{2m}\dfrac{\partial^2}{\partial x^2}$
	T	$\hat{T}$	$-\dfrac{\hbar^2}{2m}(\dfrac{\partial^2}{\partial x^2}+\dfrac{\partial^2}{\partial y^2}+\dfrac{\partial^2}{\partial z^2})$ $=-\dfrac{\hbar^2}{2m}\nabla^2$
Potential energy	$V(x)$	$\hat{V}(\hat{x})$	Multiply by $V(x)$
	$V(x,y,z)$	$\hat{V}(\hat{x},\hat{y},\hat{z})$	Multiply by $V(x,y,z)$
Total energy	E	$\hat{H}$	$-\dfrac{\hbar^2}{2m}(\dfrac{\partial^2}{\partial x^2}+\dfrac{\partial^2}{\partial y^2}+\dfrac{\partial^2}{\partial z^2})$ $+V(x,y,z)$ $=-\dfrac{\hbar^2}{2m}\nabla^2+V(x,y,z)$
Angular momentum	$l_x=yp_z-zp_y$	$\hat{l}_x$	$-i\hbar(y\dfrac{\partial}{\partial z}-z\dfrac{\partial}{\partial y})$
	$l_y=zp_x-xp_z$	$\hat{l}_y$	$-i\hbar(z\dfrac{\partial}{\partial x}-x\dfrac{\partial}{\partial z})$
	$l_z=xp_y-yp_x$	$\hat{l}_z$	$-i\hbar(x\dfrac{\partial}{\partial y}-y\dfrac{\partial}{\partial x})$

EXAMPLE 4–3
Consider the eigenvalue problem

$$\frac{d^2\Phi(\phi)}{d\phi^2}=-m^2\Phi(\phi)$$

where m is a real (not imaginary or complex) number. The two eigenfunctions of $\hat{A}=d^2/d\phi^2$ are $\Phi_m(\phi)=e^{im\phi}$ and $\Phi_{-m}(\phi)=e^{-im\phi}$. We can easily show that each of these eigenfunctions has the eigenvalue $-m^2$. Show that any linear combination of $\Phi_m(\phi)$ and $\Phi_{-m}(\phi)$ is also an eigenfunction of $\hat{A} = d^2/d\phi^2$.

SOLUTION:

$$\frac{d^2}{d\phi^2}(c_1 e^{im\phi} + c_2 e^{-im\phi}) = c_1 \frac{d^2 e^{im\phi}}{d\phi^2} + c_2 \frac{d^2 e^{-im\phi}}{d\phi^2}$$

$$= -c_1 m^2 e^{im\phi} - c_2 m^2 e^{-im\phi}$$

$$= -m^2(c_1 e^{im\phi} + c_2 e^{-im\phi})$$

Example 4–3 helps show that this result is directly due to the linear property of quantum-mechanical operators. Although we have considered only a two-fold degeneracy, the general result is obvious. We will use this property of linear operators when we discuss the hydrogen atom in Chapter 7.

4.3 Observable Quantities Must Be Eigenvalues of Quantum-Mechanical Operators

We now present our third postulate:

Postulate 3
In any measurement of the observable associated with the operator $\hat{A}$, the only values that will ever be observed are the eigenvalues a, which satisfy the eigenvalue equation

$$\hat{A}\Psi_a = a\Psi_a \tag{4.6}$$

Generally, an operator will have a set of eigenfunctions and eigenvalues, and we indicate this by writing

$$\hat{A}\Psi_n = a_n\Psi_n \tag{4.7}$$

Thus, in any experiment designed to measure the observable corresponding to $\hat{A}$, the only values we find are $a_1, a_2, a_3, \ldots$. The set of eigenvalues $\{a_n\}$ of an operator $\hat{A}$ is called the *spectrum* of $\hat{A}$.

As a specific example, consider the measurement of the energy. The operator corresponding to the energy is the Hamiltonian operator, and its eigenvalue equation is

$$\hat{H}\Psi_n = E_n\Psi_n \tag{4.8}$$

This is just the Schrödinger equation. The solution of this equation gives the Ψ_n and E_n. For the case of the particle in a box, $E_n = n^2 h^2/8ma^2$. Postulate 3 says that if we measure the energy of a particle in a box, we shall find one of these energies and no others.

Notice that the Schrödinger equation is just one of many possible eigenvalue equations because there is one for each possible operator. Equation 4.8 is the most

FIGURE 4.1
A potential well with finite potential. The potential energy $V(x)$ that describes the system illustrated in this figure is given by $V(x) = 0$ for $0 \leq x \leq a$ and $V(x) = V_0$ for $x < 0$ and $x > a$.

important and most famous one, however, because the energy spectrum of a system is one of its most important properties. Because of the equation $\Delta E = h\nu$, we see that the experimentally observed spectrum of the system is intimately related to the mathematical spectrum. This is the motivation for calling the set of eigenvalues of an operator its *spectrum,* and this is also why the Schrödinger equation is a special eigenvalue equation.

The particle in a box is a bound system, and for bound systems the spectrum is discrete. For an unbound system, the spectrum is continuous. (See Problem 3–35.) For other systems, such as a particle in a box with finite walls (Figure 4.1), the spectrum can have both discrete and continuous parts. The energies below V_0 in Figure 4.1 are discrete and those with $E > V_0$ are continuous. As the value of V_0 increases, the system becomes a particle in a box because the large potential V_0 restricts the particle to lie within the well between 0 and a.

If a system is in a state described by Ψ_n, an eigenfunction of $\hat{A}$, then a measurement of the observable corresponding to $\hat{A}$ will yield the value a_n and only a_n. It is important to realize, however, that the $\{\Psi_n\}$ are very special functions, and it is possible that a system will not be in one of these states. Postulate 1 does not say that a system must be in a state described by an eigenfunction. Any suitably well-behaved function is a possible wave function or state function. What, then, if we measure the energy of a system that is in a state described by Ψ that is not an eigenfunction of $\hat{H}$? Indeed our observed value will be one of the values E_n, but which one it will be cannot be predicted with certainty. In fact, if we were to measure the energy of each member of a collection of similarly described systems, all in the state described by Ψ, then we would observe a distribution of energies, but each member of this distribution will be one of the energies E_n. This leads to our fourth postulate.

Postulate 4

If a system is in a state described by a normalized wave function Ψ, then the average value of the observable corresponding to $\hat{A}$ is given by

$$\langle a \rangle = \int_{-\infty}^{\infty} \Psi^* \hat{A}\, \Psi\, d\tau \tag{4.9}$$

According to Postulate 4, if we were to measure the energy of each member of a collection of similarly prepared systems, each described by Ψ, then the average of the observed values is given by Equation 4.9 with $\hat{A} = \hat{H}$.

EXAMPLE 4–4

Suppose that a particle in a box is in the state represented by the normalized state function.

$$\Psi(x) = \begin{cases} \left(\dfrac{30}{a^5}\right)^{1/2} x(a - x) & 0 \le x \le a \\ 0 & \text{otherwise} \end{cases}$$

Calculate the average energy of this system.

SOLUTION: The normalization integral of $\Psi(x)$ is elementary:

$$\frac{30}{a^5} \int_0^a x^2(a - x)^2 dx = 1$$

According to Equation 4.9, the average energy associated with this state is

$$\langle E \rangle = \int_0^a \Psi^*(x)\hat{H}\ \Psi(x)dx = \int_0^a \Psi^*(x)\left(-\frac{h^2}{2m}\frac{d^2}{dx^2}\right)\Psi(x)\,dx$$

$$= -\frac{15h^2}{ma^5}\int_0^a x(a - x)\left[\frac{d^2}{dx^2}x(a - x)\right]dx$$

$$= \frac{30h^2}{ma^5}\int_0^a x(a - x)dx = \frac{5\hbar^2}{ma^2} = \frac{5h^2}{4\pi^2ma^2}$$

Note that this value is not one of the energy eigenvalues of a particle in a box (Equation 3.21). Which one is it closest to? Which eigenfunction (Equation 3.27) is $\Psi(x)$ most similar to? Does this result seem sensible?

Suppose that Ψ just happens to be an eigenfunction of $\hat{A}$; that is, suppose that $\Psi = \psi_n$ where

$$\hat{A}\psi_n = a_n\psi_n$$

Then

$$\langle a \rangle = \int_{-\infty}^{\infty} \psi_n^*\hat{A}\psi_n\,d\tau = \int_{-\infty}^{\infty} \psi_n^*a_n\psi_n\,d\tau = a_n\int_{-\infty}^{\infty} \psi_n^*\psi_n\,d\tau = a_n \quad (4.10)$$

Furthermore, if $\hat{A}\psi_n = a_n\psi_n$, then

$$\hat{A}^2\psi_n = \hat{A}(\hat{A}\psi_n) = \hat{A}(a_n\psi_n) = a_n^2\psi_n$$

and so

$$\langle a^2 \rangle = \int_{-\infty}^{\infty} \psi_n^* \hat{A}^2 \psi_n \, d\tau = a_n^2 \tag{4.11}$$

From Equations 4.10 and 4.11, we see that the variance of the measurement (cf. MathChapter B) gives

$$\sigma_n^2 = \langle a^2 \rangle - \langle a \rangle^2 = a_n^2 - a_n^2 = 0 \tag{4.12}$$

Thus, Postulate 3 says that the only value that we measure is the value a_n. Often, however, the system is not in a state described by an eigenfunction, and one measures a distribution of values whose average is given by Postulate 4.

EXAMPLE 4–5
Show by direct calculation that $\sigma_E^2 = \langle E^2 \rangle - \langle E \rangle^2 = 0$ for a particle in a box, for which

$$\psi_n(x) = \left(\frac{2}{a}\right)^{1/2} \sin\frac{n\pi x}{a} \qquad 0 \le x \le a$$

In other words, show that the only values of the energy that can be observed are the energy eigenvalues, $E_n = n^2h^2/8ma^2$ (Equation 3.21).

SOLUTION: The operator that corresponds to the observable E is the Hamiltonian operator, which for a particle in a box is Equation 3.14 with $V(x) = 0$:

$$\hat{H} = -\frac{\hbar^2}{2m}\frac{d^2}{dx^2}$$

The average energy is given by

$$\langle E \rangle = \int_0^a \psi_n^*(x) \hat{H} \psi_n(x) dx$$

$$= \frac{2}{a} \int_0^a \sin\frac{n\pi x}{a} \left(-\frac{\hbar^2}{2m}\frac{d^2}{dx^2}\right) \sin\frac{n\pi x}{a} dx$$

$$= \frac{\hbar^2}{2m} \cdot \frac{2}{a} \cdot \left(\frac{n\pi}{a}\right)^2 \int_0^a \sin^2\frac{n\pi x}{a} dx = \frac{n^2h^2}{8ma^2} = E_n$$

Similarly,

$$\langle E^2 \rangle = \int_0^a \psi_n^*(x)\hat{H}^2\psi_n(x)dx = \int_0^a \psi_n^*(x)\hat{H}[\hat{H}\psi_n(x)]dx$$

$$= \frac{2}{a}\int_0^a \sin\frac{n\pi x}{a}\left(-\frac{\hbar^2}{2m}\frac{d^2}{dx^2}\right)\left(-\frac{\hbar^2}{2m}\frac{d^2}{dx^2}\right)\sin\frac{n\pi x}{a}dx$$

$$= \frac{\hbar^4}{4m^2}\cdot\frac{2}{a}\int_0^a \sin\frac{n\pi x}{a}\left(\frac{d^4}{dx^4}\right)\sin\frac{n\pi x}{a}dx$$

$$= \frac{\hbar^4}{4m^4}\cdot\frac{2}{a}\cdot\left(\frac{n\pi}{a}\right)^4\int_0^a \sin^2\frac{n\pi x}{a}dx$$

$$= \frac{n^4 h^4}{64m^2 a^4} = \left(\frac{n^2 h^2}{8ma^2}\right)^2 = E_n^2 = \langle E \rangle^2$$

Therefore, $\sigma_E^2 = \langle E^2 \rangle - \langle E \rangle^2 = 0$, and so we find that the energies of a particle in a box in a state described by $\psi_n(x)$ above can be observed to have only the values E_1, E_2, ...

4.4 The Commutator of Two Operators Plays a Central Role in the Uncertainty Principle

When two operators act sequentially on a function, as in $\hat{A}\hat{B}f(x)$, we apply each operator in turn, working from right to left

$$\hat{A}\hat{B}f(x) = \hat{A}[\hat{B}f(x)] = \hat{A}h(x)$$

where $h(x) = \hat{B}f(x)$. An important difference between operators and ordinary algebraic quantities is that operators do not necessarily *commute*. If

$$\hat{A}\hat{B}f(x) = \hat{B}\hat{A}f(x) \qquad \text{(commutative)} \qquad (4.13)$$

for arbitrary $f(x)$, then we say that $\hat{A}$ and $\hat{B}$ *commute*. If

$$\hat{A}\hat{B}f(x) \neq \hat{B}\hat{A}f(x) \qquad \text{(noncommutative)} \qquad (4.14)$$

for arbitrary $f(x)$, then we say that $\hat{A}$ and $\hat{B}$ do not commute. For example, if $\hat{A} = d/dx$ and $\hat{B} = x$ (multiply by x), then

$$\hat{A}\hat{B}f(x) = \frac{d}{dx}[xf(x)] = f(x) + x\frac{df}{dx}$$

and

$$\hat{B}\hat{A}f(x) = x\frac{d}{dx}f(x) = x\frac{df}{dx}$$

Therefore, $\hat{A}\hat{B}f(x) \neq \hat{B}\hat{A}f(x)$, and $\hat{A}$ and $\hat{B}$ do not commute. In this particular case, we have

$$\hat{A}\hat{B}f(x) - \hat{B}\hat{A}f(x) = f(x)$$

or

$$(\hat{A}\hat{B} - \hat{B}\hat{A})f(x) = \hat{I}f(x) \tag{4.15}$$

where we have introduced the identity operator $\hat{I}$, which simply multiplies $f(x)$ by unity. Because $f(x)$ is arbitrary, we can write Equation 4.15 as an operator equation by suppressing $f(x)$ on both sides of the equation to give

$$\hat{A}\hat{B} - \hat{B}\hat{A} = \hat{I} \tag{4.16}$$

Realize that an operator equality like this is valid only if it is true for all $f(x)$. The combination of $\hat{A}$ and $\hat{B}$ appearing in Equation 4.16 occurs often and is called the *commutator*, $[\hat{A}, \hat{B}]$, of $\hat{A}$ and $\hat{B}$:

$$[\hat{A}, \hat{B}] = \hat{A}\hat{B} - \hat{B}\hat{A} \tag{4.17}$$

If $[\hat{A}, \hat{B}]f(x) = 0$ for all $f(x)$ on which the commutator acts, then we write that $[\hat{A}, \hat{B}] = 0$ and we say that $\hat{A}$ and $\hat{B}$ commute.

EXAMPLE 4–6
Let $\hat{A} = d/dx$ and $\hat{B} = x^2$. Evaluate the commutator $[\hat{A}, \hat{B}]$.

SOLUTION: We let $\hat{A}$ and $\hat{B}$ act upon an arbitrary function $f(x)$:

$$\hat{A}\hat{B}f(x) = \frac{d}{dx}[x^2 f(x)] = 2xf(x) + x^2\frac{df}{dx}$$

$$\hat{B}\hat{A}f(x) = x^2\frac{d}{dx}f(x) = x^2\frac{df}{dx}$$

By substracting these two results, we obtain

$$\hat{A}\hat{B} - \hat{B}\hat{A} = 2xf(x)$$

Because, and only because, $f(x)$ is arbitrary, we write

$$[\hat{A}, \hat{B}] = 2x\hat{I}$$

When evaluating a commutator, it is essential to include a function $f(x)$ as we have done; otherwise, we can obtain a spurious result. To this end, note well that

$$[\hat{A}, \hat{B}] = \hat{A}\hat{B} - \hat{B}\hat{A} = \frac{d}{dx}x^2 - x^2\frac{d}{dx}$$

$$\neq 2x - x^2\frac{d}{dx}$$

a result that is obtained by forgetting to include the function $f(x)$. Such errors will not occur if an arbitrary function $f(x)$ is included from the outset.

EXAMPLE 4–7

For a one-dimensional system, the momentum operator is (Table 4.1)

$$\hat{P}_x = -i\hbar\frac{d}{dx}$$

and the position operator is

$$\hat{X} = x \qquad \text{(multiply by } x\text{)}$$

Evaluate $[\hat{P}_x, \hat{X}]$.

SOLUTION: We let $[\hat{P}_x, \hat{X}]$ act upon an arbitrary function $f(x)$:

$$[\hat{P}_x, \hat{X}]f(x) = \hat{P}_x\hat{X}f(x) - \hat{X}\hat{P}_xf(x)$$

$$= -i\hbar\frac{d}{dx}[xf(x)] + xi\hbar\frac{d}{dx}f(x)$$

$$= -i\hbar f(x) - i\hbar x\frac{df}{dx} + i\hbar x\frac{df}{dx}$$

$$= -i\hbar f(x)$$

Because $f(x)$ is arbitrary, we write this result as

$$[\hat{P}_x, \hat{X}] = -i\hbar\hat{I} \tag{4.18}$$

where $\hat{I}$ is the unit operator (the multiply-by-one operator).

If two operators do not commute, then their corresponding observable quantities do not have simultaneously well-defined values, and in fact,

$$\sigma_A^2\sigma_B^2 \geq -\frac{1}{4}\left(\int \psi^*[\hat{A}, \hat{B}]\psi \, dx\right)^2 \tag{4.19}$$

where σ_A^2 and σ_B^2 are

$$\sigma_A^2 = \int \psi^*(\hat{A} - \langle a \rangle)^2 \psi \, dx \tag{4.20}$$

$$\sigma_B^2 = \int \psi^*(\hat{B} - \langle b \rangle)^2 \psi \, dx \tag{4.21}$$

and ψ is a suitably behaved state function. The proof of Equation 4.19 is left to the problems. (See Problem 4–47.)

Let's see how to use Equation 4.19 in the case where $\hat{A}$ and $\hat{B}$ are the momentum and position operators. In this case (see Equation 4.18),

$$[\hat{A}, \hat{B}] = [\hat{P}_x, \hat{X}] = -i\hbar\hat{I} \tag{4.22}$$

If we substitute Equation 4.22 into Equation 4.19, then we find that

$$\sigma_p^2 \sigma_x^2 \geq -\frac{1}{4} \left[\int \psi^*(-i\hbar)\psi \, dx \right]^2$$

or

$$\sigma_p^2 \sigma_x^2 \geq -\frac{1}{4}(-i\hbar)^2 = \frac{\hbar^2}{4}$$

By taking the square root of both sides, we have

$$\sigma_p \sigma_x \geq \frac{\hbar}{2} \tag{4.23}$$

which is the Heisenberg uncertainty principle for momentum and position.

4.5 Quantum-Mechanical Operators Must Be Hermitian Operators

Table 4.1 contains a list of some commonly occurring quantum-mechanical operators. We stated previously that these operators must have certain properties. We noticed that they all are linear, and this is one requirement that we impose. A more subtle requirement arises if we consider Postulate 3, which says that, in any measurement of the observable associated with the operator $\hat{A}$, the only values that are ever observed are the eigenvalues of $\hat{A}$. We have seen, however, that wave functions, and operators generally, are complex quantities, but certainly the eigenvalues must be real quantities if they are to correspond to the result of experimental measurement. In an equation, we have

$$\hat{A}\psi = a\psi \tag{4.24}$$

where $\hat{A}$ and ψ may be complex but a must be real. We shall insist, then, that quantum-mechanical operators have only real eigenvalues. Clearly, this places a certain restriction on the operator $\hat{A}$.

To see what this restriction is, we multiply Equation 4.24 from the left by ψ^* and integrate to obtain

$$\int \psi^* \hat{A} \, \psi \, dx = a \int \psi^* \psi \, dx = a \tag{4.25}$$

Now take the complex conjugate of Equation 4.24,

$$\hat{A}^* \psi^* = a^* \psi^* = a \psi^* \tag{4.26}$$

where the equality $a^* = a$ recognizes that a is real. Multiply Equation 4.26 from the left by ψ and integrate

$$\int \psi \hat{A}^* \psi^* dx = a \int \psi \psi^* dx = a \tag{4.27}$$

Equating the left sides of Equations 4.25 and 4.27 gives

$$\int \psi^* \hat{A} \psi \, dx = \int \psi \hat{A}^* \psi^* dx \tag{4.28}$$

The operator $\hat{A}$ must satisfy Equation 4.28 to assure that its eigenvalues are real. An operator that satisfies Equation 4.28 for *any well-behaved function* is called a *Hermitian operator*. Thus, we can write the definition of a Hermitian operator as an operator that satisfies the relation

$$\int_{-\infty}^{\infty} f^* \hat{A} f \, dx = \int_{-\infty}^{\infty} f \hat{A}^* f^* \, dx \tag{4.29}$$

where $f(x)$ is any well-behaved function. Hermitian operators have real eigenvalues. Postulate 2 should be modified to read as follows:

Postulate 2'
To every observable in classical mechanics there corresponds a linear, Hermitian operator in quantum mechanics.

All the operators in Table 4.1 are Hermitian. How do you determine if an operator is Hermitian? Consider the operator $\hat{A} = d/dx$. Does $\hat{A}$ satisfy Equation 4.29? Let's substitute $\hat{A} = d/dx$ into Equation 4.29 and integrate by parts:

$$\int_{-\infty}^{\infty} f^* \frac{d}{dx} f \, dx = \int_{-\infty}^{\infty} f^* \frac{df}{dx} \, dx = \left. f^* f \right|_{-\infty}^{\infty} - \int_{-\infty}^{\infty} f \frac{df^*}{dx} \, dx$$

For a wave function to be normalizable, it must vanish at infinity, and so the first term on the right side here is zero. Therefore, we have

$$\int_{-\infty}^{\infty} f^* \frac{d}{dx} f \, dx = - \int_{-\infty}^{\infty} f \frac{d}{dx} f^* \, dx$$

For an arbitrary function $f(x)$, d/dx does *not* satisfy Equation 4.29 and so is *not* Hermitian.

Let's consider the momentum operator $\hat{P} = -i\hbar d/dx$. Substitution into Equation 4.29 and integration by parts gives

$$\int_{-\infty}^{\infty} f^* \left(-i\hbar \frac{d}{dx} \right) f \, dx = -i\hbar \int_{-\infty}^{\infty} f^* \frac{df}{dx} \, dx = i\hbar \int_{-\infty}^{\infty} f \frac{df^*}{dx} \, dx$$

and

$$\int_{-\infty}^{\infty} f \hat{P}^* f^* \, dx = \int_{-\infty}^{\infty} f \left(-i\hbar \frac{d}{dx} \right)^* f^* \, dx = i\hbar \int_{-\infty}^{\infty} f \frac{df^*}{dx} \, dx$$

Thus, we see that $\hat{P}$ does, indeed, satisfy Equation 4.29. Therefore, the momentum operator is a Hermitian operator.

EXAMPLE 4–8
Prove that the kinetic energy operator is Hermitian.

$$\hat{T} = -\frac{\hbar^2}{2m} \frac{d^2}{dx^2}$$

SOLUTION: As always, we shall assume that f vanishes at infinity. Thus, following the procedure of integrating by parts twice,

$$-\frac{\hbar^2}{2m} \int_{-\infty}^{\infty} f^* \frac{d^2 f}{dx^2} dx = -\frac{\hbar^2}{2m} \left. f^* \frac{df}{dx} \right|_{-\infty}^{\infty} + \frac{\hbar^2}{2m} \int_{-\infty}^{\infty} \frac{df^*}{dx} \frac{df}{dx} dx$$

$$= \frac{\hbar^2}{2m} \left. \frac{df^*}{dx} f \right|_{-\infty}^{\infty} - \frac{\hbar^2}{2m} \frac{d^2 f^*}{dx^2} f \, dx$$

and so we see that

$$\int_{-\infty}^{\infty} f^* \left(-\frac{\hbar^2}{2m} \frac{d^2}{dx^2} \right) f \, dx = \int_{-\infty}^{\infty} f \left(-\frac{\hbar^2}{2m} \frac{d^2}{dx^2} \right) f^* \, dx$$

$$= \int_{-\infty}^{\infty} f \left(-\frac{\hbar^2}{2m} \frac{d^2}{dx^2} \right)^* f^* \, dx$$

Thus, Equation 4.29 is satisfied, and the kinetic energy operator is Hermitian.

The definition of a Hermitian operator that is given by Equation 4.29 is not the most general definition. A more general definition of a Hermititan operator is given by

$$\int_{-\infty}^{\infty} dx \, f_m^*(x) \hat{A} f_n(x) = \int_{-\infty}^{\infty} dx \, f_n(x) \hat{A}^* f_m^*(x) \tag{4.30}$$

where $f_n(x)$ and $f_m(x)$ are any two well-behaved functions. We shall use this definition quite often. It so happens that it is possible to prove that Equation 4.30 follows from Equation 4.29, and so the definition given by Equation 4.29 suffices if you know this. Problem 4–20 leads you through the proof.

This is a good time to introduce a notation that is extremely widely used. Let $\{f_n(x)\}$ be some set of well-behaved functions that we label by an integer n. These functions might, but not necessarily, be the eigenfunctions of some operator $\hat{O}$. In this new notational scheme, we shall express $f_n(x)$ as $| \, n \, \rangle$. Instead of denoting the complex conjugate of $f_n(x)$ by $| \, n \, \rangle^*$, we denote it by $\langle \, n \, |$. The integral of $f_n^*(x) f_n(x)$ in this notation is expressed as $\langle \, n \, | \, n \, \rangle$, or

$$\int_{-\infty}^{\infty} dx \, f_n^*(x) f_n(x) = \langle \, n \, | \, n \, \rangle$$

Normalization corresponds to writing $\langle \, n \, | \, n \, \rangle = 1$. More generally,

$$\int_{-\infty}^{\infty} dx \, f_m^*(x) f_n(x) = \langle \, m \, | \, n \, \rangle \tag{4.31}$$

You can see that $\langle \, m \, | \, n \, \rangle^* = \langle \, n \, | \, m \, \rangle$ by taking the complex conjugate of both sides of Equation 4.31. Continuing, for an operator $\hat{A}$ we write

$$\int_{-\infty}^{\infty} dx \, f_m^*(x) \hat{A} f_n(x) = \langle \, m \, | \, \hat{A} \, | \, n \, \rangle \tag{4.32}$$

Note that we enclose $\hat{A}$ between vertical lines instead of writing $\langle \, m \, | \, \hat{A} n \, \rangle$, which is not incorrect but simply not often done. In this notation, the two sides of Equation 4.30 become

$$\int_{-\infty}^{\infty} dx \, f_m^*(x) \hat{A} f_n(x) = \langle \, m \, | \, \hat{A} \, | \, n \, \rangle$$

and

$$\int_{-\infty}^{\infty} dx \, f_n(x) \hat{A}^* f_m^*(x) = \left[\int_{-\infty}^{\infty} dx \, f_n^*(x) \hat{A} f_m(x) \right]^* = \langle \, n \, | \, \hat{A} \, | \, m \, \rangle^*$$

and so $\hat{A}$ is a Hermitian operator if

$$\langle \, m \, | \, \hat{A} \, | \, n \, \rangle = \langle \, n \, | \, \hat{A} \, | \, m \, \rangle^* \tag{4.33}$$

Equation 4.33 gives the definition of a Hermitian operator (Equation 4.30) in this new notation.

This notation is due to the British physicist Paul Dirac. The quantities $| n \rangle$ are called *kets*, and the $\langle m |$ are called *bras*. This nomenclature arises from the fact that integrals such as those in Equation 4.32 are denoted by <u>bra</u> c <u>kets</u>, $\langle m | \hat{A} | n \rangle$. The notational scheme is called *Dirac notation* or *bracket notation*. This notation may be new to you, but it is very economical and we shall use it gradually more often as we go along. Problems 4–27 and 4–28 give you practice with this bracket notation.

4.6 The Eigenfunctions of Hermitian Operators Are Orthogonal

We have been led naturally to the definition and use of Hermitian operators by requiring that quantum-mechanical operators have real eigenvalues. Not only are the eigenvalues of Hermitian operators real, but their eigenfunctions satisfy a rather special condition as well. Consider the two eigenvalue equations

$$\hat{A}\psi_n = a_n \psi_n \qquad \hat{A}\psi_m = a_m \psi_m \tag{4.34}$$

We multiply the first of Equations 4.34 by ψ_m^* and integrate; then we take the complex conjugate of the second, multiply by ψ_n, and integrate to obtain

$$\int_{-\infty}^{\infty} \psi_m^* \hat{A}\psi_n \, dx = a_n \int_{-\infty}^{\infty} \psi_m^* \psi_n \, dx$$

$$\int_{-\infty}^{\infty} \psi_n \hat{A}^* \psi_m^* \, dx = \left[\int_{-\infty}^{\infty} \psi_n^* \hat{A}\psi_m \, dx \right]^* = a_m^* \int_{-\infty}^{\infty} \psi_n \psi_m^* \, dx \tag{4.35}$$

$$= a_m^* \int_{-\infty}^{\infty} \psi_m^* \psi_n \, dx$$

or

$$\langle m | \hat{A} | n \rangle = a_n \langle m | n \rangle$$

$$\langle n | \hat{A} | m \rangle^* = a_m^* \langle m | n \rangle \tag{4.36}$$

in the bracket notation. By subtracting Equations 4.35 or 4.36, we obtain

$$\int_{-\infty}^{\infty} \psi_m^* \hat{A}\psi_n \, dx - \int_{-\infty}^{\infty} \psi_n \hat{A}^* \psi_m^* \, dx = (a_n - a_m^*) \int_{-\infty}^{\infty} \psi_m^* \psi_n \, dx \tag{4.37}$$

or

$$\langle m | \hat{A} | n \rangle - \langle n | \hat{A} | m \rangle^* = (a_n - a_m^*) \langle m | n \rangle \tag{4.38}$$

Because $\hat{A}$ is Hermitian, either Equation 4.30 or 4.33 shows that the left side here is zero, and so we have

$$(a_n - a_m^*) \int_{-\infty}^{\infty} \psi_m^* \psi_n \, dx = (a_n - a_m^*)\langle m \mid n \rangle = 0 \tag{4.39}$$

There are two possibilities to consider in Equation 4.39, $n = m$ and $n \neq m$. When $n = m$, the integral is unity by normalization and so we have

$$a_n = a_n^* \tag{4.40}$$

which is just another proof that the eigenvalues are real.

When $n \neq m$, we have

$$(a_n - a_m) \int_{-\infty}^{\infty} \psi_m^* \psi_n \, dx = (a_n - a_m)\langle m \mid n \rangle = 0 \qquad m \neq n \tag{4.41}$$

Now if the system is nondegenerate, $a_n \neq a_m$, and

$$\int_{-\infty}^{\infty} \psi_m^* \psi_n \, dx = \langle m \mid n \rangle = 0 \qquad n \neq m \tag{4.42}$$

A set of eigenfunctions that satisfies the condition in Equation 4.42 is said to be *orthogonal*. We have just proved that the eigenfunctions of a Hermitian operator are orthogonal, at least for a nondegenerate system. The particle in a box is a nondegenerate system. The wave functions for this system are (Equation 3.27)

$$\psi_n(x) = \left(\frac{2}{a}\right)^{1/2} \sin \frac{n\pi x}{a} \qquad n = 1, 2, \ldots \tag{4.43}$$

It is easy to prove that these functions are orthogonal if one uses the trigonometric identity

$$\sin \alpha \sin \beta = \frac{1}{2}\cos(\alpha - \beta) - \frac{1}{2}\cos(\alpha + \beta) \tag{4.44}$$

Then

$$\frac{2}{a} \int_0^a \sin \frac{n\pi x}{a} \sin \frac{m\pi x}{a} \, dx = \frac{1}{a} \int_0^a \cos \frac{(n-m)\pi x}{a} \, dx - \frac{1}{a} \int_0^a \cos \frac{(n+m)\pi x}{a} \, dx \tag{4.45}$$

Because n and m are integers, both integrands on the right side of Equation 4.45 are of the form $\cos(N\pi x/a)$ where N is an integer. Consequently, both integrals start at

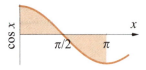

FIGURE 4.2
An illustration of the fact that the integrals of $\cos N\pi x/a$ vanish if the limits of integration start at zero and extend over a half or complete cycle of $\cos x$.

zero and go over half or complete cycles of the cosine and equal zero if $m \neq n$. (See Figure 4.2.) Thus,

$$\frac{2}{a} \int_0^a \sin\frac{n\pi x}{u} \sin\frac{m\pi x}{a} \, dx = 0 \qquad m \neq n \qquad (4.46)$$

The wave functions of a particle in a box are orthogonal.

When $n = m$ in Equation 4.45, the integrand of the first integral on the right side is equal to unity because $\cos 0 = 1$. The second integral on the right side vanishes and so we have

$$\frac{2}{a} \int_0^a \sin^2\frac{n\pi x}{a} \, dx = 1 \qquad (4.47)$$

or the particle-in-a-box wave functions are normalized. A set of functions that are both normalized and orthogonal to each other is called an *orthonormal* set. We can express the condition of orthonormality by writing

$$\int_{-\infty}^{\infty} \psi_m^* \psi_n \, dx = \langle\, m \mid n \,\rangle = \delta_{nm} \qquad (4.48)$$

where

$$\delta_{nm} = \begin{cases} 1 & m = n \\ 0 & m \neq n \end{cases} \qquad (4.49)$$

The symbol δ_{nm} occurs frequently and is called the *Kronecker delta*. (See Problem 4–26.)

EXAMPLE 4–9
According to Problem 3–33, the eigenfunctions of a particle constrained to move on a circular ring of radius a are

$$\psi_m(\theta) = (2\pi)^{-1/2} e^{im\theta} \qquad m = 0, \pm1, \pm2, \dots$$

where θ describes the angular position of the particle about the ring. Clearly, $0 \leq \theta \leq 2\pi$. Prove that these eigenfunctions form an orthonormal set.

SOLUTION: To prove that a set of functions forms an orthonormal set, we must show that they satisfy Equation 4.48. To see if they do, we have

$$\int_0^{2\pi} \psi_m^*(\theta)\psi_n(\theta)\, d\theta = \frac{1}{2\pi}\int_0^{2\pi} e^{-im\theta} e^{in\theta}\, d\theta$$

$$= \frac{1}{2\pi}\int_0^{2\pi} e^{-i(n-m)\theta}\, d\theta$$

$$= \frac{1}{2\pi}\int_0^{2\pi} \cos(n-m)\theta\, d\theta + \frac{i}{2\pi}\int_0^{2\pi} \sin(n-m)\theta\, d\theta$$

For $n \neq m$, the final two integrals vanish because they are over complete cycles of the cosine and sine. For $n = m$, the last integral vanishes because $\sin 0 = 0$, and the next to last gives 2π because $\cos 0 = 1$. Thus,

$$\int_0^{2\pi} \psi_m^*(\theta)\psi_n(\theta)\, d\theta = \delta_{mn}$$

and the $\psi_m(\theta)$ form an orthonormal set.

When we proved that the eigenfunctions of a Hermitian operator are orthogonal, we assumed that the system was nondegenerate. For simplicity, let's consider the case in which two states, described by ψ_1 and ψ_2, have the same eigenvalue a_1. By referring to Equation 4.41, we see that it does not follow that ψ_1 and ψ_2 are orthogonal because $a_1 = a_2$ in this case. The two eigenvalue equations are

$$\hat{A}\psi_1 = a_1\psi_1 \qquad \hat{A}\psi_2 = a_1\psi_2$$

Now let's consider a linear combination of ψ_1 and ψ_2, say $\phi = c_1\psi_1 + c_2\psi_2$. Then

$$\hat{A}\phi = \hat{A}(c_1\psi_1 + c_2\psi_2) = c_1\hat{A}\psi_1 + c_2\hat{A}\psi_2$$

$$= a_1c_1\psi_1 + a_1c_2\psi_2 = a_1(c_1\psi_1 + c_2\psi_2)$$

$$= a_1\phi$$

Thus, we see that if ψ_1 and ψ_2 describe a two-fold degenerate state with eigenvalue a_1, then any linear combination of ψ_1 and ψ_2 is also an eigenfunction with eigenvalue a_1. It is convenient to choose two linear combinations of ψ_1 and ψ_2, call them ϕ_1 and ϕ_2, such that

$$\int_{-\infty}^{\infty} \phi_1^*\phi_2\, dx = 0$$

To see how to do this in practice, we choose

$$\phi_1 = \psi_1 \quad \text{and} \quad \phi_2 = \psi_2 + c\psi_1$$

where ψ_1 and ψ_2 are normalized and c is a constant to be determined. Because ψ_1 is normalized, ϕ_1 is normalized. We choose c such that ϕ_1 and ϕ_2 are orthogonal:

$$\int_{-\infty}^{\infty} \phi_1^* \phi_2 \, dx = \langle \phi_1 | \phi_2 \rangle = \int_{-\infty}^{\infty} \psi_1^* (\psi_2 + c\psi_1) \, dx$$

$$= \int_{-\infty}^{\infty} \psi_1^* \psi_2 \, dx + c \int_{-\infty}^{\infty} \psi_1^* \psi_1 \, dx = \langle \psi_1 | \psi_2 \rangle + c = 0$$

If c is chosen to be

$$c = - \int_{-\infty}^{\infty} \psi_1^* \psi_2 \, dx = -\langle \psi_1 | \psi_2 \rangle$$

then ϕ_1 and ϕ_2 will be orthogonal. We can then normalize ϕ_2 by requiring that $\langle \phi_2 | \phi_2 \rangle = 1$. (This procedure can be generalized to the case of n functions and is called the *Gram-Schmidt orthonormalization procedure*.) So even if there is a degeneracy, we can construct the eigenfunctions of a Hermitian operator such that they are orthonormal and say that they form an orthonormal set (Equation 4.48).

4.7 If Two Operators Commute, They Have a Mutual Set of Eigenfunctions

Suppose that two operators $\hat{A}$ and $\hat{B}$ have the same set of eigenfunctions, so that we have

$$\hat{A}\phi_n = a_n \phi_n \quad \text{and} \quad \hat{B}\phi_n = b_n \phi_n \tag{4.50}$$

Equations 4.50 imply that the quantities corresponding to $\hat{A}$ and $\hat{B}$ have simultaneously sharply defined values. According to Equations 4.50, the values that we observe are a_n and b_n. We shall prove that if two operators have the same set of eigenfunctions, then they necessarily commute. To prove this, we must show that

$$[\hat{A}, \hat{B}]f(x) = 0 \tag{4.51}$$

for an arbitrary function $f(x)$. We can expand $f(x)$ in terms of the complete set of eigenfunctions of $\hat{A}$ and $\hat{B}$ (see Equation 4.56), and write

$$f(x) = \sum_n c_n \phi_n(x)$$

If we substitute this expansion into Equation 4.51, then we obtain

$$[\hat{A}, \hat{B}]f(x) = \sum_n c_n[\hat{A}, \hat{B}]\phi_n(x)$$

$$= \sum_n c_n(a_nb_n - b_na_n)\phi_n(x) = 0 \tag{4.52}$$

Because $f(x)$ is arbitrary, $[\hat{A}, \hat{B}] = 0$. Thus, we see that $\hat{A}$ and $\hat{B}$ commute if they have the same set of eigenfunctions.

The converse is also true; if $\hat{A}$ and $\hat{B}$ commute, then they have a mutual set of eigenfunctions. Let the eigenvalue equations of $\hat{A}$ and $\hat{B}$ be

$$\hat{A}\phi_a = a\phi_a \quad \text{and} \quad \hat{B}\phi_b = b\phi_b \tag{4.53}$$

Because $\hat{A}$ and $\hat{B}$ commute, we have

$$[\hat{A}, \hat{B}]\phi_a = 0 = \hat{A}\hat{B}\phi_a - \hat{B}\hat{A}\phi_a$$

$$= \hat{A}(\hat{B}\phi_a) - a(\hat{B}\phi_a) = 0$$

and so

$$\hat{A}(\hat{B}\phi_a) = a(\hat{B}\phi_a) \tag{4.54}$$

Equation 4.54 implies that $\hat{B}\phi_a$ is an eigenfunction of $\hat{A}$. If the system is nondenegerate, there is only one eigenfunction ϕ_a for each eigenvalue a, and so Equation 4.54 says that

$$\hat{B}\phi_a = (\text{constant})\phi_a \tag{4.55}$$

But Equation 4.54 says that ϕ_a is an eigenfunction of $\hat{B}$ as well as an eigenfunction of $\hat{A}$. Because the system is nondegenerate, there is only one eigenfunction of $\hat{B}$ for each eigenvalue b, and so Equation 4.54 implies that ϕ_a and ϕ_b are the same (within an unimportant multiplicative constant). Thus, we see that if $\hat{A}$ and $\hat{B}$ commute, then they have the same set of eigenfunctions; and because they have mutual eigenfunctions, the observables corresponding to $\hat{A}$ and $\hat{B}$ have simultaneously sharply defined values. We have proved this only in the case in which there is no degeneracy, but it is true even for a degenerate system.

4.8 The Probability of Obtaining a Certain Value of an Observable in a Measurement Is Given by a Fourier Coefficient

Consider a fairly arbitrary function $f(x)$. We assume that if $\{\psi_n(x)\}$ is some orthonormal set defined over the same interval as $f(x)$ and satisfying the same boundary conditions as $f(x)$, then it is possible to write $f(x)$ as

$$f(x) = \sum_{n=1}^{\infty} c_n \psi_n(x) \tag{4.56}$$

A set of functions such as the $\psi_n(x)$ here is said to be *complete* if Equation 4.56 holds for a suitably arbitrary function $f(x)$. It is generally difficult to prove completeness, and in practice one usually assumes that the orthonormal set associated with some Hermitian operator is complete. If we multiply both sides of Equation 4.56 by $\psi_m^*(x)$ and integrate, then we find

$$\int_{-\infty}^{\infty} \psi_m^*(x) f(x)\, dx = \sum_{n=1}^{\infty} c_n \int_{-\infty}^{\infty} \psi_m^*(x) \psi_n(x)\, dx = \sum_{n=1}^{\infty} c_n \delta_{mn} = c_m \tag{4.57}$$

All the terms in the summation equal zero except for the one term where $n = m$. The last equality in Equation 4.57 follows from the fact that $\delta_{mn} = 0$ for every term in the summation except for the term in which $m = n$, and then $\delta_{mn} = 1$ (Problem 4–26). Equation 4.57 gives us a simple formula for the coefficients in the expansion, Equation 4.56,

$$c_n = \int_{-\infty}^{\infty} \psi_n^*(x) f(x)\, dx \tag{4.58}$$

The expansion of a function in terms of an orthonormal set as in Equation 4.56 is an important and useful technique in many branches of physics and chemistry. We can illustrate the procedure by considering a particle in a box. In Example 4–4, we calculated the average energy of a particle in a box if it is in the state described by the normalized wave function $(30/a^5)^{1/2}x(a - x)$. Let's expand this function in terms of the orthonormal complete set of eigenfunctions of a particle in a box. If we substitute Equation 3.27 for $\psi_n(x)$ into Equation 4.56, then we obtain

$$f(x) = \left(\frac{2}{a}\right)^{1/2} \sum_{n=1}^{\infty} c_n \sin \frac{n\pi x}{a} \qquad 0 \leq x \leq a \tag{4.59}$$

Equation 4.58 gives

$$c_n = \left(\frac{2}{a}\right)^{1/2} \int_0^a f(x) \sin \frac{n\pi x}{a}\, dx \tag{4.60}$$

In our case,

$$f(x) = \left(\frac{30}{a^5}\right)^{1/2} x(a - x) \qquad 0 \leq x \leq a \tag{4.61}$$

If we substitute Equation 4.61 into Equation 4.60, we find that

$$c_n = \left(\frac{60}{a^6}\right)^{1/2} \int_0^a x(a-x) \sin \frac{n\pi x}{a} \, dx$$

$$= \left(\frac{60}{a^6}\right)^{1/2} \left[2 \left(\frac{a}{n\pi}\right)^3 (1 - \cos n\pi)\right]$$

$$= \frac{4(15)^{1/2}}{\pi^3 n^3}[1 - (-1)^n]$$

Therefore, we find that

$$c_n = \begin{cases} \dfrac{8(15)^{1/2}}{\pi^3 n^3} & \text{for odd values of } n \\ 0 & \text{for even values of } n \end{cases} \tag{4.62}$$

Problem 4–48 explores why $c_n = 0$ for even values of n. We shall soon show that c_n^2 is the probability that we obtain the energy $E_n = n^2 h^2 / 8ma^2$ if we measure the energy of a particle in a box described by the wave function given by Equation 4.61. The expansion of a function in terms of an orthonormal set is called a *Fourier expansion* or a *Fourier series*. Because Equation 4.59 contains only sine functions, it is called a *Fourier sine series*, in particular. The coefficients c_n in the expansion are called *Fourier coefficients*.

Postulate 3 says that the only values of the observable corresponding to $\hat{A}$ that one obtains in a measurement are the eigenvalues of $\hat{A}$. Postulate 4 tells us how to calculate the average in a series of measurements:

$$\langle a \rangle = \int_{-\infty}^{\infty} \Psi^*(x) \hat{A} \, \Psi(x) \, dx \tag{4.63}$$

If $\Psi(x)$ is an eigenstate of $\hat{A}$, then we observe only one value of a; but in the general case, when we carry out a series of measurements, we observe a distribution of the set of possible results $\{a_n\}$, whose average is given by Equation 4.63. If the system is in a state described by $\Psi(x)$, then what is the probability of obtaining the particular result a_n in a single measurement? To answer this question, we consider a measurement of the energy, so that $\hat{A} = \hat{H}$ and $\langle a \rangle = \langle E \rangle$ in Equation 4.63. Using an equation like Equation 4.56 for $\Psi(x)$, we have

$$\langle E \rangle = \int_{-\infty}^{\infty} \Psi^*(x) \hat{H} \, \Psi(x) \, dx$$

$$= \int_{-\infty}^{\infty} \left[\sum_n c_n^* \psi_n^*(x)\right] \hat{H} \left[\sum_m c_m \psi_m(x)\right] dx$$

$$= \sum_n \sum_m c_n^* c_m \int_{-\infty}^{\infty} \psi_n^*(x) \hat{H} \psi_m(x) \, dx \tag{4.64}$$

If we use the fact that $\hat{H}\psi_m(x) = E_m\psi_m(x)$, Equation 4.64 becomes

$$\langle E \rangle = \sum_n \sum_m c_n^* c_m E_m \int_{-\infty}^{\infty} \psi_n^*(x)\psi_m(x)\, dx \tag{4.65}$$

The integral here is equal to the Kronecker delta δ_{mn} and so Equation 4.65 becomes

$$\langle E \rangle = \sum_n \sum_m c_n^* c_m E_m \delta_{nm}$$

or

$$\langle E \rangle = \sum_n c_n^* c_n E_n = \sum_n |c_n|^2 E_n \tag{4.66}$$

Recall that the average of a set of energies is defined as

$$\langle E \rangle = \sum_n p_n E_n \tag{4.67}$$

where p_n is the probability of observing the value E_n. By comparing Equations 4.66 and 4.67, we see that we can interpret $|c_n|^2$ as the probability of observing E_n when carrying out a measurement on the system, or

$$\text{probability of observing } E_n = |c_n|^2 \tag{4.68}$$

Suppose the system is in an energy eigenstate, say $\psi_1(x)$, so that $\hat{H}\psi_1(x) = E_1\psi_1(x)$. Then all the c_n's are zero, except for c_1, which is equal to unity. According to Equation 4.68, the probability of observing E_1 is unity, in agreement with Postulate 3. Example 4–10 illustrates the case where the system is not in an energy eigenstate.

EXAMPLE 4–10
Consider the system discussed in Example 4–4. We have a particle in a box that is described by the normalized wave function

$$f(x) = \left(\frac{30}{a^5}\right)^{1/2} x(a - x) \qquad 0 \leq x \leq a$$

Calculate the probability that if we were to measure the energy of the particle, the value $E_n = n^2h^2/8ma^2$ would result. Show that the sum of these probabilites is unity. Using these probabilities, calculate $\langle E \rangle$ and compare the result to the value found in Example 4–4.

SOLUTION: We're going to need the values of two series in this solution:

$$\sum_{j=0}^{\infty} \frac{1}{(2j+1)^4} \qquad \text{and} \qquad \sum_{j=0}^{\infty} \frac{1}{(2j+1)^6}$$

Both series occur in numerous handbooks; for example, if you look up 'Sums of reciprocal powers' in the index of the *CRC Standard Mathematical Tables and Formulas*, you find that

$$\sum_{j=0}^{\infty} \frac{1}{(2j+1)^4} = \frac{\pi^4}{96} \quad \text{and} \quad \sum_{j=0}^{\infty} \frac{1}{(2j+1)^6} = \frac{\pi^6}{960}$$

We're now ready to go on. The energy eigenfunctions of a particle in a box are $(2/a)^{1/2} \sin(n\pi x/a)$. According to Equation 4.68, we must determine the Fourier coefficients $\{c_n\}$ in the expansion

$$f(x) = \left(\frac{2}{a}\right)^{1/2} \sum_{n=1}^{\infty} c_n \sin \frac{n\pi x}{a}$$

In Section 4.8, we showed that (Equation 4.62)

$$c_n = \begin{cases} \dfrac{8(15)^{1/2}}{\pi^3 n^3} & \text{if } n \text{ is odd} \\ 0 & \text{if } n \text{ is even} \end{cases}$$

The probability of observing the energy $n^2 h^2/8ma^2$ is

$$\text{probability of observing } E_n = |c_n|^2 = \begin{cases} \dfrac{960}{\pi^6 n^6} & \text{if } n \text{ is odd} \\ 0 & \text{if } n \text{ is even} \end{cases}$$

To show that these probabilities sum to unity, we must show that

$$\sum_{n=1}^{\infty} c_n^2 = \frac{960}{\pi^6} \sum_{n \text{ odd}}^{\infty} \frac{1}{n^6} = \frac{960}{\pi^6} \sum_{j=0}^{\infty} \frac{1}{(2j+1)^6} \stackrel{?}{=} 1$$

The value of the summation here is $\pi^6/960$, and so we see that $\displaystyle\sum_{n=1}^{\infty} c_n^2$ does indeed equal 1. The average energy is given by Equation 4.66:

$$\langle E \rangle = \sum_{n=1}^{\infty} c_n^2 E_n = \sum_{n \text{ odd}}^{\infty} \left(\frac{960}{\pi^6 n^6}\right)\left(\frac{n^2 h^2}{8ma^2}\right)$$

$$= \frac{120 h^2}{m\pi^6 a^2} \sum_{n \text{ odd}}^{\infty} \frac{1}{n^4} = \frac{120 h^2}{m\pi^6 a^2} \sum_{j=0}^{\infty} \frac{1}{(2j+1)^4}$$

The value of the summation here is $\pi^4/96$ and so we see that

$$\langle E \rangle = \frac{5 h^2}{4\pi^2 ma^2}$$

which is exactly the value that we obtained in Example 4–4.

4.9 The Time Dependence of Wave Functions Is Governed by the Time-Dependent Schrödinger Equation

We need one more postulate. Notice that the wave function in Postulate 1 contains time explicitly. This is something that we have not considered yet. Except for this one thing, we have used all the tacitly given postulates in Chapter 3, and so all our discussion so far should have been fairly familiar. Now we must discuss the time dependence of wave functions. The time dependence of wave functions is governed by the time-dependent Schrödinger equation. We cannot derive the time-dependent Schrödinger equation any more than we can derive the time-independent Schrödinger equation. It is difficult to try to justify the form of the time-dependent Schrödinger equation without using some arguments that are beyond the level of this book, and so we shall simply postulate its form and then show that it is consistent with the time-independent Schrödinger equation.

Postulate 5
The wave function or state function of a system evolves in time according to the time-dependent Schrödinger equation

$$\hat{H}\Psi(x, t) = i\hbar \frac{\partial \Psi}{\partial t} \tag{4.69}$$

Postulate 5 is the only one of the postulates presented here that we did not use in Chapter 3 and that should be new. For most systems that we shall study in this book, $\hat{H}$ does not contain time explicitly, and in this case we can apply the method of separation of variables and write

$$\Psi(x, t) = \psi(x)f(t)$$

If we substitute this into Equation 4.69 and divide both sides by $\psi(x)f(t)$, we obtain

$$\frac{1}{\psi(x)}\hat{H}\psi(x) = \frac{i\hbar}{f(t)}\frac{df}{dt} \tag{4.70}$$

If $\hat{H}$ does not contain time explicitly, then the left side of Equation 4.70 is a function of x only and the right side is a function of t only, and so both sides must equal a constant. If we denote the separation constant by E, then Equation 4.70 gives

$$\hat{H}\psi(x) = E\psi(x) \tag{4.71}$$

and

$$\frac{df}{dt} = -\frac{i}{\hbar}Ef(t) \tag{4.72}$$

The first of these two equations is what we have been calling the *Schrödinger equation*. In view of Equation 4.69, Equation 4.71 is often called the *time-independent Schrödinger equation*.

Equation 4.72 can be integrated immediately to give

$$f(t) = e^{-iEt/\hbar}$$

and so $\Psi(x, t)$ is of the form

$$\Psi(x, t) = \psi(x)e^{-iEt/\hbar} \tag{4.73}$$

If we use the relation $E = h\nu = \hbar\omega$, we can write Equation 4.73 as

$$\Psi(x, t) = \psi(x)e^{-i\omega t} \tag{4.74}$$

It is interesting to note that Equation 4.74 oscillates harmonically in time and is characteristic of wave motion. Yet the time-dependent Schrödinger equation does *not* have the same form as a classical wave equation. The Schrödinger equation has a first derivative in time, whereas the classical wave equation (Equation 2.1) has a second derivative in time. Nevertheless, the Schrödinger equation does have wavelike solutions, which is one reason why quantum mechanics is sometimes called *wave mechanics*.

There is a set of solutions to Equation 4.71, and we write Equation 4.73 as

$$\Psi_n(x, t) = \psi_n(x)e^{-iE_n/\hbar} \tag{4.75}$$

If the system happens to be in one of the eigenstates given by Equation 4.75, then

$$\Psi_n^*(x, t)\Psi_n(x, t)dx = \psi_n^*(x)\psi_n(x)dx \tag{4.76}$$

Thus, the probability density and the averages calculated from Equation 4.75 are independent of time, and the $\psi_n(x)$ are called *stationary-state* wave functions. Stationary states are of central importance in chemistry. For example, in later chapters we shall represent an atom or a molecule by a set of stationary energy states and express the spectroscopic properties of the system in terms of transitions from one stationary state to another. The Bohr model of a hydrogen atom is a simple illustration of the idea.

It is important to realize that a system is not generally in a state described by a wave function of the form of Equation 4.75. The general solution to Equation 4.69 is a superposition of Equation 4.75:

$$\Psi(x, t) = \sum_n c_n\psi_n(x)e^{-iE_n t/\hbar} \tag{4.77}$$

For simplicity, let's consider a case in which $\Psi(x, t)$ is a summation of only two terms:

$$\Psi(x, t) = c_1\psi_1(x)e^{-iE_1 t/\hbar} + c_2\psi_2(x)e^{-iE_2 t/\hbar}$$

The probability density in this case is

$$\Psi^*(x, t)\Psi(x, t)$$

$$= [c_1^*\psi_1^*(x)e^{i E_1 t/\hbar} + c_2^*\psi_2^*(x)e^{i E_2 t/\hbar}][c_1\psi_1(x)e^{-i E_1 t/\hbar} + c_2\psi_2(x)e^{-i E_2 t/\hbar}]$$

$$= |c_1|^2\psi_1^*(x)\psi_1(x) + |c_2|^2\psi_2^*(x)\psi_2(x)$$

$$+ c_1^*c_2\psi_1^*(x)\psi_2(x) \exp\left[\frac{i(E_1 - E_2)t}{\hbar}\right]$$

$$+ c_2^*c_1\psi_2^*(x)\psi_1(x) \exp\left[\frac{i(E_2 - E_1)t}{\hbar}\right]$$

The third and fourth terms here contain time explicitly, and so $\Psi^*(x, t)\Psi(x, t)$ is not independent of time. In this case, we do not have a stationary state.

Suppose, for example, that $\Psi(x, t)$ is a linear combination of the first two eigenstates of a particle in a box,

$$\Psi(x, t) = c_1 \left(\frac{2}{a}\right)^{1/2} e^{-i E_1 t/\hbar} \sin\frac{\pi x}{a} + c_2 \left(\frac{2}{a}\right)^{1/2} e^{-i E_2 t/\hbar} \sin\frac{2\pi x}{a}$$

where $E_n = n^2 h^2/8ma^2$. Suppose, furthermore, that both states occur on an equal footing, so that $c_1 = c_2$. For simplicity, let's assume that $c_1 = c_2 =$ real. We can determine c_1 and c_2 by requiring that $\Psi(x, t)$ be normalized:

$$\int_0^a dx\ \Psi^*(x, t)\Psi(x, t) = \frac{2c_1^2}{a} \int_0^a dx\ \left(e^{i E_1 t/\hbar} \sin\frac{\pi x}{a} + e^{i E_2 t/\hbar} \sin\frac{2\pi x}{a}\right)$$

$$\times \left(e^{-i E_1 t/\hbar} \sin\frac{\pi x}{a} + e^{-i E_2 t/\hbar} \sin\frac{2\pi x}{a}\right)$$

$$= \frac{2c_1^2}{a} \left[\int_0^a dx\ \sin^2\frac{\pi x}{a} + \int_0^a dx\ \sin^2\frac{2\pi x}{a}\right.$$

$$\left. + 2\cos\frac{(E_2 - E_1)t}{\hbar} \int_0^a dx\ \sin\frac{\pi x}{a} \sin\frac{2\pi x}{a}\right]$$

$$= \frac{2c_1^2}{a}\left(\frac{a}{2} + \frac{a}{2} + 0\right) = 2c_1^2 = 1$$

or $c_1 = (1/2)^{1/2}$. In going from the second line to the third line, we used the fact that

$$e^{i(E_2 - E_1)t/\hbar} + e^{-i(E_2 - E_1)t/\hbar} = 2\cos\frac{(E_2 - E_1)t}{\hbar}$$

and in going from the third line to the last line, we used the fact that $\sin(\pi x/a)$ and $\sin(2\pi x/a)$ are orthogonal over the interval $0 \leq x \leq a$.

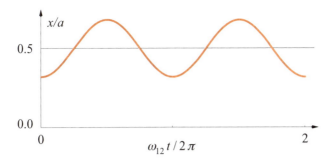

FIGURE 4.3
The time dependence of the average position of a particle in a box calculated from the state function given by Equation 4.79. The particle oscillates about the midpoint of the well from $x/a = 0.32$ to $x/a = 0.68$ with a frequency ω_{12}.

Therefore, our normalized time-dependent state function is (Problem 4–33)

$$\Psi(x, t) = \left(\frac{1}{a}\right)^{1/2} e^{-iE_1t/\hbar} \sin \frac{\pi x}{a} + \left(\frac{1}{a}\right)^{1/2} e^{-iE_2t/\hbar} \sin \frac{2\pi x}{a} \tag{4.78}$$

Let's use $\Psi(x, t)$ to calculate $\langle x \rangle$, the average position of the particle within the box.

$$\langle x \rangle = \int_0^a dx \ \Psi^*(x, t) x \Psi(x, t)$$

$$= \frac{1}{a} \int_0^a dx \ x \sin^2 \frac{\pi x}{a} + \frac{1}{a} \int_0^a dx \ x \sin^2 \frac{2\pi x}{a}$$

$$+ \frac{2 \cos \omega_{12} t}{a} \int_0^a dx \ x \sin \frac{\pi x}{a} \sin \frac{2\pi x}{a}$$

$$= \frac{a}{2} - \frac{16a}{9\pi^2} \cos \omega_{12} t \tag{4.79}$$

where $\omega_{12} = (E_2 - E_1)/\hbar$ (Problem 4–34). Figure 4.3 shows $\langle x \rangle/a$ plotted against t. Notice that the average position of the particle oscillates about the midpoint of the well with frequency ω_{12}. Problems 4–39 and 4–40 have you redo this calculation for linear combinations of other states.

We can also plot the probability density associated with $\Psi(x, t)$ in Equation 4.78. The probability density is given by

$$\text{probability density} = \Psi^*(x, t)\Psi(x, t)$$

$$= \frac{1}{a} \sin^2 \frac{\pi x}{a} + \frac{1}{a} \sin^2 \frac{2\pi x}{a} + \frac{1}{a} \sin \frac{\pi x}{a} \sin \frac{2\pi x}{a} \cos \omega_{12} t$$

$$\tag{4.80}$$

The interpretation of the measurement process that we have given here is due primarily to Bohr and Heisenberg, and is called the Copenhagen interpretation. In the quantum mechanics literature in the 1920s and '30s, there were fervent disagreements about the interpretation of quantum mechanics. For example, Einstein and Schrödinger never accepted its probabilistic interpretation and the doctrine of the Copenhagen school. Einstein, in particular, always felt that there was something missing in the formalism, and that quantum mechanics was incomplete in some sense. Questions such as "Can you claim that a particle even has a property such as momentum until you measure it?" were hotly debated in the early years of quantum mechanics, but the Copenhagen school eventually won over most scientists, possibly due to the strong personalitites of Bohr and Heisenberg. Although the Copenhagen interpretation has held sway over the years, other interpretations have become increasingly deliberated since the 1980s. There is an extensive, fascinating semipopular literature on this subject. Some references are given at the end of this chapter. The one entitled *In Search of Schrödinger's Cat* is based upon a thought experiment proposed by Schrödinger in which a cat is in a state that is a superposition of a live cat and a dead cat, and just what such a state actually means.

We should emphasize that there never was a question about the validity of the results of quantum mechanics. In fact, quantum mechanics might be the most successful calculational tool in the history of science. It is simply the interpretation of its equations that is an issue. In this book, we'll take the easy way out and just accept the Copenhagen view. Let's see how this picture can be used to describe the two-slit experiment that we discussed in Section 1.13. We start with the Schrödinger equation for a free particle of energy E. By a free particle, we mean that there are no boundaries and the potential is equal to zero everywhere.

$$-\frac{\hbar^2}{2m}\frac{\partial^2 \Psi}{\partial x^2} = i\hbar \frac{\partial \Psi}{\partial t}$$

(4.88)

Following the development in Section 4.9, we obtain

$$\Psi(x, t) = A\psi(x)e^{-iEt/\hbar}$$

where $\psi(x)$ in this case is given by

$$-\frac{\hbar^2}{2m}\frac{d^2\psi}{dx^2} = E\psi(x)$$

The solutions to this equation are $\psi(x) = e^{\pm ikx}$, where $k = (2mE/\hbar^2)^{1/2}$. Because there is no potential energy, E is simply the kinetic energy, $p^2/2m$. Thus, we see that $\hbar k = p$, and write

$$\Psi(x, t) = Ae^{ikx}e^{-i\omega t}$$

(4.89)

where $\omega = E/\hbar$, $\hbar k = p$, and A is the amplitude, which we take to be real for simplicity. The quantity k, called the wave vector, is equal to $2\pi/\lambda$, as you can see by substituting the de Broglie condition, $p = h/\lambda$, into $p = \hbar k$.

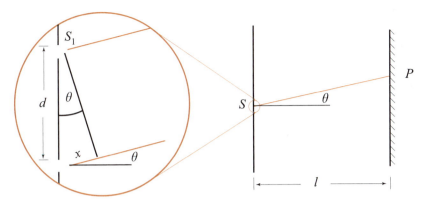

FIGURE 4.5
A schematic illustration of the geometry of a two-slit experiment.

Equation 4.89 is the same as Equation 2.54. For $k > 0$ ($k < 0$), it represents a harmonic wave of wavelength λ and frequency ω traveling to the right (left) with velocity $\omega/k = \nu\lambda$. Notice that there are no restrictions on the energy and the momentum in this case, and that they are continuous because there are no boundary conditions. (There are no boundaries, as there are for a particle in a box.) Equation 4.89 represents a *free particle*.

Let's now go back to the two-slit experiment in Section 1.13. Figure 4.5 illustrates the geometry that we use for this discussion. Let $\psi_1(x_1)$ be the wave function of a particle that goes through slit 1 and $\psi_2(x_2)$ be that for slit 2. We don't know which slit the particle has gone through, and so the wave function beyond the slits is a superposition of the two states, or

$$\Psi(x_1, x_2, t) = \Psi_1(x_1, t) + \Psi_2(x_2, t)$$

$$= A_1 e^{i(kx_1 - \omega t)} + A_2 e^{i(kx_2 - \omega t)} \tag{4.90}$$

where x_1 is the distance from slit 1 and x_2 is that from slit 2. Equation 4.90 is a fundamental tenet of quantum mechanics. The probability density for finding the particle at some point is given by $\Psi^*(x_1, x_2, t)\Psi(x_1, x_2, t)$, which, assuming for simplicity that A_1 and A_2 are real, is given by

$$\text{probability density} = (A_1 e^{-ikx_1} + A_1 e^{-ikx_2})(A_2 e^{ikx_1} + A_2 e^{ikx_2})$$

$$= A_1^2 + A_2^2 + 2A_1 A_2 \cos k(x_2 - x_1) \tag{4.91}$$

This result gives the observed interference pattern. When $|x_2 - x_1|$ is an integral multiple of λ, then $\cos k(x_2 - x_1) = \cos(2\pi|x_2 - x_1|/\lambda) = \cos(2\pi n) = 1$, and we have constructive interference. When $|x_2 - x_1|$ is an odd integral multiple of $\lambda/2$, then $\cos k(x_2 - x_1) = n\pi$, which equals -1 for $n = 1, 3, \ldots$, and so we have destructive interference.

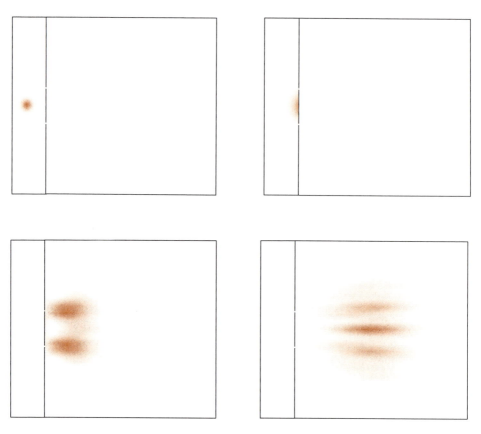

FIGURE 4.6
The wave function of a quantum-mechanical free particle approaching and then passing through an opaque screen with two narrow slits and then impinging on a second screen to produce an interference pattern.

Suppose now we use a detector and find that the particle actually goes through slit 1. In this case, the $\Psi(x_1, x_2, t)$ in Equation 4.90 collapses to $\Psi_1(x_1, t)$, and we obtain

$$\text{probability density} = A_1 e^{-i(kx_1-\omega t)} A_1 e^{i(kx_1-\omega t)}$$

$$= A_1^2 \tag{4.92}$$

This is the result that would obtain if slit 2 were closed. Thus, the measurement of which slit that the particle goes through collapses the superposition wave function to that one component that represents passage through that given slit. At least, this is according to the Copenhagen interpretation.

The above treatment is fairly simplified, taking the wave function of the particle to be one-dimensional, but it does capture the essence of the process. Figure 4.6 shows the result of a more advanced calculation of a particle incident upon the two-slit screen and then passing through it. The solution to the Schrödinger equation for this process shows the particle as a spherical wave approaching the two-slit screen and as a more complicated wave as it emerges from the screen. The emerging wave function leads to an interference pattern when it strikes the second screen.

Before concluding this chapter, we summarize our set of postulates:

Postulate 1

The state of a quantum-mechanical system is completely specified by a function $\Psi(\mathbf{r}, t)$ that depends on the coordinates of the particle and on time. This function, called the wave function or state function, has the important property that $\Psi^*(\mathbf{r}, t)\Psi(\mathbf{r}, t)dxdydz$ is the probability that the particle lies in the volume element $dxdydz$ located at $\mathbf{r}$ at time t.

Postulate 2

To every observable in classical mechanics there corresponds a linear, Hermitian operator in quantum mechanics.

Postulate 3

In any measurement of the observable associated with the operator $\hat{A}$, the only values that will ever be observed are the eigenvalues a_n, which satisfy the eigenvalue equation

$$\hat{A}\Psi_a = a\Psi_a$$

Postulate 4

If a system is in a state described by a normalized wave function Ψ, then the average value of the observable corresponding to $\hat{A}$ is given by

$$\langle a \rangle = \int_{-\infty}^{\infty} \Psi^*\hat{A}\Psi d\tau$$

Postulate 5

The wave function or state function of a system evolves in time according to the time-dependent Schrödinger equation

$$\hat{H}\Psi(x, t) = i\hbar\frac{\partial\Psi}{\partial t}$$

Problems

4–1. Which of the following candidates for wave functions are normalizable over the indicated intervals?

(a) $e^{-x^2/2}$ $(-\infty, \infty)$

(b) e^{-x} $(-\infty, \infty)$

(c) $e^{i\theta}$ $(0, 2\pi)$

(d) $\cosh x$ $(0, \infty)$

(e) xe^{-x} $(0, \infty)$

Normalize those that can be normalized. Are the others suitable wave functions?

4–2. Which of the following wave functions are normalized over the indicated two-dimensional intervals?

(a) $e^{-(x^2+y^2)/2}$ $0 \le x < \infty, 0 \le y < \infty$

(b) $e^{-(x+y)/2}$ $0 \le x < \infty, 0 \le y < \infty$

(c) $\left(\dfrac{4}{ab}\right)^{1/2} \sin \dfrac{\pi x}{a} \sin \dfrac{\pi y}{b}$ $0 \le x \le a, 0 \le y \le b$

Normalize those that aren't.

4–3. Why does $\psi^*\psi$ have to be everywhere real, nonnegative, finite, and of definite value?

4–4. In this problem, we will prove that the form of the Schrödinger equation imposes the condition that the first derivative of a wave function be continuous. The Schrödinger equation is

$$\frac{d^2\psi}{dx^2} + \frac{2m}{\hbar^2}[E - V(x)]\psi(x) = 0$$

If we integrate both sides from $a - \epsilon$ to $a + \epsilon$, where a is an arbitrary value of x and ϵ is infinitesimally small, then we have

$$\left.\frac{d\psi}{dx}\right|_{x=a+\epsilon} - \left.\frac{d\psi}{dx}\right|_{x=a-\epsilon} = \frac{2m}{\hbar^2}\int_{a-\epsilon}^{a+\epsilon}[V(x) - E]\psi(x)dx$$

Now show that $d\psi/dx$ is continuous if $V(x)$ is continuous.

Suppose now that $V(x)$ is *not* continuous at $x = a$, as in

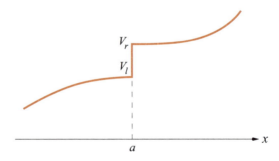

Show that

$$\left.\frac{d\psi}{dx}\right|_{x=a+\epsilon} - \left.\frac{d\psi}{dx}\right|_{x=a-\epsilon} = \frac{2m}{\hbar^2}(V_l + V_r - 2E)\psi(a)\epsilon$$

so that $d\psi/dx$ is continuous even if $V(x)$ has a *finite* discontinuity. What if $V(x)$ has an infinite discontinuity, as in the problem of a particle in a box? Are the first derivatives of the wave functions continuous at the boundaries of the box?

4–5. Determine whether the following functions are acceptable or not as state functions over the indicated intervals.

(a) $\dfrac{1}{x}$ $(0, \infty)$ (b) $e^{-2x} \sinh x$ $(0, \infty)$

(c) $e^{-x} \cos x$ $(0, \infty)$ (d) e^x $(-\infty, \infty)$

(e) $e^{-x} \sinh x$ $(0, \infty)$

4–6. Consider the linear differential equation

$$a(x)y''(x) + b(x)y'(x) + c(x)y(x) = 0$$

where $y''(x)$ and $y'(x)$ are standard notation for d^2y/dx^2 and dy/dx, respectively. Show that if $y_1(x)$ and $y_2(x)$ are each solutions to the above differential equation, then so is $y(x) = c_1 y_1(x) + c_2 y_2(x)$, where c_1 and c_2 are constants.

4–7. Calculate the values of $\sigma_E^2 = \langle E^2 \rangle - \langle E \rangle^2$ for a particle in a box in the state described by

$$\psi(x) = \left(\frac{630}{a^9} \right)^{1/2} x^2(a-x)^2 \qquad 0 \le x \le a$$

4–8. Consider a free particle constrained to move over the rectangular region $0 \le x \le a$, $0 \le y \le b$. The energy eigenfunctions of this system are

$$\psi_{n_x, n_y}(x, y) = \left(\frac{4}{ab} \right)^{1/2} \sin \frac{n_x \pi x}{a} \sin \frac{n_y \pi y}{b} \qquad \begin{cases} n_x = 1, 2, 3, \dots \\ n_y = 1, 2, 3, \dots \end{cases}$$

The Hamiltonian operator for this system is

$$\hat{H} = -\frac{\hbar^2}{2m} \left(\frac{\partial^2}{\partial x^2} + \frac{\partial^2}{\partial y^2} \right)$$

Show that if the system is one of its eigenstates, then

$$\sigma_E^2 = \langle E^2 \rangle - \langle E \rangle^2 = 0$$

4–9. The momentum operator in two dimensions is

$$\hat{P} = -i\hbar \left(\mathbf{i} \frac{\partial}{\partial x} + \mathbf{j} \frac{\partial}{\partial y} \right)$$

Using the wave function given in Problem 4–8, calculate the value of $\langle p \rangle$ and then

$$\sigma_p^2 = \langle p^2 \rangle - \langle p \rangle^2$$

Compare your result with σ_p^2 in the one-dimensional case.

4–10. Suppose that a particle in a two-dimensional box (cf. Problem 4–8) is in the state

$$\psi(x, y) = \frac{30}{(a^5 b^5)^{1/2}} x(a-x)y(b-y)$$

Show that $\psi(x, y)$ is normalized, and then calculate the value of $\langle E \rangle$ associated with the state described by $\psi(x, y)$.

4–11. Evaluate the commutator $[\hat{A}, \hat{B}]$, where $\hat{A}$ and $\hat{B}$ are given below.

	$\hat{A}$	$\hat{B}$
(a)	$\dfrac{d^2}{dx^2}$	x
(b)	$\dfrac{d}{dx} - x$	$\dfrac{d}{dx} + x$
(c)	$\displaystyle\int_0^x dx$	$\dfrac{d}{dx}$
(d)	$\dfrac{d^2}{dx^2} - x$	$\dfrac{d}{dx} + x^2$

4–12. Referring to Table 4.1 for the operator expressions for angular momentum, show that

$$[\hat{L}_x, \hat{L}_y] = i\hbar \hat{L}_z$$

$$[\hat{L}_y, \hat{L}_z] = i\hbar \hat{L}_x$$

and

$$[\hat{L}_z, \hat{L}_x] = i\hbar \hat{L}_y$$

(Do you see a pattern here to help remember these commutation relations?) What do these expressions say about the ability to measure the components of angular momentum simultaneously?

4–13. Defining

$$\hat{L}^2 = \hat{L}_x^2 + \hat{L}_y^2 + \hat{L}_z^2$$

show that $\hat{L}^2$ commutes with each component separately. What does this result tell you about the ability to measure the square of the total angular momentum and its components simultaneously?

4–14. The operators

$$\hat{L}_+ = \hat{L}_x + i\hat{L}_y \qquad \text{and} \qquad \hat{L}_- = \hat{L}_x - i\hat{L}_y$$

play a central role in the quantum-mechanical theory of angular momentum. (See the Appendix to Chapter 5.) Show that

$$\hat{L}_+ \hat{L}_- = \hat{L}^2 - \hat{L}_z^2 + \hbar \hat{L}_z$$

$$[\hat{L}_z, \hat{L}_+] = \hbar \hat{L}_+$$

and that

$$[\hat{L}_z, \hat{L}_-] = -\hbar\hat{L}_-$$

4–15. Consider a particle in a two-dimensional box. Determine $[\hat{X}, \hat{P}_y]$, $[\hat{X}, \hat{P}_x]$, $[\hat{Y}, \hat{P}_y]$, and $[\hat{Y}, \hat{P}_x]$.

4–16. Can the position and kinetic energy of an electron be measured simultaneously to arbitrary precision? (See Problem 4–42.)

4–17. Using the result of Problem 4–15, what are the "uncertainty relationships" $\Delta x \Delta p_y$ and $\Delta y \Delta p_x$ equal to?

4–18. Which of the following operators is Hermitian: d/dx, id/dx, d^2/dx^2, id^2/dx^2, xd/dx, and x? Assume that the functions on which these operators operate are appropriately well behaved at infinity.

4–19. Show that if $\hat{A}$ is Hermitian, then $\hat{A} - \langle a \rangle$ is Hermitian. Show that the sum of two Hermitian operators is Hermitian.

4–20. To prove that Equation 4.30 follows from Equation 4.29, first write Equation 4.30 with f and with g:

$$\int f^* \hat{A} f \, dx = \int f \hat{A}^* f^* \, dx \qquad \text{and} \qquad \int g^* \hat{A} g \, dx = \int g \hat{A}^* g^* \, dx$$

Now let $\psi = c_1 f + c_2 g$, where c_1 and c_2 are arbitrary complex constants, to write

$$\int (c_1^* f^* + c_2^* g^*) \hat{A} \, (c_1 f + c_2 g) \, dx = \int (c_1 f + c_2 g) \hat{A}^* \, (c_1^* f^* + c_2^* g^*) \, dx$$

If we expand both sides and use the first two equations, we find that

$$c_1^* c_2 \int f^* \hat{A} g \, dx + c_2^* c_1 \int g^* \hat{A} f \, dx = c_1 c_2^* \int f \hat{A}^* g^* \, dx + c_1^* c_2 \int g \hat{A}^* f^* \, dx$$

Rearrange this into

$$c_1^* c_2 \int (f^* \hat{A} g - g \hat{A}^* f^*) \, dx = c_1 c_2^* \int (f \hat{A}^* g^* - g^* \hat{A} f) \, dx$$

Notice that the two sides of this equation are complex conjugates of each other. If $z = x + iy$ and $z = z^*$, then show that this implies that z is real. Thus, both sides of this equation are real. But because c_1 and c_2 are arbitrary complex constants, the only way for both sides to be real is for both integrals to equal zero. Show that this implies Equation 4.30.

4–21. Show that if $\hat{A}$ is Hermitian, then

$$\int \hat{A}^* \psi^* \hat{B} \, \psi \, dx = \int \psi^* \hat{A} \, \hat{B} \, \psi \, dx$$

Hint: Use Equation 4.30.

4–22. Show that

$$\psi_0(x) = \pi^{-1/4}e^{-x^2/2}$$

$$\psi_1(x) = (4/\pi)^{1/4}xe^{-x^2/2}$$

$$\psi_2(x) = (4\pi)^{-1/4}(2x^2 - 1)e^{-x^2/2}$$

are orthonormal over the interval $-\infty < x < \infty$.

4–23. Show that the polynomials

$$P_0(x) = 1, \qquad P_1(x) = x, \qquad P_2(x) = \frac{1}{2}(3x^2 - 1) \quad \text{and} \quad P_3(x) = \frac{1}{2}(5x^3 - 3x)$$

satisfy the orthogonality relation

$$\int_{-1}^{1} P_l(x)P_n(x)dx = \frac{2\delta_{ln}}{2l + 1} \qquad l = 0, 1, 2, 3$$

4–24. Show that the set of functions $\{(2/a)^{1/2} \cos(n\pi x/a)\}$, $n = 0, 1, 2, \ldots$, is orthonormal over the interval $0 \le x \le a$.

4–25. Generate an orthogonal set of polynomials $\{\phi_j(x),\ j = 1, 2, 3\}$ over the interval $-1 \le x \le 1$ starting with $f_0(x) = 1$, $f_1(x) = x$, and $f_2(x) = x^2$. Instead of normalizing the final result, choose a multiplicative constant such that $\phi_j(1) = 1$. Compare these polynomials to those in the previous problem.

4–26. Prove that if δ_{nm} is the Kronecker delta

$$\delta_{nm} = \begin{cases} 1 & n = m \\ 0 & n \neq m \end{cases}$$

then

$$\sum_{n=1}^{\infty} c_n\delta_{nm} = c_m \qquad \text{and} \qquad \sum_n \sum_m a_n b_m \delta_{nm} = \sum_n a_n b_n$$

These results will be used often.

4–27. Express the orthonormality of the set of functions $\{\psi_n(x)\}$ in Dirac notation. Express the eigenfunction expansion $\phi(x) = \sum_n c_n\psi_n(x)$ and the coefficients c_n in Dirac notation.

4–28. A general state function, expressed in the form of a ket vector $|\phi\rangle$, can be written as a superposition of the eigenstates $|1\rangle$, $|2\rangle$, $\ldots$ of an operator $\hat{A}$ with eigenvalues $a_1, a_2, \ldots$ (in other words, $\hat{A}|n\rangle = a_n|n\rangle$):

$$|\phi\rangle = c_1|1\rangle + c_2|2\rangle + \cdots = \sum_n c_n|n\rangle$$

Show that $c_n = \langle n \mid \phi \rangle$. This quantity is called the *amplitude* of measuring a_n if a measurement of $\hat{A}$ is made in the state $\mid \phi \rangle$. The probability of obtaining a_n is $c_n^* c_n$. Show that $\mid \phi \rangle$ can be written as

$$\mid \phi \rangle = \sum_n \mid n \rangle \langle n \mid \phi \rangle$$

Similarly, the corresponding bra vector of $\mid \phi \rangle$ can be written in terms of the corresponding bra vectors of the $\mid n \rangle$ as

$$\langle \phi \mid = \sum_n c_n^* \langle n \mid$$

Show that $c_n^* = \langle \phi \mid n \rangle$. Show that $\langle \phi \mid$ can be written as

$$\langle \phi \mid = \sum_n \langle \phi \mid n \rangle \langle n \mid$$

Now show that if $\langle \phi \mid$ is normalized, then

$$\langle \phi \mid \phi \rangle = 1 = \langle \phi \mid n \rangle \langle n \mid \phi \rangle$$

and use this result to argue that

$$\sum_n \mid n \rangle \langle n \mid = 1$$

is a unit operator.

4–29. Given the three polynomials $f_0(x) = a_0$, $f_1(x) = a_1 + b_1 x$, and $f_2(x) = a_2 + b_2 x + c_2 x^2$, find the constants such that the f's form an orthonormal set over the interval $0 \leq x \leq 1$.

4–30. Using the orthogonality of the set $\{\sin(n\pi x/a)\}$ over the interval $0 \leq x \leq a$, show that if

$$f(x) = \sum_{n=1}^{\infty} b_n \sin \frac{n\pi x}{a}$$

then

$$b_n = \frac{2}{a} \int_0^a f(x) \sin \frac{n\pi x}{a} \, dx \qquad n = 1, 2, \ldots$$

Use this to show that the Fourier expansion of $f(x) = x$, $0 \leq x \leq a$, is

$$x = \frac{2a}{\pi} \sum_{n=1}^{\infty} \frac{(-1)^{n+1}}{n} \sin \frac{n\pi x}{a}$$

4–31. We can define functions of operators through their Maclaurin series (MathChapter D). For example, we define the operator $\exp(\hat{S})$ by

$$e^{\hat{S}} = \sum_{n=0}^{\infty} \frac{(\hat{S})^n}{n!}$$

Under what conditions does the equality $e^{\hat{A}+\hat{B}} \stackrel{?}{=} e^{\hat{A}}e^{\hat{B}}$ hold?

4–32. In this chapter, we learned that if ψ_n is an eigenfunction of the time-independent Schrödinger equation, then

$$\Psi_n(x, t) = \psi_n(x)e^{-iE_n t/\hbar}$$

Show that if $\psi_m(x)$ and $\psi_n(x)$ are both stationary states of $\hat{H}$, then the state

$$\Psi(x, t) = c_m\psi_m(x)e^{-iE_m t/\hbar} + c_n\psi_n(x)e^{-iE_n t/\hbar}$$

satisfies the time-dependent Schrödinger equation.

4–33. Show that $\Psi(x, t)$ given by Equation 4.78 is normalized.

4–34. Verify Equation 4.79.

4–35. What is the normalization constant for $\Psi(x, t) = \sum_{n=1}^{N} \psi_n(x)e^{-iE_n t/\hbar}$ if the $\psi_n(x)$ are normalized?

4–36. Superimpose the behavior of $\langle x \rangle$ for a classical particle moving with the same period onto Figure 4.3.

4–37. Show that the average energy of a particle described by Equation 4.78 is a constant.

4–38. What would be the form of $\Psi(x, t)$ in Equation 4.78 if it were a superposition of the three lowest states instead of two?

4–39. Derive an expression for the average position of a particle in a box in a state described by

$$\Psi(x, t) = \left(\frac{1}{a}\right)^{1/2} e^{-iE_2 t/\hbar} \sin\frac{2\pi x}{a} + \left(\frac{1}{a}\right)^{1/2} e^{-iE_3 t/\hbar} \sin\frac{3\pi x}{a}$$

With what frequency does the particle oscillate about the midpoint of the box?

4–40. Calculate the amplitude associated with the oscillation of a particle in a box in a state described by

$$\Psi(x, t) = \left(\frac{1}{a}\right)^{1/2} e^{-iE_1 t/\hbar} \sin\frac{\pi x}{a} + \left(\frac{1}{a}\right)^{1/2} e^{-iE_4 t/\hbar} \sin\frac{4\pi x}{a}$$

What is the frequency? Compare the amplitude here with that in the previous problem.

4–41. Use a program such as *MathCad* or *Mathematica* to plot the time evolution of the probability density for a particle in a box in a state described in the previous problem. Plot your result through one cycle.

4–42. In this problem, we shall develop the consequence of measuring the position of a particle in a box. If we find that the particle is located between $a/2 - \epsilon/2$ and $a/2 + \epsilon/2$, then its wave function may be ideally represented by

$$\phi_\epsilon = \begin{cases} 0 & x < a/2 - \epsilon/2 \\ 1/\sqrt{\epsilon} & a/2 - \epsilon/2 < x < a/2 + \epsilon/2 \\ 0 & x > a/2 + \epsilon/2 \end{cases}$$

Plot $\phi_\epsilon(x)$ and show that it is normalized. The parameter ϵ is in a sense a gauge of the accuracy of the measurement; the smaller the value of ϵ, the more accurate the measurement. Now let's suppose we measure the energy of the particle. The probability that we observe the value E_n is given by the value of $|c_n|^2$ in the expansion

$$\phi_\epsilon(x) = \sum_{n=1}^{\infty} c_n \psi_n(x) e^{-iE_n t/\hbar}$$

where $\psi_n(x) = (2/a)^{1/2} \sin n\pi x/a$ and $E_n = n^2 h^2/8ma^2$. Multiply both sides of this equation by $\psi_m(x)$ and integrate over x from 0 to a to get

$$c_m = e^{iE_m t/\hbar} \int_0^a \phi_\epsilon(x) \psi_m(x) \, dx = \frac{2^{3/2} a^{1/2} e^{iE_m t/\hbar}}{\epsilon^{1/2} m\pi} \sin \frac{m\pi}{2} \sin \frac{m\pi\epsilon}{2a}$$

Now show that the probability of observing E_n is given by

$$p(E_n) = \begin{cases} 0 & \text{if } n \text{ is even} \\ \dfrac{8a}{\epsilon} \left(\dfrac{1}{n\pi}\right)^2 \sin^2 \dfrac{n\pi\epsilon}{2a} & \text{if } n \text{ is odd} \end{cases}$$

Plot $p(E_n)$ against n for $\epsilon/a = 0.10$, 0.050, and 0.010. Interpret the result in terms of the uncertainty principle.

4–43. Starting with

$$\langle x \rangle = \int \Psi^*(x, t) x \Psi(x, t) dx$$

and the time-dependent Schrödinger equation, show that

$$\frac{d\langle x \rangle}{dt} = \int \Psi^* \frac{i}{\hbar} [\hat{H}, x] \Psi \, dx = \langle \hat{P}_x \rangle$$

Interpret this result.

4–44. Use Equation 4.84 to show that

$$\frac{d\langle \hat{P}_x\rangle}{dt} = \left\langle -\frac{d\hat{V}}{dx}\right\rangle$$

Interpret this result, which is known as *Ehrenfest's theorem.*

4–45. In this problem, we shall prove the Schwartz inequality, which says that if f and g are two suitably well-behaved functions, then

$$\left(\int |f|^2\,dx\right)\left(\int |g|^2\,dx\right) \geq \left|\int f^*g\,dx\right|^2$$

In order to prove the Schwartz inequality, we start with

$$\int (f + \lambda g)^*(f + \lambda g)\,dx \geq 0$$

where λ is an arbitrary complex number. Expand this to find

$$|\lambda|^2\int g^*g\,dx + \lambda\int f^*g\,dx + \lambda^*\int g^*f\,dx + \int f^*f\,dx \geq 0$$

This inequality must be true for any complex λ and, in particular, choose

$$\lambda = -\frac{\displaystyle\int g^*f\,dx}{\displaystyle\int g^*g\,dx} = -\frac{\left(\displaystyle\int gf^*\,dx\right)^*}{\displaystyle\int g^*g\,dx}$$

Show that this choice of λ gives the Schwartz inequality:

$$\left(\int f^*f\,dx\right)\left(\int g^*g\,dx\right) \geq \left|\int f^*g\,dx\right|^2$$

When does the equality hold?

4–46. In this problem, we shall prove that if $f(x)$ and $g(x)$ are suitably behaved functions, then

$$\left[\int f^*(x)f(x)\,dx\right]\left[\int g^*(x)g(x)\,dx\right] \geq \frac{1}{4}\left[\int (f^*(x)g(x) + f(x)g^*(x))\,dx\right]^2$$

We shall use this inequality to derive Equation 4.19 in the next problem. First, let

$$A = \int f^*(x)f(x)\,dx \qquad B = \int f^*(x)g(x)\,dx \qquad C = \int g^*(x)g(x)\,dx$$

Now argue that

$$\int [\lambda f^*(x) + g^*(x)][\lambda f(x) + g(x)]\, dx = A\lambda^2 + (B + B^*)\lambda + C$$

is equal to or greater than zero for real values of λ. Show that $A > 0$, $C > 0$, and that $B + B^* \geq 0$, and then argue that the roots of the quadratic form $A\lambda^2 + (B + B^*)\lambda + C$ cannot be real. Show that this can be so only if

$$AC \geq \frac{1}{4}(B + B^*)^2$$

which is the same as the above inequality.

4–47. We shall derive Equation 4.19 in this problem. You need the inequality that is derived in the previous problem to do this problem. Referring to the previous problem, let

$$f(x) = (\hat{A} - \langle a \rangle)\psi(x) \quad \text{and} \quad g(x) = i(\hat{B} - \langle b \rangle)\psi(x)$$

where $\psi(x)$ is any suitably behaved function. Substitute these into the left side of the inequality in the previous problem and use the fact that $\hat{A} - \langle a \rangle$ and $\hat{B} - \langle b \rangle$ are Hermitian (Problem 4–19) to write

$$\sigma_A^2 \sigma_B^2 \geq \frac{1}{4}\Big[i \int dx\, (\hat{A} - \langle a \rangle)^* \psi^*(x)(\hat{B} - \langle b \rangle)\psi(x)$$

$$- i \int dx\, (\hat{B} - \langle b \rangle)^* \psi^*(x)(\hat{A} - \langle a \rangle)\psi(x) \Big]^2$$

Use the Hermitian property of $\hat{A} - \langle a \rangle$ and $\hat{B} - \langle b \rangle$ again to write the right side as

$$-\frac{1}{4}\Big\{ \int dx\, \psi^*(x)\, [\hat{A}\hat{B} - \langle a \rangle\hat{B} - \hat{A}\langle b \rangle + \langle a \rangle\langle b \rangle$$

$$- \hat{B}\hat{A} + \langle b \rangle\hat{A} + \hat{B}\langle a \rangle - \langle a \rangle\langle b \rangle]\, \psi(x) \Big\}^2$$

Now use the fact that $\langle a \rangle$ and $\langle b \rangle$ are just numbers to write

$$\sigma_A^2 \sigma_B^2 \geq -\frac{1}{4}\Big\{ \int dx\, \psi^*(x)[\hat{A}, \hat{B}]\, \psi(x) \Big\}^2$$

which is Equation 4.19.

4–48. Show that $\sin(n\pi x/a)$ is an even function of x about $a/2$ if n is odd and is an odd function about $a/2$ if n is even. Use this result to show that the c_n in Equation 4.60 are zero for even values of n.

Problems 4–49 through 4–54 deal with systems with piecewise constant potentials.

4–49. Consider a particle moving in the potential energy

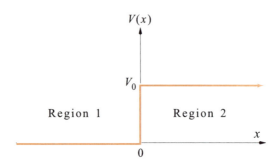

whose mathematical form is

$$V(x) = \begin{cases} 0 & x < 0 \\ V_0 & x > 0 \end{cases}$$

where V_0 is a constant. Show that if $E > V_0$, then the solutions to the Schrödinger equation in the two regions (1 and 2) are (see Problem 3–35)

$$\psi_1(x) = Ae^{ik_1x} + Be^{-ik_1x} \qquad x < 0 \tag{1}$$

and

$$\psi_2(x) = Ce^{ik_2x} + De^{-ik_2x} \qquad x > 0 \tag{2}$$

where

$$k_1 = \left(\frac{2mE}{\hbar^2}\right)^{1/2} \qquad \text{and} \qquad k_2 = \left[\frac{2m(E - V_0)}{\hbar^2}\right]^{1/2} \tag{3}$$

As we learned in Problem 3–35, e^{ikx} represents a particle traveling to the right and e^{-ikx} represents a particle traveling to the left. The physical problem we wish to set up is a particle of energy E traveling to the right and incident on a potential barrier of height V_0. If we wish to exclude the case of a particle traveling to the left in region 2, we set $D = 0$ in equation 2. The squares of the coefficients in equations 1 and 2 represent the probability that the particle is traveling in a certain direction in a given region. For example, $|A|^2$ is the probability that the particle is traveling with momentum $+\hbar k_1$ in the region $x < 0$. If we consider many particles, N_0, instead of just one, then we can interpret $|A|^2 N_0$ to be the number of particles with momentum $\hbar k_1$ in the region $x < 0$. The number of these particles that pass a given point per unit time is given by $v|A|^2 N_0$, where the velocity v is given by $\hbar k_1/m$.

Now apply the conditions that $\psi(x)$ and $d\psi/dx$ must be continuous at $x = 0$ (see Problem 4–4) to obtain

$$A + B = C$$

and

$$k_1(A - B) = k_2 C$$

Now define a quantity

$$r = \frac{\hbar k_1 |B|^2 N_0/m}{\hbar k_1 |A|^2 N_0/m} = \frac{|B|^2}{|A|^2}$$

and show that

$$r = \left(\frac{k_1 - k_2}{k_1 + k_2}\right)^2$$

Similarly, define

$$t = \frac{\hbar k_2 |C|^2 N_0/m}{\hbar k_1 |A|^2 N_0/m} = \frac{k_2 |C|^2}{k_1 |A|^2}$$

and show that

$$t = \frac{4 k_1 k_2}{(k_1 + k_2)^2}$$

The symbols r and t stand for reflection coefficient and transmission coefficient, respectively. Give a physical interpretation of these designations. Show that $r + t = 1$. Would you have expected the particle to have been reflected even though its energy, E, is greater than the barrier height, V_0? Show that $r \to 0$ and $t \to 1$ as $V_0 \to 0$.

4–50. Show that $r = 1$ for the system described in Problem 4–49 but with $E < V_0$. Discuss the physical interpretation of this result.

4–51. In this problem, we introduce the idea of *quantum-mechanical tunneling*, which plays a central role in such diverse processes as the α decay of nuclei, electron-transfer reactions, and hydrogen bonding. Consider a particle in the potential energy regions as shown below.

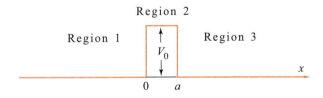

Mathematically, we have

$$V(x) = \begin{cases} 0 & x < 0 \\ V_0 & 0 < x < a \\ 0 & x > a \end{cases}$$

Show that if $E < V_0$, the solution to the Schrödinger equation in each region is given by

$$\psi_1(x) = Ae^{ik_1x} + Be^{-ik_1x} \qquad x < 0 \tag{1}$$

$$\psi_2(x) = Ce^{k_2x} + De^{-k_2x} \qquad 0 < x < a \tag{2}$$

and

$$\psi_3(x) = Ee^{ik_1x} + Fe^{-ik_1x} \qquad x > a \tag{3}$$

where

$$k_1 = \left(\frac{2mE}{\hbar^2}\right)^{1/2} \qquad \text{and} \qquad k_2 = \left[\frac{2m(V_0 - E)}{\hbar^2}\right]^{1/2} \tag{4}$$

If we exclude the situation of the particle coming from large positive values of x, then $F = 0$ in equation 3. Following Problem 4–49, argue that the transmission coefficient, the probability the particle will get past the barrier, is given by

$$t = \frac{|E|^2}{|A|^2} \tag{5}$$

Now use the fact that $\psi(x)$ and $d\psi/dx$ must be continuous at $x = 0$ and $x = a$ to obtain

$$A + B = C + D \qquad ik_1(A - B) = k_2(C - D) \tag{6}$$

and

$$Ce^{k_2a} + De^{-k_2a} = Ee^{ik_1a} \qquad k_2Ce^{k_2a} - k_2De^{-k_2a} = ik_1Ee^{ik_1a} \tag{7}$$

Eliminate B from equations 6 to get A in terms of C and D. Then solve equations 7 for C and D in terms of E. Substitute these results into the equation for A in terms of C and D to get the intermediate result

$$2ik_1A = \left[(k_1^2 - k_2^2 + 2ik_1k_2)e^{k_2a} + (k_2^2 - k_1^2 + 2ik_1k_2)e^{-k_2a}\right]\frac{Ee^{ik_1a}}{2k_2}$$

Now use the relations $\sinh x = (e^x - e^{-x})/2$ and $\cosh x = (e^x + e^{-x})/2$ (Problem A–11) to get

$$\frac{E}{A} = \frac{4ik_1k_2e^{-ik_1a}}{2(k_1^2 - k_2^2)\sinh k_2a + 4ik_1k_2\cosh k_2a}$$

Now multiply the right side by its complex conjugate and use the relation $\cosh^2 x = 1 + \sinh^2 x$ to get

$$t = \left|\frac{E}{A}\right|^2 = \frac{4}{4 + \dfrac{(k_1^2 + k_2^2)^2}{k_1^2k_2^2}\sinh^2 k_2a}$$

Finally, use the definition of k_1 and k_2 to show that the probability the particle gets through the barrier (even though it does not have enough energy!) is

$$t = \cfrac{1}{1 + \cfrac{v_0^2}{4\varepsilon(v_0 - \varepsilon)} \sinh^2(v_0 - \varepsilon)^{1/2}} \tag{8}$$

or

$$t = \cfrac{1}{1 + \cfrac{\sinh^2[v_0^{1/2}(1 - \alpha)^{1/2}]}{4\alpha(1 - \alpha)}} \tag{9}$$

where $v_0 = 2ma^2 V_0/\hbar^2$, $\varepsilon = 2ma^2 E/\hbar^2$, and $\alpha = E/V_0 = \varepsilon/v_0$. Figure 4.7 shows a plot of t versus α for $v_0 = 10$. To plot t versus α for values of $\alpha > 1$, you need to use the relation $\sinh ix = i \sin x$ (Problem A–12). What would the classical result look like?

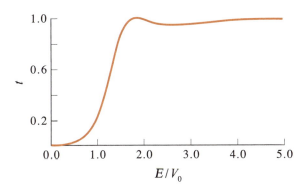

FIGURE 4.7
A plot of the probability that a particle of energy E will penetrate a barrier of height V_0 plotted against the ratio E/V_0 (equation 9 of Problem 4–51 with $v_0 = 10$).

4–52. Use the result of Problem 4–51 to determine the probability that an electron with a kinetic energy 8.0×10^{-21} J will tunnel through a 1.0 nm thick potential barrier with $V_0 = 12.0 \times 10^{-21}$ J.

4–53. Problem 4–51 shows that the probability that a particle of relative energy E/V_0 will penetrate a rectangular potential barrier of height V_0 and thickness a is

$$t = \cfrac{1}{1 + \cfrac{\sinh^2[v_0^{1/2}(1 - \alpha)^{1/2}]}{4\alpha(1 - \alpha)}}$$

where $v_0 = 2m V_0 a^2/\hbar^2$ and $\alpha = E/V_0$. What is the limit of t as $\alpha \to 1$? Plot t against α for $v_0 = 1/2$, 1, and 2. Interpret your results.

4–54. In this problem, we will consider a particle in a *finite* potential well,

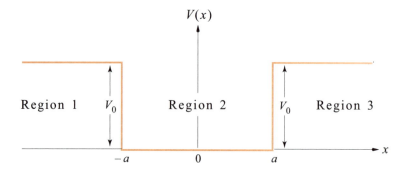

whose mathematical form is

$$V(x) = \begin{cases} V_0 & x < -a \\ 0 & -a < x < a \\ V_0 & x > a \end{cases} \tag{1}$$

Note that this potential describes what we have called a "particle in a box" if $V_0 \to \infty$. Show that if $0 < E < V_0$, the solution to the Schrödinger equation in each region is

$$\psi_1(x) = Ae^{k_1 x} \quad x < -a$$

$$\psi_2(x) = B \sin \alpha x + C \cos \alpha x \qquad -a < x < a \tag{2}$$

$$\psi_3(x) = De^{-k_1 x} \quad x > a$$

where

$$k_1 = \left[\frac{2m(V_0 - E)}{\hbar^2} \right]^{1/2} \quad \text{and} \quad \alpha = \left(\frac{2mE}{\hbar^2} \right)^{1/2} \tag{3}$$

Now apply the conditions that $\psi(x)$ and $d\psi/dx$ must be continuous at $x = -a$ and $x = a$ to obtain

$$Ae^{-k_1 a} = -B \sin \alpha a + C \cos \alpha a \tag{4}$$

$$De^{-k_1 a} = B \sin \alpha a + C \cos \alpha a \tag{5}$$

$$k_1 Ae^{-k_1 a} = \alpha B \cos \alpha a + \alpha C \sin \alpha a \tag{6}$$

and

$$-k_1 De^{-k_1 a} = \alpha B \cos \alpha a - \alpha C \sin \alpha a \tag{7}$$

Add and subtract equations 4 and 5 and add and subtract equations 6 and 7 to obtain

$$2C \cos \alpha a = (A + D)e^{-k_1 a} \tag{8}$$

$$2B \sin \alpha a = (D - A)e^{-k_1 a} \tag{9}$$

$$2\alpha C \sin \alpha a = k_1(A + D)e^{-k_1 a} \tag{10}$$

and

$$2\alpha B \cos \alpha a = -k_1(D - A)e^{-k_1 a} \tag{11}$$

Now divide equation 10 by equation 8 to get

$$\frac{\alpha \sin \alpha a}{\cos \alpha a} = \alpha \tan \alpha a = k_1 \qquad (D \neq -A \text{ and } C = 0) \tag{12}$$

and then divide equation 11 by equation 9 to get

$$\frac{\alpha \cos \alpha a}{\sin \alpha a} = \alpha \cot \alpha a = -k_1 \qquad (D \neq A \text{ and } B = 0) \tag{13}$$

Referring back to equation 3, note that equations 12 and 13 give the allowed values of E in terms of V_0. It turns out that these two equations cannot be solved simultaneously, so we have two sets of equations:

$$\alpha \tan \alpha a = k_1 \tag{14}$$

and

$$\alpha \cot \alpha a = -k_1 \tag{15}$$

Let's consider equation 14 first. Multiply both sides by a and use the definitions of α and k_1 to get

$$\left(\frac{2ma^2 E}{\hbar^2}\right)^{1/2} \tan \left(\frac{2ma^2 E}{\hbar^2}\right)^{1/2} = \left[\frac{2ma^2}{\hbar^2}(V_0 - E)\right]^{1/2} \tag{16}$$

Show that this equation simplifies to

$$\varepsilon^{1/2} \tan \varepsilon^{1/2} = (v_0 - \varepsilon)^{1/2} \tag{17}$$

where $\varepsilon = 2ma^2 E/\hbar^2$ and $v_0 = 2ma^2 V_0/\hbar^2$. Thus, if we fix v_0 (actually $2ma^2 V_0/\hbar^2$), then we can use equation 17 to solve for the allowed values of ε (actually $2ma^2 E/\hbar^2$). Equation 17 cannot be solved analytically, but if we plot both $\varepsilon^{1/2} \tan \varepsilon^{1/2}$ and $(v_0 - \varepsilon)^{1/2}$ versus ε on the same graph, then the solutions are given by the intersections of the two curves. Figure 4.8a shows such a plot for $v_0 = 12$.

The intersections occur at $\varepsilon = 2ma^2 E/\hbar^2 = 1.47$ and 11.37. The other value(s) of ε are given by the solutions to equation 15, which are obtained by finding the intersection of $-\varepsilon^{1/2} \cot \varepsilon^{1/2}$ and $(v_0 - \varepsilon)^{1/2}$ plotted against ε. Such a plot is shown in Figure 4.8b for $v_0 = 12$, giving $\varepsilon = 2ma^2 E/\hbar^2 = 5.68$. Thus, we see there are only three bound states for a well of depth $V_0 = 12\hbar^2/2ma^2$. The important point here is not the numerical values of E, but the fact that there is only a finite number of bound states. Show that there are only two bound states for $v_0 = 2ma^2 V_0/\hbar^2 = 4$.

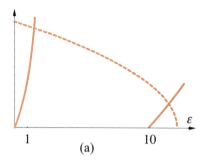
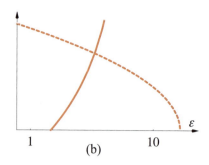

FIGURE 4.8
(a) Plots of both $\varepsilon^{1/2} \tan \varepsilon^{1/2}$ (solid curve) and $(12 - \varepsilon)^{1/2}$ (dotted curve) versus ε. The intersections of the curves give the allowed values of ε for a one-dimensional potential well of depth $V_0 = 12\hbar^2/2ma^2$. (b) Plots of both $-\varepsilon^{1/2} \cot \varepsilon^{1/2}$ (solid curve) and $(12 - \varepsilon)^{1/2}$ (dotted curve) plotted against ε. The intersection gives an allowed value of ε for a one-dimensional potential well of depth $V_0 = 12\hbar^2/2ma^2$.

References

Cassidy, D. *Uncertainty: The Life and Science of Werner Heisenberg*. W. H. Freeman: New York, 1993.

Whitaker, A. *Einstein, Bohr, and the Quantum Dilemma*. Cambridge University Press: Cambridge, UK, 1996.

Al-Khalili, J. *Quantum: A Guide for the Perplexed*. Weidenfeld & Nicolson: London, 2004.

Herbert, N. *Quantum Reality Beyond the New Physics*. Anchor/Doubleday: New York, 1987.

Gribbin, J. *In Search of Schrödinger's Cat: Quantum Physics and Reality*. Bantam Books: New York, 1984.

Davies, P. C. W., Brown, J. R., editors. *The Ghost in the Atom: A Discussion of the Mysteries of Quantum Physics*. Cambridge University Press: New York, 1993.

Greenstein, G., Zajone, A. *The Quantum Challenge,* 2nd ed. Jones and Bartett Publishers: Sudbury, MA, 2005.

Kahn, P. A., Hansen, E. H. The Dirac (Bracket) Notation in the Undergraduate Physical Curriculum: A Pictorial Introduction. *Chem. Educator,* **5**, 113 (2000).

Series and Limits

Frequently, we need to investigate the behavior of an equation for small values (or perhaps large values) of one of the variables in the equation. For example, in Chapter 1 we derived the low-frequency behavior of the Planck blackbody distribution law (Equation 1.2):

$$\rho_\nu(T)d\nu = \frac{8\pi h}{c^3}\frac{\nu^3 d\nu}{e^{\beta h\nu} - 1} \tag{D.1}$$

To do this, we used the fact that e^x can be written as the infinite series (i.e., a series containing an unending number of terms)

$$e^x = \sum_{n=0}^{\infty}\frac{x^n}{n!} = 1 + x + \frac{x^2}{2!} + \frac{x^3}{3!} + \cdots \tag{D.2}$$

and then realized if x is small, then x^2, x^3, etc., are even smaller. We can express this result by writing

$$e^x = 1 + x + O(x^2)$$

where $O(x^2)$ is a bookkeeping device that reminds us we are neglecting terms involving x^2 and higher powers of x. If we apply this result to Equation D.1, we have

$$\rho_\nu(T)d\nu = \frac{8\pi h}{c^3}\frac{\nu^3 d\nu}{1 + \beta h\nu + O[(\beta h\nu)^2] - 1}$$

$$\approx \frac{8\pi h}{c^3}\frac{\nu^3 d\nu}{\beta h\nu}$$

$$= \frac{8\pi k_B T}{c^3}\nu^2 d\nu$$

Thus, we see that $\rho_v(T)$ goes as v^2 for small values of v. In this MathChapter, we will review some useful series and apply them to some physical problems.

One of the most useful series we will use is the geometric series:

$$\frac{1}{1-x} = \sum_{n=0}^{\infty} x^n = 1 + x + x^2 + x^3 + \cdots \qquad |x| < 1 \qquad (D.3)$$

This result can be derived by algebraically dividing 1 by $1 - x$, or by the following trick. Consider the finite series (i.e., a series with a finite number of terms)

$$S_N = 1 + x + x^2 + \cdots + x^N$$

Now multiply S_N by x:

$$x S_N = x + x^2 + \cdots + x^{N+1}$$

Now notice that

$$S_N - x S_N = 1 - x^{N+1}$$

or that

$$S_N = \frac{1 - x^{N+1}}{1 - x} \qquad (D.4)$$

If $|x| < 1$, then $x^{N+1} \rightarrow 0$ as $N \rightarrow \infty$, so we recover Equation D.3.

Recovering Equation D.3 from Equation D.4 brings us to an important point regarding infinite series: Equation D.3 is valid only if $|x| < 1$. It makes no sense at all if $|x| \geq 1$. We say that the infinite series in Equation D.3 converges for $|x| < 1$ and diverges for $|x| \geq 1$. How can we tell whether a given infinite series converges or diverges? There are a number of so-called convergence tests, but one simple and useful one is the *ratio test*. To apply the ratio test, we form the ratio of the $(n + 1)$th term, u_{n+1}, to the nth term, u_n, and then let n become very large:

$$r = \lim_{n \to \infty} \left| \frac{u_{n+1}}{u_n} \right| \qquad (D.5)$$

If $r < 1$, the series converges; if $r > 1$, the series diverges; and if $r = 1$, the test is inconclusive. Let's apply this test to the geometric series (Equation D.3):

$$r = \lim_{n \to \infty} \left| \frac{x^{n+1}}{x^n} \right| = |x|$$

Thus, we see that the series converges if $|x| < 1$ and diverges if $|x| > 1$. It actually diverges at $x = 1$, but the ratio test does not tell us that. We would have to use a more sophisticated convergence test to determine the behavior at $x = 1$.

For the exponential series (Equation D.2), we have

$$r = \lim_{n\to\infty} \left| \frac{x^{n+1}/(n+1)!}{x^n/n!} \right| = \lim_{n\to\infty} \left| \frac{x}{n+1} \right|$$

Thus, we conclude that the exponential series converges for all values of x.

In Chapter 5, we encounter the summation

$$S = \sum_{n=0}^{\infty} e^{-nh\nu/k_BT} \tag{D.6}$$

where ν represents the vibrational frequency of a diatomic molecule and the other symbols have their usual meanings. We can sum this series by letting

$$x = e^{-h\nu/k_BT}$$

in which case we have

$$S = \sum_{n=0}^{\infty} x^n \qquad |x| < 1$$

According to Equation D.3, $S = 1/(1-x)$, or

$$S = \frac{1}{1 - e^{-h\nu/k_BT}} \tag{D.7}$$

We say that S has been evaluated in closed form because its numerical evaluation requires only a finite number of steps, in contrast to Equation D.6, which would require an infinite number of steps.

A practical question that arises is how we find the infinite series that corresponds to a given function. For example, how do we derive Equation D.2? First, assume that the function $f(x)$ can be expressed as a power series (i.e., a series in powers of x):

$$f(x) = c_0 + c_1 x + c_2 x^2 + c_3 x^3 + \cdots$$

where the c_j are to be determined. Then let $x = 0$ and find that $c_0 = f(0)$. Now differentiate once with respect to x,

$$\frac{df}{dx} = c_1 + 2c_2 x + 3c_3 x_2 + \cdots$$

and let $x = 0$ to find that $c_1 = (df/dx)_{x=0}$. Differentiate again,

$$\frac{d^2 f}{dx^2} = 2c_2 + 3 \cdot 2c_3 x + \cdots$$

and let $x = 0$ to get $c_2 = (d^2 f/dx^2)_{x=0}/2$. Differentiate once more,

$$\frac{d^3 f}{dx^3} = 3 \cdot 2c_3 + 4 \cdot 3 \cdot 2x + \cdots$$

and let $x = 0$ to get $c_3 = (d^3 f/dx^3)_{x=0}/3!$. The general result is

$$c_n = \frac{1}{n!} \left(\frac{d^n f}{dx^n} \right)_{x=0} \tag{D.8}$$

so we can write

$$f(x) = f(0) + \left(\frac{df}{dx} \right)_{x=0} x + \frac{1}{2!} \left(\frac{d^2 f}{dx^2} \right)_{x=0} x^2 + \frac{1}{3!} \left(\frac{d^3 f}{dx^3} \right)_{x=0} x^3 + \cdots \tag{D.9}$$

Equation D.9 is called the Maclaurin series of $f(x)$. If we apply Equation D.9 to $f(x) = e^x$, we find that

$$\left(\frac{d^n e^x}{dx^n} \right)_{x=0} = 1$$

so

$$e^x = 1 + x + \frac{x^2}{2!} + \frac{x^3}{3!} + \cdots$$

Some other important Maclaurin series, which can be obtained from a straightforward application of Equation D.9 (Problem D–13) are

$$\sin x = x - \frac{x^3}{3!} + \frac{x^5}{5!} - \frac{x^7}{7!} + \cdots \tag{D.10}$$

$$\cos x = 1 - \frac{x^2}{2!} + \frac{x^4}{4!} - \frac{x^6}{6!} + \cdots \tag{D.11}$$

$$\ln(1 + x) = x - \frac{x^2}{2} + \frac{x^3}{3} - \frac{x^4}{4} + \cdots \qquad -1 < x \le 1 \tag{D.12}$$

and

$$(1 + x)^n = 1 + nx + \frac{n(n-1)}{2!} x^2 + \frac{n(n-1)(n-2)}{3!} x^3 + \cdots \qquad x^2 < 1 \tag{D.13}$$

Series D.10 and D.11 converge for all values of x, but as indicated, Series D.12 converges only for $-1 < x \le 1$ and Series D.13 converges only for $x^2 < 1$. Note that if n is a positive integer in Series D.13, the series truncates. For example, if $n = 2$ or 3, we have

$$(1 + x)^2 = 1 + 2x + x^2$$

and

$$(1+x)^3 = 1 + 3x + 3x^2 + x^3$$

Equation D.13 for a positive integer is called the binomial expansion. If n is not a positive integer, the series continues indefinitely, and Equation D.13 is called the binomial series. For example,

$$(1+x)^{1/2} = 1 + \frac{x}{2} - \frac{1}{8}x^2 + O(x^3) \tag{D.14}$$

$$(1+x)^{-1/2} = 1 - \frac{x}{2} + \frac{3}{8}x^2 + O(x^3) \tag{D.15}$$

Any handbook of mathematical tables will have the Maclaurin series for many functions. Problem D–20 discusses a Taylor series, which is an extension of a Maclaurin series.

We can use the series presented here to derive a number of results used throughout the book. For example, the limit

$$\lim_{x \to 0} \frac{\sin x}{x}$$

occurs several times. Because this limit gives 0/0, we could use l'Hôpital's rule, which tells us that

$$\lim_{x \to 0} \frac{\sin x}{x} = \lim_{x \to 0} \frac{\dfrac{d \sin x}{dx}}{\dfrac{dx}{dx}} = \lim_{x \to 0} \cos x = 1$$

We could derive the same result by dividing Equation D.10 by x and then letting $x \to 0$. (These two methods are really equivalent. See Problem D–21.)

We will do one final example involving series and limits. Einstein's theory of the temperature dependence of the molar heat capacity of a crystal is given by

$$\overline{C}_V = 3R \left(\frac{\Theta_E}{T} \right)^2 \frac{e^{-\Theta_E/T}}{(1 - e^{-\Theta_E/T})^2} \tag{D.16}$$

where R is the molar gas constant and Θ_E is a constant, called the Einstein constant, that is characteristic of the solid (cf. Section 1.4). We'll now show that this equation gives the Dulong and Petit limit ($\overline{C}_V \to 3R$) at high temperatures. First let $x = \Theta_E/T$ in Equation D.16 to obtain

$$\overline{C}_V = 3Rx^2 \frac{e^{-x}}{(1 - e^{-x})^2} \tag{D.17}$$

When T is large, x is small, and so we shall use

$$e^{-x} = 1 - x + O(x^2)$$

Equation D.17 becomes

$$\overline{C}_V = 3Rx^2 \frac{1 - x + O(x^2)}{(x + O(x^2))^2} \longrightarrow 3R$$

as $x \to 0$ ($T \to \infty$). This result is called the law of Dulong and Petit; the molar heat capacity of a crystal becomes $3R = 24.9$ J·K^{-1}·mol^{-1} for a monatomic crystal at high temperatures. By "high temperatures" we actually mean that $T \gg \Theta_E$, which for many substances is less than 1000 K.

Problems

D–1. Calculate the percentage difference between e^x and $1 + x$ for $x = 0.0050, 0.0100, 0.0150, \ldots, 0.1000$.

D–2. Calculate the percentage difference between $\ln(1 + x)$ and x for $x = 0.0050, 0.0100, 0.0150, \ldots, 0.1000$.

D–3. Write out the expansion of $(1 + x)^{1/2}$ through the quadratic term.

D–4. Write out the expansion of $(1 + x)^{-1/2}$ through the quadratic term.

D–5. Show that

$$\frac{1}{(1 - x)^2} = 1 + 2x + 3x^2 + 4x^3 + \cdots$$

D–6. Evaluate the series

$$S = \frac{1}{2} + \frac{1}{4} + \frac{1}{8} + \frac{1}{16} + \cdots$$

D–7. Evaluate the series

$$S = \sum_{n=0}^{\infty} \frac{1}{3^n}$$

D–8. Evaluate the series

$$S = \sum_{n=1}^{\infty} \frac{(-1)^{n+1}}{2^n}$$

D–9. Numbers whose decimal formula are recurring decimals such as $0.272\,727\ldots$ are rational numbers, meaning that they can be expressed as the ratio of two numbers (in other words, as a fraction). Show that $0.272\,727\ldots = 27/99$.

D–10. Show that $0.142\,857\,142\,857\,142\,857\ldots = 1/7$. (See the previous problem.)

D–11. Series of the form

$$S(x) = \sum_{n=0}^{\infty} nx^n$$

occur frequently in physical problems. To find a closed expression for $S(x)$, we start with

$$\frac{1}{1-x} = \sum_{n=0}^{\infty} x^n$$

Notice now that $S(x)$ can be expressed as

$$x\frac{d}{dx}\sum_{n=0}^{\infty} x^n = \sum_{n=0}^{\infty} nx^n$$

and show that $S(x) = x/(1-x)^2$.

D–12. Using the method introduced in the previous problem, show that

$$S(x) = \sum_{n=0}^{\infty} n^2x^n = \frac{x(1+x)}{(1-x)^3}$$

D–13. Use Equation D.9 to derive Equations D.10 and D.11.

D–14. Show that Equations D.2, D.10, and D.11 are consistent with the relation $e^{ix} = \cos x + i\sin x$.

D–15. Use Equation D.2 and the definitions

$$\sinh x = \frac{e^x - e^{-x}}{2} \qquad \text{and} \qquad \cosh x = \frac{e^x + e^{-x}}{2}$$

to show that

$$\sinh x = x + \frac{x^3}{3!} + \frac{x^5}{5!} + \cdots$$

$$\cosh x = 1 + \frac{x^2}{2!} + \frac{x^4}{4!} + \cdots$$

D–16. Show that Equations D.10 and D.11 and the results of the previous problem are consistent with the relations

$$\sin ix = i\sinh x \qquad\qquad \cos ix = \cosh x$$

$$\sinh ix = i\sin x \qquad\qquad \cosh ix = \cos x$$

D–17. Evaluate the limit of

$$f(x) = \frac{e^{-x} \sin^2 x}{x^2}$$

as $x \to 0$.

D–18. Evaluate the integral

$$I = \int_0^a x^2 e^{-x} \cos^2 x \, dx$$

for small values of a by expanding I in powers of a through quadratic terms.

D–19. Prove that the series for $\sin x$ converges for all values of x.

D–20. A Maclaurin series is an expansion about the point $x = 0$. A series of the form

$$f(x) = c_0 + c_1(x - x_0) + c_2(x - x_0)^2 + \cdots$$

is an expansion about the point x_0 and is called a Taylor series. First show that $c_0 = f(x_0)$. Now differentiate both sides of the above expansion with respect to x and then let $x = x_0$ to show that $c_1 = (df/dx)_{x=x_0}$. Now show that

$$c_n = \frac{1}{n!} \left(\frac{d^n f}{dx^n} \right)_{x=x_0}$$

and so

$$f(x) = f(x_0) + \left(\frac{df}{dx} \right)_{x=x_0} (x - x_0) + \frac{1}{2} \left(\frac{d^2 f}{dx^2} \right)_{x=x_0} (x - x_0)^2 + \cdots$$

D–21. Show that l'Hôpital's rule amounts to forming a Taylor expansion of both the numerator and the denominator. Evaluate the limit

$$\lim_{x \to 0} \frac{\ln(1 + x) - x}{x^2}$$

both ways.

D–22. Start with

$$\frac{1}{1 - x} = 1 + x + x^2 + \cdots$$

Now let $x = 1/x$ to write

$$\frac{1}{1 - \frac{1}{x}} = \frac{x}{x - 1} = 1 + \frac{1}{x} + \frac{1}{x^2} + \cdots$$

Now add these two expressions to get

$$1 = \cdots + \frac{1}{x^2} + \frac{1}{x} + 2 + x + x^2 + \cdots$$

Does this make sense? What went wrong?

D–23. The energy of a quantum-mechanical harmonic oscillator is given by $\varepsilon_n = (n + \frac{1}{2})h\nu$, $n = 0, 1, 2, \ldots$, where h is the Planck constant and ν is the fundamental frequency of the oscillator. The average vibrational energy of a harmonic oscillator in an ideal gas is given by

$$\varepsilon_{\text{vib}} = (1 - e^{-h\nu/k_{\text{B}}T}) \sum_{n=0}^{\infty} \varepsilon_n e^{-nh\nu/k_{\text{B}}T}$$

where k_{B} is the Boltzmann constant and T is the kelvin temperature. Show that

$$\varepsilon_{\text{vib}} = \frac{h\nu}{2} + \frac{h\nu e^{-h\nu/k_{\text{B}}T}}{1 - e^{-h\nu/k_{\text{B}}T}}$$

E. Bright Wilson, Jr. was born on December 18, 1908, in Gallatin, Tennessee, and died in 1992. Wilson received his Ph.D. in 1933 from the California Institute of Technology, where he studied with Linus Pauling. In 1934, he went to Harvard University as a junior fellow and became a full professor just three years later. He was the Theodore Richards Professor of Chemistry from 1948 until his formal retirement in 1979. Wilson's experimental and theoretical work in microwave spectroscopy contributed to the understanding of the structure and dynamics of molecules. During World War II, he directed underwater explosives research at Woods Hole, Massachusetts. In the early 1950s, he spent a year at the Pentagon as a research director of the Weapons System Evaluation Group. In later years, he served on and chaired committees of the National Research Council, seeking solutions to various environmental problems. Wilson wrote three books, all of which became classics. His book *Introduction to Quantum Mechanics*, written with Linus Pauling in 1935, was used by almost all physical chemistry graduate students for 20 years, and *Molecular Vibrations: The Theory of Infrared and Raman Vibrational Spectra*, written with J. C. Decius and Paul Cross, was a standard reference for most of a generation of physical chemists. His *An Introduction to Scientific Research* is a model for both substance and clarity. One of his sons, Kenneth, was awarded the Nobel Prize in Physics in 1982.

CHAPTER 5

The Harmonic Oscillator and Vibrational Spectroscopy

The vibration of a diatomic molecule can be described by a harmonic oscillator. In this chapter, we shall first study a classical harmonic oscillator and then present and discuss the energies and the corresponding wave functions of a quantum-mechanical harmonic oscillator. We shall use the quantum-mechanical energies to describe the infrared spectrum of a diatomic molecule and learn how to determine molecular force constants. Then we shall discuss selection rules for a harmonic oscillator, and finally normal coordinates, which describe the vibrational motion of polyatomic molecules.

5.1 A Harmonic Oscillator Obeys Hooke's Law

Consider a mass m connected to a wall by a spring, as shown in Figure 5.1.

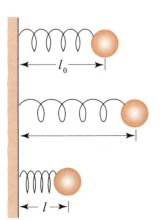

FIGURE 5.1
A mass connected to a wall by a spring. If the force acting upon the mass is directly proportional to the displacement of the spring from its undistorted length, then the force law is called Hooke's law.

Suppose further that no gravitational force is acting on m so that the only force is due to the spring. If we let l_0 be the equilibrium, or undistorted, length of the spring, then the restoring force must be some function of the displacement of the spring from its equilibrium length. Let this displacement be denoted by $x = l - l_0$, where l is the length of the spring. The simplest assumption we can make about the force on m as a function of the displacement is that the force is directly proportional to the displacement and to write

$$f = -k(l - l_0) = -kx \tag{5.1}$$

The negative sign indicates that the force points to the right in Figure 5.1 if the spring is compressed ($l < l_0$) and points to the left if the spring is stretched ($l > l_0$). Equation 5.1 is called *Hooke's law* and the (positive) proportionality constant k is called the *force constant* of the spring. A small value of k implies a weak or loose spring, and a large value of k implies a stiff spring.

Newton's equation with a Hooke's law force is

$$m\frac{d^2l}{dt^2} = -k(l - l_0) \tag{5.2}$$

If we let $x = l - l_0$, then $d^2l/dt^2 = d^2x/dt^2$ (l_0 is a constant) and we have

$$m\frac{d^2x}{dt^2} + kx = 0 \tag{5.3}$$

According to Section 2.3, the general solution to this equation is (Problem 5–2)

$$x(t) = c_1 \sin \omega t + c_2 \cos \omega t \tag{5.4}$$

where

$$\omega = \left(\frac{k}{m}\right)^{1/2} \tag{5.5}$$

EXAMPLE 5–1
Show that Equation 5.4 can be written in the form

$$x(t) = A \sin(\omega t + \phi) \tag{5.6}$$

SOLUTION: The easiest way to prove this is to write

$$\sin(\omega t + \phi) = \sin \omega t \cos \phi + \cos \omega t \sin \phi$$

and substitute this into Equation 5.6 to obtain

$$x(t) = A \cos \phi \sin \omega t + A \sin \phi \cos \omega t$$

$$= c_1 \sin \omega t + c_2 \cos \omega t$$

where

$$c_1 = A \cos \phi \qquad \text{and} \qquad c_2 = A \sin \phi$$

Equation 5.6 shows that the displacement oscillates sinusoidally, or *harmonically*, with a natural frequency $\omega = (k/m)^{1/2}$. In Equation 5.6, A, the maximum displacement, is the *amplitude* of the vibration and ϕ is the *phase angle*.

Suppose we stretch the spring so that its displacement is A and then let go. The initial velocity in this case is zero and so from Equation 5.4, we have

$$x(0) = c_2 = A$$

and

$$\left(\frac{dx}{dt} \right)_{t=0} = 0 = c_1 \omega$$

These two equations imply that $c_1 = 0$ and $c_2 = A$ in Equation 5.4, and so

$$x(t) = A \cos \omega t \qquad (5.7)$$

The displacement versus time is plotted in Figure 5.2, which shows that the mass oscillates back and forth between A and $-A$ with a frequency ω radians per second, or $\nu = \omega/2\pi$ cycles per second. The quantity A is called the *amplitude* of the vibration.

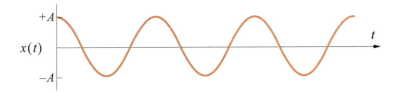

FIGURE 5.2
An illustration of the displacement of a harmonic oscillator versus time.

5.2 The Energy of a Harmonic Oscillator Is Conserved

Let's look at the total energy of a harmonic oscillator. The force is given by Equation 5.1. Recall from physics that a force can be expressed as a derivative of a potential energy or that

$$f(x) = -\frac{dV}{dx} \tag{5.8}$$

so that the potential energy is

$$V(x) = -\int f(x)dx + \text{constant} \tag{5.9}$$

Using Equation 5.1 for $f(x)$, we see that

$$V(x) = \frac{k}{2}x^2 + \text{constant} \tag{5.10}$$

The constant term here is an arbitrary constant that can be used to fix the zero of energy. If we choose the potential energy of the system to be zero when the spring is undistorted ($x = 0$), then we have

$$V(x) = \frac{k}{2}x^2 \tag{5.11}$$

for the potential energy associated with a simple harmonic oscillator.

The kinetic energy is

$$T = \frac{1}{2}m\left(\frac{dl}{dt}\right)^2 = \frac{1}{2}m\left(\frac{dx}{dt}\right)^2 \tag{5.12}$$

Using Equation 5.7 for $x(t)$, we see that

$$T = \frac{1}{2}m\omega^2 A^2 \sin^2 \omega t \tag{5.13}$$

and

$$V = \frac{1}{2}kA^2 \cos^2 \omega t \tag{5.14}$$

Both T and V are plotted in Figure 5.3. The total energy is

$$E = T + V = \frac{1}{2}m\omega^2 A^2 \sin^2 \omega t + \frac{1}{2}kA^2 \cos^2 \omega t$$

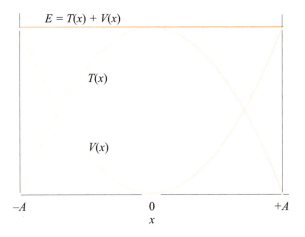

FIGURE 5.3
The kinetic energy [curve labeled $T(x)$] and the potential energy [curve labeled $V(x)$] of a harmonic oscillator during one oscillation. The spring is fully compressed at $-A$ and fully stretched at $+A$. The equilibrium length is $x = 0$. The total energy is the horizontal curve labelled E, which is the sum of $T(x)$ and $V(x)$.

If we recall that $\omega = (k/m)^{1/2}$, we see that the coefficient of the first term is $kA^2/2$, so that the total energy becomes

$$E = \frac{kA^2}{2}(\sin^2 \omega t + \cos^2 \omega t)$$

$$= \frac{kA^2}{2} \tag{5.15}$$

Thus, we see that the total energy is a constant and, in particular, is equal to the potential energy at its largest displacement, where the kinetic energy is zero. Figure 5.3 shows how the total energy is distributed between the kinetic energy and the potential energy. Each oscillates between $+A$ and $-A$ but in such a way that their sum is always a constant. We say that the total energy is conserved and that the system is a *conservative system*.

EXAMPLE 5–2
Using the more general equation

$$x(t) = C \sin(\omega t + \phi)$$

prove that the total energy of a harmonic oscillator is

$$E = \frac{k}{2}C^2$$

SOLUTION:

$$E = T + V = \frac{1}{2}m\left(\frac{dx}{dt}\right)^2 + \frac{1}{2}kx^2$$

$$= \frac{m}{2}\omega^2 C^2 \cos^2(\omega t + \phi) + \frac{k}{2}C^2 \sin^2(\omega t + \phi)$$

Using the fact that $\omega^2 = k/m$, we have

$$E = \frac{k}{2}C^2[\cos^2(\omega t + \phi) + \sin^2(\omega t + \phi)] = \frac{k}{2}C^2$$

The concept of a conservative system is important and is worth discussing in more detail. For a system to be conservative, the force must be derivable from a potential energy function that is a function of only the spatial coordinates describing the system. In the case of a simple harmonic oscillator, $V(x, y, z)$ is given by Equation 5.11 in three dimensions, and the force is given by Equation 5.8. For a single particle in three dimensions, we have $V = V(x, y, z)$ and

$$f_x(x, y, z) = -\frac{\partial V}{\partial x}$$

$$f_y(x, y, z) = -\frac{\partial V}{\partial y}$$

$$f_z(x, y, z) = -\frac{\partial V}{\partial z}$$

or, in vector notation

$$\mathbf{f}(x, y, z) = -\nabla\, V(x, y, z) \tag{5.16}$$

where ∇ is the gradient operator, defined by (MathChapter C)

$$\nabla = \mathbf{i}\frac{\partial}{\partial x} + \mathbf{j}\frac{\partial}{\partial y} + \mathbf{k}\frac{\partial}{\partial z} \tag{5.17}$$

To prove that Equation 5.16 implies that the system is conservative, consider the one-dimensional case for simplicity. In this case, Newton's equation is

$$m\frac{d^2x}{dt^2} = -\frac{dV}{dx} \tag{5.18}$$

If we integrate both sides of Equation 5.18, then the right side becomes

$$\int -\frac{dV}{dx}\,dx = -V(x) + \text{constant} \tag{5.19}$$

and the left side becomes

$$\int m\frac{d^2x}{dt^2}\,dx = m\int \frac{d^2x}{dt^2}\frac{dx}{dt}\,dt$$
$$= \frac{m}{2}\int \frac{d}{dt}\left(\frac{dx}{dt}\right)^2\,dt = \frac{m}{2}\left(\frac{dx}{dt}\right)^2 + \text{constant}$$

(5.20)

By equating Equations 5.19 and 5.20, we find that

$$\frac{m}{2}\left(\frac{dx}{dt}\right)^2 + V(x) = \text{constant} \tag{5.21}$$

or that the total energy is conserved. Thus, we see that if the force can be expressed as the derivative of a potential energy that is a function of the spatial coordinates only, then the system is conservative.

5.3 The Equation for a Harmonic-Oscillator Model of a Diatomic Molecule Contains the Reduced Mass of the Molecule

The simple harmonic oscillator is a good model for a vibrating diatomic molecule. A diatomic molecule, however, does not look like the system pictured in Figure 5.1, but more like two masses connected by a spring, as in Figure 5.4. In this case we have two equations of motion, one for each mass:

$$m_1\frac{d^2x_1}{dt^2} = k(x_2 - x_1 - l_0) \tag{5.22}$$

and

$$m_2\frac{d^2x_2}{dt^2} = -k(x_2 - x_1 - l_0) \tag{5.23}$$

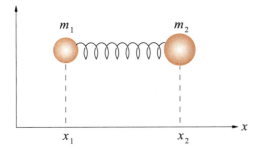

FIGURE 5.4
Two masses connected by a spring, which is a model used to describe the vibrational motion of a diatomic molecule.

where l_0 is the undistorted length of the spring. Note that if $x_2 - x_1 > l_0$, the spring is stretched and the force on mass m_1 is toward the right and that on mass m_2 is toward the left. This is why the force term in Equation 5.22 is positive and that in Equation 5.23 is negative. Note also that the force on m_1 is equal and opposite to the force on m_2, as it should be according to Newton's third law (action and reaction).

If we add Equations 5.22 and 5.23, we find that

$$\frac{d^2}{dt^2}(m_1 x_1 + m_2 x_2) = 0 \tag{5.24}$$

This form suggests that we introduce a *center-of-mass coordinate*

$$X = \frac{m_1 x_1 + m_2 x_2}{M} \tag{5.25}$$

where $M = m_1 + m_2$, so that we can write Equation 5.24 in the form

$$M\frac{d^2 X}{dt^2} = 0 \tag{5.26}$$

There is no force term here, so Equation 5.26 shows that the center of mass moves uniformly in time with a constant momentum (Problem 5–1).

The vibrational motion of the two-mass or two-body system in Figure 5.4 must depend upon only the *relative* separation of the two masses, or upon the *relative coordinate*

$$x = x_2 - x_1 - l_0 \tag{5.27}$$

If we divide Equation 5.23 by m_2 and subtract Equation 5.22 divided by m_1, we find that

$$\frac{d^2 x_2}{dt^2} - \frac{d^2 x_1}{dt^2} = -\frac{k}{m_2}(x_2 - x_1 - l_0) - \frac{k}{m_1}(x_2 - x_1 - l_0)$$

or

$$\frac{d^2}{dt^2}(x_2 - x_1) = -k\left(\frac{1}{m_1} + \frac{1}{m_2}\right)(x_2 - x_1 - l_0)$$

If we let

$$\frac{1}{m_1} + \frac{1}{m_2} = \frac{m_1 + m_2}{m_1 m_2} = \frac{1}{\mu}$$

and introduce $x = x_2 - x_1 - l_0$ from Equation 5.27, then we have

$$\mu\frac{d^2 x}{dt^2} + kx = 0 \tag{5.28}$$

The quantity μ that we have defined is called the *reduced mass*.

Equation 5.28 is an important result with a nice physical interpretation. If we compare Equation 5.28 with Equation 5.3, we see that Equation 5.28 is the same except for the substitution of the reduced mass μ. Thus, the two-body system in Figure 5.4 can be treated as easily as the one-body problem in Figure 5.1 by using the reduced mass of the two-body system. In particular, the motion of the system is governed by Equation 5.6 but with $\omega = (k/\mu)^{1/2}$. Generally, if the potential energy depends upon only the *relative* distance between two bodies, then we can introduce relative coordinates such as $x_2 - x_1$ and reduce a two-body problem to a one-body problem. This important and useful theorem of classical mechanics is discussed in Problems 5–6 and 5–7.

5.4 The Harmonic-Oscillator Approximation Results from the Expansion of an Internuclear Potential Around Its Minimum

Before we discuss the quantum-mechanical treatment of a harmonic oscillator, we should discuss how good an approximation it is for a vibrating diatomic molecule. The internuclear potential for a diatomic molecule is illustrated by the solid line in Figure 5.5. Notice that the curve rises steeply to the left of the minimum due to the difficulty of pushing the two nuclei closer together. The curve to the right side of the equilibrium position rises initially but eventually levels off. The potential energy at large separations is essentially the bond energy. The dashed line shows the potential $\frac{1}{2}k(l - l_0)^2$ associated with Hooke's law.

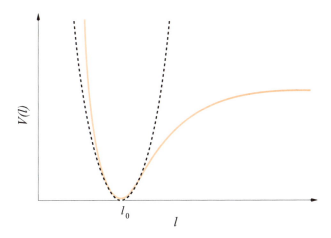

FIGURE 5.5

A comparison of the harmonic-oscillator potential ($kl^2/2$; dashed line) with the complete internuclear potential (solid line) of a diatomic molecule. The harmonic-oscillator potential is a satisfactory approximation at small displacements from l_0.

Although the harmonic-oscillator potential may appear to be a terrible approximation to the experimental curve, note that it is, indeed, a good approximation in the region of the minimum. This region is the physically important region for many molecules at room temperature. Although the harmonic oscillator unrealistically allows the displacement to vary from $-\infty$ to $+\infty$, these large displacements produce potential energies that are so large that they do not often occur in practice. The harmonic oscillator is a good approximation for vibrations with small amplitudes.

We can put the previous discussion into mathematical terms by considering the Taylor expansion (see MathChapter D) of the potential energy $V(l)$ about the equilibrium bond length $l = l_0$. The first few terms in this expansion are

$$V(l) = V(l_0) + \left(\frac{dV}{dl}\right)_{l=l_0}(l - l_0) + \frac{1}{2!}\left(\frac{d^2V}{dl^2}\right)_{l=l_0}(l - l_0)^2$$

$$+ \frac{1}{3!}\left(\frac{d^3V}{dl^3}\right)_{l=l_0}(l - l_0)^3 + \cdots \tag{5.29}$$

The first term in Equation 5.29 is a constant and depends upon where we choose the zero of energy. It is convenient to choose the zero of energy such that $V(l_0)$ equals zero and relate $V(l)$ to this convention. The second term on the right side of Equation 5.29 involves the quantity $(dV/dl)_{l=l_0}$. Because the point $l = l_0$ is the minimum of the potential energy curve, dV/dl vanishes there, so there is no linear term in the displacement in Equation 5.29. Note that dV/dl is the force acting between the two nuclei, and the fact that dV/dl vanishes at $l = l_0$ means that the force acting between the nuclei is zero at this point. This is why $l = l_0$ is called the *equilibrium bond length*.

If we denote $l - l_0$ by x, $(d^2V/dl^2)_{l=l_0}$ by k, and $(d^3V/dl^3)_{l=l_0}$ by γ_3, Equation 5.29 becomes

$$V(x) = \frac{1}{2}k(l - l_0)^2 + \frac{1}{6}\gamma_3(l - l_0)^3 + \cdots$$

$$= \frac{1}{2}kx^2 + \frac{1}{6}\gamma_3 x^3 + \cdots \tag{5.30}$$

If we restrict ourselves to small displacements, then x will be small and we can neglect the terms beyond the quadratic term in Equation 5.30, showing that the general potential energy function $V(l)$ can be approximated by a harmonic-oscillator potential. Note that the force constant is equal to the curvature of $V(l)$ at the minimum. We can consider corrections or extensions of the harmonic-oscillator model by the higher-order terms in Equation 5.30. These are called *anharmonic terms* and will be considered in Section 5.7.

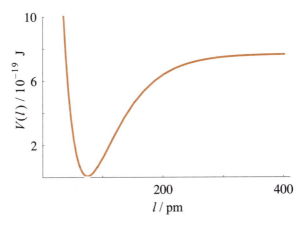

FIGURE 5.6
The Morse potential-energy curve $V(l) = D_e(1 - e^{-\beta(l-l_0)})^2$ plotted against the internuclear displacement l for H_2. The values of the parameters for H_2 are $D_e = 7.61 \times 10^{-19}$ J, $\beta = 0.0193$ pm^{-1}, and $l_0 = 74.1$ pm.

EXAMPLE 5–3
An analytic expression that is a good approximation to an intermolecular potential energy curve is a *Morse potential:*

$$V(l) = D_e(1 - e^{-\beta(l-l_0)})^2$$

First let $x = l - l_0$ so that we can write

$$V(x) = D_e(1 - e^{-\beta x})^2$$

where D_e and β are parameters that depend upon the molecule. The parameter D_e is the dissociation energy of the molecule measured from the minimum of $V(l)$, and β is a measure of the curvature of $V(l)$ at its minimum. Figure 5.6 shows $V(l)$ plotted against l for H_2. Derive a relation between the force constant and the parameters D_e and β.

SOLUTION: We now expand $V(x)$ about $x = 0$ (Equation 5.30), using

$$V(0) = 0 \qquad \left(\frac{dV}{dx}\right)_{x=0} = [2D_e\beta(e^{-\beta x} - e^{-2\beta x})]_{x=0} = 0$$

and

$$\left(\frac{d^2V}{dx^2}\right)_{x=0} = [-2D_e\beta(\beta e^{-\beta x} - 2\beta e^{-2\beta x})]_{x=0} = 2D_e\beta^2$$

Therefore, we can write

$$V(x) = D_e \beta^2 x^2 + \cdots$$

Comparing this result with Equation 5.11 gives

$$k = 2D_e \beta^2$$

5.5 The Energy Levels of a Quantum-Mechanical Harmonic Oscillator Are $E_v = h\nu(v + \frac{1}{2})$ with $v = 0, 1, 2, \ldots$

The Schrödinger equation for a one-dimensional harmonic oscillator is

$$-\frac{\hbar^2}{2\mu}\frac{d^2\psi}{dx^2} + V(x)\psi(x) = E\psi(x)$$

with $V(x) = \frac{1}{2}kx^2$. Thus, we must solve the second-order differential equation

$$\frac{d^2\psi}{dx^2} + \frac{2\mu}{\hbar^2}\left(E - \frac{1}{2}kx^2\right)\psi(x) = 0 \qquad -\infty < x < \infty \qquad (5.31)$$

This differential equation, however, does not have constant coefficients, so we cannot use the method we developed in Section 2.2. In fact, when a differential equation does not have constant coefficients, there is no simple, general technique for solving it, and each case must be considered individually. One method that often works, however laborious it is, is to substitute a power series into the differential equation and then determine each coefficient in the series sequentially. This method is called the power series method and is discussed in any book on differential equations. We present an alternative method of solving Equation 5.31 using operator methods in the appendix at the end of the chapter. This method involves no knowledge of differential equations and involves only algebraic manipulations of certain operators.

When Equation 5.31 is solved, well-behaved, finite solutions can be obtained only if the energy is restricted to the quantized values

$$E_v = h\nu\left(v + \frac{1}{2}\right) \qquad v = 0, 1, 2, \ldots \qquad (5.32)$$

where

$$\nu = \frac{1}{2\pi}\left(\frac{k}{\mu}\right)^{1/2} \qquad (5.33)$$

The energies are plotted in Figure 5.7. Note that the energy levels are equally spaced, with a separation $h\nu$. This uniform spacing between energy levels is a property peculiar

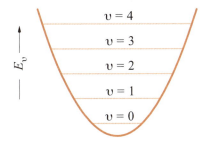

FIGURE 5.7
The energy levels of a quantum-mechanical harmonic oscillator. The curve is a parabola representing the potential energy, $V(x) = kx^2/2$.

to the quadratic potential of a harmonic oscillator. Note also that the energy of the ground state, the state with $v = 0$, is $\frac{1}{2}hv$ and is not zero as the lowest classical energy is. This energy is called the *zero-point energy* of the harmonic oscillator and is a direct result of the uncertainty principle. The energy of a harmonic oscillator can be written in the form $(p^2/2\mu) + (kx^2/2)$, and so we see that a zero value for the energy would require that both p and x or, more precisely, the expectation values of $\hat{P}^2$ and $\hat{X}^2$ be simultaneously zero, in violation of the uncertainty principle.

5.6 The Harmonic Oscillator Accounts for the Infrared Spectrum of a Diatomic Molecule

We now discuss the spectroscopic predictions of a harmonic oscillator. If we model the potential energy function of a diatomic molecule as a harmonic oscillator, then, according to Equation 5.32, the vibrational energy levels of the diatomic molecule are given by

$$E_v = hv \left(v + \frac{1}{2} \right) \qquad v = 0, 1, 2, \ldots \tag{5.34}$$

A diatomic molecule can make a transition from one vibrational energy state to another by absorbing or emitting electromagnetic radiation whose observed frequency satisfies the Bohr frequency condition

$$\Delta E = hv_{\text{obs}} \tag{5.35}$$

We will prove in Chapter 8 that the harmonic-oscillator model allows transitions only between adjacent energy states, so that we have the condition that $\Delta v = \pm 1$. Such a condition is called a *selection rule*.

For absorption to occur, $\Delta v = +1$ and so

$$\Delta E = E_{v+1} - E_v = hv \tag{5.36}$$

Thus, the observed frequency of the radiation absorbed is

$$\nu_{obs} = \frac{1}{2\pi}\left(\frac{k}{\mu}\right)^{1/2} \quad \text{(Hz)} \tag{5.37}$$

Furthermore, because successive energy states of a harmonic oscillator are separated by the same energy, ΔE is the same for all allowed transitions, so this model predicts that the spectrum consists of just one line whose frequency is given by Equation 5.37. This prediction is in good accord with experiment, and this line is called the *fundamental vibrational frequency*. For diatomic molecules, these lines occur at around 10^{14} Hz, which is in the infrared region.

It is customary in discussions of vibrational spectroscopy to write Equation 5.34 in the form

$$G(v) = \tilde{\omega}\left(v + \frac{1}{2}\right) \tag{5.38}$$

where $G(v) = E_v/hc$ and

$$\tilde{\omega} = \frac{1}{2\pi c}\left(\frac{k}{\mu}\right)^{1/2} \quad (\text{cm}^{-1}) \tag{5.39}$$

Both $G(v)$, which is called the *vibrational term*, and $\tilde{\omega}$ have units of cm^{-1}, called *wave numbers*. The tilde on ω emphasizes that $\tilde{\omega}$ has units of wave numbers.

Equation 5.39 enables us to determine force constants if the fundamental vibrational frequency is known. For example, for $H^{35}Cl$, $\tilde{\omega}_{obs}$ is 2.886×10^3 cm^{-1} and so, according to Equation 5.39, the force constant of $H^{35}Cl$ is

$$k = (2\pi c\tilde{\omega})^2\mu$$

$$= [2\pi(2.998 \times 10^8 \text{ m·s}^{-1})(2.886 \times 10^3 \text{ cm}^{-1})(100 \text{ cm·m}^{-1})]^2$$

$$\times \frac{(35.0 \text{ amu})(1.00 \text{ amu})}{(35.0 + 1.00) \text{ amu}}(1.661 \times 10^{-27} \text{ kg·amu}^{-1})$$

$$= 4.78 \times 10^2 \text{ kg·s}^{-1} = 4.78 \times 10^2 \text{ N·m}^{-1}$$

EXAMPLE 5–4
The infrared spectrum of $^{75}Br^{19}F$ consists of an intense line at 380 cm^{-1}. Calculate the force constant of $^{75}Br^{19}F$.

SOLUTION: The force constant is given by

$$k = (2\pi c\tilde{\omega})^2\mu$$

The reduced mass is

$$\mu = \frac{(75.0 \text{ amu})(19.0 \text{ amu})}{(75.0 + 19.0) \text{ amu}}(1.661 \times 10^{-27} \text{ kg·amu}^{-1}) = 2.52 \times 10^{-26} \text{ kg}$$

and so

$$k = [2\pi(2.998 \times 10^8 \text{ m·s}^{-1})(380 \text{ cm}^{-1})(100 \text{ cm·m}^{-1})]^2(2.52 \times 10^{-26} \text{ kg})$$

$$= 129 \text{ kg·s}^{-2} = 129 \text{ N·m}^{-1}$$

Force constants for diatomic molecules are of the order of 10^2 to 10^3 N·m^{-1}. Table 5.1 lists the vibrational spectroscopic parameters of some diatomic molecules. We will also see in Section 5.12 that not only must $\Delta v = \pm 1$ in the harmonic-oscillator model but that the dipole moment of the molecule must change as the molecule vibrates if the molecule is to absorb infrared radiation. Thus, the harmonic-oscillator model predicts that HCl absorbs in the infrared but N_2 does not. We will see that this prediction

TABLE 5.1
Vibrational Spectroscopic Parameters of Some Diatomic Molecules

Molecule	$\tilde{\omega}_e$/cm^{-1}	$\tilde{x}_e\tilde{\omega}_e$/cm^{-1} [b]	k/N·m^{-1}	$\bar{l}(v=0)$/pm [a]	D_0/kJ·mol^{-1} [b]
H_2	4403.56	123.86	510	74.14	432.1
D_2	3116.33	62.51	527	74.15	439.6
$H^{19}F$	4138.32	89.88	920	91.68	566.2
$H^{35}Cl$	2990.94	52.819	478	127.46	427.8
$H^{79}Br$	2648.97	45.218	381	141.44	362.6
$H^{127}I$	2309.01	39.644	291	160.92	294.7
$^{12}C^{16}O$	2169.81	13.288	1857	112.83	1070.2
$^{14}N^{16}O$	1904.20	14.075	1550	115.08	626.8
$^{14}N^{14}N$	2358.57	14.324	2243	109.77	941.6
$^{16}O^{16}O$	1580.19	11.98	1142	120.75	493.6
$^{19}F^{19}F$	916.64	11.236	454	141.19	154.6
$^{35}Cl^{35}Cl$	559.72	2.675	319	198.79	239.2
$^{79}Br^{79}Br$	325.321	1.0774	240	228.11	190.1
$^{127}I^{127}I$	214.502	0.6147	170	266.63	148.8
$^{23}Na^{23}Na$	159.125	0.7255	17	307.89	69.5
$^{39}K^{39}K$	92.021	0.2829	9.7	390.51	49.6

a. The quantity $\bar{l}(v=0)$ is the average internuclear separation in the $v=0$ state.
b. The meanings of $\tilde{x}_e\tilde{\omega}_e$, and D_0 are explained in Section 5.7.

is in good agreement with experiment. There are, indeed, deviations from the harmonic-oscillator model, but we will see in the next section that not only are they fairly small but that we can systematically introduce corrections and extensions to account for them.

5.7 Overtones Are Observed in Vibrational Spectra

Thus far we have treated the vibrational motion of a diatomic molecule by means of a harmonic-oscillator model. We saw in Section 5.4, however, that the internuclear potential energy is not a simple parabola but is more like that illustrated in Figure 5.5. The dashed line in Figure 5.5 depicts the harmonic oscillator. Recall from Equation 5.29 that the potential energy $V(l)$ may be expanded in a Taylor series about l_0, the value of l at the minimum of $V(l)$, to give [recall that $(dV/dl)_{l_0} = 0$]

$$V(l) - V(l_0) = \frac{1}{2!}\left(\frac{d^2V}{dl^2}\right)_{l=l_0}(l-l_0)^2 + \frac{1}{3!}\left(\frac{d^3V}{dl^3}\right)_{l=l_0}(l-l_0)^3 + \cdots$$

$$= \frac{k}{2}x^2 + \frac{\gamma_3}{6}x^3 + \frac{\gamma_4}{24}x^4 + \cdots \tag{5.40}$$

where x is the displacement of the nuclei from their equilibrium separation, k is the (Hooke's law) force constant, and $\gamma_j = (d^jV/dl^j)_{l=l_0}$. The harmonic-oscillator approximation consists of keeping only the quadratic term in Equation 5.40, and it predicts that there will be only one line in the vibrational spectrum of a diatomic molecule. Experimental data show there is, indeed, one dominant line (the *fundamental*), but there are also lines of weaker intensity at almost integral multiples of the fundamental. These lines are called *overtones* (Table 5.2). If the anharmonic terms in Equation 5.40 are included in the Hamiltonian operator for the vibrational motion of a diatomic molecule, the Schrödinger equation can be solved by a technique called perturbation theory (see Chapter 8) to give

TABLE 5.2
The Vibrational Spectrum of $H^{35}Cl$

| | | $\tilde{\omega}_{obs}/cm^{-1}$ | | |
| | | Harmonic oscillator | Anharmonic oscillator | $\tilde{\omega}(0 \to v)/$ |
Transition	$\tilde{\omega}_{obs}/cm^{-1}$	$\tilde{\omega} = 2885.90v$	$\tilde{\omega} = 2988.90v - 51.60v(v+1)$	$v\tilde{\omega}(0 \to 1)$
$0 \to 1$ (fundamental)	2885.9	2885.9	2885.7	1.000
$0 \to 2$ (first overtone)	5668.0	5771.8	5668.2	0.982
$0 \to 3$ (second overtone)	8347.0	8657.7	8347.5	0.964
$0 \to 4$ (third overtone)	10 923.1	11 543.6	10 923.6	0.946
$0 \to 5$ (fourth overtone)	13 396.5	14 429.5	13 396.5	0.928

$$G(v) = \tilde{\omega}_e(v + \tfrac{1}{2}) - \tilde{x}_e\tilde{\omega}_e(v + \tfrac{1}{2})^2 + \cdots \qquad v = 0, 1, 2, \ldots \qquad (5.41)$$

where $\tilde{x}_e$ is called the *anharmonicity constant*. The anharmonic correction in Equation 5.41 is much smaller than the harmonic term because $\tilde{x}_e \ll 1$ (cf. Table 5.2).

Figures 5.8 and 5.9 show the levels given by Equation 5.41. Notice that the levels are not equally spaced as they are for a harmonic oscillator and, in fact, that their separation decreases with increasing v. This is reflected by the numbers in the last column of Table 5.2. The values of $\tilde{\omega}_e$ and $\tilde{\omega}_e\tilde{x}_e$ extracted from the data are slightly different from those in Table 5.1 because of higher-order effects. Notice from Figure 5.9 that the harmonic-oscillator approximation is best for small values of v, which are the most important values at room temperature (Problem 5–40).

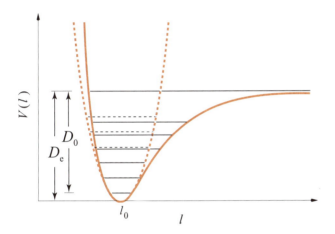

FIGURE 5.8
The energy levels of a harmonic oscillator (dotted horizontal lines) and the energy levels of an anharmonic oscillator (solid horizontal lines) superimposed on the harmonic-oscillator potential (dotted color) and the anharmonic-oscillator potential (solid color).

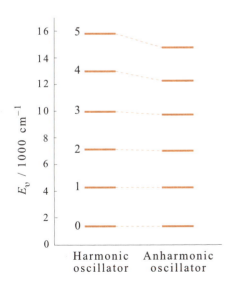

FIGURE 5.9
The vibrational energy states of $H^{35}Cl(g)$ calculated in the harmonic-oscillator approximation and with a correction for anharmonicity.

Figure 5.8 also shows two quantities D_e and D_0. The energy difference between the minimum of the potential well at the dissociation limit is D_e; the energy difference between the zero-point vibrational energy and the dissociation limit is D_0. They are related by

$$D_0 = D_e - \frac{1}{2}h\nu \tag{5.42}$$

It is D_0 that is measured experimentally.

The selection rule for an anharmonic oscillator is that Δv can have any integral value, although the intensities of the $\Delta v = \pm 2, \pm 3, \ldots$ transitions are much less than for the $\Delta v = \pm 1$ transitions. If we recognize that most diatomic molecules are in the ground vibrational state at room temperature (Problem 5–40), the frequencies of the observed $0 \rightarrow v$ transitions will be given by

$$\tilde{\omega}_{obs} = G(v) - G(0) = \tilde{\omega}_e v - \tilde{x}_e \tilde{\omega}_e v(v+1) \qquad v = 1, 2, \ldots \tag{5.43}$$

The application of Equation 5.43 to the spectrum of $H^{35}Cl$ is given in Table 5.2. You can see that the agreement with experimental data is a substantial improvement over the harmonic-oscillator approximation.

EXAMPLE 5–5

Given that $\tilde{\omega}_e = 536.10$ cm^{-1} and $x_e \omega_e = 3.4$ cm^{-1} for $^{23}Na^{19}F(g)$, calculate the frequencies of the first and second vibrational overtone transitions.

SOLUTION: We use Equation 5.43:

$$\tilde{\omega}_{obs} = \tilde{\omega}_e v - \tilde{x}_e \tilde{\omega}_e v(v+1) \qquad v = 1, 2, \ldots$$

The fundamental is given by letting $v = 1$, and the first two overtones are given by letting $v = 2$ and 3.

$$\text{Fundamental: } \tilde{\omega}_{obs} = \tilde{\omega}_e - 2\tilde{x}_e\tilde{\omega}_e = 529.3 \text{ cm}^{-1}$$

$$\text{First overtone: } \tilde{\omega}_{obs} = 2\tilde{\omega}_e - 6\tilde{x}_e\tilde{\omega}_e = 1051.8 \text{ cm}^{-1}$$

$$\text{Second overtone: } \tilde{\omega}_{obs} = 3\tilde{\omega}_e - 12\tilde{x}_e\tilde{\omega}_e = 1567.5 \text{ cm}^{-1}$$

Note that the overtones are not quite integral multiples of the fundamental frequency, and the fundamental frequency is less than the frequency for pure harmonic motion.

5.8 The Harmonic-Oscillator Wave Functions Involve Hermite Polynomials

The wave functions corresponding to the E_v for a harmonic oscillator are nondegenerate and are given by

$$\psi_v(x) = |v\rangle = N_v H_v(\alpha^{1/2}x)e^{-\alpha x^2/2} \tag{5.44}$$

where

$$\alpha = \left(\frac{k\mu}{\hbar^2}\right)^{1/2} \tag{5.45}$$

and the middle entry in Equation 5.44 represents $\psi_v(x)$ in the bracket notation. The normalization constant N_v is

$$N_v = \frac{1}{(2^v v!)^{1/2}} \left(\frac{\alpha}{\pi}\right)^{1/4} \tag{5.46}$$

and the $H_v(\alpha^{1/2}x)$ are polynomials called *Hermite polynomials*. The first few Hermite polynomials are listed in Table 5.3. Note that $H_v(\xi)$ is a vth-degree polynomial in ξ. The first few harmonic-oscillator wave functions are listed in Table 5.4 and plotted in Figure 5.10.

Although we have not actually solved the Schrödinger equation for a harmonic oscillator (Equation 5.31), we can at least show that the functions given by Equation 5.44

TABLE 5.3
The First Few Hermite Polynomials [a]

$H_0(\xi) = 1$	$H_1(\xi) = 2\xi$
$H_2(\xi) = 4\xi^2 - 2$	$H_3(\xi) = 8\xi^3 - 12\xi$
$H_4(\xi) = 16\xi^4 - 48\xi^2 + 12$	$H_5(\xi) = 32\xi^5 - 160\xi^3 + 120\xi$

a. The variable ξ is equal to $\alpha^{1/2}x$, where $\alpha = (k\mu)^{1/2}/\hbar$.

TABLE 5.4
The First Few Harmonic-Oscillator Wave Functions, Equation 5.44 [a]

$\psi_0(x) = |0\rangle = \left(\dfrac{\alpha}{\pi}\right)^{1/4} e^{-\alpha x^2/2}$	$\psi_2(x) = |2\rangle = \left(\dfrac{\alpha}{4\pi}\right)^{1/4} (2\alpha x^2 - 1)e^{-\alpha x^2/2}$
$\psi_1(x) = |1\rangle = \left(\dfrac{4\alpha^3}{\pi}\right)^{1/4} xe^{-\alpha x^2/2}$	$\psi_3(x) = |3\rangle = \left(\dfrac{\alpha^3}{9\pi}\right)^{1/4} (2\alpha x^3 - 3x)e^{-\alpha x^2/2}$

a. The parameter $\alpha = (k\mu)^{1/2}/\hbar$.

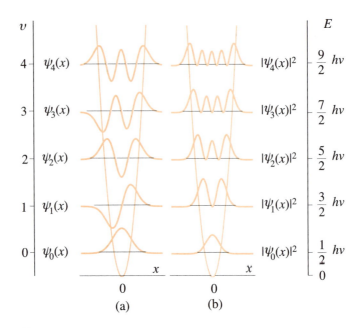

FIGURE 5.10
(a) The normalized harmonic-oscillator wave functions. (b) The probability densities for a harmonic oscillator. As in Figure 5.7, the potential energy is indicated by the parabolas in (a) and (b).

are solutions. For example, let's consider $\psi_0(x)$, which according to Table 5.4 is

$$\psi_0(x) = |0\rangle = \left(\frac{\alpha}{\pi}\right)^{1/4} e^{-\alpha x^2/2}$$

Substitution of this equation into Equation 5.31 with $E_0 = \frac{1}{2}\hbar\omega$ yields

$$\frac{d^2\psi_0}{dx^2} + \frac{2\mu}{\hbar^2}\left(E_0 - \frac{1}{2}kx^2\right)\psi_0(x) = 0$$

$$\left(\frac{\alpha}{\pi}\right)^{1/4}(\alpha^2 x^2 e^{-\alpha x^2/2} - \alpha e^{-\alpha x^2/2}) + \frac{2\mu}{\hbar^2}\left(\frac{\hbar\omega}{2} - \frac{kx^2}{2}\right)\left(\frac{\alpha}{\pi}\right)^{1/4} e^{-\alpha x^2/2} \stackrel{?}{=} 0$$

or

$$(\alpha^2 x^2 - \alpha) + \left(\frac{\mu\omega}{\hbar} - \frac{\mu k}{\hbar^2}x^2\right) \stackrel{?}{=} 0$$

Using the relations $\alpha = (k\mu/\hbar^2)^{1/2}$ and $\omega = (k/\mu)^{1/2}$, we see that everything cancels on the left side of the above expression. Thus, $\psi_0(x)$ is a solution to Equation 5.31.

Problem 5–21 involves proving explicitly that $\psi_1(x)$ and $\psi_2(x)$ are solutions of Equation 5.31.

We can also show explicitly that the $\psi_v(x)$ are normalized, or that N_v given by Equation 5.46 is the normalization constant.

EXAMPLE 5–6
Show that $\psi_0(x)$ and $\psi_1(x)$ are normalized.

SOLUTION: According to Table 5.4,

$$\psi_0(x) = \left(\frac{\alpha}{\pi}\right)^{1/4} e^{-\alpha x^2/2} \quad \text{and} \quad \psi_1(x) = \left(\frac{4\alpha^3}{\pi}\right)^{1/4} x e^{-\alpha x^2/2}$$

Then

$$\int_{-\infty}^{\infty} \psi_0^2(x)dx = \langle 0|0\rangle = \left(\frac{\alpha}{\pi}\right)^{1/2} \int_{-\infty}^{\infty} e^{-\alpha x^2}dx = \left(\frac{\alpha}{\pi}\right)^{1/2}\left(\frac{\pi}{\alpha}\right)^{1/2} = 1$$

and

$$\int_{-\infty}^{\infty} \psi_1^2(x)dx = \langle 1|1\rangle = \left(\frac{4\alpha^3}{\pi}\right)^{1/2} \int_{-\infty}^{\infty} x^2 e^{-\alpha x^2}dx = \left(\frac{4\alpha^3}{\pi}\right)^{1/2}\left[\frac{1}{2\alpha}\left(\frac{\pi}{\alpha}\right)^{1/2}\right] = 1$$

The integrals here are given inside the front cover of the book and evaluated in Problem 5–23.

We can appeal to the general results of Chapter 4 to argue that the harmonic-oscillator wave functions are orthogonal. The energy eigenvalues are nondegenerate, so

$$\int_{-\infty}^{\infty} dx\,\psi_v(x)\psi_{v'}(x) = \langle v|v'\rangle = 0 \qquad v \neq v'$$

or, more explicitly, that

$$\int_{-\infty}^{\infty} dx\, H_v(\alpha^{1/2}x) H_{v'}(\alpha^{1/2}x) e^{-\alpha x^2} = \int_{-\infty}^{\infty} d\xi\, H_v(\xi) H_{v'}(\xi) e^{-\xi^2} = 0 \qquad v \neq v'$$

We say that the Hermite polynomials are orthogonal with respect to the weighting function $e^{-\xi^2}$.

The even-odd property of a function is a rather special and important property. Figure 5.11b shows that if $f(x)$ is an odd function, then

$$\int_{-A}^{A} f(x)\,dx = 0 \qquad f(x)\ \text{odd} \qquad (5.49)$$

because the areas on each side of the y axis cancel. This is a useful property. According to Equation 5.44, the harmonic-oscillator wave functions are

$$\psi_v(x) = |v\rangle = N_v H_v(\alpha^{1/2}x)e^{-\alpha x^2/2}$$

Because the $\psi_v(x)$ are even when v is an even integer and odd when v is an odd integer, $\psi_v^2(x)$ is an even function for any value of v. According to Equation 5.49, then,

$$\langle x \rangle = \langle v|x|v\rangle = \int_{-\infty}^{\infty} \psi_v(x)x\psi_v(x)\,dx = 0 \qquad (5.50)$$

because the integrand is an odd function for any value of v. Thus, the average displacement of a harmonic oscillator is zero for all the quantum states of a harmonic oscillator, or the average internuclear separation is the equilibrium bond length l_0.

The average momentum is given by

$$\langle p \rangle = \langle v|\hat{P}_x|v\rangle = \int_{-\infty}^{\infty} \psi_v(x)\left(-i\hbar\frac{d}{dx}\right)\psi_v(x)\,dx \qquad (5.51)$$

The derivative of an odd (even) function is even (odd), so this integral vanishes because the integrand is the product of an odd and even function and hence is overall odd. Thus, $\langle p \rangle = 0$ for a harmonic operator.

As Figure 5.11 shows, even functions have the property that

$$\int_{-A}^{A} dx f(x) = 2\int_{0}^{A} dx f(x) \qquad f(x)\ \text{even} \qquad (5.52)$$

We shall use such symmetry arguments often.

5.10 There Are Many Useful Relations Among Hermite Polynomials

We can evaluate $\langle x^2 \rangle$ for the ground state of a harmonic oscillator by using $\psi_0(x)$ in Table 5.4 and evaluating the integral

$$\langle x^2 \rangle = \left(\frac{\alpha}{\pi}\right)^{1/2}\int_{-\infty}^{\infty} dx\, x^2 e^{-\alpha x^2}$$

Letting $\xi = \alpha^{1/2}x$ and using Equation 5.52, we have

$$\langle x^2 \rangle = \frac{2}{\alpha \pi^{1/2}} \int_0^\infty d\xi \, \xi^2 e^{-\xi^2} = \frac{1}{2\alpha} = \frac{\hbar}{2(\mu k)^{1/2}} \tag{5.53}$$

In general, we would have to evaluate

$$\langle x^2 \rangle = \langle v|x^2|v \rangle = \int_{-\infty}^\infty dx \, \psi_v^*(x) x^2 \psi_v(x)$$

$$= N_v^2 \int_{-\infty}^\infty dx \, e^{-\alpha x^2} H_v(\alpha^{1/2}x) x^2 H_v(\alpha^{1/2}x)$$

$$= \frac{N_v^2}{\alpha^{3/2}} \int_{-\infty}^\infty d\xi \, e^{-\xi^2} H_v(\xi) \xi^2 H_v(\xi) \tag{5.54}$$

where we have used Equation 5.44 for $\psi_v(x)$ and have let $\xi = \alpha^{1/2}x$ in the last step.

There are a number of relations involving the Hermite polynomials that can be used to evaluate integrals like Equation 5.54. A particularly useful one is

$$H_{v+1}(\xi) - 2\xi H_v(\xi) + 2v H_{v-1}(\xi) = 0 \tag{5.55}$$

Equation 5.55 is a *recursion formula* for the $H_v(\xi)$.

EXAMPLE 5–10
Show explicitly that the first few Hermite polynomials satisfy the recursion formula given by Equation 5.55.

SOLUTION: To do this, we shall use the recursion formula to generate Hermite polynomials starting with $H_0(\xi)$ and $H_1(\xi)$. Setting $v = 1$ in Equation 5.55, we have

$$H_2(\xi) = 2\xi H_1(\xi) - 2H_0(\xi)$$

Substituting $H_0(\xi) = 1$ and $H_1(\xi) = 2\xi$ into this equation gives

$$H_2(\xi) = 4\xi^2 - 2$$

Now let $v = 2$ in Equation 5.55:

$$H_3(\xi) = 2\xi H_2(\xi) - 4H_1(\xi)$$

$$= 2\xi(4\xi^2 - 2) - 4(2\xi) = 8\xi^3 - 12\xi$$

With $v = 3$, we have

$$H_4(\xi) = 2\xi H_3(\xi) - 6H_2(\xi)$$

$$= 2\xi(8\xi^3 - 12\xi) - 6(4\xi^2 - 2)$$

$$= 16\xi^4 - 48\xi^2 + 12$$

All these results agree with the entries in Table 5.3.

Recursion formulas can be used to evaluate integrals involving Hermite polynomials. For example, consider

$$\langle \xi \rangle = \int_{-\infty}^{\infty} \psi_v(\xi)\xi\psi_v(\xi)\,d\xi \tag{5.56}$$

We have shown earlier that this integral is equal to zero because of the even-odd character of the $\psi_v(x)$, but let's show this using Equation 5.55. Using Equation 5.44, Equation 5.56 becomes

$$\langle \xi \rangle = N_v^2 \int_{-\infty}^{\infty} d\xi\, H_v(\xi)\xi H_v(\xi)e^{-\xi^2} \tag{5.57}$$

According to Equation 5.55,

$$\xi H_v(\xi) = v H_{v-1}(\xi) + \frac{1}{2}H_{v+1}(\xi) \tag{5.58}$$

and when this is substituted into Equation 5.57, we have

$$\langle \xi \rangle = v N_v^2 \int_{-\infty}^{\infty} d\xi\, H_v(\xi)H_{v-1}(\xi)e^{-\xi^2} + \frac{N_v^2}{2}\int_{-\infty}^{\infty} d\xi\, H_v(\xi)H_{v+1}(\xi)e^{-\xi^2}$$

But both of these integrals are equal to zero because of orthogonality. Notice that we did not have to know the form of the Hermite polynomials to show that $\langle \xi \rangle = 0$.

This example did not give us anything new because we could have evaluated $\langle \xi \rangle$ by symmetry. Let's consider the more difficult case,

$$\langle \xi^2 \rangle = N_v^2 \int_{-\infty}^{\infty} d\xi\, H_v(\xi)\xi^2 H_v(\xi)e^{-\xi^2} \tag{5.59}$$

Multiply Equation 5.58 by ξ to write

$$\xi^2 H_v(\xi) = v\xi H_{v-1}(\xi) + \frac{1}{2}\xi H_{v+1}(\xi)$$

Now apply Equation 5.58 to each of the two terms on the right to obtain

$$\xi^2 H_v(\xi) = v\left[(v-1)H_{v-2}(\xi) + \frac{1}{2}H_v(\xi)\right] + \frac{1}{2}\left[(v+1)H_v(\xi) + \frac{1}{2}H_{v+2}(\xi)\right]$$

$$= v(v-1)H_{v-2}(\xi) + \left(v + \frac{1}{2}\right)H_v(\xi) + \frac{1}{4}H_{v+2}(\xi)$$

When we substitute this into Equation 5.59, we obtain

$$\langle\xi^2\rangle = \langle v|\xi^2|v\rangle$$

$$= N_v^2 \int_{-\infty}^{\infty} d\xi\, H_v(\xi)\left[v(v-1)H_{v-2}(\xi) + \left(v + \frac{1}{2}\right)H_v(\xi) + \frac{1}{4}H_{v+2}(\xi)\right]$$

The two integrals involving $H_{v-2}(\xi)$ and $H_{v+2}(\xi)$ vanish by orthogonality, and we are left with

$$\langle\xi^2\rangle = \left(v + \frac{1}{2}\right)N_v^2 \int_{-\infty}^{\infty} d\xi\, H_v(\xi)H_v(\xi)e^{-\xi^2}$$

$$= v + \frac{1}{2}$$

because of normalization. Because $\xi = \alpha^{1/2}x$, we have

$$\langle x^2\rangle = \frac{1}{\alpha}\langle\xi^2\rangle = \frac{1}{\alpha}\left(v + \frac{1}{2}\right) = \frac{\hbar}{(\mu k)^{1/2}}\left(v + \frac{1}{2}\right) \qquad (5.60)$$

Notice that Equation 5.60 agrees with Equation 5.53 when $v = 0$. The evaluation of $\langle p\rangle$ and $\langle p^2\rangle$ by this method is left to Problem 5–27.

5.11 The Vibrations of Polyatomic Molecules Are Represented by Normal Coordinates

The vibrational spectra of polyatomic molecules turn out to be easily understood in the harmonic-oscillator approximation. The key point is the introduction of normal coordinates, which we discuss in this section.

Consider a molecule containing N nuclei. A complete specification of this molecule in space requires $3N$ coordinates, three cartesian coordinates for each nucleus. We say that the N-atomic molecule has a total of $3N$ *degrees of freedom*. Of these $3N$ coordinates, three can be used to specify the center of mass of the molecule. Motion along these three coordinates corresponds to translational motion of the center of mass of the molecule, and so we call these three coordinates *translational degrees of freedom*. It requires two coordinates to specify the orientation of a linear molecule about its center

TABLE 5.5
The Number of Various Degrees of Freedom of a Polyatomic
Molecule Containing N Atoms

	Linear	Nonlinear
Translational degrees of freedom	3	3
Rotational degrees of freedom	2	3
Vibrational degrees of freedom	$3N - 5$	$3N - 6$

of mass and three coordinates to specify the orientation of a nonlinear molecule about its center of mass. Because motion along these coordinates corresponds to rotational motion, we say that a linear molecule has two *degrees of rotational freedom* and that a nonlinear molecule has three degrees of rotational freedom. The remaining coordinates ($3N - 5$ for a linear molecule and $3N - 6$ for a nonlinear molecule) specify the relative positions of the N nuclei. Because motion along these coordinates corresponds to vibrational motion, we say that a linear molecule has $3N - 5$ *vibrational degrees of freedom* and that a nonlinear molecule has $3N - 6$ vibrational degrees of freedom. These results are summarized in Table 5.5.

EXAMPLE 5–11
Determine the number of various degrees of freedom of HCl, CO_2, H_2O, NH_3, and CH_4.

SOLUTION:

	Total	Translational	Rotational	Vibrational
HCl	6	3	2	1
CO_2 (linear)	9	3	2	4
H_2O	9	3	3	3
NH_3	12	3	3	6
CH_4	15	3	3	9

In the absence of external fields, the energy of a molecule does not depend upon the position of its center of mass or its orientation. The potential energy of a polyatomic molecule is therefore a function of only the $3N - 5$ or $3N - 6$ vibrational coordinates. If we let the displacements about the equilibrium values of these coordinates be denoted by $x_1, x_2, \ldots, x_{N_{vib}}$, where N_{vib} is the number of vibrational degrees of freedom, then the potential energy is given by the multidimensional generalization of the one-dimensional case given by Equation 5.29:

$$\Delta V = V(x_1, x_2, \ldots, x_{N_{\text{vib}}}) - V(0, 0, \ldots, 0) = \frac{1}{2} \sum_{i=1}^{N_{\text{vib}}} \sum_{j=1}^{N_{\text{vib}}} \left(\frac{\partial^2 V}{\partial x_i \partial x_j} \right) x_i x_j + \cdots$$

$$= \frac{1}{2} \sum_{i=1}^{N_{\text{vib}}} \sum_{j=1}^{N_{\text{vib}}} f_{ij} x_i x_j + \cdots \tag{5.61}$$

In general, there are other terms that contain higher powers of x_i, but these anharmonic terms are neglected here. The presence of the cross terms in Equation 5.61 makes the solution of the corresponding Schrödinger equation very difficult to obtain. A theorem of classical mechanics, however, allows us to eliminate all the cross terms in Equation 5.61. The details are too specialized to go into here, but a straightforward procedure using matrix algebra can be used to find a new set of coordinates $\{Q_j\}$, such that

$$\Delta V = \frac{1}{2} \sum_{j=1}^{N_{\text{vib}}} F_j Q_j^2 \tag{5.62}$$

Note the lack of cross terms in this expression. These new coordinates are called *normal coordinates* or *normal modes*. In terms of normal coordinates, the vibrational Hamiltonian operator is

$$\hat{H}_{\text{vib}} = -\sum_{j=1}^{N_{\text{vib}}} \frac{\hbar^2}{2\mu_j} \frac{d^2}{dQ_j^2} + \frac{1}{2} \sum_{j=1}^{N_{\text{vib}}} F_j Q_j^2 \tag{5.63}$$

Recall from Section 3.9 that if a Hamiltonian operator can be written as a sum of independent terms, the total wave function is a product of individual wave functions and the energy is a sum of independent energies. Applying this theorem to Equation 5.63, we have

$$\hat{H}_{\text{vib}} = \sum_{j=1}^{N_{\text{vib}}} \hat{H}_{\text{vib},j} = \sum_{j=1}^{N_{\text{vib}}} \left(-\frac{\hbar^2}{2\mu_j} \frac{d^2}{dQ_j^2} + \frac{1}{2} F_j Q_j^2 \right) \tag{5.64}$$

$$\psi_{\text{vib}}(Q_1, Q_2, \ldots, Q_{\text{vib}}) = \psi_{\text{vib},1}(Q_1) \psi_{\text{vib},2}(Q_2) \cdots \psi_{\text{vib},N_{\text{vib}}}(Q_{N_{\text{vib}}})$$

and

$$E_{\text{vib}} = \sum_{j=1}^{N_{\text{vib}}} h\nu_j (\nu_j + \tfrac{1}{2}) \qquad \text{each } \nu_j = 0, 1, 2, \ldots \tag{5.65}$$

The practical consequence of Equations 5.64 and 5.65 is that under the harmonic-oscillator approximation, the vibrational motion of a polyatomic molecule appears as N_{vib} independent harmonic oscillators. In the absence of degeneracies, each will have its own characteristic fundamental frequency v_j. The normal modes of two molecules are shown in Figure 5.12.

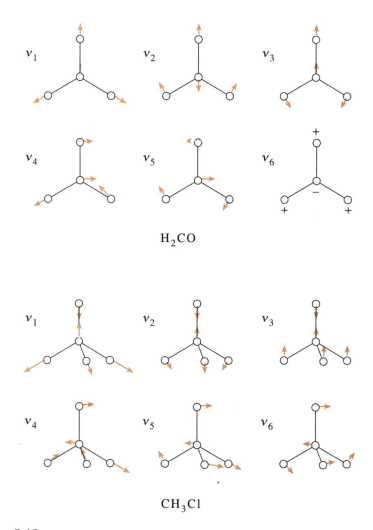

FIGURE 5.12

The normal modes of a formaldehyde molecule (H_2CO) and a chloromethane molecule (CH_3Cl). For a given normal mode, the arrows indicate how the atoms move. Each atom oscillates about its equilibrium position with the same frequency and phase, but different atoms have different amplitudes of oscillation. Although specific molecules are indicated, the normal modes are characteristic of the symmetry of the molecules and so are more general. Each of the v_4, v_5, and v_6 modes of CH_3Cl is doubly degenerate, so the total number of normal modes is 9.

5.12 The Harmonic-Oscillator Selection Rule Is $\Delta v = \pm 1$

We said in Section 5.6 that transitions between harmonic-oscillator states can occur only between adjacent levels. This condition, which we shall discuss in this section, is called a *selection rule*. We shall show in Section 8.6 that if a harmonic oscillator is irradiated with electromagnetic radiation propagating in the z direction, then the probability that the oscillator makes a transition from state v to state v' is proportional to the square of the integral

$$\langle v | \mu_z | v' \rangle = \int_{-\infty}^{\infty} dx \, \psi_v(x) \mu_z(x) \psi_{v'}(x) \tag{5.66}$$

where μ_z is the z component of the dipole moment. This type of integral, which occurs frequently in spectroscopy, is called a *dipole transition moment*.

We now expand $\mu_z(x)$ about the equilibrium nuclear separation:

$$\mu_z(x) = \mu_0 + \left(\frac{d\mu}{dx}\right)_0 x + \cdots \tag{5.67}$$

where μ_0 is the dipole moment at the equilibrium bond length and x is the displacement from that equilibrium value. Thus, when $x = 0$, $\mu_z = \mu_0$. If we substitute Equation 5.67 into Equation 5.66, we have two terms:

$$\langle v | \mu_z | v' \rangle = \mu_0 \int_{-\infty}^{\infty} dx \, \psi_v(x) \psi_{v'}(x) + \left(\frac{d\mu}{dx}\right)_0 \int_{-\infty}^{\infty} dx \, \psi_v(x) x \psi_{v'}(x) \tag{5.68}$$

The first integral here vanishes if $v \neq v'$ due to the orthogonality of the harmonic-oscillator wave functions. The second integral can be evaluated in general by using Equation 5.58:

$$\xi H_v(\xi) = v H_{v-1}(\xi) + \tfrac{1}{2} H_{v+1}(\xi) \tag{5.69}$$

If we substitute Equation 5.69 into Equation 5.68, letting $\alpha^{1/2} x = \xi$, we obtain

$$\langle v | \mu_z | v' \rangle = \frac{N_v N_{v'}}{\alpha} \left(\frac{d\mu}{d\xi}\right)_0 \int_{-\infty}^{\infty} H_{v'}(\xi) \left[v H_{v-1}(\xi) + \frac{1}{2} H_{v+1}(\xi) \right] e^{-\xi^2} d\xi \tag{5.70}$$

Using now the orthogonality property of the Hermite polynomials, we see that $\langle v | \mu_z | v' \rangle$ vanishes unless $v' = v \pm 1$. Thus, the selection rule for vibrational transitions under the harmonic-oscillator approximation is that $\Delta v = \pm 1$. In addition, the factor $(d\mu/d\xi)_0$ in front of the transition moment integral reminds us that the dipole moment of the molecule must vary during a vibration (Equation 5.67), or the transition will not take place.

EXAMPLE 5–12

Using the explicit formulas for the Hermite polynomials given in Table 5.3, show that a $0 \rightarrow 1$ vibrational transition is allowed and that a $0 \rightarrow 2$ transition is forbidden.

SOLUTION: Letting $\xi = \alpha^{1/2}x$ in Table 5.3, we have

$$\psi_0(\xi) = \left(\frac{\alpha}{\pi}\right)^{1/4} e^{-\xi^2/2}$$

$$\psi_1(\xi) = \sqrt{2}\left(\frac{\alpha}{\pi}\right)^{1/4} \xi e^{-\xi^2/2}$$

$$\psi_2(\xi) = \frac{1}{\sqrt{2}}\left(\frac{\alpha}{\pi}\right)^{1/4} (2\xi^2 - 1)e^{-\xi^2/2}$$

The dipole transition moment is given by the integral

$$I_{0\rightarrow v} \propto \int_{-\infty}^{\infty} \psi_v(\xi)\xi\psi_0(\xi)d\xi$$

The transition is allowed if $I_{0\rightarrow v} \neq 0$ and is forbidden if $I_{0\rightarrow v} = 0$. For $v = 1$, we have

$$I_{0\rightarrow 1} \propto \left(\frac{2\alpha}{\pi}\right)^{1/2} \int_{-\infty}^{\infty} \xi^2 e^{-\xi^2} d\xi \neq 0$$

because the integrand is everywhere positive. For $v = 2$,

$$I_{0\rightarrow 2} \propto \left(\frac{\alpha}{2\pi}\right)^{1/2} \int_{-\infty}^{\infty} (2\xi^3 - \xi)e^{-\xi^2} d\xi = 0$$

because the integrand is an odd function and the limits go from $-\infty$ to $+\infty$.

A selection rule for vibrational absorption spectroscopy is that the dipole moment of the molecule must vary during the normal mode motion. When this is so, the normal mode is said to be *infrared active*. Otherwise, it is *infrared inactive*. The normal modes of H_2O are as follows:

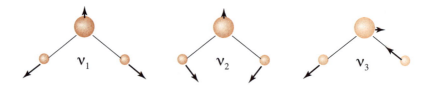

Note that the dipole moment changes during the motion of all three normal modes, so all three normal modes of H_2O are infrared active. Therefore, H_2O has three bands in its infrared spectrum. For CO_2, there are four normal modes $(3N - 5)$:

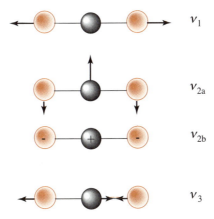

v_1

v_{2a}

v_{2b}

v_3

There is no change in the dipole moment during the symmetric stretch of CO_2 (v_1), so this mode is infrared inactive. The other modes are infrared active, but the bending mode (v_2) is doubly degenerate, so it leads to only one infrared band.

Appendix: Operator Method Solution to the Schrödinger Equation for a Harmonic Oscillator

The Schrödinger equation for a harmonic operator is

$$-\frac{\hbar^2}{2\mu}\frac{d^2\psi}{dx^2} + \frac{1}{2}kx^2\psi = E\psi \tag{1}$$

As we mentioned in Section 5.5, this is a linear differential equation in $\psi(x)$, but it does not have constant coefficients. A standard method to solve such equations is to assume that $\psi(x)$ can be expressed as a power series in x:

$$\psi(x) = \sum_{n=0}^{\infty} a_n x^n$$

We substitute this expression into the differential equation and obtain a set of algebraic equations for the a_n. This method is called the *power series method*, and is at best tedious.

There is an alternative method of solving equation 1 that is based on *operator methods*, which is not only much easier than the power series method, but is also much more elegant. In this appendix, we shall use operator methods to determine the eigenvalues and eigenfunctions of a harmonic oscillator. Operator methods are used frequently in quantum mechanics and they are well worth assimilating.

The Hamiltonian operator of a harmonic oscillator is given by

$$\hat{H} = -\frac{\hbar^2}{2\mu}\frac{d^2}{dx^2} + \frac{1}{2}kx^2 = \frac{\hat{P}^2}{2\mu} + \frac{k}{2}\hat{X}^2$$

where $k = \mu\omega^2$ is the force constant and $\hat{P}$ and $\hat{X}$ are the momentum and position operators, respectively. If we introduce new operators

$$\hat{p} = (\mu\hbar\omega)^{-1/2}\hat{P} \quad \text{and} \quad \hat{x} = (\mu\omega/\hbar)^{1/2}\hat{X} \tag{2}$$

then the Hamiltonian operator can be written as

$$\hat{H} = \frac{\hat{P}^2}{2\mu} + \frac{k}{2}\hat{X}^2 = \frac{\hbar\omega}{2}(\hat{p}^2 + \hat{x}^2) \tag{3}$$

Using the fact that $[\hat{P}, \hat{X}] = -i\hbar$, we have

$$[\hat{p}, \hat{x}] = \hat{p}\hat{x} - \hat{x}\hat{p} = \frac{1}{\hbar}(\hat{P}\hat{X} - \hat{X}\hat{P}) = \frac{1}{\hbar}(-i\hbar) = -i \tag{4}$$

We now define the (not necessarily Hermitian) operators $\hat{a}_+$ and $\hat{a}_-$ by

$$\hat{a}_+ = \frac{1}{\sqrt{2}}(\hat{x} - i\hat{p}) \quad \text{and} \quad \hat{a}_- = \frac{1}{\sqrt{2}}(\hat{x} + i\hat{p}) \tag{5}$$

Using these definitions, we have

$$\hat{a}_-\hat{a}_+ = \frac{1}{2}(\hat{x} + i\hat{p})(\hat{x} - i\hat{p}) = \frac{1}{2}[\hat{x}^2 + i(\hat{p}\hat{x} - \hat{x}\hat{p}) + \hat{p}^2]$$

$$= \frac{1}{2}(\hat{x}^2 + \hat{p}^2 + 1) \tag{6}$$

Similarly,

$$\hat{a}_+\hat{a}_- = \frac{1}{2}(\hat{p}^2 + \hat{x}^2 - 1) \tag{7}$$

Equations 6 and 7 give us

$$\hat{a}_+\hat{a}_- - \hat{a}_-\hat{a}_+ = [\hat{a}_+, \hat{a}_-] = -1 \tag{8}$$

Note that the Hamiltonian operator, equation 3, can be written as

$$\hat{H} = \hbar\omega\left(\hat{a}_+\hat{a}_- + \frac{1}{2}\right) \tag{9}$$

To make equation 9 more transparent, we denote the operator $\hat{a}_+\hat{a}_-$ by $\hat{v}$ and write equation 9 as

$$\hat{H} = \hbar\omega\left(\hat{v} + \frac{1}{2}\right) \tag{10}$$

Equation 10 is an operator form of the eigenvalues, $E_v = \hbar\omega(v + \frac{1}{2})$, of a harmonic oscillator. We might expect that the eigenvalues of $\hat{v}$ are $0, 1, 2, \ldots$, but we don't know that at this point.

So far all we've done is to express $\hat{H}$ in a new notation [although the form of equation 10 suggests the eigenvalues $\hbar\omega(v + \frac{1}{2})$]. Now let's explore some of the properties of $\hat{a}_+$ and $\hat{a}_-$. Start with the Schrödinger equation written in the form

$$\hat{H}|v\rangle = \hbar\omega\left(\hat{a}_+\hat{a}_- + \frac{1}{2}\right)|v\rangle = E_v|v\rangle \tag{11}$$

If we multiply from the left by $\hat{a}_-$, then we obtain

$$\hbar\omega\left(\hat{a}_-\hat{a}_+\hat{a}_- + \frac{\hat{a}_-}{2}\right)|v\rangle = \hbar\omega\left(\hat{a}_-\hat{a}_+ + \frac{1}{2}\right)\hat{a}_-|v\rangle = E_v\hat{a}_-|v\rangle$$

Now use equation 6 to write this equation as

$$\frac{\hbar\omega}{2}(\hat{p}^2 + \hat{x}^2 + 2)\,|\,\hat{a}_-v\,\rangle = E_v\,|\,\hat{a}_-v\,\rangle$$

or, using equation 3,

$$\hat{H}\,|\,\hat{a}_-v\,\rangle = (E_v - \hbar\omega)\,|\,\hat{a}_-v\,\rangle \tag{12}$$

Equation 12 has an important interpretation: the function $|\hat{a}_-v\rangle$ is an eigenfunction corresponding to an energy of one unit (in terms of $\hbar\omega$) less than that of $|v\rangle$. We can express this result by writing

$$|\,\hat{a}_-v\,\rangle \propto |\,v-1\,\rangle \tag{13}$$

Similarly, by starting with

$$\hat{H}|v\rangle = \hbar\omega\left(\hat{a}_-\hat{a}_+ - \frac{1}{2}\right)|v\rangle = E_v|v\rangle$$

instead of equation 11 and then multiplying from the left by $\hat{a}_+$, we obtain

$$\hbar\omega\left(\hat{a}_+\hat{a}_- - \frac{1}{2}\right)\hat{a}_+|v\rangle = E_v\hat{a}_+|v\rangle$$

Using equation 7 for $\hat{a}_+\hat{a}_-$ gives

$$\hat{H}\,|\,\hat{a}_+v\rangle = (E_v + \hbar\omega)\,|\,\hat{a}_+v\,\rangle$$

or

$$|\,\hat{a}_+v\,\rangle \propto |\,v+1\,\rangle \tag{14}$$

Thus, $\hat{a}_+$ operating on ψ_v gives ψ_{v+1} (to within a multiplicative constant) and $\hat{a}_-\psi_v$ gives ψ_{v-1} (to within a multiplicative constant). The operators $\hat{a}_+$ and $\hat{a}_-$ are called *raising* or *lowering operators,* or simply *ladder operators.* If we think of each rung of a ladder as a quantum state, then the operators $\hat{a}_+$ and $\hat{a}_-$ enable us to move up and down the ladder once we know the wave function of a single rung.

Although we know that $\hat{a}_+$ and $\hat{a}_-$ are raising and lowering operators, we still don't know what values v can take on, so let's look at the operator $\hat{v} = \hat{a}_+\hat{a}_-$. Let the eigenvalues of $\hat{v}$ be denoted by v, so that we have

$$\hat{v}|v\rangle = v|v\rangle \tag{15}$$

Don't be put off by the notation here; $\hat{v}$ represents an operator, $|v\rangle$ is its eigenvector, and v is its eigenvalue. Using this notation, equation 10 becomes

$$\hat{H}|v\rangle = \hbar\omega\left(v + \frac{1}{2}\right)|v\rangle$$

where once again, we do not yet know the possible values that v can take on.

It's easy to show, however, that v, the eigenvalue of the number operator $\hat{v}$, must be equal to or greater than zero. To prove this, multiply equation 15 from the left by $\langle v|$ to get

$$v = \langle\,v\,|\,\hat{v}\,|\,v\,\rangle = \langle\,v\,|\,\hat{a}_+\hat{a}_-\,|\,v\,\rangle$$

$$= \langle\,v\,|\,\tfrac{1}{\sqrt{2}}(\hat{x} - i\,\hat{p})\hat{a}_-\,|\,v\,\rangle$$

Using the fact that $\hat{x}$ and $\hat{p}$ are Hermitian, we have

$$v = \langle\,v\,|\,\tfrac{1}{\sqrt{2}}(\hat{x} - i\,\hat{p})\hat{a}_-\,|\,v\,\rangle = \langle\,\tfrac{1}{\sqrt{2}}(\hat{x} + i\,\hat{p})v\,|\,\hat{a}_-v\,\rangle = \langle\,\hat{a}_-v\,|\,\hat{a}_-v\,\rangle \geq 0$$

because this last expression represents the integral of the magnitude of $|\hat{a}_-v\rangle$ squared. Thus,

$$v \geq 0 \tag{16}$$

Now, because $|\hat{a}_- v\rangle \propto |v-1\rangle$ and $v \geq 0$, there must be some minimum value of v. Let this value be v_{min}. By definition, then, we must have

$$\hat{a}_-|v_{min}\rangle = 0$$

Multiply from the left by $\hat{a}_+$ to get

$$\hat{a}_+\hat{a}_-|v_{min}\rangle = \hat{v}|v_{min}\rangle = v_{min}|v_{min}\rangle = 0$$

which says that $v_{min} = 0$. Furthermore, repeated application of the raising operator $\hat{a}_+$ shows that $v = 0, 1, 2, \ldots$, so the energy eigenvalues of a harmonic oscillator are given by $\hbar\omega(v + \frac{1}{2})$, $v = 0, 1, 2, \ldots$.

Now that we know the allowed energies, we can determine the corresponding wave functions. Start with

$$\hat{a}_-|0\rangle = \hat{a}_-\psi_0 = 0 \tag{17}$$

Using the definition of $\hat{a}_-$, we have

$$\frac{1}{\sqrt{2}}(\hat{x} + i\hat{p})\psi_0 = 0$$

which, using equations 2, becomes

$$\frac{1}{\sqrt{2}}\left[\left(\frac{\mu\omega}{\hbar}\right)^{1/2}x + \left(\frac{\hbar}{\mu\omega}\right)^{1/2}\frac{d}{dx}\right]\psi_0 = 0$$

or

$$\frac{d\psi_0}{dx} = -\frac{\mu\omega}{\hbar}x\psi_0 \tag{18}$$

This equation can be integrated to give

$$\psi_0(x) = |0\rangle = ce^{-\mu\omega x^2/2\hbar}$$

where c is just a normalization constant. This result for $\psi_0(x)$ is the same as that given in Table 5.4 when you use the relation $k = \mu\omega^2$. All the other eigenfunctions can be obtained from $\psi_0(x)$ by a repeated application of the raising operator $\hat{a}_+$ (Problem 5–45).

Problems

5–1. Show that the equation $Md^2X/dt^2 = 0$ (Equation 5.26) implies that the motion is uniform.

5–2. Verify that $x(t) = A\sin\omega t + B\cos\omega t$, where $\omega = (k/m)^{1/2}$ is a solution to Newton's equation for a harmonic oscillator.

5–3. Verify that $x(t) = C \sin(\omega t + \phi)$ is a solution to Newton's equation for a harmonic oscillator.

5–4. The general solution for the classical harmonic oscillator is $x(t) = C \sin(\omega t + \phi)$. Show that the displacement oscillates between $+C$ and $-C$ with a frequency of ω radian·s^{-1} or $\nu = \omega/2\pi$ cycle·s^{-1}. What is the period of the oscillations; that is, how long does it take to undergo one cycle?

5–5. From Problem 5–4, we see that the period of a harmonic vibration is $\tau = 1/\nu$. The average of the kinetic energy over one cycle is given by

$$\langle T \rangle = \frac{1}{\tau} \int_0^\tau \frac{m\omega^2 C^2}{2} \cos^2(\omega t + \phi) dt$$

Show that $\langle T \rangle = E/2$, where E is the total energy. Show also that $\langle V \rangle = E/2$, where the instantaneous potential energy is given by

$$V = \frac{kC^2}{2} \sin^2(\omega t + \phi)$$

Interpret the result $\langle T \rangle = \langle V \rangle$.

5–6. Consider two masses m_1 and m_2 in one dimension, interacting through a potential that depends only upon their relative separation $(x_1 - x_2)$, so that $V(x_1, x_2) = V(x_1 - x_2)$. Given that the force acting upon the jth particle is $f_j = -(\partial V/\partial x_j)$, show that $f_1 = -f_2$. What law is this?

Newton's equations for m_1 and m_2 are

$$m_1 \frac{d^2 x_1}{dt^2} = -\frac{\partial V}{\partial x_1} \quad \text{and} \quad m_2 \frac{d^2 x_2}{dt^2} = -\frac{\partial V}{\partial x_2}$$

Now introduce center-of-mass and relative coordinates by

$$X = \frac{m_1 x_1 + m_2 x_2}{M} \qquad x = x_1 - x_2$$

where $M = m_1 + m_2$, and solve for x_1 and x_2 to obtain

$$x_1 = X + \frac{m_2}{M} x \qquad \text{and} \qquad x_2 = X + \frac{m_1}{M} x$$

Show that Newton's equations in these coordinates are

$$m_1 \frac{d^2 X}{dt^2} + \frac{m_1 m_2}{M} \frac{d^2 x}{dt^2} = -\frac{\partial V}{\partial x}$$

and

$$m_2 \frac{d^2 X}{dt^2} - \frac{m_1 m_2}{M} \frac{d^2 x}{dt^2} = +\frac{\partial V}{\partial x}$$

Now add these two equations to find

$$M \frac{d^2 X}{dt^2} = 0$$

Interpret this result. Now divide the first equation by m_1 and the second by m_2 and subtract to obtain

$$\frac{d^2 x}{dt^2} = -\left(\frac{1}{m_1} + \frac{1}{m_2} \right) \frac{\partial V}{\partial x}$$

or

$$\mu \frac{d^2 x}{dt^2} = -\frac{\partial V}{\partial x}$$

where $\mu = m_1 m_2 / (m_1 + m_2)$ is the reduced mass. Interpret this result, and discuss how the original two-body problem has been reduced to two one-body problems.

5–7. Extend the results of Problem 5–6 to three dimensions. Realize that in three dimensions the relative separation is given by

$$r_{12} = [(x_1 - x_2)^2 + (y_1 - y_2)^2 + (z_1 - z_2)^2]^{1/2}$$

5–8. Show that the reduced mass of two equal masses, m, is $m/2$.

5–9. Example 5–3 shows that a Maclaurin expansion of a Morse potential leads to

$$V(x) = D\beta^2 x^2 + \cdots$$

Given that $D = 7.31 \times 10^{-19}$ J·molecule^{-1} and $\beta = 1.81 \times 10^{10}$ m^{-1} for HCl, calculate the force constant of HCl. Plot the Morse potential for HCl, and plot the corresponding harmonic oscillator potential on the same graph (cf. Figure 5.5).

5–10. Use the result of Example 5–3 and Equation 5.39 to show that

$$\beta = 2\pi c \tilde{\omega}_{obs} \left(\frac{\mu}{2D} \right)^{1/2}$$

Given that $\tilde{\omega}_{obs} = 2886$ cm^{-1} and $D = 440.2$ kJ·mol^{-1} for H^{35}Cl, calculate β. Compare your result with that in Problem 5–9.

5–11. Carry out the Maclaurin expansion of the Morse potential in Example 5–3 through terms in x^4. Express γ_3 in Equation 5.30 in terms of D and β.

5–12. It turns out that the solution of the Schrödinger equation for the Morse potential can be expressed as

$$G(v) = \tilde{\omega}_e \left(v + \frac{1}{2} \right) - \tilde{\omega}_e \tilde{x}_e \left(v + \frac{1}{2} \right)^2$$

where

$$\tilde{x}_e = \frac{hc\tilde{\omega}_e}{4D}$$

Given that $\tilde{\omega}_e = 2886 \text{ cm}^{-1}$ and $D = 440.2 \text{ kJ} \cdot \text{mol}^{-1}$ for $H^{35}Cl$, calculate $\tilde{x}_e$ and $\tilde{\omega}_e \tilde{x}_e$.

5–13. In the infrared spectrum of $H^{127}I$, there is an intense line at 2309 cm^{-1}. Calculate the force constant of $H^{127}I$ and the period of vibration of $H^{127}I$.

5–14. The force constant of $^{35}Cl^{35}Cl$ is $319 \text{ N} \cdot \text{m}^{-1}$. Calculate the fundamental vibrational frequency and the zero-point energy of $^{35}Cl^{35}Cl$.

5–15. The fundamental line in the infrared spectrum of $^{12}C^{16}O$ occurs at 2143.0 cm^{-1}, and the first overtone occurs at 4260.0 cm^{-1}. Calculate the values of $\tilde{\omega}_e$ and $\tilde{x}_e \tilde{\omega}_e$ for $^{12}C^{16}O$.

5–16. Using the parameters given in Table 5.1, calculate the fundamental and the first three overtones of $H^{79}Br$.

5–17. The frequencies of the vibrational transitions in the anharmonic-oscillator approximation are given by Equation 5.43. Show how the values of both $\tilde{\omega}_e$ and $\tilde{x}_e \tilde{\omega}_e$ may be obtained by plotting $\tilde{\omega}_{obs}/v$ versus $(v + 1)$. Use this method and the data in Table 5.1 to determine the values of $\tilde{\omega}_e$ and $\tilde{x}_e \tilde{\omega}_e$ for $H^{35}Cl$.

5–18. The following data are obtained from the infrared spectrum of $^{127}I^{35}Cl$. Using the method of Problem 5–17, determine the values of $\tilde{\omega}_e$ and $\tilde{x}_e \tilde{\omega}_e$ from these data.

Transition	Frequency/cm^{-1}
$0 \rightarrow 1$	381.20
$0 \rightarrow 2$	759.60
$0 \rightarrow 3$	1135.00
$0 \rightarrow 4$	1507.40
$0 \rightarrow 5$	1877.00

5–19. The vibrational term of a diatomic molecule is given to a good approximation by

$$G(v) = \left(v + \frac{1}{2}\right)\tilde{\omega}_e - \left(v + \frac{1}{2}\right)^2 \tilde{x}_e \tilde{\omega}_e$$

where v is the vibrational quantum number. Show that the spacing between the adjacent levels ΔG is given by

$$\Delta G = G(v + 1) - G(v) = \tilde{\omega}_e\{1 - 2\tilde{x}_e(v + 1)\} \tag{1}$$

The diatomic molecule dissociates in the limit that $\Delta G \rightarrow 0$. Show that the maximum vibrational quantum number, v_{max}, is given by

$$v_{max} = \frac{1}{2\tilde{x}_e} - 1$$

Use this result to show that the dissociation energy D_e of the diatomic molecule can be written as

$$\tilde{D}_e = \frac{\tilde{\omega}_e(1 - \tilde{x}_e^2)}{4\tilde{x}_e} \approx \frac{\tilde{\omega}_e}{4\tilde{x}_e} \tag{2}$$

Referring to equation 1, explain how the constants $\tilde{\omega}_e$ and $\tilde{x}_e$ can be evaluated from a plot of ΔG versus $v + 1$. This type of plot is called a *Birge–Sponer plot*. Once the values of $\tilde{\omega}_e$ and $\tilde{x}_e$ are known, equation 2 can be used to determine the dissociation energy of the molecule. Use the following experimental data for H_2 to calculate the dissociation energy, $\tilde{D}_e$.

v	$G(v)/\text{cm}^{-1}$	v	$G(v)/\text{cm}^{-1}$
0	4161.12	7	26 830.97
1	8087.11	8	29 123.93
2	11 782.35	9	31 150.19
3	15 250.36	10	32 886.85
4	18 497.92	11	34 301.83
5	21 505.65	12	35 351.01
6	24 287.83	13	35 972.97

Explain why your Birge–Sponer plot is not linear for high values of v. How does the value of $\tilde{D}_e$ obtained from the Birge–Sponer analysis compare with the experimental value of $38\,269.48\ \text{cm}^{-1}$?

5–20. An analysis of the vibrational spectrum of the ground-electronic-state homonuclear diatomic molecule C_2 gives $\tilde{\omega}_e = 1854.71\ \text{cm}^{-1}$ and $\tilde{\omega}_e \tilde{x}_e = 13.34\ \text{cm}^{-1}$. Suggest an experimental method that can be used to determine these spectroscopic parameters. Use the expression derived in Problem 5–19 to determine the number of vibrational levels for the ground state of C_2.

5–21. Verify that $\psi_1(x)$ and $\psi_2(x)$ given in Table 5.4 satisfy the Schrödinger equation for a harmonic oscillator.

5–22. Show explicitly for a harmonic oscillator that $\psi_0(\xi)$ is orthogonal to $\psi_1(\xi)$, $\psi_2(\xi)$, and $\psi_3(\xi)$ and that $\psi_1(\xi)$ is orthogonal to $\psi_2(\xi)$ and $\psi_3(\xi)$ (see Table 5.4).

5–23. To normalize the harmonic-oscillator wave functions and calculate various expectation values, we must be able to evaluate integrals of the form

$$I_v(a) = \int_{-\infty}^{\infty} x^{2v} e^{-ax^2} dx \qquad v = 0, 1, 2, \ldots$$

We can simply either look them up in a table of integrals or continue this problem. First, show that

$$I_v(a) = 2 \int_0^{\infty} x^{2v} e^{-ax^2} dx$$

The case $v = 0$ can be handled by the following trick. Show that the square of $I_0(a)$ can be written in the form

$$I_0^2(a) = 4 \int_0^\infty \int_0^\infty dx\,dy\,e^{-a(x^2+y^2)}$$

Now convert to plane polar coordinates, letting

$$r^2 = x^2 + y^2 \qquad \text{and} \qquad dx\,dy = r\,dr\,d\theta$$

Show that the appropriate limits of integration are $0 \le r < \infty$ and $0 \le \theta \le \pi/2$ and that

$$I_0^2(a) = 4 \int_0^{\pi/2} d\theta \int_0^\infty dr\,r\,e^{-ar^2}$$

which is elementary and gives

$$I_0^2(a) = 4 \cdot \frac{\pi}{2} \cdot \frac{1}{2a} = \frac{\pi}{a}$$

or

$$I_0(a) = \left(\frac{\pi}{a}\right)^{1/2}$$

Now prove that the $I_v(a)$ may be obtained by repeated differentiation of $I_0(a)$ with respect to a and, in particular, that

$$\frac{d^v I_0(a)}{da^v} = (-1)^v I_v(a)$$

Use this result and the fact that $I_0(a) = (\pi/a)^{1/2}$ to generate $I_1(a)$, $I_2(a)$, and so forth.

5–24. Prove that the product of two even functions is even, that the product of two odd functions is even, and that the product of an even and an odd function is odd.

5–25. Prove that the derivative of an even (odd) function is odd (even).

5–26. Show that

$$\langle x^2 \rangle = \int_{-\infty}^\infty \psi_2(x)x^2\psi_2(x)dx = \frac{5}{2}\frac{\hbar}{(\mu k)^{1/2}}$$

for a harmonic oscillator. Note that $\langle x^2 \rangle^{1/2}$ is the square root of the mean of the square of the displacement (the *root-mean-square displacement*) of the oscillator.

5–27. Show that $\langle p \rangle = 0$ and that

$$\langle p^2 \rangle = \int_{-\infty}^\infty \psi_2(x)\hat{P}^2\psi_2(x)dx = \frac{5}{2}\hbar(\mu k)^{1/2}$$

for a harmonic oscillator.

5–28. Use the result of the previous two problems to show that $E_2 = 5h\nu/2$.

5–29. Prove that

$$\langle T \rangle = \langle V(x) \rangle = \frac{E_v}{2}$$

for a one-dimensional harmonic oscillator for $v = 0$ and $v = 1$.

5–30. Show that the eigenfunctions and eigenvalues of a three-dimensional harmonic oscillator whose potential energy is

$$V(x, y, z) = \frac{1}{2}k_x x^2 + \frac{1}{2}k_y y^2 + \frac{1}{2}k_z z^2$$

are

$$\psi_{v_x, v_y, v_z}(x, y, z) = \psi_{v_x}(x)\psi_{v_y}(y)\psi_{v_z}(z)$$

where

$$\psi_{v_u}(u) = \left[\frac{(\alpha_u/\pi)^{1/2}}{2^{v_u} v_u!}\right]^{1/2} H_{v_u}(\alpha_u^{1/2} u) e^{-\alpha_u u^2/2} \qquad (u = x, y, \text{ or } z)$$

$$\alpha_u^2 = \frac{\mu k_u}{\hbar^2}$$

and

$$E_{v_x v_y v_z} = h\nu_x(v_x + \tfrac{1}{2}) + h\nu_y(v_y + \tfrac{1}{2}) + h\nu_z(v_z + \tfrac{1}{2})$$

where

$$\nu_u = \frac{1}{2\pi}\left(\frac{k_u}{\mu}\right)^{1/2}$$

Discuss the degeneracy of this system when the oscillator is isotropic—that is, when $k_x = k_y = k_z$. Make a diagram like that shown in Figure 3.6.

5–31. There are a number of general relations between the Hermite polynomials and their derivatives (which we will not derive). Some of these are

$$\frac{dH_v(\xi)}{d\xi} = 2\xi H_v(\xi) - H_{v+1}(\xi)$$

$$H_{v+1}(\xi) - 2\xi H_v(\xi) + 2v H_{v-1}(\xi) = 0$$

and

$$\frac{dH_v(\xi)}{d\xi} = 2v H_{v-1}(\xi)$$

Such connecting relations are called *recursion formulas*. Verify these formulas explicitly using the first few Hermite polynomials given in Table 5.3.

5–32. Use the recursion formulas for the Hermite polynomials given in Problem 5–31 to show that $\langle p \rangle = 0$ and $\langle p^2 \rangle = \hbar(\mu k)^{1/2}(v + \frac{1}{2})$. Remember that the momentum operator involves a differentiation with respect to x, not ξ.

5–33. It can be proved generally that

$$\langle x^2 \rangle = \frac{1}{\alpha}\left(v + \frac{1}{2}\right) = \frac{\hbar}{(\mu k)^{1/2}}\left(v + \frac{1}{2}\right)$$

and that

$$\langle x^4 \rangle = \frac{3}{4\alpha^2}(2v^2 + 2v + 1) = \frac{3\hbar^2}{4\mu k}(2v^2 + 2v + 1)$$

for a harmonic oscillator. Verify these formulas explicitly for the first two states of a harmonic oscillator.

5–34. This problem is similar to Problem 3–39. Show that the harmonic-oscillator wave functions are alternately even and odd functions of x because the Hamiltonian operator obeys $\hat{H}(x) = \hat{H}(-x)$. Define a reflection operator $\hat{R}$ by

$$\hat{R}u(x) = u(-x)$$

Show that $\hat{R}$ is linear and that it commutes with $\hat{H}$. Show also that the eigenvalues of $\hat{R}$ are ± 1. What are its eigenfunctions? Show that the harmonic-oscillator wave functions are eigenfunctions of $\hat{R}$. Note that they are eigenfunctions of both $\hat{H}$ and $\hat{R}$. What does this observation say about $\hat{H}$ and $\hat{R}$?

5–35. Use Ehrenfest's theorem (Problem 4–44) to show that $\langle p_x \rangle$ does not depend upon time for a one-dimensional harmonic oscillator.

5–36. Figure 5.13 compares the probability distribution associated with $\psi_{10}(\xi)$ to the classical distribution. (See also Problem 3–18.) This problem illustrates what is meant by the classical distribution. Consider

$$x(t) = A \sin(\omega t + \phi)$$

which can be written as

$$\omega t = \sin^{-1}\left(\frac{x}{A}\right) - \phi$$

Now

$$dt = \frac{\omega^{-1}dx}{\sqrt{A^2 - x^2}} \tag{1}$$

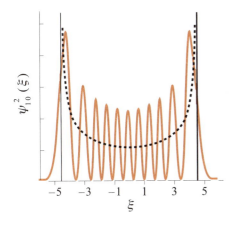

FIGURE 5.13
The probability distribution function of a harmonic oscillator in the $v = 10$ state. The dashed line is that for a classical harmonic oscillator with the same energy. The vertical lines at $\xi \approx \pm 4.6$ represent the extreme limits of the classical harmonic motion.

This equation gives the time that the oscillator spends between x and $x + dx$. We can convert equation 1 to a probability distribution in x by dividing by the time that it takes for the oscillator to go from $-A$ to A. Show that this time is π/ω and that the probability distribution in x is

$$p(x)dx = \frac{dx}{\pi\sqrt{A^2 - x^2}} \tag{2}$$

Show that $p(x)$ is normalized. Why does $p(x)$ achieve its maximum value at $x = \pm A$? Now use the fact that $\xi = \alpha^{1/2}x$, where $\alpha = (k\mu/\hbar^2)^{1/2}$, to show that

$$p(\xi)d\xi = \frac{d\xi}{\pi\sqrt{\alpha A^2 - \xi^2}} \tag{3}$$

Show that the limits of ξ are $\pm(\alpha A^2)^{1/2} = \pm(21)^{1/2}$, and compare this result to the vertical lines shown in Figure 5.13. *Hint:* You need to use the fact that $kA^2/2 = E_{10}$ ($n = 10$). Finally, plot equation 3 and compare your result with the curve in Figure 5.13.

5–37. We can use the harmonic-oscillator wave function to illustrate tunneling, a strictly quantum-mechancial property. The probability that the displacement of a harmonic oscillator in its ground state lies between x and $x + dx$ is given by

$$P(x)\, dx = \psi_0^2(x)\, dx = \left(\frac{\alpha}{\pi}\right)^{1/2} e^{-\alpha x^2}\, dx \tag{1}$$

The energy of the oscillator in its ground state is $(\hbar/2)(k/\mu)^{1/2}$. Show that the greatest displacement that this oscillator can have classically is its amplitude

$$A = \left[\frac{\hbar}{(k\mu)^{1/2}}\right]^{1/2} = \frac{1}{\alpha^{1/2}} \tag{2}$$

According to equation 1, however, there is a nonzero probability that the displacement of the oscillator will exceed this classical value and *tunnel* into the classically forbidden region. Show that this probability is given by

$$\int_{-\infty}^{-\alpha^{1/2}} P(x)\, dx + \int_{-\alpha^{1/2}}^{\infty} P(x)\, dx = \frac{2}{\pi^{1/2}} \int_{1}^{\infty} e^{-z^2}\, dz \tag{3}$$

The integral here cannot be evaluated in closed form but occurs so frequently in a number of different fields (such as the kinetic theory of gases and statistics) that it is well tabulated under the name *complementary error function*, erfc(x), which is defined as

$$\text{erfc}(x) = \frac{2}{\pi^{1/2}} \int_{x}^{\infty} e^{-z^2}\, dz \tag{4}$$

By referring to tables, it can be found that the probability that the displacement of the molecule will exceed its classical amplitude is 0.16.

5–38. Problem 5–37 shows that although a classical harmonic oscillator has a fixed amplitude, a quantum-mechanical harmonic oscillator does not. We can, however, use $\langle x^2 \rangle$ as a measure of the square of the amplitude, and, in particular, we can use the square root of $\langle x^2 \rangle$, the root-mean-square value of x, $A_{\text{rms}} = x_{\text{rms}} = \langle x^2 \rangle^{1/2}$, as a measure of the amplitude. Show that

$$A_{\text{rms}} = x_{\text{rms}} = \left[\frac{\hbar}{2(\mu k)^{1/2}} \right]^{1/2} = \left(\frac{\hbar}{4\pi c \tilde{\omega}_{\text{obs}} \mu} \right)^{1/2}$$

for the ground state of a harmonic oscillator. Thus, we can calculate $\langle x^2 \rangle^{1/2}$ in terms of $\tilde{\omega}_{\text{obs}}$. Show that $A_{\text{rms}} = 7.58$ pm for $H^{35}Cl$. The bond length of HCl is 127 pm, and so we can see that the root-mean-square amplitude is only about 5% of the bond length. This is typical of diatomic molecules.

5–39. The fundamental vibrational frequency of $H^{79}Br$ is 2.63×10^3 cm^{-1} and $\bar{l}$ is 141 pm. Calculate the root-mean-square displacement in the ground state and compare your result to $\bar{l}$.

5–40. In this problem, we'll calculate the fraction of diatomic molecules in a particular vibrational state at a temperature T using the harmonic-oscillator approximation. A fundamental equation of physical chemistry is the *Boltzmann distribution*, which says that the number of molecules with an energy E_j is proportional to $e^{-E_j/k_B T}$, where k_B is the Boltzmann constant and T is the kelvin temperature. Thus, we write

$$N_j \propto e^{-E_j/k_B T}$$

Using the fact that $\sum_j N_j = N$, show that the fraction of molecules with an energy E_j is given by

$$f_j = \frac{e^{-E_j/k_B T}}{\sum_j e^{-E_j/k_B T}} = \frac{e^{-E_j/k_B T}}{Q(T)}$$

The denominator here, $Q(T)$, is called a partition function. Show that the partition function of a harmonic oscillator $[E_j = (j + \frac{1}{2})h\nu]$ is

$$Q(T) = e^{-h\nu/2k_BT}(1 - e^{-h\nu/k_BT})^{-1}$$

Hint: You need to use the geometric series $\sum_{j=0}^{\infty} x^j = (1 - x)^{-1}$ for $|x| < 1$. Now show that

$$f_j = (1 - e^{-h\nu/k_BT})e^{-jh\nu/k_BT}$$

and, in particular, that

$$f_0 = 1 - e^{-h\nu/k_BT}$$

Now show that $f_0 \approx 1$ at 300 K for a typical molecule, with $\tilde{\omega}_e = 1000$ cm^{-1}. Plot the fraction of molecules in the jth vibrational state at $T = 300$ K.

5–41. Calculate the fraction of HBr molecules in the ground vibrational state at 300 K and 2000 K. Take $\tilde{\omega}_e$ to be 2650 cm^{-1}.

5–42. Determine the number of translational, rotational, and vibrational degrees of freedom in

(a) CH$_3$Cl   (b) OCS

(c) C$_6$H$_6$ (d) H$_2$CO

5–43. In this problem, we will prove the so-called *quantum-mechanical virial theorem*. Start with $\hat{H}\psi = E\psi$, where

$$\hat{H} = -\frac{\hbar^2}{2m}\nabla^2 + V(x, y, z)$$

Using the fact that $\hat{H}$ is a Hermitian operator, show that

$$\int \psi^*[\hat{H}, \hat{A}]\psi \, d\tau = 0 \tag{1}$$

where $\hat{A}$ is any linear operator. Choose $\hat{A}$ to be

$$\hat{A} = -i\hbar\left(x\frac{\partial}{\partial x} + y\frac{\partial}{\partial y} + z\frac{\partial}{\partial z}\right) \tag{2}$$

and show that

$$[\hat{H}, \hat{A}] = i\hbar\left(x\frac{\partial V}{\partial x} + y\frac{\partial V}{\partial y} + z\frac{\partial V}{\partial z}\right) - \frac{i\hbar}{m}(\hat{P}_x^2 + \hat{P}_y^2 + \hat{P}_z^2)$$

$$= i\hbar\left(x\frac{\partial V}{\partial x} + y\frac{\partial V}{\partial y} + z\frac{\partial V}{\partial z}\right) - 2i\hbar\hat{T}$$

where $\hat{T}$ is the kinetic energy operator. Now use equation 1 and show that

$$\left\langle x\frac{\partial V}{\partial x} + y\frac{\partial V}{\partial y} + z\frac{\partial V}{\partial z} \right\rangle = 2\langle \hat{T} \rangle \tag{3}$$

Equation 3 is the quantum-mechanical *virial theorem*.

5–44. Use the virial theorem (Problem 5–43) to prove that $\langle \hat{T} \rangle = \langle V \rangle = E/2$ for a harmonic oscillator.

5–45. Use the fact that

$$\hat{a}_+\psi_v = \frac{1}{\sqrt{2}}\left[\left(\frac{\mu\omega}{\hbar}\right)^{1/2}\hat{x} - \left(\frac{\hbar}{\mu\omega}\right)^{1/2}\hat{p}\right]\psi_v \propto \psi_{v+1}$$

and that $\psi_0(x) = (\alpha/\pi)^{1/4}e^{-\alpha x^2/2}$ to generate $\psi_1(x)$ and $\psi_2(x)$.

References

Barrow, G. *Molecular Spectroscopy*. McGraw-Hill: New York, 1962.

Brown, J. M. *Molecular Spectroscopy*. Oxford University Press: New York, 1998.

Wilson, E. B., Decius, J. C., Cross, P. C. *Molecular Vibrations*. Dover Publications: Mineola, NY, 1955.

Banwell, C. N., McCash, E. M. *Fundamentals of Molecular Spectroscopy*, 4th ed. McGraw-Hill: New York, 1944.

Hollas, J. Michael. *Molecular Spectroscopy*, 4th ed. Wiley & Sons: New York, 2004.

Hollas, J. M. *Basic Atomic and Molecular Spectroscopy*. Wiley & Sons: New York, 2002.

Herzberg, G. *Molecular Spectra and Molecular Structure: Spectra of Diatomic Molecules*, 2nd ed. Krieger Publishing: Malabar, FL, 1989.

Harris, D. C., Bertolucci, M. D. *Symmetry and Spectroscopy: An Introduction to Vibrational and Electronic Spectroscopy*. Dover Publications: Mineola, NY, 1989.

Spherical Coordinates

Although cartesian coordinates (x, y, and z) are suitable for many problems, there are many other problems for which they prove to be cumbersome. A particularly important type of such a problem occurs when the system being described has some sort of a natural center, as in the case of an atom, where the (heavy) nucleus serves as one. In describing atomic systems, as well as many other systems, it is most convenient to use spherical coordinates (Figure E.1).

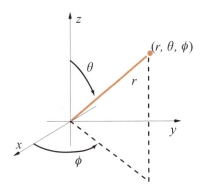

FIGURE E.1
A representation of a spherical coordinate system. A point is specified by the spherical coordinates r, θ, and ϕ.

Instead of locating a point in space by specifying the cartesian coordinates x, y, and z, we can equally well locate the same point by specifying the spherical coordinates r, θ, and ϕ. From Figure E.1, we can see that the relations between the two sets of coordinates are given by

$$x = r \sin \theta \cos \phi$$

$$y = r \sin \theta \sin \phi \qquad \text{(E.1)}$$

$$z = r \cos \theta$$

This coordinate system is called a *spherical coordinate system* because the graph of the equation $r = c =$ constant is a sphere of radius c centered at the origin.

Occasionally, we need to know r, θ, and ϕ in terms of x, y, and z. These relations are given by (Problem E–1)

$$r = \left(x^2 + y^2 + z^2\right)^{1/2}$$

$$\cos\theta = \frac{z}{(x^2 + y^2 + z^2)^{1/2}} \tag{E.2}$$

$$\tan\phi = \frac{y}{x}$$

Any point on the surface of a sphere of unit radius can be specified by the values of θ and ϕ. The angle θ represents the declination from the north pole, and hence $0 \le \theta \le \pi$. The angle ϕ represents the angle about the equator, and so $0 \le \phi \le 2\pi$. Although there is a natural zero value for θ (along the north pole), there is none for ϕ. Conventionally, the angle ϕ is measured from the x axis, as illustrated in Figure E.1. Note that r, being the distance from the origin, is intrinsically a positive quantity. In mathematical terms, $0 \le r < \infty$.

In Chapter 6, we will encounter integrals involving spherical coordinates. The differential volume element in cartesian coordinates is $dxdydz$, but it is not quite so simple in spherical coordinates. Figure E.2 shows a differential volume element in spherical coordinates, which can be seen to be

$$dV = (r\sin\theta d\phi)(rd\theta)dr = r^2\sin\theta dr d\theta d\phi \tag{E.3}$$

Let's use Equation E.3 to evaluate the volume of a sphere of radius a. In this case, $0 \le r \le a$, $0 \le \theta \le \pi$, and $0 \le \phi \le 2\pi$. Therefore,

$$V = \int_0^a r^2 dr \int_0^\pi \sin\theta d\theta \int_0^{2\pi} d\phi = \left(\frac{a^3}{3}\right)(2)(2\pi) = \frac{4\pi a^3}{3}$$

Similarly, if we integrate only over θ and ϕ, then we obtain

$$dV = r^2 dr \int_0^\pi \sin\theta d\theta \int_0^{2\pi} d\phi = 4\pi r^2 dr \tag{E.4}$$

This quantity is the volume of a spherical shell of radius r and thickness dr (Figure E.3). The factor $4\pi r^2$ represents the surface area of the spherical shell and dr is its thickness. The quantity

$$dA = r^2\sin\theta d\theta d\phi \tag{E.5}$$

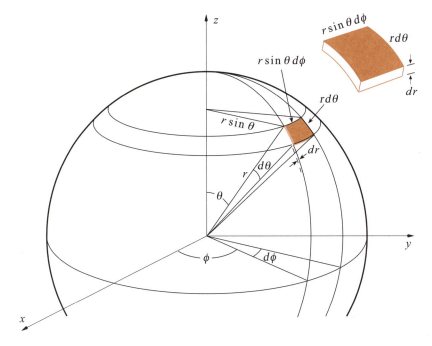

FIGURE E.2
A geometrical construction of the differential volume element in spherical coordinates.

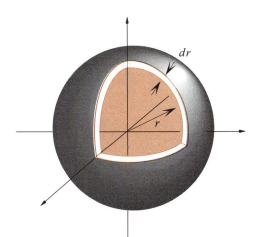

FIGURE E.3
A spherical shell of radius r and thickness dr. The volume of such a shell is $4\pi r^2 dr$, which is its area $(4\pi r^2)$ times its thickness (dr).

is the differential area on the surface of a sphere of radius r. (See Figure E.2.) If we integrate Equation E.5 over all values of θ and ϕ, then we obtain $A = 4\pi r^2$, the area of the surface of a sphere of radius r.

Often, the integral we need to evaluate will be of the form

$$I = \int_0^\infty \int_0^\pi \int_0^{2\pi} F(r, \theta, \phi) r^2 \sin\theta \, dr \, d\theta \, d\phi \tag{E.6}$$

When writing multiple integrals, for convenience we use a notation that treats an integral like an operator. To this end, we write the triple integral in Equation E.6 in the form

$$I = \int_0^\infty dr \, r^2 \int_0^\pi d\theta \, \sin\theta \int_0^{2\pi} d\phi \, F(r, \theta, \phi) \tag{E.7}$$

In Equation E.7, each integral "acts on" everything that lies to its right; in other words, we first integrate $F(r, \theta, \phi)$ over ϕ from 0 to 2π, then multiply the result by $\sin\theta$ and integrate over θ from 0 to π, and finally multiply that result by r^2 and integrate over r from 0 to ∞. The advantage of the notation in Equation E.7 is that the integration variable and its associated limits are always unambiguous. As an example of the application of this notation, let's evaluate Equation E.7 with

$$F(r, \theta, \phi) = \frac{1}{32\pi} r^2 e^{-r} \sin^2\theta \cos^2\phi$$

(We will learn in Chapter 7 that this function is the square of a $2p_x$ hydrogen atomic orbital.) If we substitute $F(r, \theta, \phi)$ into Equation E.7, we obtain

$$I = \frac{1}{32\pi} \int_0^\infty dr \, r^2 \int_0^\pi d\theta \, \sin\theta \int_0^{2\pi} d\phi \, r^2 e^{-r} \sin^2\theta \cos^2\phi$$

The integral over ϕ gives

$$\int_0^{2\pi} d\phi \, \cos^2\phi = \pi$$

so that

$$I = \frac{1}{32} \int_0^\infty dr \, r^2 \int_0^\pi d\theta \, \sin\theta \, r^2 e^{-r} \sin^2\theta \tag{E.8}$$

The integral over θ, I_θ, is

$$I_\theta = \int_0^\pi d\theta \, \sin^3\theta$$

It is often convenient to perform a transformation of variables and let $x = \cos\theta$ in integrals involving θ. Then $\sin\theta \, d\theta$ becomes $-dx$ and the limits become $+1$ to -1, so in this case we have

$$I_\theta = \int_0^\pi d\theta \, \sin^3\theta = -\int_1^{-1} dx(1-x^2) = \int_{-1}^1 dx(1-x^2) = 2 - \frac{2}{3} = \frac{4}{3}$$

Using this result in Equation E.8 gives

$$I = \frac{1}{24} \int_0^\infty dr\, r^4 e^{-r} = \frac{1}{24}(4!) = 1$$

where we have used the general integral

$$\int_0^\infty x^n e^{-x} dx = n!$$

This final result for I simply shows that our above expression for a $2p_x$ hydrogen atomic orbital is normalized.

Frequently, the integrand in Equation E.7 will be a function only of r, in which case we say that the integrand is spherically symmetric. Let's look at Equation E.7 when $F(r, \theta, \phi) = f(r)$:

$$I = \int_0^\infty dr\, r^2 \int_0^\pi d\theta \sin\theta \int_0^{2\pi} d\phi\, f(r) \tag{E.9}$$

Because $f(r)$ is independent of θ and ϕ, we can integrate over ϕ to get 2π and then integrate over θ to get 2:

$$\int_0^\pi \sin\theta\, d\theta = \int_{-1}^1 dx = 2$$

Therefore, Equation E.9 becomes

$$I = \int_0^\infty f(r) 4\pi r^2 dr \tag{E.10}$$

The point here is that if $F(r, \theta, \phi) = f(r)$, then Equation E.7 becomes effectively a one-dimensional integral with a factor of $4\pi r^2 dr$ multiplying the integrand. The quantity $4\pi r^2 dr$ is the volume of a spherical shell of radius r and thickness dr.

EXAMPLE E–1
We will learn in Chapter 7 that a 1s hydrogen atomic orbital is given by

$$f(r) = \frac{1}{(\pi a_0^3)^{1/2}} e^{-r/a_0}$$

Show that the square of this function is normalized.

SOLUTION: Realize that $f(r)$ is a spherically symmetric function of x, y, and z, where $r = (x^2 + y^2 + z^2)^{1/2}$. Therefore, we use Equation E.10 and write

$$I = \int_0^\infty f^2(r) 4\pi r^2 dr = \frac{4\pi}{\pi a_0^3} \int_0^\infty r^2 e^{-2r/a_0} dr$$

$$= \frac{4}{a_0^3} \cdot \frac{2}{(2/a_0)^3} = 1$$

If we restrict ourselves to the surface of a sphere of unit radius, then the angular part of Equation E.5 gives us the differential surface area

$$dA = \sin\theta d\theta d\phi \tag{E.11}$$

If we integrate over the entire spherical surface ($0 \leq \theta \leq \pi$, $0 \leq \phi \leq 2\pi$), then

$$A = \int_0^\pi \sin\theta d\theta \int_0^{2\pi} d\phi = 4\pi \tag{E.12}$$

which is the area of a sphere of unit radius.

We call the solid enclosed by the surface that connects the origin and the area dA a *solid angle*, as shown in Figure E.4. Because of Equation E.12, we say that a complete solid angle is 4π, just as we say that a complete angle of a circle is 2π. We often denote a solid angle by $d\Omega$, so that we sometimes write

$$d\Omega = \sin\theta d\theta d\phi \tag{E.13}$$

and Equation E.12 becomes

$$\int_{\text{sphere}} d\Omega = 4\pi \tag{E.14}$$

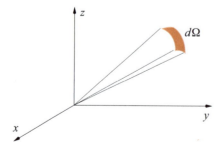

FIGURE E.4
The solid angle, $d\Omega$, subtended by the differential area element $dA = \sin\theta d\theta d\phi$.

In discussing the quantum theory of a hydrogen atom in Chapter 7, we will frequently encounter angular integrals of the form

$$I = \int_0^\pi d\theta \sin\theta \int_0^{2\pi} d\phi \, F(\theta, \phi) \tag{E.15}$$

Note that we are integrating $F(\theta, \phi)$ over the surface of a sphere. For example, we will encounter the integral

$$I = \frac{15}{8\pi} \int_0^{2\pi} d\phi \int_0^\pi d\theta (\sin^2\theta \cos^2\theta) \sin\theta$$

The value of this integral is

$$I = \left(\frac{15}{8\pi}\right) \int_0^\pi d\theta \sin^2\theta \cos^2\theta \sin\theta \int_0^{2\pi} d\phi$$

$$= \frac{15}{4} \int_{-1}^1 (1 - x^2)x^2 dx = \frac{15}{4}\left[\frac{2}{3} - \frac{2}{5}\right] = 1$$

EXAMPLE E–2
Show that

$$I = \int_0^\pi d\theta \sin\theta \int_0^{2\pi} d\phi \, Y_1^1(\theta, \phi)^* Y_1^{-1}(\theta, \phi) = 0$$

where

$$Y_1^1(\theta, \phi) = -\left(\frac{3}{8\pi}\right)^{1/2} e^{i\phi} \sin\theta$$

and

$$Y_1^{-1}(\theta, \phi) = \left(\frac{3}{8\pi}\right)^{1/2} e^{-i\phi} \sin\theta$$

SOLUTION:

$$I = -\frac{3}{8\pi} \int_0^\pi d\theta \sin^3\theta \int_0^{2\pi} d\phi \, e^{-2i\phi}$$

The integral over ϕ is an integral over a complete cycle of $\sin 2\phi$ and $\cos 2\phi$ and therefore $I = 0$. We say that $Y_1^1(\theta, \phi)$ and $Y_1^{-1}(\theta, \phi)$ are orthogonal over the surface of a unit sphere.

There is one final topic involving spherical coordinates that we should discuss here. The operator

$$
\left(\frac{\partial^2}{\partial x^2} + \frac{\partial^2}{\partial y^2} + \frac{\partial^2}{\partial z^2} \right) f = \frac{\partial^2 f}{\partial x^2} + \frac{\partial^2 f}{\partial y^2} + \frac{\partial^2 f}{\partial z^2}
$$

occurs frequently in physical problems. The operator

$$
\nabla^2 = \frac{\partial^2}{\partial x^2} + \frac{\partial^2}{\partial y^2} + \frac{\partial^2}{\partial z^2} \tag{E.16}
$$

is called the *Laplacian operator*. When dealing with problems involving a center of symmetry, so that we use spherical coordinates, we express ∇^2 in terms of spherical coordinates rather than cartesian coordinates. The conversion of ∇^2 from cartesian coordinates to spherical coordinates can be carried out starting with Equation E.1, but it is a long, tedious exercise involving partial derivatives that perhaps you should do once, but probably never again. The final result is (see Problems E–13 and E–14)

$$
\nabla^2 = \frac{1}{r^2} \frac{\partial}{\partial r} \left(r^2 \frac{\partial}{\partial r} \right) + \frac{1}{r^2 \sin \theta} \frac{\partial}{\partial \theta} \left(\sin \theta \frac{\partial}{\partial \theta} \right) + \frac{1}{r^2 \sin^2 \theta} \frac{\partial^2}{\partial \phi^2} \tag{E.17}
$$

EXAMPLE E–3
Show that $u(r, \theta, \phi) = 1/r$ is a solution to $\nabla^2 u = 0$. (This equation is called *Laplace's equation*.)

SOLUTION: The fact that u depends only upon r means that $\nabla^2 u$ reduces to

$$
\nabla^2 u = \frac{1}{r^2} \frac{\partial}{\partial r} \left(r^2 \frac{\partial u}{\partial r} \right)
$$

If we substitute $u = 1/r$ into this expression, we find that $r^2 \partial u / \partial r = -1$ and that $\nabla^2 u = 0$.

EXAMPLE E–4
Show that $u(\theta, \phi) = Y_1^1(\theta, \phi)$ given in Example E–2 satisfies the equation $\nabla^2 u = \frac{c}{r^2} u$, where c is a constant. What is the value of c?

SOLUTION: Because $u(\theta, \phi)$ is independent of r, we start with

$$
\nabla^2 u = \frac{1}{r^2 \sin \theta} \frac{\partial}{\partial \theta} \left(\sin \theta \frac{\partial u}{\partial \theta} \right) + \frac{1}{r^2 \sin^2 \theta} \frac{\partial^2 u}{\partial \phi^2}
$$

Substituting

$$
u(\theta, \phi) = -\left(\frac{3}{8\pi} \right)^{1/2} e^{i\phi} \sin \theta
$$

into $\nabla^2 u$ gives

$$\nabla^2 u = -\left(\frac{3}{8\pi}\right)^{1/2}\left[\frac{e^{i\phi}}{r^2 \sin\theta}(\cos^2\theta - \sin^2\theta) - \frac{\sin\theta}{r^2 \sin^2\theta}e^{i\phi}\right]$$

$$= -\left(\frac{3}{8\pi}\right)^{1/2}\frac{e^{i\phi}}{r^2}\left(\frac{1 - 2\sin^2\theta}{\sin\theta} - \frac{1}{\sin\theta}\right)$$

$$= 2\left(\frac{3}{8\pi}\right)^{1/2}\frac{e^{i\phi}\sin\theta}{r^2}$$

or $c = -2$.

Problems

E–1. Derive Equations E.2 from Equations E.1.

E–2. Express the following points given in cartesian coordinates in terms of spherical coordinates: (x, y, z): $(1, 0, 0)$; $(0, 1, 0)$; $(0, 0, 1)$; $(0, 0, -1)$.

E–3. Describe the graphs of the following equations:

(a) $r = 5$ (b) $\theta = \pi/4$ (c) $\phi = \pi/2$

E–4. Use Equation E.3 to determine the volume of a hemisphere of radius a.

E–5. Use Equation E.5 to determine the surface area of a hemisphere of radius a.

E–6. Evaluate the integral

$$I = \int_0^\pi \cos^2\theta \sin^3\theta \, d\theta$$

by letting $x = \cos\theta$.

E–7. We will learn in Chapter 7 that a $2p_y$ hydrogen atom orbital is given by

$$\psi_{2p_y} = \frac{1}{4\sqrt{2\pi}}re^{-r/2}\sin\theta\sin\phi$$

Show that ψ_{2p_y} is normalized. (Don't forget to square ψ_{2p_y} first.)

E–8. We will learn in Chapter 7 that a $2s$ hydrogen atomic orbital is given by

$$\psi_{2s} = \frac{1}{4\sqrt{2\pi}}(2 - r)e^{-r/2}$$

Show that ψ_{2s} is normalized.

E–9. Show that

$$
Y_1^0(\theta, \phi) = \left(\frac{3}{4\pi}\right)^{1/2} \cos\theta
$$

$$
Y_1^1(\theta, \phi) = -\left(\frac{3}{8\pi}\right)^{1/2} e^{i\phi} \sin\theta
$$

and

$$
Y_1^{-1}(\theta, \phi) = \left(\frac{3}{8\pi}\right)^{1/2} e^{-i\phi} \sin\theta
$$

are orthonormal over the surface of a sphere.

E–10. Evaluate the average of $\cos\theta$ and $\cos^2\theta$ over the surface of a sphere.

E–11. We shall frequently use the notation $d\mathbf{r}$ to represent the volume element in spherical coordinates. Evaluate the integral

$$
I = \int d\mathbf{r}\, e^{-r} \cos^2\theta
$$

where the integral is over all space (in other words, over all possible values of r, θ, and ϕ).

E–12. Show that the two functions

$$
f_1(r) = e^{-r} \cos\theta \quad \text{and} \quad f_2(r) = (2 - r)e^{-r/2} \cos\theta
$$

are orthogonal over all space (in other words, over all possible values of r, θ, and ϕ).

E–13. Consider the transformation from cartesian coordinates to plane polar coordinates,

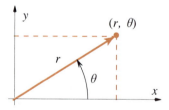

where

$$
x = r \cos\theta \qquad r = (x^2 + y^2)^{1/2}
$$

$$
y = r \sin\theta \qquad \theta = \tan^{-1}\left(\frac{y}{x}\right)
$$

(1)

If a function $f(r, \theta)$ depends upon the polar coordinates r and θ, then the chain rule of partial differentiation says that

$$
\left(\frac{\partial f}{\partial x}\right)_y = \left(\frac{\partial f}{\partial r}\right)_\theta \left(\frac{\partial r}{\partial x}\right)_y + \left(\frac{\partial f}{\partial \theta}\right)_r \left(\frac{\partial \theta}{\partial x}\right)_y
$$

(2)

and that

$$\left(\frac{\partial f}{\partial y}\right)_x = \left(\frac{\partial f}{\partial r}\right)_\theta \left(\frac{\partial r}{\partial y}\right)_x + \left(\frac{\partial f}{\partial \theta}\right)_r \left(\frac{\partial \theta}{\partial y}\right)_x \tag{3}$$

For simplicity, we will assume that r is equal to a constant, l, so that we can ignore terms involving derivatives with respect to r. In other words, we will consider a particle that is constrained to move on the circumference of a circle. This system is sometimes called a *particle on a ring*. Using equations 1 and 2, show that

$$\left(\frac{\partial f}{\partial x}\right)_y = -\frac{\sin\theta}{l}\left(\frac{\partial f}{\partial \theta}\right)_r \qquad \text{and} \qquad \left(\frac{\partial f}{\partial y}\right)_x = \frac{\cos\theta}{l}\left(\frac{\partial f}{\partial \theta}\right)_r \tag{4}$$

Now apply equation 2 again to show that

$$\left(\frac{\partial^2 f}{\partial x^2}\right)_y = \left[\frac{\partial}{\partial x}\left(\frac{\partial f}{\partial x}\right)_y\right] = \left[\frac{\partial}{\partial \theta}\left(\frac{\partial f}{\partial x}\right)_y\right]_l \left(\frac{\partial \theta}{\partial x}\right)_y$$

$$= \left\{\frac{\partial}{\partial \theta}\left[-\frac{\sin\theta}{l}\left(\frac{\partial f}{\partial \theta}\right)_r\right]\right\}_l \left(-\frac{\sin\theta}{l}\right)$$

$$= \frac{\sin\theta\cos\theta}{l^2}\left(\frac{\partial f}{\partial \theta}\right) + \frac{\sin^2\theta}{l^2}\left(\frac{\partial^2 f}{\partial \theta^2}\right)$$

Similarly, show that

$$\left(\frac{\partial^2 f}{\partial y^2}\right)_x = -\frac{\sin\theta\cos\theta}{l^2}\left(\frac{\partial f}{\partial \theta}\right) + \frac{\cos^2\theta}{l^2}\left(\frac{\partial^2 f}{\partial \theta^2}\right)$$

and that

$$\nabla^2 f = \frac{\partial^2 f}{\partial x^2} + \frac{\partial^2 f}{\partial y^2} \longrightarrow \frac{1}{l^2}\left(\frac{\partial^2 f}{\partial \theta^2}\right)$$

E–14. Generalize Problem E–13 to the case of a particle moving in a plane under the influence of a central force; in other words, convert

$$\nabla^2 = \frac{\partial^2}{\partial x^2} + \frac{\partial^2}{\partial y^2}$$

to plane polar coordinates, this time without assuming that r is a constant. Use the method of separation of variables to separate the equation for this problem. Solve the angular equation.

E–15. Show that $u(r, \theta, \phi) = r\sin\theta\cos\phi$ satisfies Laplace's equation, $\nabla^2 u = 0$.

E–16. Show that $u(r, \theta, \phi) = r\sin^2\theta\cos 2\phi$ satisfies Laplace's equation, $\nabla^2 u = 0$.

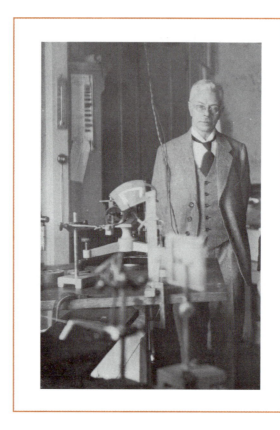

Pieter Zeeman was born on May 25, 1865, in a small village on the isle of Schouwen, Zeeland, the Netherlands, and died in Amsterdam in 1943. Although he showed early signs of excellence in science, he had to leave his small village and move to Delft to study Greek and Latin, which were required for acceptance to the University of Leyden. In 1885, he became a student at Leyden, where he became an assistant to the prominent physicist, Hendrik Lorentz. After obtaining his doctorate in 1893, he remained at Leyden except for a brief period of time to study in Strasbourg. In 1896, he announced his discovery of the magnetic splitting of spectral lines, and the following year he received a position at the University of Amsterdam. He received many offers of prestigous chairs in other countries but chose to remain at Amsterdam, where he remained until he reached the mandatory retirement age in 1935. In 1923, the university had a new laboratory specially built for his work, which is now called the Zeeman Laboratory. Zeeman was an excellent teacher, remembered fondly by his many students. Outside of science, he was interested in literature and drama and he enjoyed entertaining his collaborators and students at his home. He shared the Nobel Prize in Physics in 1902 (the second one awarded) with Lorentz for "their researches into the influence of magnetism upon radiation phenomena."

CHAPTER 6

The Rigid Rotator and Rotational Spectroscopy

The rotation of a diatomic molecule can be described by a rigid rotator, which is like a rotating dumbbell. Using the concept of a reduced mass, we can picture the rotator as a body of reduced mass μ rotating about a fixed center. This problem lends itself naturally to spherical coordinates, and the wave functions are functions of θ and ϕ, the two angles in spherical coordinates (MathChapter E). The wave functions are called spherical harmonics and occur in many problems involving a fixed center. In fact, we shall see in the next chapter that the s, p, d, etc., orbitals of a hydrogen atom are described by spherical harmonics.

Solving the Schrödinger equation for a rigid rotator is a fairly lengthy process, and so in most of this chapter we shall simply present and discuss the energies and the corresponding wave functions. We shall use the allowed energies to describe the microwave spectrum of a diatomic molecule and see how you can determine bond lengths.

6.1 The Energy Levels of a Rigid Rotator Are $E = \hbar^2 J(J+1)/2I$

In this section, we will discuss a simple model for a rotating diatomic molecule. The model consists of two point masses m_1 and m_2 at fixed distances l_1 and l_2 from their center of mass (cf. Figure 6.1). Because the distance between the two masses is fixed, this model is referred to as the *rigid-rotator model*. Even though a diatomic molecule vibrates as it rotates, the vibrational amplitude is small compared with the bond length, so considering the bond length fixed is a good approximation (see Problem 5–38).

Let the molecule rotate about its center of mass at a frequency of ν_{rot} cycles per second. The velocities of the two masses are $v_1 = 2\pi l_1 \nu_{rot}$ and $v_2 = 2\pi l_2 \nu_{rot}$, which we write as $v_1 = l_1 \omega$ and $v_2 = l_2 \omega$, where ω (radians per second) $= 2\pi \nu_{rot}$ and is called the

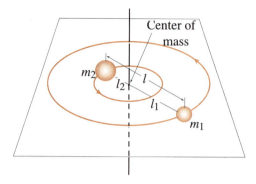

FIGURE 6.1
Two masses, m_1 and m_2, rotating about
their center of mass.

angular velocity (Section 1.7). The kinetic energy of the rigid rotator is

$$T = \frac{1}{2}m_1v_1^2 + \frac{1}{2}m_2v_2^2 = \frac{1}{2}(m_1l_1^2 + m_2l_2^2)\omega^2$$

$$= \frac{1}{2}I\omega^2 \tag{6.1}$$

where I, the *moment of inertia*, is given by

$$I = m_1l_1^2 + m_2l_2^2 \tag{6.2}$$

Using the fact that the location of the center of mass is given by $m_1l_1 = m_2l_2$, the moment of inertia can be rewritten as (Problem 6–1)

$$I = \mu l^2 \tag{6.3}$$

where $l = l_1 + l_2$ (the fixed separation of the two masses) and μ is the *reduced mass* (Section 5.3). In Section 1.7, we discussed a single body of mass m rotating at a distance l from a fixed center. In that case, the moment of inertia, I, was equal to ml^2. By comparing Equation 6.3 with this result, we may consider Equation 6.3 to be an equation for the moment of inertia of a single body of mass μ rotating at a distance l from a fixed center. Thus, we have transformed a two-body problem into an equivalent one-body problem, just as we did for a harmonic oscillator in Section 5.3.

Following Equations 1.17 and 1.19, the angular momentum L is

$$L = I\omega \tag{6.4}$$

and the kinetic energy is

$$T = \frac{L^2}{2I} \tag{6.5}$$

There is no potential energy term because in the absence of any external forces (e.g., electric or magnetic forces) the energy of the molecule does not depend on its orientation in space. The Hamiltonian operator of a rigid rotator is therefore just the kinetic energy operator. Using the operator $\hat{T}$ given in Table 4.1 and the correspondences between linear and rotating systems given in Table 1.2, we can write the Hamiltonian operator of a rigid rotator as

$$\hat{H} = \hat{T} = -\frac{\hbar^2}{2\mu}\nabla^2 \quad (r \text{ constant}) \tag{6.6}$$

where ∇^2 is the Laplacian operator. We encountered ∇^2 in cartesian coordinates in Section 3.9, but if the system has a natural center of symmetry, such as one particle revolving around one fixed at the origin, then using spherical coordinates (MathChapter E) is much more convenient. Therefore, we must convert ∇^2 from cartesian coordinates to spherical coordinates. This conversion involves a tedious exercise in the chain rule of partial differentiation, which is best left as a problem (see Problems 6–2 and 6–3). The final result is

$$\nabla^2 = \frac{1}{r^2}\frac{\partial}{\partial r}\left(r^2\frac{\partial}{\partial r}\right)_{\theta,\phi} + \frac{1}{r^2 \sin\theta}\frac{\partial}{\partial\theta}\left(\sin\theta\frac{\partial}{\partial\theta}\right)_{r,\phi} + \frac{1}{r^2 \sin^2\theta}\left(\frac{\partial^2}{\partial\phi^2}\right)_{r,\theta} \tag{6.7}$$

where the subscripts indicate which variables are held constant. The rigid rotator is a special case where r is equal to a constant, l, so Equation 6.7 becomes

$$\nabla^2 = \frac{1}{l^2}\frac{1}{\sin\theta}\frac{\partial}{\partial\theta}\left(\sin\theta\frac{\partial}{\partial\theta}\right) + \frac{1}{l^2}\frac{1}{\sin^2\theta}\frac{\partial^2}{\partial\phi^2} \quad (r \text{ constant}) \tag{6.8}$$

If we use this result in Equation 6.6, we obtain

$$\hat{H} = -\frac{\hbar^2}{2I}\left[\frac{1}{\sin\theta}\frac{\partial}{\partial\theta}\left(\sin\theta\frac{\partial}{\partial\theta}\right) + \frac{1}{\sin^2\theta}\left(\frac{\partial^2}{\partial\phi^2}\right)\right] \tag{6.9}$$

Because $\hat{H} = \hat{L}^2/2I$, we see we can make the correspondence

$$\hat{L}^2 = -\hbar^2\left[\frac{1}{\sin\theta}\frac{\partial}{\partial\theta}\left(\sin\theta\frac{\partial}{\partial\theta}\right) + \frac{1}{\sin^2\theta}\left(\frac{\partial^2}{\partial\phi^2}\right)\right] \tag{6.10}$$

Note that the square of the angular momentum is a naturally occurring operator in quantum mechanics. Both θ and ϕ are unitless, so Equation 6.10 shows that the natural units of angular momentum are $\hbar$ for atomic and molecular systems. We will make use of this fact later.

The orientation of a rigid rotator is completely specified by the two angles θ and ϕ, so rigid-rotator wave functions depend upon only these two variables. The rigid-rotator wave functions are customarily denoted by $Y(\theta, \phi)$, so the Schrödinger equation for a rigid rotator reads

$$\hat{H}Y(\theta, \phi) = EY(\theta, \phi)$$

or

$$-\frac{\hbar^2}{2I}\left[\frac{1}{\sin\theta}\frac{\partial}{\partial\theta}\left(\sin\theta\frac{\partial}{\partial\theta}\right) + \frac{1}{\sin^2\theta}\left(\frac{\partial^2}{\partial\phi^2}\right)\right]Y(\theta, \phi) = EY(\theta, \phi) \quad (6.11)$$

If we multiply Equation 6.11 by $\sin^2\theta$ and let

$$\beta = \frac{2IE}{\hbar^2} \tag{6.12}$$

we find the partial differential equation

$$\sin\theta\frac{\partial}{\partial\theta}\left(\sin\theta\frac{\partial Y}{\partial\theta}\right) + \frac{\partial^2 Y}{\partial\phi^2} + (\beta\sin^2\theta)Y = 0 \tag{6.13}$$

To solve Equation 6.13, we again use the method of separation of variables and let

$$Y(\theta, \phi) = \Theta(\theta)\Phi(\phi) \tag{6.14}$$

If we substitute Equation 6.14 into Equation 6.13 and divide by $\Theta(\theta)\Phi(\phi)$, we find

$$\frac{\sin\theta}{\Theta(\theta)}\frac{d}{d\theta}\left(\sin\theta\frac{d\Theta}{d\theta}\right) + \beta\sin^2\theta + \frac{1}{\Phi(\phi)}\frac{d^2\Phi}{d\phi^2} = 0 \tag{6.15}$$

Because θ and ϕ are independent variables, we must have

$$\frac{\sin\theta}{\Theta(\theta)}\frac{d}{d\theta}\left(\sin\theta\frac{d\Theta}{d\theta}\right) + \beta\sin^2\theta = m^2 \tag{6.16}$$

and

$$\frac{1}{\Phi(\phi)}\frac{d^2\Phi}{d\phi^2} = -m^2 \tag{6.17}$$

where m^2 is a constant. We use m^2 as a separation constant in anticipation of using the square of the separation constant in later equations.

Because Equation 6.17 contains only constant coefficients, it is relatively easy to solve. Its solutions are

$$\Phi(\phi) = A_m e^{im\phi} \qquad \text{and} \qquad \Phi(\phi) = A_{-m}e^{-im\phi} \tag{6.18}$$

The requirement that $\Phi(\phi)$ be a single-valued function of ϕ is

$$\Phi(\phi + 2\pi) = \Phi(\phi) \tag{6.19}$$

By substituting Equation 6.18 into Equation 6.19, we see that

$$A_m e^{im(\phi+2\pi)} = A_m e^{im\phi} \tag{6.20}$$

and that

$$A_{-m} e^{-im(\phi+2\pi)} = A_{-m} e^{-im\phi} \tag{6.21}$$

Equations 6.20 and 6.21 together imply that

$$e^{\pm i2\pi m} = 1 \tag{6.22}$$

In terms of sines and cosines, Equation 6.22 is (Equation A.6)

$$\cos(2\pi m) \pm i \sin(2\pi m) = 1$$

which implies that $m = 0, \pm 1, \pm 2, \ldots$, because $\cos 2\pi m = 1$ and $\sin 2\pi m = 0$ for $m = 0, \pm 1, \pm 2, \ldots$. Thus, Equations 6.18 can be written as one equation:

$$\Phi_m(\phi) = A_m e^{im\phi} \qquad m = 0, \pm 1, \pm 2, \ldots \tag{6.23}$$

We can find the value of A_m by requiring that the $\Phi_m(\phi)$ be normalized.

EXAMPLE 6–1
Determine the value of A_m in Equation 6.23.

SOLUTION: The A_m in Equation 6.23 are determined by the requirement that the $\Phi_m(\phi)$ are normalized. The normalization condition is that

$$\int_0^{2\pi} \Phi_m^*(\phi)\Phi_m(\phi)d\phi = 1$$

Using Equation 6.20 for the $\Phi_m(\phi)$, we have

$$|A_m|^2 \int_0^{2\pi} d\phi = 1$$

or

$$|A_m|^2 2\pi = 1$$

or

$$A_m = (2\pi)^{-1/2}$$

Thus, the normalized version of Equation 6.23 is

$$\Phi_m(\phi) = \frac{1}{(2\pi)^{1/2}} e^{im\phi} \qquad m = 0, \pm1, \pm2, \ldots \qquad (6.24)$$

When we solve Equation 6.16, it turns out naturally that β, given by Equation 6.12, must obey the condition

$$\beta = J(J+1) \qquad J = 0, 1, 2, \ldots$$

Substituting this result into Equation 6.12 gives

$$E_J = \frac{\hbar^2}{2I} J(J+1) = BJ(J+1) \qquad J = 0, 1, 2, \ldots \qquad (6.25)$$

where $B = \hbar^2/2I$ is called the *rotational constant* of the molecule. Once again, we obtain a set of discrete energy levels. In addition to the allowed energies given by Equation 6.25, we also find that each energy level has a degeneracy g_J given by

$$g_J = 2J + 1 \qquad (6.26)$$

We'll see later that this degeneracy is due to the fact that J can take on the values $-m$ through $+m$.

A standard method to solve Equation 6.16 and obtain the result $\beta = J(J+1)$ for $J = 0, 1, 2, \ldots$ is by the power series method, which we mentioned in Section 5.5 for a harmonic oscillator. We also pointed out there that we could use a more elegant method using operators to obtain the eigenvalues and eigenfunctions, and this is also true for a rigid rotator. We present this operator method for a rigid rotator in the appendix at the end of the chapter.

6.2 The Rigid Rotator Is a Model for a Rotating Diatomic Molecule

The allowed energies of a rigid rotator are given by Equation 6.25. We will show in Section 6.7 that the selection rule for the rigid rotator says that transitions are allowed only from adjacent states or that

$$\Delta J = \pm 1 \qquad (6.27)$$

In addition to the requirement that $\Delta J = \pm 1$, the molecule must also possess a permanent dipole moment to absorb electromagnetic radiation. Thus, HCl has a pure rotational spectrum, but N_2 does not. In the case of absorption of electromagnetic radiation, the molecule goes from a state with a quantum number J to one with $J + 1$. The energy difference, then, is

$$\Delta E = E_{J+1} - E_J = \frac{\hbar^2}{2I}[(J + 1)(J + 2) - J(J + 1)]$$

$$= \frac{\hbar^2}{I}(J + 1) = \frac{h^2}{4\pi^2 I}(J + 1) \qquad J = 0, 1, 2, \ldots \qquad (6.28)$$

The energy levels and the absorption transitions are shown in Figure 6.2.

Using the Bohr frequency condition $\Delta E = h\nu_{obs}$, the frequencies at which the absorption transitions occur are

$$\nu_{obs} = \frac{h}{4\pi^2 I}(J + 1) \qquad J = 0, 1, 2, \ldots \qquad (6.29)$$

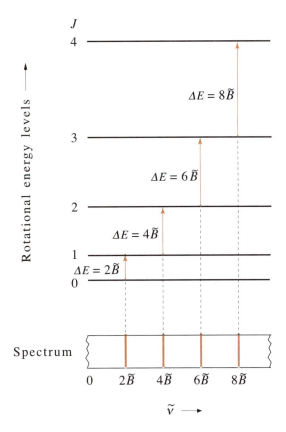

FIGURE 6.2

The energy levels and absorption transitions of a rigid rotator. The absorption transitions occur between adjacent levels, so the absorption spectrum shown below the energy levels consists of a series of equally spaced lines. The quantity $\tilde{B}$ is $h/8\pi^2 cI$ (Equation 6.33).

The reduced mass of a diatomic molecule is typically around 10^{-25} to 10^{-26} kg, and a typical bond distance is approximately 10^{-10} m (100 pm), so the moment of inertia of a diatomic molecule typically ranges from 10^{-45} to 10^{-46} kg·m^2. Substituting $I = 5 \times 10^{-46}$ kg·m^2 into Equation 6.29 shows that the absorption frequencies are about 2×10^{10} to 10^{11} Hz (cf. Problem 6–4). By referring to Figure 1.11 in Problem 1–1, we see that these frequencies lie in the microwave region. Consequently, rotational transitions of diatomic molecules occur in the microwave region, and the study of rotational transitions in molecules is called *microwave spectroscopy*.

It is common practice in microwave spectroscopy to write Equation 6.29 as

$$\nu_{obs} = 2B(J + 1) \qquad J = 0, 1, 2, \ldots \tag{6.30}$$

where

$$B = \frac{h}{8\pi^2 I} \quad \text{(Hz)} \tag{6.31}$$

is the rotational constant expressed in units of hertz. Furthermore, the transition frequency in Equation 6.29 is commonly expressed in terms of wave numbers (cm^{-1}) rather than hertz (Hz). If we use the relation $\tilde{\omega} = \nu/c$, then Equation 6.30 becomes

$$\tilde{\omega}_{obs} = 2\tilde{B}(J + 1) \qquad J = 0, 1, 2, \ldots \tag{6.32}$$

where $\tilde{B}$ is the rotational constant expressed in units of wave numbers:

$$\tilde{B} = \frac{h}{8\pi^2 c I} \quad \text{(cm}^{-1}\text{)} \tag{6.33}$$

The tilde here emphasizes that $\tilde{B}$ has units of wave numbers. From either Equation 6.30 or 6.32, we see that the rigid-rotator model predicts that the microwave spectrum of a diatomic molecule consists of a series of equally spaced lines with a separation of $2B$ Hz or $2\tilde{B}$ cm^{-1}, as shown in Figure 6.2. From the separation between the absorption frequencies, we can determine the rotational constant and hence the moment of inertia of the molecule. Furthermore, because $I = \mu l^2$, where l is the internuclear distance or bond length, we can determine the bond length. This procedure is illustrated in Example 6–2.

EXAMPLE 6–2
To a good approximation, the microwave spectrum of H^{35}Cl consists of a series of equally spaced lines, separated by 6.26×10^{11} Hz. Calculate the bond length of H^{35}Cl.

SOLUTION: According to Equation 6.30, the spacing of the lines in the microwave spectrum of H^{35}Cl is given by

$$2B = \frac{h}{4\pi^2 I}$$

and so

$$\frac{h}{4\pi^2 I} = 6.26 \times 10^{11} \text{ Hz}$$

Solving this equation for I, we have

$$I = \frac{6.626 \times 10^{-34} \text{ J} \cdot \text{s}}{4\pi^2(6.26 \times 10^{11} \text{ s}^{-1})} = 2.68 \times 10^{-47} \text{ kg} \cdot \text{m}^2$$

The reduced mass of $H^{35}Cl$ is

$$\mu = \frac{(1.00 \text{ amu})(35.0 \text{ amu})}{36.0 \text{ amu}}(1.661 \times 10^{-27} \text{ kg} \cdot \text{amu}^{-1}) = 1.66 \times 10^{-27} \text{ kg}$$

Using the fact that $I = \mu l^2$, we obtain

$$l = \left(\frac{2.68 \times 10^{-47} \text{ kg} \cdot \text{m}^2}{1.661 \times 10^{-27} \text{ kg}}\right)^{1/2} = 1.29 \times 10^{-10} \text{ m} = 129 \text{ pm}$$

Problems 6–5 through 6–8 give other examples of the determination of bond lengths from microwave data.

6.3 Rotational Transitions Accompany Vibrational Transitions

In the previous chapter, we showed that the vibrational motion of a diatomic molecule can be well approximated by a harmonic oscillator, and in this chapter, we have just seen that the rotational motion of a diatomic molecule can be well represented by a rigid rotator. We combine these ideas and say that the rotational-vibrational motion of a diatomic molecule can be well approximated by the rigid rotator–harmonic oscillator model.

Within the rigid rotator–harmonic oscillator approximation, the rotational and vibrational energy of a diatomic molecule is given by the sum of Equations 5.32 and 6.25:

$$E_{v,J} = \left(v + \frac{1}{2}\right)h\nu + BJ(J+1) \tag{6.34}$$

If we express $E_{v,J}$ in terms of wave numbers (cm^{-1}) by dividing by hc, we obtain

$$\tilde{E}_{v,J} = (v + \tfrac{1}{2})\tilde{\omega} + \tilde{B}J(J+1) \qquad v = 0, 1, 2, \ldots \qquad J = 0, 1, 2, \ldots \tag{6.35}$$

where

$$\tilde{\omega} = \frac{1}{2\pi c}\left(\frac{k}{\mu}\right)^{1/2} \qquad \text{and} \qquad \tilde{B} = \frac{h}{8\pi^2 cI} \tag{6.36}$$

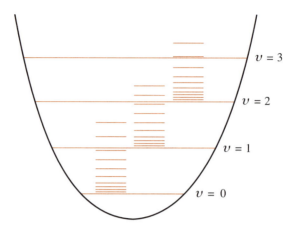

FIGURE 6.3
An energy diagram showing the rotational levels associated with each vibrational state for a diatomic molecule.

Typical values of $\tilde{\omega}$ and $\tilde{B}$ are of the order of $10^3\ cm^{-1}$ and $1\ cm^{-1}$, respectively, so the spacing between vibrational energy levels is about 100 to 1000 times the spacing between rotational levels. This result is shown schematically in Figure 6.3.

When a molecule absorbs infrared radiation, the vibrational transition is accompanied by a rotational transition. The selection rules for absorption of infrared radiation in the rigid rotator–harmonic oscillator approximation are

$$\begin{array}{c} \Delta v = +1 \\ \Delta J = \pm 1 \end{array} \quad \text{absorption} \tag{6.37}$$

For the case $\Delta J = +1$, Equation 6.35 gives

$$\tilde{\nu}_{obs}(\Delta J = +1) = \tilde{E}_{v+1,J+1} - \tilde{E}_{v,J}$$

$$= \left(v + \frac{3}{2}\right)\tilde{\omega} + \tilde{B}(J+1)(J+2) - \left(v + \frac{1}{2}\right)\tilde{\omega} + \tilde{B}J(J+1)$$

$$= \tilde{\omega} + 2\tilde{B}(J+1) \qquad J = 0, 1, 2, \ldots \tag{6.38}$$

and likewise for the case $\Delta J = -1$, we have

$$\tilde{\nu}_{obs}(\Delta J = -1) = \tilde{E}_{v+1,J-1} - \tilde{E}_{v,J} = \tilde{\omega} - 2\tilde{B}J \qquad J = 1, 2, \ldots \tag{6.39}$$

In both Equations 6.38 and 6.39, J is the initial rotational quantum number. Typically, $\tilde{\omega} \approx 10^3\ cm^{-1}$ and $\tilde{B} \approx 1\ cm^{-1}$, so the spectrum predicted by Equations 6.38 and 6.39 contains lines at $10^3\ cm^{-1} \pm$ integral multiples of $\approx 1\ cm^{-1}$. Notice that

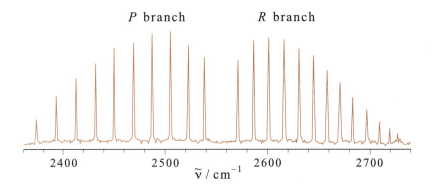

FIGURE 6.4
The rotation-vibration spectrum of the $0 \to 1$ vibrational transition of HBr(g). The R branch and the P branch are indicated in the figure. (Adapted from Barrow, G., *Introduction to Molecular Spectroscopy*; McGraw-Hill: New York, 1962. Used with permission of McGraw-Hill Book Company.)

there is no line at $\tilde{\omega}$ because the transition $\Delta J = 0$ is forbidden. The rotation-vibration spectrum of HBr(g) is shown in Figure 6.4.

The gap centered around 2560 cm^{-1} corresponds to the missing line at $\tilde{\omega}$. On each side of the gap is a series of lines whose spacing is about 10 cm^{-1}. The series toward the high-frequency side is called the R *branch* and is due to rotational transitions with $\Delta J = +1$. The series toward the low frequencies is called the P *branch* and is due to rotational transitions with $\Delta J = -1$.

EXAMPLE 6–3
The bond length of ^{12}C^{14}N is 117 pm and its force constant is 1630 N·m^{-1}. Predict the rotation-vibration spectrum of ^{12}C^{14}N.

SOLUTION: First we must calculate the fundamental frequency $\tilde{\omega}$ and the rotational constant $\tilde{B}$ (Equations 6.36). Both quantities require the reduced mass, which is

$$\mu = \frac{(12.0 \text{ amu})(14.0 \text{ amu})}{(12.0 + 14.0) \text{ amu}}(1.661 \times 10^{-27} \text{ kg·amu}^{-1}) = 1.07 \times 10^{-26} \text{ kg}$$

Using Equation 6.36 for $\tilde{\omega}$,

$$\tilde{\omega} = \frac{1}{2\pi c}\left(\frac{k}{\mu}\right)^{1/2} = \frac{1}{2\pi(2.998 \times 10^8 \text{ m})}\left(\frac{1630 \text{ N·m}^{-1}}{1.07 \times 10^{-26} \text{ kg}}\right)^{1/2}$$

$$= 2.07 \times 10^5 \text{ m}^{-1} = 2.07 \times 10^3 \text{ cm}^{-1}$$

and for $\tilde{B}$,

$$\tilde{B} = \frac{h}{8\pi^2 cI} = \frac{h}{8\pi^2 c\mu l^2}$$

$$= \frac{6.626 \times 10^{-34}\text{ J·s}}{8\pi^2(2.998 \times 10^8\text{ m·s}^{-1})(1.07 \times 10^{-26}\text{ kg})(117 \times 10^{-12}\text{ m})^2}$$

$$= 191\text{ m}^{-1} = 1.91\text{ cm}^{-1}$$

The rotation-vibration spectrum will consist of lines at $\tilde{\omega} \pm 2\tilde{B}J$ where $J = 1, 2,$ 3, ... There will be no line at $\tilde{\omega}$, and the separation of the lines in the P and R branches will be $2\tilde{B} = 3.82\text{ cm}^{-1}$ (cf. Figure 6.4 for HBr).

If we compare the results of Example 6–3 with experimental data, or look closely at Figure 6.4, we see several features we cannot explain. Close examination shows that the lines in the R branch are more closely spaced with increasing frequency and that the lines of the P branch become further apart with decreasing frequency. We will discuss the spacing of the lines in the R and P branches in the next section.

6.4 Rotation-Vibration Interaction Accounts for the Unequal Spacing of the Lines in the P and R Branches of a Rotation-Vibration Spectrum

The energies of a rigid rotator–harmonic oscillator are given by (Equation 6.35)

$$\tilde{E}_{v,J} = \tilde{\omega}(v + \tfrac{1}{2}) + \tilde{B}J(J + 1)$$

where $\tilde{B} = h/8\pi^2 cI = h/8\pi^2 c\mu l^2$. Because the vibrational amplitude increases with the vibrational state (cf. Figure 6.3), we expect that l should increase slightly with v, causing $\tilde{B}$ to decrease with increasing v. We will indicate the dependence of $\tilde{B}$ upon v by using $\tilde{B}_v$ in place of $\tilde{B}$:

$$\tilde{E}_{v,J} = \tilde{\omega}(v + \tfrac{1}{2}) + \tilde{B}_v J(J + 1) \tag{6.40}$$

The dependence of $\tilde{B}$ on v is called *rotation-vibration interaction*. If we consider a $v = 0 \to 1$ transition, then the frequencies of the P and R branches will be given by

$$\tilde{\nu}_R(\Delta J = +1) = E_{1,J+1} - E_{0,J}$$

$$= \frac{3}{2}\tilde{\omega} + \tilde{B}_1(J + 1)(J + 2) - \frac{1}{2}\tilde{\omega} - \tilde{B}_0 J(J + 1)$$

$$= \tilde{\omega} + 2\tilde{B}_1 + (3\tilde{B}_1 - \tilde{B}_0)J + (\tilde{B}_1 - \tilde{B}_0)J^2 \qquad J = 0, 1, 2, \ldots$$

$$\tag{6.41}$$

and

$$\tilde{v}_P(\Delta J = -1) = E_{1,J-1} - E_{0,J}$$

$$= \tilde{\omega} - (\tilde{B}_1 + \tilde{B}_0)J + (\tilde{B}_1 - \tilde{B}_0)J^2 \qquad J = 1, 2, 3, \ldots$$
(6.42)

In both cases, J corresponds to the initial rotational quantum number. Note that Equations 6.41 and 6.42 reduce to Equations 6.38 and 6.39 if $\tilde{B}_1 = \tilde{B}_0$. Because the bond length increases with increasing v, $\tilde{B}_1 < \tilde{B}_0$, and therefore the spacing between the lines in the R branch decreases and the spacing between the lines in the P branch increases with increasing J. This behavior is reflected in Figure 6.4.

EXAMPLE 6–4
The lines in the R and P branches are customarily labeled by the initial value of the rotational quantum number giving rise to the lines. Thus, the lines given by Equation 6.41 are $R(0)$, $R(1)$, $R(2)$, ..., and those given by Equation 6.42 are $P(1)$, $P(2)$, ... Given the following data for $^1H^{127}I$

Line	Frequency/cm^{-1}
$R(0)$	2242.087
$R(1)$	2254.257
$P(1)$	2216.723
$P(2)$	2203.541

calculate $\tilde{B}_0$ and $\tilde{B}_1$ and $l(v = 0)$ and $l(v = 1)$. Take the reduced mass of the molecule to be 1.660×10^{-27} kg.

SOLUTION: Using Equation 6.41 with $J = 0$ and 1 and Equation 6.42 with $J = 1$ and 2, we have

$$\left. \begin{array}{l} 2242.087 \text{ cm}^{-1} = \tilde{\omega}_0 + 2\tilde{B}_1 \\ 2254.257 \text{ cm}^{-1} = \tilde{\omega}_0 + 6\tilde{B}_1 - 2\tilde{B}_0 \end{array} \right\} R \text{ branch}$$

and

$$\left. \begin{array}{l} 2216.723 \text{ cm}^{-1} = \tilde{\omega}_0 - 2\tilde{B}_0 \\ 2203.541 \text{ cm}^{-1} = \tilde{\omega}_0 + 2\tilde{B}_1 - 6\tilde{B}_0 \end{array} \right\} P \text{ branch}$$

If we subtract the first line of the P branch from the second line of the R branch, we find

$$37.534 \text{ cm}^{-1} = 6\tilde{B}_1$$

or $\tilde{B}_1 = 6.256$ cm^{-1}. If we subtract the second line of the P branch from the first line of the R branch, we find

$$38.546 \text{ cm}^{-1} = 6\tilde{B}_0$$

or $\tilde{B}_0 = 6.424$ cm^{-1}. Using the fact that $\tilde{B}_v = h/8\pi^2 c\mu l^2(v)$, we obtain $l(v=0) = 162.0$ pm and $l(v=1) = 164.1$ pm.

The dependence of $\tilde{B}_v$ on v is usually expressed as

$$\tilde{B}_v = \tilde{B}_e - \tilde{\alpha}_e(v + \tfrac{1}{2}) \tag{6.43}$$

Using the values of $\tilde{B}_0$ and $\tilde{B}_1$ from the above example, we find that $\tilde{B}_e = 6.508$ cm^{-1} and that $\tilde{\alpha}_e = 0.168$ cm^{-1}. Values of $\tilde{B}_e$ and $\tilde{\alpha}_e$ as well as other spectroscopic parameters are given in Table 6.1.

TABLE 6.1
Some Rotational Spectroscopic Parameters of Some Diatomic Molecules in the Ground Electronic State

Molecule	$\tilde{B}_e$/cm^{-1}	$\tilde{\alpha}_e$/cm^{-1}	$\tilde{D}$/cm^{-1}	$\bar{l}(v=0)$/pm	D_0/kJ·mol^{-1}
H_2	60.8530	3.0622	4.71×10^{-2}	74.14	432.1
$H^{19}F$	20.9557	0.798	2.15×10^{-3}	91.68	566.2
$H^{35}Cl$	10.5934	0.3072	5.319×10^{-4}	127.46	427.8
$H^{79}Br$	8.4649	0.2333	3.458×10^{-4}	141.44	362.6
$H^{127}I$	6.510	0.1689	2.069×10^{-4}	160.92	294.7
$^{12}C^{16}O$	1.9313	0.0175	6.122×10^{-6}	112.83	1070.2
$^{14}N^{16}O$	1.6719	0.0171	5.4×10^{-6}	115.08	626.8
$^{14}N^{14}N$	1.9982	0.0173	5.76×10^{-6}	109.77	941.6
$^{16}O^{16}O$	1.4456	0.0159	4.84×10^{-6}	120.75	493.6
$^{19}F^{19}F$	0.8902	0.1385	3.3×10^{-6}	141.19	154.6
$^{35}Cl^{35}Cl$	0.2440	0.00149	1.86×10^{-7}	198.79	239.2
$^{79}Br^{79}Br$	0.0821	0.00319	2.09×10^{-8}	228.11	190.1
$^{127}I^{127}I$	0.03737	0.00011	4.25×10^{-9}	266.63	148.8
$^{35}Cl^{19}F$	0.5165	0.00437	8.77×10^{-7}	162.83	252.5
$^{23}Na^{23}Na$	0.1547	0.00087	5.81×10^{-7}	307.89	69.5
$^{39}K^{39}K$	0.05674	0.000165	8.63×10^{-8}	390.51	49.6

TABLE 6.2
The Rotational Absorption Spectrum of $H^{35}Cl$

Transition	$\tilde{\omega}_{obs}/cm^{-1}$	$\Delta\tilde{\omega}_{obs}/cm^{-1}$	$\tilde{\omega}_{calc} = 2\tilde{B}(J+1)$ $\tilde{B} = 10.340\ cm^{-1}$	$\tilde{\omega}_{calc} = 2\tilde{B}(J+1) - 4\tilde{D}(J+1)^3$ $\tilde{B} = 10.395\ cm^{-1}$ $\tilde{D} = 0.0004\ cm^{-1}$
$3 \to 4$	83.03		82.72	83.06
		21.07		
$4 \to 5$	104.10		103.40	103.75
		20.20		
$5 \to 6$	124.30		124.08	124.39
		20.73		
$6 \to 7$	145.03		144.76	144.98
		20.48		
$7 \to 8$	165.51		165.44	165.50
		20.35		
$8 \to 9$	185.86		186.12	185.94
		20.52		
$9 \to 10$	206.38		206.80	206.30
		20.12		
$10 \to 11$	226.50		227.48	226.56

6.5 The Lines in a Pure Rotational Spectrum Are Not Equally Spaced

Table 6.2 lists some of the observed lines in the pure rotational spectrum (no vibrational transitions) of $H^{35}Cl$. The differences listed in the third column clearly show that the lines are not exactly equally spaced as the rigid-rotator approximation predicts. The discrepancy can be resolved by realizing that a chemical bond is not truly rigid. As the molecule rotates more energetically (increasing J), the centrifugal force causes the bond to stretch slightly (Problem 6–26). This small effect can be treated by perturbation theory (Chapter 8), and the end result is that the energy can be written as

$$E_J = \tilde{B}J(J+1) - \tilde{D}J^2(J+1)^2 \tag{6.44}$$

where $\tilde{D}$ is called the *centrifugal distortion constant*. Rigid-rotator and nonrigid-rotator energy levels are sketched in Figure 6.5.

The frequencies of the absorption due to $J \to J+1$ transitions are given by

$$\tilde{\omega} = E_{J+1} - E_J$$

$$= 2\tilde{B}(J+1) - 4\tilde{D}(J+1)^3 \qquad J = 0, 1, 2, \ldots \tag{6.45}$$

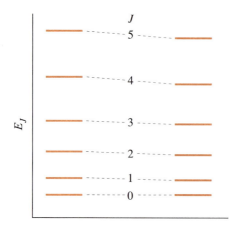

FIGURE 6.5
The rotational energy levels of a rigid rotator
(left) and a nonrigid rotator (right).

The predictions of this equation are given in Table 6.2, where we obtain $\tilde{B} = 10.395 \text{ cm}^{-1}$ and $\tilde{D} = 0.0004 \text{ cm}^{-1}$ for $H^{35}Cl$ by fitting Equation 6.45 to the experimental data. These values differ slightly from those in Table 6.1 because of higher-order effects. The inclusion of centrifugal distortion alters the extracted value of $\tilde{B}$ (see Problems 6–22 and 6–23).

6.6 The Wave Functions of a Rigid Rotator Are Called Spherical Harmonics

The wave functions of a rigid rotator are given by the solutions to Equation 6.11. To solve Equation 6.11, we assumed separation of variables and wrote $Y(\theta, \phi) = \Theta(\theta)\Phi(\phi)$ (Equation 6.14). The resulting differential equation for $\Phi(\phi)$ (Equation 6.17) is relatively easy to solve, and we showed that its solutions are

$$\Phi(\phi) = \frac{1}{(2\pi)^{1/2}} e^{im\phi} \qquad m = 0, \pm 1, \pm 2, \ldots \tag{6.46}$$

The differential equation for $\Theta(\theta)$, Equation 6.16, is not as easy to solve because it does not have constant coefficients. It is convenient to let $x = \cos\theta$ and $\Theta(\theta) = P(x)$ in Equation 6.16. (This x should not be confused with the cartesian coordinate, x.) Because $0 \le \theta \le \pi$, the range of x is $-1 \le x \le +1$. Under the change of variable, $x = \cos\theta$, Equation 6.16 becomes (Problem 6–27)

$$(1 - x^2)\frac{d^2 P}{dx^2} - 2x\frac{dP}{dx} + \left[\beta - \frac{m^2}{1 - x^2}\right] P(x) = 0 \tag{6.47}$$

with $m = 0, \pm 1, \pm 2, \ldots$. Equation 6.47 for $P(x)$ is called *Legendre's equation* and is a well-known equation in classical physics. It occurs in a variety of problems formulated in spherical coordinates. When Equation 6.47 is solved, it is found that β must equal

$J(J+1)$ with $J = 0, 1, 2, \ldots$, and that $|m| \leq J$, where $|m|$ denotes the magnitude of m, if the solutions are to remain finite. Thus, Equation 6.47 can be written as

$$(1 - x^2)\frac{d^2 P}{dx^2} - 2x\frac{dP}{dx} + \left[J(J+1) - \frac{m^2}{1 - x^2} \right] P(x) = 0 \qquad (6.48)$$

with $J = 0, 1, 2, \ldots$, and $m = 0, \pm 1, \pm 2, \ldots, \pm J$.

The solutions to Equation 6.48 are most easily discussed by considering the $m = 0$ case first. When $m = 0$, the solutions to Equation 6.48 are called *Legendre polynomials* and are denoted by $P_J(x)$. Legendre polynomials arise in a number of physical problems. The first few Legendre polynomials are listed in Table 6.3.

EXAMPLE 6–5
Prove that the first few Legendre polynomials satisfy Equation 6.48 when $m = 0$.

SOLUTION: Equation 6.48 with $m = 0$ is

$$(1 - x^2)\frac{d^2 P}{dx^2} - 2x\frac{dP}{dx} + J(J+1)P(x) = 0 \qquad (1)$$

The first Legendre polynomial, $P_0(x) = 1$, is clearly a solution of equation 1 with $J = 0$. When we substitute $P_1(x) = x$ into equation 1 with $J = 1$, we obtain

$$-2x + 1(2)x = 0$$

and so $P_1(x)$ is a solution. For $P_2(x)$, equation 1 is

$$(1 - x^2)(3) - 2x(3x) + 2(3)[\tfrac{1}{2}(3x^2 - 1)] = (3 - 3x^2) - 6x^2 + (9x^2 - 3) = 0$$

TABLE 6.3
The First Few Legendre Polynomials [a]

$P_0(x) = 1$
$P_1(x) = x$
$P_2(x) = \tfrac{1}{2}(3x^2 - 1)$
$P_3(x) = \tfrac{1}{2}(5x^3 - 3x)$
$P_4(x) = \tfrac{1}{8}(35x^4 - 30x^2 + 3)$

a. Legendre polynomials are the solutions to Equation 6.48 with $m = 0$. The subscript indexing the Legendre polynomials is the value of J in Equation 6.48.

Notice from Table 6.3 that $P_J(x)$ is an even function of x if J is even and an odd function of x if J is odd. The factors in front of the $P_J(x)$ are chosen such that $P_J(1) = 1$. In addition, although we will not prove it, it can be shown generally that the $P_J(x)$ in Table 6.3 are orthogonal, or that

$$\int_{-1}^{1} P_J(x) P_{J'}(x) dx = \langle J \mid J' \rangle = 0 \qquad J \neq J' \tag{6.49}$$

Keep in mind here that the limits on x correspond to the natural, physical limits on θ (0 to π) in spherical coordinates because $x = \cos \theta$. The Legendre polynomials are normalized by the general relation

$$\int_{-1}^{1} [P_J(x)]^2 dx = \langle J \mid J \rangle = \frac{2}{2J + 1} \tag{6.50}$$

Equation 6.50 shows that the normalization constant of $P_J(x)$ is $[(2J + 1)/2]^{1/2}$.

Although the Legendre polynomials arise only in the case $m = 0$, they are customarily studied first because the solutions for the $m \neq 0$ case, called *associated Legendre functions*, are defined in terms of the Legendre polynomials. If we denote the associated Legendre functions by $P_J^{|m|}(x)$, then their defining relation is

$$P_J^{|m|}(x) = (1 - x^2)^{|m|/2} \frac{d^{|m|}}{dx^{|m|}} P_J(x) \tag{6.51}$$

Note that only the magnitude of m is relevant here because the defining differential equation, Equation 6.48, depends on only m^2. Because the leading term in $P_J(x)$ is x^J (see Table 6.3), Equation 6.51 shows that $P_J^{|m|}(x) = 0$ if $m > J$. The first few associated Legendre functions (Problem 6–31) are given in Table 6.4.

Before we discuss a few of the properties of the associated Legendre functions, let's be sure to realize that θ and not x is the variable of physical interest. Table 6.4 also lists the associated Legendre functions in terms of $\cos \theta$ and $\sin \theta$. Note that the factors $(1 - x^2)^{1/2}$ in Table 6.4 become $\sin \theta$ when the associated Legendre functions are expressed in the variable θ. Because $x = \cos \theta$, then $dx = -\sin \theta \, d\theta$ and Equations 6.49 and 6.50 can be written as

$$\int_{-1}^{1} P_J(x) P_{J'}(x) dx = \int_{0}^{\pi} P_J(\cos \theta) P_{J'}(\cos \theta) \sin \theta d\theta = \frac{2\delta_{JJ'}}{2J + 1} \tag{6.52}$$

Because the differential volume element in spherical coordinates is $d\tau = r^2 \sin \theta dr d\theta d\phi$, we see that the factor $\sin \theta d\theta$ in Equation 6.52 is the "θ" part of $d\tau$ in spherical coordinates.

TABLE 6.4
The First Few Associated Legendre Functions $P_J^{|m|}(x)$

$P_0^0(x) = 1$

$P_1^0(x) = x = \cos\theta$

$P_1^1(x) = (1 - x^2)^{1/2} = \sin\theta$

$P_2^0(x) = \frac{1}{2}(3x^2 - 1) = \frac{1}{2}(3\cos^2\theta - 1)$

$P_2^1(x) = 3x(1 - x^2)^{1/2} = 3\cos\theta\sin\theta$

$P_2^2(x) = 3(1 - x^2) = 3\sin^2\theta$

$P_3^0(x) = \frac{1}{2}(5x^3 - 3x) = \frac{1}{2}(5\cos^3\theta - 3\cos\theta)$

$P_3^1(x) = \frac{3}{2}(5x^2 - 1)(1 - x^2)^{1/2} = \frac{3}{2}(5\cos^2\theta - 1)\sin\theta$

$P_3^2(x) = 15x(1 - x^2) = 15\cos\theta\sin^2\theta$

$P_3^3(x) = 15(1 - x^2)^{3/2} = 15\sin^3\theta$

The associated Legendre functions satisfy the relation

$$\int_{-1}^{1} P_J^{|m|}(x) P_{J'}^{|m|}(x)dx = \int_0^{\pi} P_J^{|m|}(\cos\theta) P_{J'}^{|m|}(\cos\theta)\sin\theta d\theta$$

$$= \frac{2}{(2J + 1)}\frac{(J + |m|)!}{(J - |m|)!}\delta_{JJ'} \tag{6.53}$$

[Remember that $0! = 1$ (Problem 6–47).] Equation 6.53 can be used to show that the normalization constant of the associated Legendre functions is

$$N_{Jm} = \left[\frac{(2J + 1)}{2}\frac{(J - |m|)!}{(J + |m|)!}\right]^{1/2} \tag{6.54}$$

Thus, the (normalized) $\Theta(\theta)$ part of Equation 6.14 is given by

$$\Theta_J^m(\theta) = \left[\frac{2J + 1}{2}\frac{(J - |m|)!}{(J + |m|)!}\right]^{1/2} P_J^{|m|}(\cos\theta) \tag{6.55}$$

EXAMPLE 6–6
Use Equation 6.53 in both the x and θ variables and Table 6.4 to prove that $P_1^1(x)$ and $P_2^1(x)$ are orthogonal.

SOLUTION: According to Equation 6.53, we must prove that

$$\int_{-1}^{1} P_1^1(x) P_2^1(x)dx = 0$$

From Table 6.4, we have

$$\int_{-1}^{1} (1 - x^2)^{1/2} [3x(1 - x^2)^{1/2}] \, dx = 3 \int_{-1}^{1} x(1 - x^2) \, dx = 0$$

In terms of θ, we have (from Equation 6.53 and Table 6.4)

$$\int_{0}^{\pi} (\sin \theta)(3 \cos \theta \sin \theta) \sin \theta \, d\theta = 3 \int_{0}^{\pi} \sin^3 \theta \cos \theta \, d\theta = 0$$

Returning to the original problem now, the solutions to Equation 6.11, which are not only the rigid-rotator wave functions but also the angular part of the hydrogen atomic orbitals, are given by $\Theta_J^m(\theta) \Phi_m(\phi)$ (Equation 6.14). Using Equations 6.55 and 6.46, we see that the normalized functions

$$Y_J^m(\theta, \phi) = i^{m+|m|} \left[\frac{(2J + 1)}{4\pi} \frac{(J - |m|)!}{(J + |m|)!} \right]^{1/2} P_J^{|m|}(\cos \theta) e^{im\phi} \tag{6.56}$$

with $J = 0, 1, 2, \ldots$ and $m = 0, \pm 1, \pm 2, \ldots, \pm J$ satisfy Equation 6.11. The peculiar-looking factor of $i^{m+|m|}$ in Equation 6.56 is simply a convention that is used by most authors. Note that this factor is equal to 1 when m is odd and negative and is equal to -1 when m is odd and positive (Problem 6–34). Note that the spherical harmonics given in Table 6.5 display this convention. The $Y_J^m(\theta, \phi)$ form an orthonormal set

$$\int_{0}^{\pi} d\theta \sin \theta \int_{0}^{2\pi} d\phi \, Y_J^m(\theta, \phi)^* Y_{J'}^k(\theta, \phi) = \delta_{JJ'} \delta_{mk} \tag{6.57}$$

Note that the $Y_J^m(\theta, \phi)$ are orthonormal with respect to $\sin \theta \, d\theta d\phi$ and not just $d\theta d\phi$. MathChapter E shows that the factor $\sin \theta \, d\theta d\phi$ has a simple physical inter-

TABLE 6.5
The First Few Spherical Harmonics, $Y_J^m(\theta, \phi)$ [a]

$$Y_0^0 = \frac{1}{(4\pi)^{1/2}} \qquad\qquad Y_1^0 = \left(\frac{3}{4\pi} \right)^{1/2} \cos \theta$$

$$Y_1^1 = -\left(\frac{3}{8\pi} \right)^{1/2} \sin \theta e^{i\phi} \qquad Y_1^{-1} = \left(\frac{3}{8\pi} \right)^{1/2} \sin \theta e^{-i\phi}$$

$$Y_2^0 = \left(\frac{5}{16\pi} \right)^{1/2} (3 \cos^2 \theta - 1) \qquad Y_2^1 = -\left(\frac{15}{8\pi} \right)^{1/2} \sin \theta \cos \theta e^{i\phi}$$

$$Y_2^{-1} = \left(\frac{15}{8\pi} \right)^{1/2} \sin \theta \cos \theta e^{-i\phi} \qquad Y_2^2 = \left(\frac{15}{32\pi} \right)^{1/2} \sin^2 \theta e^{2i\phi}$$

$$Y_2^{-2} = \left(\frac{15}{32\pi} \right)^{1/2} \sin^2 \theta e^{-2i\phi}$$

a. The negative signs in $Y_1^1(\theta, \phi)$ and $Y_2^1(\theta, \phi)$ are simply a convention.

pretation. The differential volume element in spherical coordinates is $r^2 \sin\theta\, d\theta d\phi$. If r is a constant, as it is in the case of a rigid rotator, and set equal to unity for convenience, then the spherical coordinate volume element becomes a surface element $dA = \sin\theta\, d\theta d\phi$. If this surface element is integrated over θ and ϕ, we obtain 4π, the surface area of a sphere of unit radius (Problem 6–35). Thus, $\sin\theta\, d\theta d\phi$ is an area element on the surface of a sphere of unit radius. According to Equation 6.56, the $Y_J^m(\theta, \phi)$ are orthonormal over a spherical surface and so are called *spherical harmonics*.

EXAMPLE 6–7

Show that $Y_1^{-1}(\theta, \phi)$ is normalized and that it is orthogonal to $Y_2^1(\theta, \phi)$.

SOLUTION: Using $Y_1^{-1}(\theta, \phi)$ from Table 6.5, the normalization condition is

$$\int_0^\pi d\theta \sin\theta \int_0^{2\pi} d\phi\, Y_1^{-1}(\theta, \phi)^* Y_1^{-1}(\theta, \phi) = \frac{3}{8\pi}\int_0^\pi d\theta \sin\theta \sin^2\theta \int_0^{2\pi} d\phi \overset{?}{=} 1$$

Letting $x = \cos\theta$, we have

$$\frac{3}{8\pi}\cdot 2\pi \int_{-1}^1 (1-x^2)dx = \frac{3}{4}\left(2 - \frac{2}{3}\right) = 1$$

The orthogonality condition is

$$\int_0^\pi d\theta \sin\theta \int_0^{2\pi} d\phi\, Y_2^1(\theta, \phi)^* Y_1^{-1}(\theta, \phi)$$

$$= -\left(\frac{15}{8\pi}\right)^{1/2}\left(\frac{3}{8\pi}\right)^{1/2}\int_0^\pi d\theta \sin\theta \int_0^{2\pi} d\phi\, (e^{-i\phi}\sin\theta\cos\theta)(e^{-i\phi}\sin\theta)$$

$$= -\left(\frac{45}{64\pi^2}\right)^{1/2}\int_0^\pi d\theta \sin^3\theta \cos\theta \int_0^{2\pi} d\phi\, e^{-2i\phi}$$

The integral over ϕ is zero because it is the integral of $\cos 2\phi$ and $\sin 2\phi$ over complete cycles. Thus, we see that $Y_1^{-1}(\theta, \phi)$ and $Y_2^1(\theta, \phi)$ are orthogonal.

In summary, the Schrödinger equation for a rigid rotator is

$$\hat{H}Y_J^m(\theta, \phi) = \frac{\hbar^2 J(J+1)}{2I}Y_J^m(\theta, \phi) \qquad J = 0, 1, 2, \ldots \qquad (6.58)$$

where $\hat{H}$ is given by Equation 6.9. Because $\hat{H} = \hat{L}^2/2I$, $\hat{L}^2$ is given by

$$\hat{L}^2 = -\hbar^2\left[\frac{1}{\sin\theta}\frac{\partial}{\partial\theta}\left(\sin\theta\frac{\partial}{\partial\theta}\right) + \frac{1}{\sin^2\theta}\frac{\partial^2}{\partial\phi^2}\right] \qquad (6.59)$$

The eigenvalue equation for $\hat{L}^2$ is

$$\hat{L}^2 Y_J^m(\theta, \phi) = \hbar^2 J(J+1) Y_J^m(\theta, \phi) \tag{6.60}$$

Thus, we see that the spherical harmonics are also eigenfunctions of $\hat{L}^2$ and that the square of the angular momentum can have only the values given by

$$L^2 = \hbar^2 J(J+1) \qquad J = 0, 1, 2, \ldots \tag{6.61}$$

EXAMPLE 6–8
Show that $Y_1^1(\theta, \phi)$ is an eigenfunction of $\hat{L}^2$ with eigenvalue $L^2 = 2\hbar$. (See Equation 6.60.)

SOLUTION: From Table 6.5, we have

$$Y_1^1(\theta, \phi) = \left(\frac{3}{8\pi} \right)^{1/2} \sin\theta\, e^{i\phi}$$

The "θ" part of the differential operator in brackets in Equation 6.59 gives $e^{i\phi}(\cos^2\theta - \sin^2\theta)/\sin\theta$, and the "$\phi$" part gives $-e^{i\phi}/\sin\theta$. If we add these two results, we get

$$\frac{(\cos^2\theta - \sin^2\theta - 1)e^{i\phi}}{\sin\theta} = -\frac{2e^{i\phi}\sin^2\theta}{\sin\theta} = -2e^{i\phi}\sin\theta$$

Therefore,

$$\hat{L}^2 Y_1^1(\theta, \phi) = 2\hbar^2 Y_1^1(\theta, \phi)$$

which is Equation 6.60 with $J = 1$.

Angular momentum plays an important role in quantum mechanics, as it does in classical mechanics, and so we shall discuss angular momentum more fully in Section 6.8.

6.7 The Selection Rule in the Rigid-Rotator Approximation Is $\Delta J = \pm 1$

As we said when we discussed the selection rule for a harmonic oscillator in Section 5.12, the probability that a molecule makes a transition from one vibrational state to another when it is irradiated with electromagnetic radiation in the z direction is given by the square of an integral of the form

$$\langle v' | \mu_z | v \rangle = \int_{-\infty}^{\infty} dx\, \psi_{v'}(x) \mu_z(x) \psi_v(x)$$

We called this integral a transition dipole moment. In the case of a rigid rotator, the transition dipole moment integral takes the form (we shall derive this result in Section 8.6)

$$\langle J', m' \mid \mu_z \mid J, m \rangle = \int_0^{2\pi} \int_0^{\pi} Y_{J'}^{m'}(\theta, \phi)^* \mu_z Y_J^m(\theta, \phi) \sin \theta d\theta d\phi$$

Using the fact that $\mu_z = \mu_0 \cos \theta$ gives

$$\langle J', m' \mid \mu_z \mid J, m \rangle = \mu_0 \int_0^{2\pi} \int_0^{\pi} Y_{J'}^{m'}(\theta, \phi)^* Y_J^m(\theta, \phi) \cos \theta \sin \theta d\theta d\phi \quad (6.62)$$

Notice that μ_0 must be nonzero for the transition moment to be nonzero. Thus, we have now proven our earlier assertion that a molecule must have a permanent dipole moment for it to have a pure rotational spectrum, at least in the rigid-rotator approximation. Now let's prove that $\Delta J = \pm 1$. Recall that (Equation 6.56)

$$Y_J^m(\theta, \phi) = i^{m+|m|} N_{Jm} P_J^{|m|}(\cos \theta) e^{im\phi} \quad (6.63)$$

where N_{Jm} is given by Equation 6.54. Substitute Equation 6.63 into Equation 6.62 and let $x = \cos \theta$ to obtain

$$\langle J', m' \mid \mu_z \mid J, m \rangle = \mu_0 N_{J'm'} N_{Jm} \int_0^{2\pi} d\phi e^{i(m-m')\phi} \int_{-1}^1 dx\, x P_{J'}^{|m'|}(x) P_J^{|m|}(x)$$

$$(6.64)$$

The integral over ϕ is zero unless $m = m'$, so we find that $\Delta m = 0$ is part of the rigid-rotator selection rule. Integration over ϕ for $m = m'$ gives a factor of 2π, so we have

$$\langle J', m \mid \mu_z \mid J, m \rangle = 2\pi \mu_0 N_{J'm} N_{Jm} \int_{-1}^1 dx\, P_{J'}^{|m|}(x) x P_J^{|m|}(x) \quad (6.65)$$

We can evaluate this integral in general by using the recursion formula (Problem 6–33)

$$(2J + 1)x P_J^{|m|}(x) = (J - |m| + 1) P_{J+1}^{|m|}(x) + (J + |m|) P_{J-1}^{|m|}(x) \quad (6.66)$$

By using this relation in Equation 6.65, we obtain

$$\langle J', m \mid \mu_z \mid J, m \rangle = 2\pi \mu_0 N_{J'm} N_{Jm} \int_{-1}^1 dx\, P_{J'}^{|m|}(x)$$

$$\times \left[\frac{(J - |m| + 1)}{2J + 1} P_{J+1}^{|m|}(x) + \frac{(J + |m|)}{2J + 1} P_{J-1}^{|m|}(x) \right]$$

Using the orthogonality relation for the $P_J^m(x)$ (Equation 6.53), we find that the above integral will vanish unless $J' = J + 1$ or $J' = J - 1$. This finding leads to the selection rule $J' = J \pm 1$, or $\Delta J = \pm 1$. Thus, we have shown that the selection rule for pure rotational spectra in the rigid-rotator approximation is that the molecule must have a permanent dipole moment and that $\Delta J = \pm 1$ and $\Delta m = 0$.

EXAMPLE 6–9
Using the explicit formulas for the spherical harmonics given in Table 6.3, show that the rotational transition $J = 0 \rightarrow J = 1$ is allowed, but $J = 0 \rightarrow J = 2$ is forbidden in microwave spectroscopy (in the rigid-rotator approximation).

SOLUTION: Referring to Equation 6.62, we see that we must show that the integral

$$I_{0 \rightarrow 1} = \int_0^{2\pi} \int_0^{\pi} Y_1^m(\theta, \phi)^* Y_0^0(\theta, \phi) \cos \theta \sin \theta d\theta d\phi$$

is nonzero and that

$$I_{0 \rightarrow 2} = \int_0^{2\pi} \int_0^{\pi} Y_2^m(\theta, \phi)^* Y_0^0(\theta, \phi) \cos \theta \sin \theta d\theta d\phi$$

is equal to zero. In either case, we can easily see that the integral over ϕ will be zero unless $m = 0$, so we will concentrate only on the θ integration. For $I_{0 \rightarrow 1}$, we have

$$I_{0 \rightarrow 1} = 2\pi \int_0^{\pi} \left(\frac{3}{4\pi} \right)^{1/2} \cos \theta \left(\frac{1}{4\pi} \right)^{1/2} \cos \theta \sin \theta d\theta$$

$$= \frac{\sqrt{3}}{2} \int_{-1}^{1} dx x^2 = \frac{1}{\sqrt{3}} \neq 0$$

For $I_{0 \rightarrow 2}$, we have

$$I_{0 \rightarrow 2} = 2\pi \int_0^{\pi} \left(\frac{5}{16\pi} \right)^{1/2} (3 \cos^2 \theta - 1) \left(\frac{1}{4\pi} \right)^{1/2} \cos \theta \sin \theta d\theta$$

$$= \frac{\sqrt{5}}{4} \int_{-1}^{1} dx (3x^3 - x) = 0$$

because the integrand is an odd function of x.

6.8 The Precise Values of the Three Components of Angular Momentum Cannot Be Measured Simultaneously

In this section, we will explore some of the quantum-mechanical properties of angular momentum. Angular momentum plays a key role not only in the theory of the rigid rotator, but in the theory of the hydrogen atom as well. In fact, we're going to see in the next chapter that the spherical harmonics are the angular wave functions of a hydrogen atom as well as the rigid-rotator wave function. To keep things general, we'll use l for angular momentum and m for its z component in this section.

Recall that angular momentum is a vector quantity. The quantum-mechanical operators corresponding to the three components of angular momentum are given in Table 4.1. These operators are obtained from the classical expressions (Equation C.20) by replacing the classical momenta by their quantum-mechanical equivalents to obtain

$$\hat{L}_x = y\hat{P}_z - z\hat{P}_y = -i\hbar \left(y\frac{\partial}{\partial z} - z\frac{\partial}{\partial y} \right)$$

$$\hat{L}_y = z\hat{P}_x - x\hat{P}_z = -i\hbar \left(z\frac{\partial}{\partial x} - x\frac{\partial}{\partial z} \right)$$

$$\hat{L}_z = x\hat{P}_y - y\hat{P}_x = -i\hbar \left(x\frac{\partial}{\partial y} - y\frac{\partial}{\partial x} \right)$$

Through a straightforward, but somewhat tedious, exercise in partial differentiation, we can convert these equations into spherical coordinates (Problems 6–38 and 6–39) to obtain

$$\hat{L}_x = -i\hbar \left(-\sin\phi \frac{\partial}{\partial \theta} - \cot\theta \cos\phi \frac{\partial}{\partial \phi} \right)$$

$$\hat{L}_y = -i\hbar \left(\cos\phi \frac{\partial}{\partial \theta} - \cot\theta \sin\phi \frac{\partial}{\partial \phi} \right) \tag{6.67}$$

$$\hat{L}_z = -i\hbar \frac{\partial}{\partial \phi}$$

Using the definition of the spherical harmonics (Equations 6.56), we see that all the ϕ dependence of the spherical harmonics occurs in the factor $e^{im\phi}$ in $\Phi_m(\phi)$, and so the spherical harmonics are eigenfunctions of $\hat{L}_z$:

$$\hat{L}_z Y_l^m(\theta, \phi) = N_{lm}\hat{L}_z P_l^{|m|}(\cos\theta)e^{im\phi}$$

$$= N_{lm} P_l^{|m|}(\cos\theta)\hat{L}_z e^{im\phi}$$

$$= \hbar m Y_l^m(\theta, \phi) \tag{6.68}$$

Equation 6.68 shows that measured values of L_z are integral multiples of $\hbar$. Notice that $\hbar$ is a fundamental measure of the angular momentum of a quantum-mechanical system.

The spherical harmonics are not eigenfunctions of $\hat{L}_x$ or $\hat{L}_y$, however, as the following example shows.

EXAMPLE 6–10
Use Equation 6.67 to show that $Y_1^{-1}(\theta, \phi)$ is not an eigenfunction of $\hat{L}_x$.

SOLUTION: From Table 6.5, $Y_1^{-1}(\theta, \phi) = (3/8\pi)^{1/2} \sin \theta e^{-i\phi}$. Using the first of Equations 6.67, we have

$$\hat{L}_x Y_1^{-1}(\theta, \phi) = -i\hbar \left(\frac{3}{8\pi} \right)^{1/2} (-\sin \phi \cos \theta e^{-i\phi} + i \cot \theta \cos \phi \sin \theta e^{-i\phi})$$

$$= -i\hbar \left(\frac{3}{8\pi} \right)^{1/2} \cos \theta (-\sin \phi + i \cos \phi)e^{-i\phi}$$

But the term in parentheses is

$$-\sin \phi + i \cos \phi = -\frac{(e^{i\phi} - e^{-i\phi})}{2i} + i\frac{(e^{i\phi} + e^{-i\phi})}{2}$$

$$= +\frac{i}{2}(e^{i\phi} - e^{-i\phi}) + \frac{i}{2}(e^{i\phi} + e^{-i\phi}) = ie^{i\phi}$$

Therefore,

$$\hat{L}_x Y_1^{-1}(\theta, \phi) = \hbar \left(\frac{3}{8\pi} \right)^{1/2} \cos \theta = \frac{\hbar}{2^{1/2}} Y_1^0(\theta, \phi)$$

Note that

$$\langle \hat{L}_x \rangle = \int_0^\pi d\theta \sin \theta \int_0^{2\pi} d\phi \, Y_1^{-1}(\theta, \phi)^* \hat{L}_x Y_1^{-1}(\theta, \phi)$$

$$= \frac{\hbar}{2^{1/2}} \int_0^{2\pi} d\phi \int_0^\pi d\theta \sin \theta \, Y_1^{-1}(\theta, \phi)^* Y_1^0(\theta, \phi) = 0$$

because of the orthogonality of $Y_1^{-1}(\theta, \phi)$ and $Y_1^0(\theta, \phi)$.

In the notation that we are using in this section, Equation 6.60 reads

$$\hat{L}^2 Y_l^m(\theta, \phi) = \hbar^2 l(l + 1)Y_l^m(\theta, \phi) \tag{6.69}$$

Because the spherical harmonics are simultaneous eigenfunctions of both $\hat{L}^2$ and $\hat{L}_z$, we can determine precise values of L^2 and L_z simultaneously (Section 4.4), which implies that the operators $\hat{L}^2$ and $\hat{L}_z$ commute.

EXAMPLE 6–11

Prove that the operators $\hat{L}^2$ and $\hat{L}_z$ commute.

SOLUTION: Using $\hat{L}^2$ from Equation 6.59 and $\hat{L}_z$ from Equation 6.67, we have

$$\hat{L}^2\hat{L}_z f = -\hbar^2 \left[\frac{1}{\sin\theta} \frac{\partial}{\partial\theta} \left(\sin\theta \frac{\partial}{\partial\theta} \right) + \frac{1}{\sin^2\theta} \frac{\partial^2}{\partial\phi^2} \right] \left(-i\hbar \frac{\partial f}{\partial\phi} \right)$$

$$= i\hbar^3 \left[\frac{1}{\sin\theta} \frac{\partial}{\partial\theta} \left(\sin\theta \frac{\partial^2 f}{\partial\theta\partial\phi} \right) + \frac{1}{\sin^2\theta} \frac{\partial^3 f}{\partial\phi^3} \right]$$

and

$$\hat{L}_z\hat{L}^2 f = \left(-i\hbar \frac{\partial}{\partial\phi} \right) \left\{ -\hbar^2 \left[\frac{1}{\sin\theta} \frac{\partial}{\partial\theta} \left(\sin\theta \frac{\partial}{\partial\theta} \right) + \frac{1}{\sin^2\theta} \frac{\partial^2}{\partial\phi^2} \right] \right\} f$$

$$= i\hbar^3 \left[\frac{1}{\sin\theta} \frac{\partial}{\partial\theta} \left(\sin\theta \frac{\partial^2 f}{\partial\phi\partial\theta} \right) + \frac{1}{\sin^2\theta} \frac{\partial^3 f}{\partial\phi^3} \right]$$

where in writing the last line here we have recognized that $(\partial/\partial\phi)$ does not affect terms involving θ. Because

$$\frac{\partial^2 f}{\partial\theta\partial\phi} = \frac{\partial^2 f}{\partial\phi\partial\theta}$$

for any function well enough behaved to be a wave function, we see that

$$\hat{L}^2\hat{L}_z f = \hat{L}_z\hat{L}^2 f$$

or that

$$[\hat{L}^2, \hat{L}_z] = 0$$

because f is arbitrary.

We can use Equations 6.60 and 6.68 to prove that $|m| \leq l$, or that $m = 0, \pm 1, \pm 2, \ldots, \pm l$. It follows from Equation 6.68 that

$$\hat{L}_z^2 Y_l^m(\theta, \phi) = m^2\hbar^2 Y_l^m(\theta, \phi) \tag{6.70}$$

Subtracting Equation 6.70 from 6.60 gives

$$(\hat{L}^2 - \hat{L}_z^2)Y_l^m(\theta, \phi) = \hbar^2[l(l+1) - m^2]Y_l^m(\theta, \phi)$$

Furthermore, because

$$\hat{L}^2 = \hat{L}_x^2 + \hat{L}_y^2 + \hat{L}_z^2$$

then

$$(\hat{L}^2 - \hat{L}_z^2)Y_l^m(\theta, \phi) = (\hat{L}_x^2 + \hat{L}_y^2)Y_l^m(\theta, \phi) = \hbar^2[l(l+1) - m^2]Y_l^m(\theta, \phi) \quad (6.71)$$

Thus, the observed values of $L_x^2 + L_y^2$ are $[l(l+1) - m^2]\hbar^2$. But because $L_x^2 + L_y^2$ is the sum of two squared terms, it cannot be negative, and so we have

$$[l(l+1) - m^2]\hbar^2 \geq 0$$

or

$$l(l+1) \geq m^2 \qquad (6.72)$$

Because l and m are integers, Equation 6.72 says that

$$|m| \leq l$$

or that the only possible values of the integer m are

$$m = 0, \pm 1, \pm 2, \ldots, \pm l \qquad (6.73)$$

This result might be familiar as the condition of the magnetic quantum number associated with the hydrogen atom.

Equation 6.73 shows that there are $2l + 1$ values of m for each value of l. Let's look at the case of $l = 1$ for which $l(l+1) = 2$. Because $l = 1$, m can have only the values 0 and ± 1. Using the equations

$$\hat{L}^2 Y_1^m(\theta, \phi) = 1(1+1)\hbar^2 Y_1^m(\theta, \phi) = 2\hbar^2 Y_1^m(\theta, \phi) \qquad m = 0, \pm 1$$

and

$$\hat{L}_z Y_1^m(\theta, \phi) = m\hbar Y_1^m(\theta, \phi) \qquad m = 0, \pm 1$$

we see that

$$|\mathbf{L}| = (L^2)^{1/2} = \sqrt{2}\hbar$$

and

$$L_z = -\hbar, 0, +\hbar$$

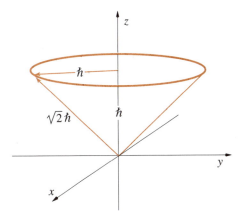

FIGURE 6.6
The $m = +1$ component of the angular-momentum state, $l = 1$. The angular momentum describes a cone because the x and y components cannot be specified. The projection of the motion onto the x–y plane is a circle of radius $\hbar$ centered at the origin (Example 6–12).

where $|\mathbf{L}|$ is the magnitude of the angular-momentum vector. Note that the maximum value of L_z is less than $|\mathbf{L}|$, which implies that $\mathbf{L}$ and L_z cannot point in the same direction. This is illustrated in Figure 6.6, which shows L_z with a value $+\hbar$ and $|\mathbf{L}|$ with its value $\sqrt{2}\hbar$.

Now let's try to specify L_x and L_y. Problem 6–41 has you prove that $\hat{L}_x$, $\hat{L}_y$, and $\hat{L}_z$ commute with $\hat{L}^2$.

$$[\hat{L}^2, \hat{L}_x] = [\hat{L}^2, \hat{L}_y] = [\hat{L}^2, \hat{L}_z] = 0 \tag{6.74}$$

But they do not commute among themselves. In particular, we have

$$[\hat{L}_x, \hat{L}_y] = i\hbar\hat{L}_z$$

$$[\hat{L}_y, \hat{L}_z] = i\hbar\hat{L}_x \tag{6.75}$$

$$[\hat{L}_z, \hat{L}_x] = i\hbar\hat{L}_y$$

(Do you see a cyclic pattern in these three commutator equations?) Equations 6.74 and 6.75 imply that although it is possible to observe precise values of L^2 and L_z simultaneously, it is not possible to observe precise values of L_x and L_y simultaneously because they do not commute with each other. Even though L_x and L_y do not have precise values, they do have an average value, and Problem 6–42 shows that $\langle L_x \rangle = \langle L_y \rangle = 0$. These results are illustrated in Figure 6.6, which shows L_z with a value of $+\hbar$ and $|\mathbf{L}|$ with a value of $\hbar\sqrt{2}$. A nice classical interpretation of these results is that $\mathbf{L}$ precesses about the z axis, mapping out the surface of the cone shown there. The average values of $\langle L_x \rangle$ and $\langle L_y \rangle$ are zero. This picture is in nice accord with the uncertainty principle: by specifying L_z exactly, we have a complete uncertainty in the angle ϕ associated with L_z.

EXAMPLE 6–12

Show that the projection onto the $x–y$ plane of the motion of the angular-momentum vector with $L^2 = 2\hbar^2$ and $L_z = \hbar$ is a circle of radius $\hbar$ in the $x–y$ plane.

SOLUTION: From the cone in Figure 6.6, we see that the $x–y$ projection will be a circle. To determine the radius, r, of the circle, consider the x, z cross section:

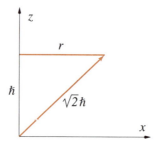

Because we have a right triangle, $r^2 + \hbar^2 = 2\hbar^2$ and so $r = \hbar$. Thus, we also see that while we know the magnitude of the angular momentum and its z component, we do not know the direction in which the vector $L_x \mathbf{i} + L_y \mathbf{j}$ points.

Before leaving this section, we should address the following question: What is so special about the z direction? The answer is that nothing at all is special about the z direction. We could have chosen either the x or y direction as the unique direction and all the above results would be the same, except for exchanging x or y for z. For example, we can know both L^2 and L_x precisely simultaneously, in which case L_y and L_z do not have precise values. It is customary to choose the z direction because the expression for $\hat{L}_z$ in spherical coordinates is so much simpler than for $\hat{L}_x$ or $\hat{L}_y$ (cf. Equation 6.67). The rotating system does not know x from y from z and, in fact, this inability to distinguish between the three directions explains the $(2l + 1)$-fold degeneracy.

Appendix: Determination of the Eigenvalues of $\hat{L}^2$ and $\hat{L}_z$ by Operator Methods

In the appendix to Chapter 5, we determined the energies and the wave functions of a harmonic oscillator using operator methods. The key step was expressing the Hamiltonian operator in terms of a raising operator and a lowering operator and then using the commutation relation $[\hat{P}_x, \hat{X}] = -i\hbar$. In this appendix, we shall use an analogous procedure to determine the eigenfunctions and eigenvalues of $\hat{L}^2$ and $\hat{L}_z$. This procedure is very commonly used in general discussions of angular momentum.

We learned in this chapter that the eigenvalues of $\hat{L}^2$ are $\hbar^2 l(l+1)$ with $l = 0, 1, 2, \ldots$, and that the eigenvalues of $\hat{L}_z$ are $m\hbar$ with $m = 0, \pm 1, \ldots, \pm l$. Furthermore, because $\hat{L}^2$ and $\hat{L}_z$ commute, they have mutual eigenfunctions, which are the spherical harmonics, $Y_l^m(\theta, \phi)$, but we don't need to know this here. All we need to know is that $\hat{L}^2$ and $\hat{L}_z$ commute, so that they have mutual eigenfunctions. To emphasize this point, let $\psi_{\alpha\beta}$ be the mutual (normalized) eigenfunctions of $\hat{L}^2$ and $\hat{L}_z$ such that

$$\hat{L}^2 \psi_{\beta\alpha} = \hbar^2 \beta^2 \psi_{\beta\alpha} \qquad \text{and} \qquad \hat{L}_z \psi_{\beta\alpha} = \hbar\alpha\psi_{\beta\alpha} \tag{1}$$

or

$$\hat{L}^2 | \beta\,\alpha \rangle = \hbar^2\beta^2 | \beta\,\alpha \rangle \tag{2a}$$

and

$$\hat{L}_z | \beta\,\alpha \rangle = \hbar\alpha | \beta\,\alpha \rangle \tag{2b}$$

in the bracket notation. We operate on equation 2b with $\hat{L}_z$ and subtract the result from equation 2a to obtain

$$(\hat{L}^2 - \hat{L}_z^2) | \beta\,\alpha \rangle = \hbar^2(\beta^2 - \alpha^2) | \beta\,\alpha \rangle$$

Because $\hat{L}^2 - \hat{L}_z^2 = \hat{L}_x^2 + \hat{L}_y^2$ corresponds to a nonnegative quantity, we know that

$$\langle \beta\,\alpha | \hat{L}_x^2 + \hat{L}_y^2 | \beta\,\alpha \rangle = \langle \beta\,\alpha | \hat{L}^2 - \hat{L}_z^2 | \beta\,\alpha \rangle$$

$$= \hbar^2(\beta^2 - \alpha^2)\langle \beta\,\alpha | \beta\,\alpha \rangle$$

$$= \hbar^2(\beta^2 - \alpha^2) \geq 0$$

or that

$$-\beta \leq \alpha \leq \beta \tag{3}$$

Because β is fixed, the possible values of α must be finite in number.

To determine the possible values of α, we start with Equations 6.74 and 6.75, which say that

$$[\hat{L}^2, \hat{L}_x] = [\hat{L}^2, \hat{L}_y] = [\hat{L}^2, \hat{L}_z] = 0 \tag{4}$$

$$[\hat{L}_x, \hat{L}_y] = i\hbar\hat{L}_z \tag{5a}$$

$$[\hat{L}_y, \hat{L}_z] = i\hbar\hat{L}_x \tag{5b}$$

$$[\hat{L}_z, \hat{L}_x] = i\hbar\hat{L}_y \tag{5c}$$

where $\hat{L}^2 = \hat{L}_x^2 + \hat{L}_y^2 + \hat{L}_z^2$. We now introduce the (not necessarily Hermitian) operators

$$\hat{L}_+ = \hat{L}_x + i\hat{L}_y \tag{6}$$

and

$$\hat{L}_- = \hat{L}_x - i\hat{L}_y \tag{7}$$

Using equations 4 and 5, it's easy to show that

$$[\hat{L}^2, \hat{L}_+] = [\hat{L}^2, \hat{L}_-] = 0 \tag{8}$$

$$[\hat{L}_z, \hat{L}_+] = \hbar\hat{L}_+ \tag{9}$$

and

$$[\hat{L}_z, \hat{L}_-] = \hbar\hat{L}_- \tag{10}$$

For example,

$$[\hat{L}_z, \hat{L}_+] = \hat{L}_z(\hat{L}_x + i\hat{L}_y)) - (\hat{L}_x + i\hat{L}_y)\hat{L}_z$$

$$= \hat{L}_z\hat{L}_x - \hat{L}_x\hat{L}_z + i(\hat{L}_z\hat{L}_y - \hat{L}_y\hat{L}_z)$$

$$= i\hbar\hat{L}_y + i(-i\hbar\hat{L}_x) = \hbar(\hat{L}_x + i\hat{L}_y) = \hbar\hat{L}_+$$

Now let

$$\psi_{\beta\alpha}^{+1} = \hat{L}_+\psi_{\beta\alpha} \tag{11}$$

Operate on both sides of equation 11 with $\hat{L}_z$ and use equation 9 to write

$$\hat{L}_z\psi_{\beta\alpha}^{+1} = \hat{L}_z\hat{L}_+\psi_{\beta\alpha}$$

$$= (\hbar\hat{L}_+ + \hat{L}_+\hat{L}_z)\psi_{\beta\alpha}$$

$$= \hbar\hat{L}_+\psi_{\beta\alpha} + \hbar\alpha\hat{L}_+\psi_{\beta\alpha}$$

$$= \hbar(\alpha + 1)\hat{L}_+\psi_{\beta\alpha}$$

$$= \hbar(\alpha + 1)\psi_{\beta\alpha}^{+1} \tag{12}$$

If we operate on both sides of equation 11 with $\hat{L}^2$ and use the fact that $\hat{L}^2$ commutes with $\hat{L}_z$ (equation 4), then we obtain

$$\hat{L}^2\hat{L}_+\psi_{\beta\alpha} = \hat{L}_+\hat{L}^2\psi_{\beta\alpha} = \hbar^2\beta^2\hat{L}_+\psi_{\beta\alpha} = \hbar^2\beta^2\psi_{\beta\alpha}^{+1} \tag{13}$$

Thus, the result of operating on $\psi_{\beta\,\alpha}$ by $\hat{L}_+$ is to produce $\psi_{\beta\,\alpha+1}$ from $\psi_{\beta\,\alpha}$, or in bracket notation,

$$\hat{L}_+|\,\beta\,\alpha\rangle = c_+|\,\beta\,\alpha+1\rangle \qquad (14)$$

where c_+ is just a normalization constant. The operator $\hat{L}_+$ is a *raising operator*. Repeated application of $\hat{L}_+$ on $|\,\beta\,\alpha\rangle$ produces a sequence of eigenfunctions of $\hat{L}_z$, $|\,\beta\,\alpha\rangle$, $|\,\beta\,\alpha+1\rangle$, $|\beta\,\alpha+2\rangle$, ..., so long as the result is nonzero.

Similarly, it is easy to show that $\hat{L}_-$ is a *lowering operator*, in the sense that

$$\hat{L}_-|\,\beta\,\alpha\rangle = c_-|\,\beta\,\alpha-1\rangle \qquad (15)$$

Consecutive application of $\hat{L}_-$ on $|\,\beta\,\alpha\rangle$ produces a sequence of eigenfunctions of $\hat{L}_z$, $|\,\beta\,\alpha\rangle$, $|\,\beta\,\alpha-1\rangle$, $|\beta\,\alpha-2\rangle$, ..., so long as the result is nonzero. Notice that consecutive applications of $\hat{L}_+$ and $\hat{L}_-$ on $|\,\beta\,\alpha\rangle$ produces the set of eigenfunctions of $\hat{L}_z$, $|\,\beta\,\alpha\rangle$, $|\,\beta\,\alpha\pm1\rangle$, $|\,\beta\,\alpha\pm2\rangle$, ..., so long as none of these is nonzero. The α index of the eigenfunctions increases or decreases in unit steps, forming a ladder of eigenvalues. For this reason, $\hat{L}_+$ and $\hat{L}_-$ are called *ladder operators*.

According to equation 3, $-\beta \leq \alpha \leq \beta$. Let α_{max} be the largest possible value of α. By definition, then, we have

$$\hat{L}_+|\,\beta\,\alpha_{\text{max}}\rangle = 0 \qquad (16)$$

Operate on equation 16 with $\hat{L}_-$ to obtain

$$\hat{L}_-\hat{L}_+|\,\beta\,\alpha_{\text{max}}\rangle = 0 \qquad (17)$$

Now express $\hat{L}_-\hat{L}_+$ in terms of $\hat{L}^2$ and $\hat{L}_z$ by writing

$$\hat{L}_-\hat{L}_+ = (\hat{L}_x - i\hat{L}_y)(\hat{L}_x + i\hat{L}_y)$$

$$= \hat{L}_x^2 + \hat{L}_y^2 + i(\hat{L}_x\hat{L}_y - \hat{L}_y\hat{L}_x)$$

$$= \hat{L}^2 - \hat{L}_z^2 - \hbar\hat{L}_z$$

Substitute this result into equation 17 to obtain

$$\hat{L}_-\hat{L}_+|\,\beta\,\alpha_{\text{max}}\rangle = (\hat{L}^2 - \hat{L}_z^2 - \hbar\hat{L}_z)\,|\,\beta\,\alpha_{\text{max}}\rangle$$

$$= \hbar^2(\beta^2 - \alpha_{\text{max}}^2 - \alpha_{\text{max}})\,|\,\beta\,\alpha_{\text{max}}\rangle = 0$$

from which we see that

$$\beta^2 = \alpha_{\text{max}}^2 + \alpha_{\text{max}} \qquad (18)$$

Similarly, we start with

$$\hat{L}_-|\beta\,\alpha_{min}\rangle = 0$$

and then operate with $\hat{L}_+$ to obtain

$$\beta^2 = \alpha^2_{min} - \alpha_{min} \tag{19}$$

We can find a relation between α_{max} and α_{min} by equating equations 18 and 19:

$$\alpha^2_{max} + \alpha_{max} = \alpha^2_{min} - \alpha_{min}$$

Solving for α_{max} gives

$$\alpha_{max} = -\frac{1}{2} \pm \frac{1}{2}(1 - 4\alpha_{min} + 4\alpha^2_{min})^{1/2}$$

$$= -\frac{1}{2} \pm \frac{1}{2}(1 - 2\alpha_{min})$$

$$= \alpha_{min} - 1 \quad \text{or} \quad -\alpha_{min}$$

Clearly, α_{max} cannot be equal to $\alpha_{min} - 1$ (by definition), and so we see that $\alpha_{max} = -\alpha_{min}$. Furthermore, because as we have seen above, the eigenvalues of $\hat{L}_z$ vary in unit steps, the eigenvalues of $\hat{L}_z$ extend from $+\alpha_{max}$ to $-\alpha_{max}$ in unit steps. This is possible only if α_{max} is itself an integer (or possibly a half-integer). Thus, if we let $\alpha_{max} = l =$ an integer, then we find that

$$\alpha = 0, \pm 1, \pm 2, \pm 3, \ldots, \pm l \tag{20}$$

If we substitute $\alpha_{max} = l$ into equation 18, we see that

$$\beta^2 = l(l+1) \quad l = 0, 1, 2, \ldots \tag{21}$$

in accord with our results in the chapter.

We can use operator methods to determine the eigenfunctions also. We did this for a harmonic oscillator in the appendix to Chapter 5, but here we shall just refer to the reference to Townsend at the end of the chapter.

Problems

6–1. Show that the moment of inertia for a rigid rotator can be written as $I = \mu l^2$, where $l = l_1 + l_2$ (the fixed separation of the two masses) and μ is the reduced mass.

6–2. Consider the transformation from cartesian coordinates to plane polar coordinates,

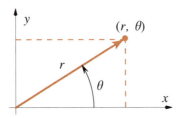

where

$$x = r \cos \theta \qquad r = (x^2 + y^2)^{1/2}$$

$$y = r \sin \theta \qquad \theta = \tan^{-1}\left(\frac{y}{x}\right) \tag{1}$$

If a function $f(r, \theta)$ depends upon the polar coordinates r and θ, then the chain rule of partial differentiation says that

$$\left(\frac{\partial f}{\partial x}\right)_y = \left(\frac{\partial f}{\partial r}\right)_\theta \left(\frac{\partial r}{\partial x}\right)_y + \left(\frac{\partial f}{\partial \theta}\right)_r \left(\frac{\partial \theta}{\partial x}\right)_y \tag{2}$$

and that

$$\left(\frac{\partial f}{\partial y}\right)_x = \left(\frac{\partial f}{\partial r}\right)_\theta \left(\frac{\partial r}{\partial y}\right)_x + \left(\frac{\partial f}{\partial \theta}\right)_r \left(\frac{\partial \theta}{\partial y}\right)_x \tag{3}$$

For simplicity, we will assume that r is equal to a constant, l, so that we can ignore terms involving derivatives with respect to r. In other words, we will consider a particle that is constrained to move on the circumference of a circle. This system is sometimes called a *particle on a ring*. Using equations 1 and 2, show that

$$\left(\frac{\partial f}{\partial x}\right)_y = -\frac{\sin \theta}{l}\left(\frac{\partial f}{\partial \theta}\right)_r \qquad \text{and} \qquad \left(\frac{\partial f}{\partial y}\right)_x = \frac{\cos \theta}{l}\left(\frac{\partial f}{\partial \theta}\right)_r \tag{4}$$

Now apply equation 2 again to show that

$$\left(\frac{\partial^2 f}{\partial x^2}\right)_y = \left[\frac{\partial}{\partial x}\left(\frac{\partial f}{\partial x}\right)_y\right] = \left[\frac{\partial}{\partial \theta}\left(\frac{\partial f}{\partial x}\right)_y\right]_l \left(\frac{\partial \theta}{\partial x}\right)_y$$

$$= \left\{\frac{\partial}{\partial \theta}\left[-\frac{\sin \theta}{l}\left(\frac{\partial f}{\partial \theta}\right)_r\right]\right\}_l \left(-\frac{\sin \theta}{l}\right)$$

$$= \frac{\sin \theta \cos \theta}{l^2}\left(\frac{\partial f}{\partial \theta}\right) + \frac{\sin^2 \theta}{l^2}\left(\frac{\partial^2 f}{\partial \theta^2}\right)$$

Similarly, show that

$$\left(\frac{\partial^2 f}{\partial y^2}\right)_x = -\frac{\sin \theta \cos \theta}{l^2}\left(\frac{\partial f}{\partial \theta}\right) + \frac{\cos^2 \theta}{l^2}\left(\frac{\partial^2 f}{\partial \theta^2}\right)$$

and that

$$\nabla^2 f = \frac{\partial^2 f}{\partial x^2} + \frac{\partial^2 f}{\partial y^2} \longrightarrow \frac{1}{l^2}\left(\frac{\partial^2 f}{\partial \theta^2}\right)$$

Now show that the Schrödinger equation for a particle of mass m constrained to move on a circle of radius r is (see Problem 3–34)

$$-\frac{\hbar^2}{2I}\frac{\partial^2 \psi(\theta)}{\partial \theta^2} = E\psi(\theta) \qquad 0 \le \theta \le 2\pi$$

where $I = ml^2$ is the moment of inertia. Solve this equation and determine the allowed values of E.

6–3. Generalize Problem 6–2 to the case of a particle moving in a plane under the influence of a central force; in other words, convert

$$\nabla^2 = \frac{\partial^2}{\partial x^2} + \frac{\partial^2}{\partial y^2}$$

to plane polar coordinates, this time without assuming that r is a constant. Use the method of separation of variables to separate the equation for this problem. Solve the angular equation.

6–4. Show that rotational transitions of a diatomic molecule occur in the microwave region or the far infrared region of the spectrum.

6–5. In the far infrared spectrum of $H^{79}Br$, there is a series of lines separated by 16.72 cm^{-1}. Calculate the values of the moment of inertia and the internuclear separation in $H^{79}Br$.

6–6. The $J = 0$ to $J = 1$ transition for carbon monoxide ($^{12}C^{16}O$) occurs at $1.153 \times 10^5 \text{ MHz}$. Calculate the value of the bond length in carbon monoxide.

6–7. The spacing between the lines in the microwave spectrum of $H^{35}Cl$ is $6.350 \times 10^{11} \text{ Hz}$. Calculate the bond length of $H^{35}Cl$.

6–8. The microwave spectrum of $^{39}K^{127}I$ consists of a series of lines whose spacing is almost constant at 3634 MHz. Calculate the bond length of $^{39}K^{127}I$.

6–9. The equilibrium internuclear distance of $H^{127}I$ is 160.4 pm. Calculate the value of B in wave numbers and megahertz.

6–10. Assuming the rotation of a diatomic molecule in the $J = 10$ state may be approximated by classical mechanics, calculate how many revolutions per second $^{23}Na^{35}Cl$ makes in the $J = 10$ rotational state. The rotational constant of $^{23}Na^{35}Cl$ is 6500 MHz.

6–11. The results we derived for a rigid rotator apply to linear polyatomic molecules as well as to diatomic molecules. Given that the moment of inertia I for $H^{12}C^{14}N$ is $1.89 \times 10^{-46} \text{ kg·m}^2$ (cf. Problem 6–12), predict the microwave spectrum of $H^{12}C^{14}N$.

6–12. This problem involves the calculation of the moment of inertia of a linear triatomic molecule such as $H^{12}C^{14}N$ (see Problem 6–11). The moment of inertia of any set of point masses is

$$I = \sum_j m_j l_j^2$$

where l_j is the distance of the jth mass from the center of mass. Thus, the moment of inertia of $H^{12}C^{14}N$ is

$$I = m_H l_H^2 + m_C l_C^2 + m_N l_N^2 \tag{1}$$

Show that equation 1 can be written as

$$I = \frac{m_H m_C r_{HC}^2 + m_H m_N r_{HN}^2 + m_C m_N r_{CN}^2}{m_H + m_C + m_N}$$

where the r's are the various internuclear distances. Given that $r_{HC} = 106.8$ pm and $r_{CN} = 115.6$ pm, calculate the value of I and compare the result with that given in Problem 6–11.

6–13. The following lines were observed in the microwave absorption spectrum of $H^{127}I$ and $D^{127}I$ between 60 cm^{-1} and 90 cm^{-1}.

	$\tilde{\nu}_{obs}/$cm^{-1}			
$H^{127}I$	64.275	77.130	89.985	
$D^{127}I$	65.070	71.577	78.094	84.591

Use the rigid-rotator approximation to determine the values of $\tilde{B}$, I, and $l(v = 0)$ for each molecule. Take the mass of ^{127}I to be 126.904 amu and the mass of D to be 2.013 amu.

6–14. Given that $B = 56\,000$ MHz and $\tilde{\omega} = 2143.0$ cm^{-1} for CO, calculate the frequencies of the first few lines of the R and P branches in the rotation-vibration spectrum of CO.

6–15. Given that $l = 156$ pm and $k = 250$ N·m^{-1} for ^{6}LiF, use the rigid rotator–harmonic oscillator approximation to construct to scale an energy-level diagram for the first five rotational levels in the $v = 0$ and $v = 1$ vibrational states. Indicate the allowed transitions in an absorption experiment, and calculate the frequencies of the first few lines in the R and P branches of the rotation-vibration spectrum of ^{6}LiF.

6–16. Using the values of $\tilde{\omega}_e$, $\tilde{x}_e\tilde{\omega}_e$, $\tilde{B}_e$, and $\tilde{\alpha}_e$ given in Tables 5.1 and 6.1, construct to scale an energy-level diagram for the first five rotational levels in the $v = 0$ and $v = 1$ vibrational states for $H^{35}Cl$. Indicate the allowed transitions in an absorption experiment, and calculate the frequencies of the first few lines in the R and P branches.

6–17. The following data are obtained for the rotation-vibration spectrum of $H^{127}I$. Determine $\tilde{B}_0$, $\tilde{B}_1$, $\tilde{B}_e$, and $\tilde{\alpha}_e$ from these data.

Line	Frequency/cm^{-1}
$R(0)$	2242.6
$R(1)$	2254.8
$P(1)$	2217.1
$P(2)$	2203.8

6–18. The following spectroscopic constants were determined for pure samples of $^{74}Ge^{32}S$ and $^{72}Ge^{32}S$:

Molecule	B_e/MHz	α_e/MHz	D/kHz	$l(v=0)$/pm
$^{74}Ge^{32}S$	5593.08	22.44	2.349	0.20120
$^{72}Ge^{32}S$	5640.06	22.74	2.388	0.20120

Determine the frequency of the $J = 0$ to $J = 1$ transition for $^{74}Ge^{32}S$ and $^{72}Ge^{32}S$ in their ground vibrational state. The width of a microwave absorption line is on the order of 1 KHz. Could you distinguish a pure sample of $^{74}Ge^{32}S$ from a 50/50 mixture of $^{74}Ge^{32}S$ and $^{72}Ge^{32}S$ using microwave spectroscopy?

6–19. An analysis of the rotational spectrum of $^{12}C^{32}S$ gives the following results:

v	$\tilde{B}_v$/cm^{-1}
0	0.81708
1	0.81116
2	0.80524
3	0.79932

Determine the values of $\tilde{B}_e$ and $\tilde{\alpha}_e$ from these data.

6–20. How many degrees of vibrational freedom are there for NH_3, CO_2, CH_4, and C_2H_6?

6–21. The following data are obtained for the rotation-vibration spectrum of $H^{79}Br$. Determine $\tilde{B}_0$, $\tilde{B}_1$, $\tilde{B}_e$, and $\tilde{\alpha}_e$ from these data.

Line	Frequency/cm^{-1}
$R(0)$	2642.60
$R(1)$	2658.36
$P(1)$	2609.67
$P(2)$	2592.51

6–22. The frequencies of the rotational transitions in the nonrigid-rotator approximation are given by Equation 6.45. Show how both $\tilde{B}$ and $\tilde{D}$ may be obtained by plotting $\tilde{v}/(J + 1)$ versus $(J + 1)^2$. Use this method and the data in Table 6.2 to determine both $\tilde{B}$ and $\tilde{D}$ for $H^{35}Cl$.

6–23. The following data are obtained in the microwave spectrum of $^{12}C^{16}O$. Use the method of Problem 6–22 to determine the values of $\tilde{B}$ and $\tilde{D}$ from these data.

Transitions	Frequency/cm^{-1}
$0 \to 1$	3.84540
$1 \to 2$	7.69060
$2 \to 3$	11.53550
$3 \to 4$	15.37990

$4 \to 5$	19.22380
$5 \to 6$	23.06685

6–24. Using the parameters given in Table 6.1, calculate the frequencies (in cm^{-1}) of the $0 \to 1$, $1 \to 2$, $2 \to 3$, and $3 \to 4$ rotational transitions in the ground vibrational state of H^{35}Cl in the nonrigid-rotator approximation.

6–25. Use the data in Table 6.1 to calculate the ratio of centrifugal distortion energy to the total rotational energy of H^{35}Cl and ^{35}Cl^{35}Cl in the $J = 10$ state.

6–26. In this problem, we shall develop a semiclassical derivation of Equation 6.45. As usual, we'll consider one atom to be fixed at the origin with the other atom rotating about it with a reduced mass μ. Because the molecule is nonrigid, as the molecule rotates faster and faster (increasing J), the centrifugal force causes the bond to stretch. The extent of the stretching of the bond can be determined by balancing the Hooke's law force $[k(l - l_0)]$ and the centrifugal force ($\mu v^2 / l = \mu l \omega^2$, Equation 1.20):

$$k(l - l_0) = \mu l \omega^2 \tag{1}$$

In equation 1, l_0 is the bond length when there is no rotation ($J = 0$). The total energy of the rotator is made up of a kinetic energy part and a potential energy part, which we write as

$$E = \frac{1}{2} I \omega^2 + \frac{1}{2} k(l - l_0)^2 \tag{2}$$

Substitute equation 1 into equation 2 and show that the result can be written as

$$E = \frac{L^2}{2\mu l^2} + \frac{L^4}{2\mu^2 k l^6} \tag{3}$$

where $L = I\omega$. Now use the quantum condition $L^2 = \hbar^2 J(J + 1)$ to obtain

$$E_J = \frac{\hbar^2}{2\mu l^2} J(J + 1) + \frac{\hbar^4 J^2 (J + 1)^2}{2\mu^2 k l^6} \tag{4}$$

Finally, solve equation 1 for $1/l^2$, and eliminate ω by using $\omega = L/I = L/\mu l^2$ to obtain

$$\frac{1}{l^2} = \frac{1}{l_0} \left(1 - \frac{2L^2}{\mu k l^4} + \frac{L^4}{\mu^2 k^2 l^8} \right)$$

Now substitute this result into equation 4 to obtain

$$E_J = \frac{\hbar^2}{2\mu l_0^2} J(J + 1) - \frac{\hbar^4}{2\mu^2 k l_0^6} J^2 (J + 1)^2$$

6–27. In terms of the variable θ, Legendre's equation is

$$\sin \theta \frac{d}{d\theta} \left[\sin \theta \frac{d\Theta(\theta)}{d\theta} \right] + (\beta^2 \sin^2 \theta - m^2) \Theta(\theta) = 0$$

Let $x = \cos\theta$ and $P(x) = \Theta(\theta)$ and show that

$$(1 - x^2)\frac{d^2 P(x)}{dx^2} - 2x\frac{d P(x)}{dx} + \left(\beta - \frac{m^2}{1 - x^2}\right)P(x) = 0$$

6–28. Show that the Legendre polynomials given in Table 6.3 satisfy Equation 6.48 with $m = 0$.

6–29. Show that the orthogonality integral for the Legendre polynomials, Equation 6.49, is equivalent to

$$\int_0^\pi P_l(\cos\theta)P_n(\cos\theta)\sin\theta d\theta = 0 \qquad l \neq n$$

6–30. Show that the Legendre polynomials given in Table 6.3 satisfy the orthogonality and normalization conditions given by Equations 6.49 and 6.50.

6–31. Use Equation 6.51 to generate the associated Legendre functions in Table 6.4.

6–32. Show that the first few associated Legendre functions given in Table 6.4 are solutions to Equation 6.48 and that they satisfy the orthonormality condition (Equation 6.53).

6–33. There are a number of recursion formulas for the associated Legendre functions. Show that the first few associated Legendre functions in Table 6.4 satisfy the recursion formula (Equation 6.66)

$$(2l + 1)x P_l^{|m|}(x) = (l - |m| + 1)P_{l+1}^{|m|}(x) + (l + |m|)P_{l-1}^{|m|}(x)$$

6–34. Show that the factor $i^{m+|m|} = 1$ when m is odd and negative and that it equals -1 when m is odd and positive.

6–35. Show that the integral of $\sin\theta\, d\theta d\phi$ over the surface of a sphere is equal to 4π.

6–36. Show that the first few spherical harmonics in Table 6.5 satisfy the orthonormality condition (Equation 6.57).

6–37. Using explicit expressions for $Y_l^m(\theta, \phi)$, show that

$$|Y_1^1(\theta, \phi)|^2 + |Y_1^0(\theta, \phi)|^2 + |Y_{-1}^1(\theta, \phi)|^2 = \text{constant}$$

This is a special case of the general theorem

$$\sum_{m=-l}^{+l} |Y_l^m(\theta, \phi)|^2 = \text{constant}$$

known as Unsöld's theorem. What is the physical significance of this result?

6–38. In cartesian coordinates,

$$\hat{L}_z = -i\hbar\left(x\frac{\partial}{\partial y} - y\frac{\partial}{\partial x}\right)$$

Convert this equation to spherical coordinates, showing that

$$\hat{L}_z = -i\hbar \frac{\partial}{\partial \phi}$$

6–39. Convert $\hat{L}_x$ and $\hat{L}_y$ from cartesian coordinates to spherical coordinates.

6–40. Compute the value of $\hat{L}^2 Y(\theta, \phi)$ for the following functions:

(a) $1/(4\pi)^{1/2}$

(b) $(3/4\pi)^{1/2} \cos\theta$

(c) $(3/8\pi)^{1/2} \sin\theta e^{i\phi}$

(d) $(3/8\pi)^{1/2} \sin\theta e^{-i\phi}$

Do you find anything interesting about the results?

6–41. Prove that $\hat{L}^2$ commutes with $\hat{L}_x$, $\hat{L}_y$, and $\hat{L}_z$ but that

$$[\hat{L}_x, \hat{L}_y] = i\hbar \hat{L}_z \qquad [\hat{L}_y, \hat{L}_z] = i\hbar \hat{L}_x \qquad [\hat{L}_z, \hat{L}_x] = i\hbar \hat{L}_y$$

(*Hint:* Use cartesian coordinates.) Do you see a pattern in these formulas?

6–42. It is a somewhat advanced exercise to prove generally that $\langle L_x \rangle = \langle L_y \rangle = 0$, but prove that they are zero at least for the first few l, m states by using the spherical harmonics given in Table 6.5.

6–43. Calculate the ratio of the dipole transition moments for the $0 \to 1$ and $1 \to 2$ rotational transitions in the rigid-rotator approximation.

6–44. In this problem, we'll calculate the fraction of diatomic molecules in a particular rotational level at a temperature T using the rigid-rotator approximation. A fundamental equation of physical chemistry is the *Boltzmann distribution*, which says that the number of molecules with an energy E_J is proportional to $e^{-E_J/k_B T}$, where k_B is the Boltzmann constant and T is the kelvin temperature. Furthermore, because the degeneracy of the Jth rotational level is $2J + 1$, we write

$$N_J \propto (2J + 1)e^{-E_J/k_B T} = (2J + 1)e^{-BJ(J+1)/k_B T}$$

or

$$N_J = c(2J + 1)e^{-BJ(J+1)/k_B T}$$

where c is a proportionality constant. Plot N_J/N_0 versus J for $H^{35}Cl$ ($\tilde{B} = 10.60 \text{ cm}^{-1}$) and $^{127}I^{35}Cl$ ($\tilde{B} = 0.114 \text{ cm}^{-1}$) at 300 K. Treating J as a continuous parameter, show that the value of J in the most populated rotational state is the nearest integer to

$$J_{max} = \frac{1}{2}\left[\left(\frac{2k_B T}{\tilde{B}}\right)^{1/2} - 1\right]$$

Calculate J_{max} for $H^{35}Cl$ ($\tilde{B} = 10.60 \text{ cm}^{-1}$) and $^{127}I^{35}Cl$ ($\tilde{B} = 0.114 \text{ cm}^{-1}$) at 300 K.

6–45. The summation that occurs in the rotational Boltzmann distribution (previous problem) can be evaluated approximately by converting the summation to an integral. Show that

$$\sum_{J=0}^{\infty} (2J+1) \exp\left[\frac{-\tilde{B}J(J+1)}{k_B T}\right] \approx \int_0^{\infty} \exp\left[\frac{-\tilde{B}J(J+1)}{k_B T}\right] d[J(J+1)]$$

$$= \frac{k_B T}{\tilde{B}} = \frac{8\pi^2 c I k_B T}{h}$$

This is an excellent approximation for values of $\tilde{B}/k_B T$ less than 0.05 or so. Using this result, calculate and plot the fraction of $^{127}I^{35}Cl$ molecules in the Jth rotational state versus J at 25°C. ($\tilde{B} = 0.114 \text{ cm}^{-1}$.)

6–46. Can you use the result of the previous problem to rationalize the envelope of the lines in the P and R branches in the rotation-vibration spectrum in Figure 6.4?

6–47. Many students are mystified when they see that $0! = 1$. The standard formula $n! = n(n-1)\cdots 1$ is applicable *only* for $n = 1, 2, 3, \ldots$. Euler showed that we could extend the idea of a factorial to other numbers through the definite integral

$$I_n = \int_0^{\infty} x^n e^{-x} dx$$

Integration repeatedly by parts shows that $I_n = n(n-1)\cdots 1$ when $n = 1, 2, 3, \ldots$, so that I_n gives our "standard" result in that case. However, there is no reason for n to be restricted in the above integral, and we can use it to *define* $n!$ for other values of n. Use this extended definition to show that $0! = 1$. What about $(1/2)!$? Even $(-1/2)!$? For a more complete discussion of $n!$ for a general value of n, read about the gamma function in any applied mathematics book.

References

Gordy, G., Cook, R. L. *Techniques of Chemistry: Microwave Molecular Spectra,* 3rd ed. Wiley & Sons: New York, 1984.

Barrow, G. *Molecular Spectroscopy*. McGraw-Hill: New York, 1962.

Brown, J. M. *Molecular Spectroscopy*. Oxford University Press: New York, 1998.

Banwell, C. N., McCash, E. M. *Fundamentals of Molecular Spectroscopy,* 4th ed. McGraw-Hill: New York, 1944.

Hollas, J. Michael. *Molecular Spectroscopy,* 4th ed. Wiley & Sons: New York, 2004.

Hollas, J. M. *Basic Atomic and Molecular Spectroscopy*. Wiley & Sons: New York, 2002.

Herzberg, G. *Molecular Spectra and Molecular Structure: Spectra of Diatomic Molecules,* 2nd ed. Krieger Publishing: Malabar, FL, 1989.

Harris, D. C., Bertolucci, M. D. *Symmetry and Spectroscopy: An Introduction to Vibrational and Electronic Spectroscopy*. Dover Publications: Mineola, NY, 1989.

Townsend, J. S. *A Modern Approach to Quantum Mechanics*. University Science Books: Sausalito, CA, 2000.

Determinants

In the next chapter we will encounter n linear algebraic equations in n unknowns. Such equations can be solved by means of determinants, which we discuss in this MathChapter. Consider the pair of linear algebraic equations

$$
\begin{aligned}
a_{11}x + a_{12}y &= d_1 \\
a_{21}x + a_{22}y &= d_2
\end{aligned}
\tag{F.1}
$$

If we multiply the first of these equations by a_{22} and the second by a_{12} and then subtract, we obtain

$$(a_{11}a_{22} - a_{12}a_{21})x = d_1a_{22} - d_2a_{12}$$

or

$$x = \frac{a_{22}d_1 - a_{12}d_2}{a_{11}a_{22} - a_{12}a_{21}} \tag{F.2}$$

Similarly, if we multiply the first by a_{21} and the second by a_{11} and then subtract, we get

$$y = \frac{a_{11}d_2 - a_{21}d_1}{a_{11}a_{22} - a_{12}a_{21}} \tag{F.3}$$

Notice that the denominators in both Equations F.2 and F.3 are the same. We represent $a_{11}a_{22} - a_{12}a_{21}$ by the quantity $\begin{vmatrix} a_{11} & a_{12} \\ a_{21} & a_{22} \end{vmatrix}$, which equals $a_{11}a_{22} - a_{12}a_{21}$ and is called a 2×2 *determinant*. The reason for introducing this notation is that it readily generalizes to the treatment of n linear algebraic equations in n unknowns. Generally, an $n \times n$

determinant is a square array of n^2 elements arranged in n rows and n columns. A 3×3 determinant is given by

$$\begin{vmatrix} a_{11} & a_{12} & a_{13} \\ a_{21} & a_{22} & a_{23} \\ a_{31} & a_{32} & a_{33} \end{vmatrix} = \begin{matrix} a_{11}a_{22}a_{33} + a_{21}a_{32}a_{13} + a_{12}a_{23}a_{31} \\ - a_{31}a_{22}a_{13} - a_{21}a_{12}a_{33} - a_{11}a_{23}a_{32} \end{matrix} \tag{F.4}$$

(We will prove this soon.) Notice that the element a_{ij} occurs at the intersection of the ith row and the jth column.

Equation F.4 and the corresponding equations for evaluating higher-order determinants can be obtained in a systematic manner. First we define a cofactor. The *cofactor*, A_{ij}, of an element a_{ij} is an $(n-1) \times (n-1)$ determinant obtained by deleting the ith row and the jth column, multiplied by $(-1)^{i+j}$. For example, A_{12}, the cofactor of element a_{12} of

$$D = \begin{vmatrix} a_{11} & a_{12} & a_{13} \\ a_{21} & a_{22} & a_{23} \\ a_{31} & a_{32} & a_{33} \end{vmatrix}$$

is

$$A_{12} = (-1)^{1+2} \begin{vmatrix} a_{21} & a_{23} \\ a_{31} & a_{33} \end{vmatrix}$$

EXAMPLE F–1

Evaluate the cofactor of each of the first-row elements in

$$D = \begin{vmatrix} 2 & -1 & 1 \\ 0 & 3 & -1 \\ 2 & -2 & 1 \end{vmatrix}$$

SOLUTION: The cofactor of a_{11} is

$$A_{11} = (-1)^{1+1} \begin{vmatrix} 3 & -1 \\ -2 & 1 \end{vmatrix} = 3 - 2 = 1$$

The cofactor of a_{12} is

$$A_{12} = (-1)^{1+2} \begin{vmatrix} 0 & -1 \\ 2 & 1 \end{vmatrix} = -2$$

and the cofactor of a_{13} is

$$A_{13} = (-1)^{1+3} \begin{vmatrix} 0 & 3 \\ 2 & -2 \end{vmatrix} = -6$$

We can use cofactors to evaluate determinants. The value of the 3×3 determinant in Equation F.4 can be obtained from the formula

$$\begin{vmatrix} a_{11} & a_{12} & a_{13} \\ a_{21} & a_{22} & a_{23} \\ a_{31} & a_{32} & a_{33} \end{vmatrix} = a_{11}A_{11} + a_{12}A_{12} + a_{13}A_{13} \qquad \text{(F.5)}$$

Thus, the value of D in Example F–1 is

$$D = (2)(1) + (-1)(-2) + (1)(-6) = -2$$

EXAMPLE F–2
Evaluate D in Example F–1 by expanding in terms of the first *column* of elements instead of the first *row*.

SOLUTION: We will use the formula

$$D = a_{11}A_{11} + a_{21}A_{21} + a_{31}A_{31}$$

The various cofactors are

$$A_{11} = (-1)^2 \begin{vmatrix} 3 & -1 \\ -2 & 1 \end{vmatrix} = 1$$

$$A_{21} = (-1)^3 \begin{vmatrix} -1 & 1 \\ -2 & 1 \end{vmatrix} = -1$$

and

$$A_{31} = (-1)^4 \begin{vmatrix} -1 & 1 \\ 3 & -1 \end{vmatrix} = -2$$

and so

$$D = (2)(1) + (0)(-1) + (2)(-2) = -2$$

Notice that we obtained the same answer for D as we did for Example F–1. This result illustrates the general fact that a determinant may be evaluated by expanding in terms of the cofactors of the elements of any row or any column. If we choose the second row of D, then we obtain

$$D = (0)(-1)^3 \begin{vmatrix} -1 & 1 \\ -2 & 1 \end{vmatrix} + (3)(-1)^4 \begin{vmatrix} 2 & 1 \\ 2 & 1 \end{vmatrix} + (-1)(-1)^5 \begin{vmatrix} 2 & -1 \\ 2 & -2 \end{vmatrix} = -2$$

Although we have discussed only 3×3 determinants, the procedure is readily extended to determinants of any order.

EXAMPLE F–3

In Chapter 11 we will meet the *determinantal equation*

$$\begin{vmatrix} x & 1 & 0 & 0 \\ 1 & x & 1 & 0 \\ 0 & 1 & x & 1 \\ 0 & 0 & 1 & x \end{vmatrix} = 0$$

Expand this determinantal equation into a quartic equation for x.

SOLUTION: Expand about the first row of elements to obtain

$$x \begin{vmatrix} x & 1 & 0 \\ 1 & x & 1 \\ 0 & 1 & x \end{vmatrix} - \begin{vmatrix} 1 & 1 & 0 \\ 0 & x & 1 \\ 0 & 1 & x \end{vmatrix} = 0$$

Now expand about the first column of each of the 3×3 determinants to obtain

$$(x)(x) \begin{vmatrix} x & 1 \\ 1 & x \end{vmatrix} - (x)(1) \begin{vmatrix} 1 & 0 \\ 1 & x \end{vmatrix} - (1) \begin{vmatrix} x & 1 \\ 1 & x \end{vmatrix} = 0$$

or

$$x^2(x^2 - 1) - x(x) - (1)(x^2 - 1) = 0$$

or

$$x^4 - 3x^2 + 1 = 0$$

Note that because we can choose any row or column to expand the determinant, it is easiest to take the one with the most zeroes!

A number of properties of determinants are useful to know:

1. The value of a determinant is unchanged if the rows are made into columns in the same order; in other words, first row becomes first column, second row becomes second column, and so on. For example,

$$\begin{vmatrix} 1 & 2 & 5 \\ -1 & 0 & -1 \\ 3 & 1 & 2 \end{vmatrix} = \begin{vmatrix} 1 & -1 & 3 \\ 2 & 0 & 1 \\ 5 & -1 & 2 \end{vmatrix}$$

2. If any two rows or columns are the same, the value of the determinant is zero. For example,

$$\begin{vmatrix} 4 & 2 & 4 \\ -1 & 0 & -1 \\ 3 & 1 & 3 \end{vmatrix} = 0$$

3. If any two rows or columns are interchanged, the sign of the determinant is changed. For example,

$$\begin{vmatrix} 3 & 1 & -1 \\ -6 & 4 & 5 \\ 1 & 2 & 2 \end{vmatrix} = - \begin{vmatrix} 1 & 3 & -1 \\ 4 & -6 & 5 \\ 2 & 1 & 2 \end{vmatrix}$$

4. If every element in a row or column is multiplied by a factor k, the value of the determinant is multiplied by k. For example,

$$\begin{vmatrix} 6 & 8 \\ -1 & 2 \end{vmatrix} = 2 \begin{vmatrix} 3 & 4 \\ -1 & 2 \end{vmatrix}$$

5. If any row or column is written as the sum or difference of two or more terms, the determinant can be written as the sum or difference of two or more determinants according to

$$\begin{vmatrix} a_{11} \pm a_{11}' & a_{12} & a_{13} \\ a_{21} \pm a_{21}' & a_{22} & a_{23} \\ a_{31} \pm a_{31}' & a_{32} & a_{33} \end{vmatrix} = \begin{vmatrix} a_{11} & a_{12} & a_{13} \\ a_{21} & a_{22} & a_{23} \\ a_{31} & a_{32} & a_{33} \end{vmatrix} \pm \begin{vmatrix} a_{11}' & a_{12} & a_{13} \\ a_{21}' & a_{22} & a_{23} \\ a_{31}' & a_{32} & a_{33} \end{vmatrix}$$

For example,

$$\begin{vmatrix} 3 & 3 \\ 2 & 6 \end{vmatrix} = \begin{vmatrix} 2+1 & 3 \\ -2+4 & 6 \end{vmatrix} = \begin{vmatrix} 2 & 3 \\ -2 & 6 \end{vmatrix} + \begin{vmatrix} 1 & 3 \\ 4 & 6 \end{vmatrix}$$

6. The value of a determinant is unchanged if one row or column is added or subtracted to another, as in

$$\begin{vmatrix} a_{11} & a_{12} & a_{13} \\ a_{21} & a_{22} & a_{23} \\ a_{31} & a_{32} & a_{33} \end{vmatrix} = \begin{vmatrix} a_{11}+a_{12} & a_{12} & a_{13} \\ a_{21}+a_{22} & a_{22} & a_{23} \\ a_{31}+a_{32} & a_{32} & a_{33} \end{vmatrix}$$

For example,

$$\begin{vmatrix} 1 & -1 & 3 \\ 4 & 0 & 2 \\ 1 & 2 & 1 \end{vmatrix} = \begin{vmatrix} 0 & -1 & 3 \\ 4 & 0 & 2 \\ 3 & 2 & 1 \end{vmatrix} = \begin{vmatrix} 0 & -1 & 3 \\ 4 & 0 & 2 \\ 7 & 2 & 3 \end{vmatrix}$$

In the first case we add column 2 to column 1, and in the second case we added row 2 to row 3. This procedure may be repeated n times to obtain

$$\begin{vmatrix} a_{11} & a_{12} & a_{13} \\ a_{21} & a_{22} & a_{23} \\ a_{31} & a_{32} & a_{33} \end{vmatrix} = \begin{vmatrix} a_{11} + na_{12} & a_{12} & a_{13} \\ a_{21} + na_{22} & a_{22} & a_{23} \\ a_{31} + na_{32} & a_{32} & a_{33} \end{vmatrix} \tag{F.6}$$

This result is easy to prove:

$$\begin{vmatrix} a_{11} + na_{12} & a_{12} & a_{13} \\ a_{21} + na_{22} & a_{22} & a_{23} \\ a_{31} + na_{32} & a_{32} & a_{33} \end{vmatrix} = \begin{vmatrix} a_{11} & a_{12} & a_{13} \\ a_{21} & a_{22} & a_{23} \\ a_{31} & a_{32} & a_{33} \end{vmatrix} + n \begin{vmatrix} a_{12} & a_{12} & a_{13} \\ a_{22} & a_{22} & a_{23} \\ a_{32} & a_{32} & a_{33} \end{vmatrix}$$

$$= \begin{vmatrix} a_{11} & a_{12} & a_{13} \\ a_{21} & a_{22} & a_{23} \\ a_{31} & a_{32} & a_{33} \end{vmatrix} + 0$$

where we used rule 5 to write the first line. The second determinant on the right side equals zero because two columns are the same.

We provided these rules because simultaneous linear algebraic equations can be solved in terms of determinants. For simplicity, we will consider only a pair of equations, but the final result is easy to generalize. Consider the two equations

$$\begin{aligned} a_{11}x + a_{12}y &= d_1 \\ a_{21}x + a_{22}y &= d_2 \end{aligned} \tag{F.7}$$

If $d_1 = d_2 = 0$, the equations are said to be *homogeneous*. Otherwise, they are called *inhomogeneous*. Let's assume at first that they are inhomogeneous. The determinant of the coefficients of x and y is

$$D = \begin{vmatrix} a_{11} & a_{12} \\ a_{21} & a_{22} \end{vmatrix}$$

According to rule 4,

$$\begin{vmatrix} a_{11}x & a_{12} \\ a_{21}x & a_{22} \end{vmatrix} = xD$$

Furthermore, according to rule 6,

$$\begin{vmatrix} a_{11}x + a_{12}y & a_{12} \\ a_{21}x + a_{22}y & a_{22} \end{vmatrix} = xD \tag{F.8}$$

If we substitute Equation F.7 into Equation F.8, then we have

$$\begin{vmatrix} d_1 & a_{12} \\ d_2 & a_{22} \end{vmatrix} = xD$$

Solving for x gives

$$x = \frac{\begin{vmatrix} d_1 & a_{12} \\ d_2 & a_{22} \end{vmatrix}}{\begin{vmatrix} a_{11} & a_{12} \\ a_{21} & a_{22} \end{vmatrix}} \tag{F.9}$$

Similarly, we get

$$y = \frac{\begin{vmatrix} a_{11} & d_1 \\ a_{21} & d_2 \end{vmatrix}}{\begin{vmatrix} a_{11} & a_{12} \\ a_{21} & a_{22} \end{vmatrix}} \tag{F.10}$$

Notice that Equations F.9 and F.10 are identical to Equations F.2 and F.3. The solution for x and y in terms of determinants is called *Cramer's rule*. Note that the determinant in the numerator is obtained by replacing the column in D that is associated with the unknown quantity and replacing it with the column associated with the right sides of Equations F.7. This result is readily extended to more than two simultaneous equations.

EXAMPLE F–4
Solve the equations

$$x + y + z = 2$$

$$2x - y - z = 1$$

and

$$x + 2y - z = -3$$

SOLUTION: The extension of Equations F.9 and F.10 is

$$x = \frac{\begin{vmatrix} 2 & 1 & 1 \\ 1 & -1 & -1 \\ -3 & 2 & -1 \end{vmatrix}}{\begin{vmatrix} 1 & 1 & 1 \\ 2 & -1 & -1 \\ 1 & 2 & -1 \end{vmatrix}} = \frac{9}{9} = 1$$

Similarly,

$$y = \frac{\begin{vmatrix} 1 & 2 & 1 \\ 2 & 1 & -1 \\ 1 & -3 & -1 \end{vmatrix}}{\begin{vmatrix} 1 & 1 & 1 \\ 2 & -1 & -1 \\ 1 & 2 & -1 \end{vmatrix}} = \frac{-9}{9} = -1$$

and

$$z = \frac{\begin{vmatrix} 1 & 1 & 2 \\ 2 & -1 & 1 \\ 1 & 2 & -3 \end{vmatrix}}{\begin{vmatrix} 1 & 1 & 1 \\ 2 & -1 & -1 \\ 1 & 2 & -1 \end{vmatrix}} = \frac{18}{9} = 2$$

What happens if $d_1 = d_2 = 0$ in Equation F.7? In that case, we find that $x = y = 0$, which is an obvious solution called a *trivial solution*. The only way that we could obtain a nontrivial solution for a set of homogeneous equations is for the denominator in Equations F.9 and F.10 to be zero, or for

$$D = \begin{vmatrix} a_{11} & a_{12} \\ a_{21} & a_{22} \end{vmatrix} = 0 \tag{F.11}$$

In Chapter 8, we will meet equations such as

$$c_1(H_{11} - ES_{11}) + c_2(H_{12} - ES_{12}) = 0$$

and

$$c_1(H_{12} - ES_{12}) + c_2(H_{22} - ES_{22}) = 0$$

where the H_{ij} and S_{ij} are known quantities and c_1, c_2, and E are to be determined. We can appeal to Equation F.11, which says that for a nontrivial solution (in other words, one for which both c_1 and c_2 are not equal to zero) to exist, we must have

$$\begin{vmatrix} H_{11} - ES_{11} & H_{12} - ES_{12} \\ H_{12} - ES_{12} & H_{22} - ES_{22} \end{vmatrix} = 0 \tag{F.12}$$

When this determinant is expanded, we obtain a quadratic equation in E, yielding two roots. The determinant in Equation F.12 is called a *secular determinant* and Equation F.12 itself constitutes a *secular determinantal equation*.

EXAMPLE F–5

Find the roots of the determinantal equation

$$\begin{vmatrix} 2 - \lambda & 3 \\ 3 & 4 - \lambda \end{vmatrix} = 0$$

SOLUTION: Expand the determinant to obtain $(2 - \lambda)(4 - \lambda) - 9 = 0$ or $\lambda^2 - 6\lambda - 1 = 0$. The two roots are

$$\lambda = \frac{6}{2} \pm \frac{\sqrt{40}}{2} = 3 \pm \sqrt{10}$$

Although we considered only two simultaneous homogeneous algebraic equations, Equation F.11 is readily extended to any number. We will use this result in Chapter 8.

Problems

F–1. Evaluate the determinant

$$D = \begin{vmatrix} 2 & 1 & 1 \\ -1 & 3 & 2 \\ 2 & 0 & 1 \end{vmatrix}$$

Add column 2 to column 1 to get

$$\begin{vmatrix} 3 & 1 & 1 \\ 2 & 3 & 2 \\ 2 & 0 & 1 \end{vmatrix}$$

and evaluate it. Compare your result with the value of D. Now add row 2 to row 1 of D to get

$$\begin{vmatrix} 1 & 4 & 3 \\ -1 & 3 & 2 \\ 2 & 0 & 1 \end{vmatrix}$$

and evaluate it. Compare your result with the value of D above.

George Uhlenbeck (at left) was born on December 6, 1900, in Batavia, Java (now Jakarta, Indonesia), and died in 1988 in Boulder, Colorado. When he was six years old, his family moved to The Hague, the Netherlands. After graduating from the University of Leiden in 1920, Uhlenbeck began his graduate studies there with Paul Ehrenfest. While an assistant to Ehrenfest in 1925, Uhlenbeck, working with a fellow graduate student, Samuel Goudsmit, made his most important discovery—electron spin. He eventually wrote his dissertation in Copenhagen in 1927, and then received a position at the University of Michigan. He returned to the Netherlands in 1935, but left shortly before World War II for the United States, where he taught at the University of Michigan and later at Rockefeller University. Uhlenbeck was an inspiring teacher with organized and extremely clear lectures.

Samuel Goudsmit (at right) was born on July 11, 1902, in The Hague, the Netherlands, and died in 1978 in Reno, Nevada. In 1919, he entered the University of Leiden, and later did experimental work from 1923 to 1926 at the University of Amsterdam with Pieter Zeeman. He received his Ph.D. from the University of Leiden in 1927, and then immigrated to the United States with Uhlenbeck to accept a position at the University of Michigan. During World War II, he served with the Alsos Mission in Europe, where he traveled with the U.S. Army through newly occupied territory to assess the progress of the German atomic bomb project. Goudsmit wrote an account of this mission in a book titled *Alsos*. He served as the editor-in-chief of the American Physical Society for over twenty years, founding *Physical Review Letters* in 1958.

In spite of the importance of their discovery of electron spin, Uhlenbeck and Goudsmit never received the Nobel Prize.

CHAPTER 7

The Hydrogen Atom

We are now ready to study the hydrogen atom, which is of particular interest to chemists because it serves as the prototype for more complex atoms and, therefore, molecules. In addition, probably every chemistry student has studied the results of a quantum-mechanical treatment of the hydrogen atom in general chemistry, and in this chapter we will see that the familiar hydrogen atomic orbitals and their properties emerge naturally as solutions to the Schrödinger equation.

In the first three sections, we discuss the wave functions of an electron in a hydrogen atom, or the hydrogen atomic orbitals, quantitatively. We use the equations for the orbitals to calculate a number of properties of a hydrogen atom. Then in Section 7.4, we discuss the Zeeman effect, which describes the spectrum of a hydrogen atom in an external magnetic field. The Zeeman effect is not of direct interest to most chemists (although he received the 1902 Nobel Prize in Physics for this work), but the discussion in Section 7.4 leads directly to the introduction of electron spin, which certainly is of interest to most chemists. After incorporating electron spin into our quantum-mechanical formalism, we then show how the electronic states of a hydrogen atom can be described by way of term symbols, which depend upon the interaction between the spin of the electron and its orbital angular momentum. The final section, which is optional, completes our discussion of the Zeeman effect and shows how it depends upon electron spin.

7.1 The Schrödinger Equation for a Hydrogen Atom Can Be Solved Exactly

The electron and the proton in a hydrogen atom interact through a coulombic potential:

$$V(r) = -\frac{e^2}{4\pi\epsilon_0 r} \tag{7.1}$$

where e is the charge on the proton, ϵ_0 is the permittivity of free space, and r is the distance between the electron and the proton. The factor $4\pi\epsilon_0$ arises because we are using SI units. The spherical geometry of the model suggests that we use a spherical coordinate system with the proton at the origin. The Hamiltonian operator for a hydrogen atom is

$$\hat{H} = -\frac{\hbar^2}{2m_e}\nabla^2 - \frac{e^2}{4\pi\epsilon_0 r} \tag{7.2}$$

where m_e is the mass of the electron, and the Schrödinger equation is

$$-\frac{\hbar^2}{2m_e}\nabla^2\psi(r,\theta,\phi) + V(r)\psi(r,\theta,\phi) = E\psi(r,\theta,\phi) \tag{7.3}$$

where ∇^2 is the Laplacian operator in spherical coordinates (Equation 6.7):

$$\nabla^2 = \frac{1}{r^2}\frac{\partial}{\partial r}\left(r^2\frac{\partial}{\partial r}\right) + \frac{1}{r^2\sin\theta}\frac{\partial}{\partial\theta}\left(\sin\theta\frac{\partial}{\partial\theta}\right) + \frac{1}{r^2\sin^2\theta}\frac{\partial^2}{\partial\phi^2} \tag{7.4}$$

If we substitute Equation 7.4 into Equation 7.3, we obtain

$$-\frac{\hbar^2}{2m_e}\left[\frac{1}{r^2}\frac{\partial}{\partial r}\left(r^2\frac{\partial\psi}{\partial r}\right) + \frac{1}{r^2\sin\theta}\frac{\partial}{\partial\theta}\left(\sin\theta\frac{\partial\psi}{\partial\theta}\right) + \frac{1}{r^2\sin^2\theta}\frac{\partial^2\psi}{\partial\phi^2}\right] \tag{7.5}$$
$$+ V(r)\psi(r,\theta,\phi) = E\psi(r,\theta,\phi)$$

At first sight, this partial differential equation looks exceedingly complicated. To bring Equation 7.5 into a more manageable form, first multiply through by $2m_e r^2$ to obtain

$$-\hbar^2\frac{\partial}{\partial r}\left(r^2\frac{\partial\psi}{\partial r}\right) - \hbar^2\left[\frac{1}{\sin\theta}\left(\frac{\partial}{\partial\theta}\sin\theta\frac{\partial\psi}{\partial\theta}\right) + \frac{1}{\sin^2\theta}\frac{\partial^2\psi}{\partial\phi}\right] \tag{7.6}$$
$$+ 2m_e r^2[V(r) - E]\psi(r,\theta,\phi) = 0$$

The second term here, the one containing all the θ and ϕ dependence, is nothing but $\hat{L}^2\psi$ according to Equation 6.59. Thus, we can write the Schrödinger equation in the form

$$-\hbar^2\frac{\partial}{\partial r}\left(r^2\frac{\partial\psi}{\partial r}\right) + \hat{L}^2\psi + 2m_e r^2[V(r) - E]\,\psi(r,\theta,\phi) = 0 \tag{7.7}$$

Notice now that if we consider the entire left side of Equation 7.7 to be an operator acting upon $\psi(r,\theta,\phi)$, then this operator consists of a part that depends upon only r (the first and third terms) and a part that depends upon only θ and ϕ (the second term). According to Section 3.9, $\psi(r,\theta,\phi)$ factors into the product of a function that depends upon only r and one that depends upon only θ and ϕ. Furthermore, the θ, ϕ factor must be an eigenfunction of $\hat{L}^2$, which, according to Equation 6.60, we know to be the

TABLE 7.1
The First Few Spherical Harmonics

$$Y_0^0 = \frac{1}{(4\pi)^{1/2}} \qquad\qquad Y_1^0 = \left(\frac{3}{4\pi}\right)^{1/2}\cos\theta$$

$$Y_1^1 = -\left(\frac{3}{8\pi}\right)^{1/2}\sin\theta\, e^{i\phi} \qquad\qquad Y_1^{-1} = \left(\frac{3}{8\pi}\right)^{1/2}\sin\theta\, e^{-i\phi}$$

$$Y_2^0 = \left(\frac{5}{16\pi}\right)^{1/2}(3\cos^2\theta - 1) \qquad\qquad Y_2^1 = -\left(\frac{15}{8\pi}\right)^{1/2}\sin\theta\cos\theta\, e^{i\phi}$$

$$Y_2^{-1} = \left(\frac{15}{8\pi}\right)^{1/2}\sin\theta\cos\theta\, e^{-i\phi} \qquad\qquad Y_2^2 = \left(\frac{15}{32\pi}\right)^{1/2}\sin^2\theta\, e^{2i\phi}$$

$$Y_2^{-2} = \left(\frac{15}{32\pi}\right)^{1/2}\sin^2\theta\, e^{-2i\phi}$$

spherical harmonics $Y_l^{m_l}(\theta, \phi)$ (Section 6.6):

$$\hat{L}^2 Y_l^{m_l}(\theta, \phi) = \hbar^2 l\,(l+1)Y_l^{m_l}(\theta, \phi) \qquad \begin{array}{l} l = 0, 1, 2, \ldots \\ -l \le m_l \le +l \end{array} \qquad (7.8)$$

It is customary to write Equation 7.8 in terms of l and m_l instead of J and m when applying it to a hydrogen atom. The first few spherical harmonics are given in Table 6.5, but we reproduce them here in Table 7.1 for convenience.

Consequently, if we let

$$\psi(r, \theta, \phi) = R(r)Y_l^{m_l}(\theta, \phi) \qquad (7.9)$$

and use Equation 7.8, Equation 7.7 becomes (Problem 7–2)

$$-\frac{\hbar^2}{2m_e r^2}\frac{d}{dr}\left(r^2\frac{dR}{dr}\right) + \left[\frac{\hbar^2 l(l+1)}{2m_e r^2} + V(r) - E\right]R(r) = 0 \qquad (7.10)$$

Equation 7.10 is called the *radial equation* for the hydrogen atom and is the only new equation that we have to study in order to have a complete solution to the hydrogen atom.

Notice that the square of the angular momentum is quantized and conserved in a central field, just as it is conserved classically in a central field. Equation 7.10 has the direct physical interpretation that the total energy E is the sum of a radial kinetic energy, an angular kinetic energy, and the potential energy. Equation 7.10 is an ordinary differential equation in r. It is somewhat tedious to solve, but once solved, we find that for solutions to be acceptable as wave functions, the energy must be quantized according to

$$E_n = -\frac{m_e e^4}{8\epsilon_0^2 h^2 n^2} = -\frac{m_e e^4}{32\pi^2\epsilon_0^2\hbar^2 n^2} \qquad n = 1, 2, \ldots \qquad (7.11)$$

If we introduce the Bohr radius from Section 1.8, $a_0 = \epsilon_0 h^2 / \pi m_e e^2 = 4\pi \epsilon_0 \hbar^2 / m_e e^2$, then Equation 7.11 becomes

$$E_n = -\frac{e^2}{8\pi^2 \epsilon_0 a_0 n^2} \qquad n = 1, 2, \ldots \tag{7.12}$$

It is surely remarkable that these are the same energies obtained from the Bohr model of the hydrogen atom. Of course, the electron now is not restricted to the sharply defined orbits of Bohr, but is described by its wave function, $\psi(r, \theta, \phi)$.

In the course of solving Equation 7.10, we find not only that an integer occurs naturally but that n must satisfy the condition that $n \geq l + 1$, which is usually written as

$$0 \leq l \leq n - 1 \qquad n = 1, 2, \ldots \tag{7.13}$$

because we have already seen in the previous chapter that the smallest possible value of l is zero. (Equation 7.13 might be familiar from general chemistry.) The solutions to Equation 7.10, called the *radial wave functions*, depend on two quantum numbers n and l and are given by

$$R_{nl}(r) = -\left\{ \frac{(n - l - 1)!}{2n\,[(n + 1)!]^3} \right\}^{1/2} \left(\frac{2}{na_0} \right)^{l+3/2} r^l e^{-r/na_0} L_{n+l}^{2l+1}\left(\frac{2r}{na_0} \right) \tag{7.14}$$

where the L_{n+l}^{2l+1} are polynomials called *associated Laguerre polynomials*. The first few radial wave functions are given in Table 7.2.

TABLE 7.2
The Hydrogen-like Radial Wave Functions, $R_{nl}(r)$, for $n = 1, 2$, and 3 [a]

$$R_{10}(r) = 2\left(\frac{Z}{a_0} \right)^{3/2} e^{-\rho}$$

$$R_{20}(r) = \left(\frac{Z}{2a_0} \right)^{3/2} (2 - \rho)e^{-\rho/2}$$

$$R_{21}(r) = \frac{1}{\sqrt{3}} \left(\frac{Z}{2a_0} \right)^{3/2} \rho e^{-\rho/2}$$

$$R_{30}(r) = \frac{2}{27} \left(\frac{Z}{3a_0} \right)^{3/2} (27 - 18\rho + 2\rho^2)e^{-\rho/3}$$

$$R_{31}(r) = \frac{1}{27} \left(\frac{2Z}{3a_0} \right)^{3/2} \rho(6 - \rho)e^{-\rho/3}$$

$$R_{32}(r) = \frac{4}{27\sqrt{10}} \left(\frac{Z}{3a_0} \right)^{3/2} \rho^2 e^{-\rho/3}$$

a. The quantity Z is the nuclear charge, and $\rho = Zr/a_0$, where a_0 is the Bohr radius.

The radial wave functions given by Equation 7.14 may look complicated, but notice that each one is just a polynomial multiplied by an exponential. The combinatorial factor in front assures that the $R_{nl}(r)$ are normalized with respect to an integration over r, or that $R_{nl}(r)$ satisfy

$$\int_0^\infty dr\, r^2 R_{nl}^*(r) R_{nl}(r) = 1 \tag{7.15}$$

Note that the volume element here is $r^2 dr$, which is the "r" part of the spherical coordinate volume element $r^2 \sin\theta dr d\theta d\phi$. Problem 7–3 has you show that the radial wave functions in Table 7.2 are normalized.

The complete hydrogen atomic wave functions are

$$\psi_{nlm_l}(r, \theta, \phi) = R_{nl}(r) Y_l^{m_l}(\theta, \phi) \tag{7.16}$$

The first few hydrogen atomic wave functions are given in Table 7.3. The normalization

TABLE 7.3
The Complete Hydrogen-like Atomic Wave Functions
for $n = 1$, 2, and 3 [a]

$$\psi_{100} = \frac{1}{\sqrt{\pi}} \left(\frac{Z}{a_0}\right)^{3/2} e^{-\rho}$$

$$\psi_{200} = \frac{1}{4\sqrt{2\pi}} \left(\frac{Z}{a_0}\right)^{3/2} (2 - \rho) e^{-\rho/2}$$

$$\psi_{210} = \frac{1}{4\sqrt{2\pi}} \left(\frac{Z}{a_0}\right)^{3/2} \rho e^{-\rho/2} \cos\theta$$

$$\psi_{21\pm1} = \frac{1}{8\sqrt{\pi}} \left(\frac{Z}{a_0}\right)^{3/2} \rho e^{-\rho/2} \sin\theta e^{\pm i\phi}$$

$$\psi_{300} = \frac{1}{81\sqrt{3\pi}} \left(\frac{Z}{a_0}\right)^{3/2} (27 - 18\rho + 2\rho^2) e^{-\rho/3}$$

$$\psi_{310} = \frac{\sqrt{2}}{81\sqrt{\pi}} \left(\frac{Z}{a_0}\right)^{3/2} \rho(6 - \rho) e^{-\rho/3} \cos\theta$$

$$\psi_{31\pm1} = \frac{1}{81\sqrt{\pi}} \left(\frac{Z}{a_0}\right)^{3/2} \rho(6 - \rho) e^{-\rho/3} \sin\theta e^{\pm i\phi}$$

$$\psi_{320} = \frac{1}{81\sqrt{6\pi}} \left(\frac{Z}{a_0}\right)^{3/2} \rho^2 e^{-\rho/3} (3\cos^2\theta - 1)$$

$$\psi_{32\pm1} = \frac{1}{81\sqrt{\pi}} \left(\frac{Z}{a_0}\right)^{3/2} \rho^2 e^{-\rho/3} \sin\theta \cos\theta e^{\pm i\phi}$$

$$\psi_{32\pm2} = \frac{1}{162\sqrt{\pi}} \left(\frac{Z}{a_0}\right)^{3/2} \rho^2 e^{-\rho/3} \sin^2\theta e^{\pm 2i\phi}$$

a. The quantity Z is the nuclear charge, and $\rho = Zr/a_0$, where a_0 is the Bohr radius.

condition for hydrogen atomic wave functions is

$$\int_0^\pi d\theta \ \sin \theta \int_0^{2\pi} d\phi \int_0^\infty dr \ r^2 \psi_{n l m_l}^*(r, \theta, \phi) \psi_{n l m_l}(r, \theta, \phi) = 1$$

Because $\hat{H}$ is Hermitian (Section 4.5), the functions $\psi_{n l m_l}$ must also be orthogonal. This orthonormality relationship is given by

$$\int_0^{2\pi} d\phi \int_0^\pi d\theta \ \sin \theta \int_0^\infty dr \ r^2 \psi_{n'l'm_l'}^*(r, \theta, \phi) \psi_{n l m_l}(r, \theta, \phi) = \delta_{nn'} \cdot \delta_{ll'} \cdot \delta_{m_l m_l'}$$

$$(7.17)$$

where the δ's are Kronecker deltas—that is, $\delta_{ij} = 0$ when $i \neq j$ and $= 1$ when $i = j$.
Equation 7.17 in the bracket notation is

$$\langle n \ l \ m_l \ | \ n' \ l' \ m_l' \rangle = \delta_{nn'} \delta_{ll'} \delta_{m_l m_l'} \tag{7.18}$$

Note that $\psi_{n l m_l}(r, \theta, \phi) = | \ n \ l \ m_l \rangle$ depends upon three quantum numbers. These functions are simultaneous eigenfunctions of $\hat{H}$, $\hat{L}^2$, and $\hat{L}_z$, which mutually commute:

$$[\hat{H}, \hat{L}^2] = [\hat{H}, \hat{L}_z] = [\hat{L}^2, \hat{L}_z] = 0$$

The three mutually commuting operators assure the existence of the simultaneous eigenfunctions depending upon three quantum numbers. This is why the hydrogen atomic wave functions $\psi_{n l m_l}(r, \theta, \phi)$ depend upon n, l, and m_l; n for the Hamiltonian operator (the energy), l for the $\hat{L}^2$ operator (the orbital angular momentum), and m_l for the $\hat{L}_z$ operator (the z component of the orbital angular momentum) of the commuting set of operators.

Because $\hat{H}$, $\hat{L}^2$, and $\hat{L}_z$ commute with $\hat{H}$, their corresponding physical observable quantities (energy, orbital angular momentum and its z component) are conserved (see Equation 4.85). Their corresponding quantum numbers, n, l, and m_l, which are fixed and so can be used to label the wave function, are said to be *good quantum numbers*. Good quantum numbers are those associated with conserved quantities.

EXAMPLE 7–1
Show that the hydrogen-like atomic wave function ψ_{210} in Table 7.3 is normalized and that it is orthogonal to ψ_{200}.

SOLUTION: The orthonormality condition is given by Equation 7.18. Using ψ_{210} from Table 7.3,

$$\langle 210 \,|\, 210 \rangle = \int_0^\infty dr\, r^2 \int_0^\pi d\theta \sin\theta \int_0^{2\pi} d\phi \left[\frac{1}{\sqrt{32\pi}} \left(\frac{Z}{a_0} \right)^{3/2} \rho e^{-\rho/2} \cos\theta \right]^2$$

$$= \frac{1}{32\pi} \left(\frac{Z}{a_0} \right)^5 \int_0^\infty dr\, r^4 e^{-Zr/a_0} \int_0^\pi d\theta \sin\theta \cos^2\theta \int_0^{2\pi} d\phi$$

$$= \frac{1}{32\pi} \left(\frac{Z}{a_0} \right)^5 \left(\frac{a_0}{Z} \right)^5 (24) \left(\frac{2}{3} \right) (2\pi) = 1$$

and so ψ_{210} is normalized. To show that it is orthogonal to ψ_{200},

$$\langle 210 \,|\, 200 \rangle = \int_0^\infty dr\, r^2 \int_0^\pi d\theta \sin\theta \int_0^{2\pi} d\phi \left[\frac{1}{\sqrt{32\pi}} \left(\frac{Z}{a_0} \right)^{3/2} \left(\frac{Zr}{a_0} \right) e^{-Zr/2a_0} \cos\theta \right]$$

$$\times \left[\frac{1}{\sqrt{32\pi}} \left(\frac{Z}{a_0} \right)^{3/2} \left(2 - \frac{Zr}{a_0} \right) e^{-Zr/2a_0} \right]$$

$$= \frac{1}{32\pi} \left(\frac{Z}{a_0} \right)^4 \int_0^\infty dr\, r^3 \left(2 - \frac{Zr}{a_0} \right) e^{-Zr/a_0} \int_0^\pi d\theta \sin\theta \cos\theta \int_0^{2\pi} d\phi$$

The integral over θ here vanishes, so ψ_{210} and ψ_{200} are orthogonal.

7.2 *s* Orbitals Are Spherically Symmetric

The hydrogen atomic wave functions depend upon three quantum numbers, n, l, and m_l. The quantum number n is called the *principal quantum number* and has the values $1, 2, \ldots$. The energy of the hydrogen atom depends upon only the principal quantum number through the equation $E_n = -e^2/8\pi\epsilon_0 a_0 n^2$. The quantum number l is called the *angular-momentum quantum number* and has the values $0, 1, \ldots, n-1$. The magnitude of the angular momentum of the electron about the proton is determined completely by l through $|L| = \hbar\sqrt{l(l+1)}$. Note that the form of the radial wave functions depends upon both n and l. The value of l is customarily denoted by a letter, with $l = 0$ being denoted by s, $l = 1$ by p, $l = 2$ by d, $l = 3$ by f, with higher values of l denoted by the alphabetic sequence following f. The origin of the letters s, p, d, f is historic and has to do with the designation of the observed spectral lines of atomic sodium. (The letters s, p, d, and f stand for *sharp, principal, diffuse,* and *fundamental*.) A wave function with $n = 1$ and $l = 0$ is called a $1s$ wave function; one with $n = 2$ and $l = 0$ is called a $2s$ wave function, and so on.

The third quantum number m_l is called the *magnetic quantum number* and takes on the $2l + 1$ values $m_l = 0, \pm 1, \pm 2, \ldots \pm l$. The z component of the angular momentum is determined completely by m_l through $L_z = m_l \hbar$. The quantum number m_l is called the magnetic quantum number because the energy of a hydrogen atom in a magnetic field depends on m_l. In the absence of a magnetic field, each energy level has a degeneracy

FIGURE 7.1
The splitting of the $2p$ state of a hydrogen atom into three components in a magnetic field.

of $2l + 1$. In the presence of a magnetic field, these levels split, and the energy depends upon the particular value of m_l (Section 7.4). This splitting is illustrated in Figure 7.1 and is called the *Zeeman effect*, which is discussed in Section 7.4. In this case, E is a function of both the quantum numbers n and m_l.

The complete hydrogen atomic wave functions depend on three variables, so plotting or displaying them is difficult. The radial and angular parts are commonly considered separately. The state of lowest energy of a hydrogen atom is the $1s$ state. The radial function associated with the $1s$ state is (Table 7.2)

$$R_{1s}(r) = \frac{2}{a_0^{3/2}} e^{-r/a_0}$$

As mentioned above, the radial wave functions are normalized with respect to integration over r, so

$$\int_0^\infty [R_{1s}(r)]^2 r^2 dr = \frac{4}{a_0^3} \int_0^\infty r^2 e^{-2r/a_0} dr = 1 \qquad (7.19)$$

From Equation 7.19, we see that the probability that the electron lies between r and $r + dr$ is $[R_{nl}(r)]^2 r^2 dr$, and plots of $r^2 R_{nl}^2(r)$ are shown in Figure 7.2 for $Z = 1$. An important observation from the plots in Figure 7.2 is that the number of nodes in the radial function is equal to $n - l - 1$. (The point $r = 0$ is not considered to be a node.)

For the $1s$ state, the probability that the electron lies between r and $r + dr$ is

$$\text{Prob} = \frac{4}{a_0^3} r^2 e^{-2r/a_0} dr \qquad (7.20)$$

This result is contrary to the Bohr model in which the electron is incorrectly restricted to lie in fixed, well-defined orbits. Figure 7.3 shows surface plots of both the wave functions, $\psi(r)$, and the associated probability densities, $\psi^2(r)$, for the $1s$, $2s$, and $3s$ states of atomic hydrogen. The nucleus lies in the center of the horizontal plane in each case.

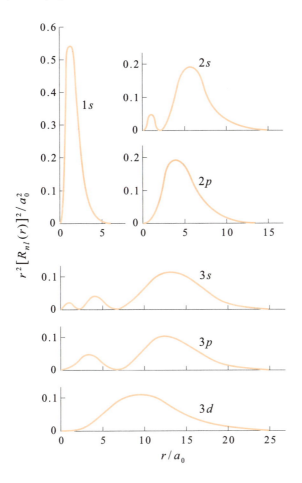

FIGURE 7.2
The probability densities $r^2[R_{nl}(r)]^2$ associated with the radial parts of the hydrogen atomic wave functions.

EXAMPLE 7–2
Calculate the probability that an electron described by a hydrogen atomic 1*s* wave function will be found within one Bohr radius of the nucleus.

SOLUTION: The probability that the electron will be found within one Bohr radius of the nucleus is obtained by integrating Equation 7.20 from 0 to a_0:

$$\text{Prob}(0 \le r \le a_0) = \frac{4}{a_0^3} \int_0^{a_0} r^2 e^{-2r/a_0} dr$$

$$= 4 \int_0^1 x^2 e^{-2x} dx$$

$$= 1 - 5e^{-2} = 0.323$$

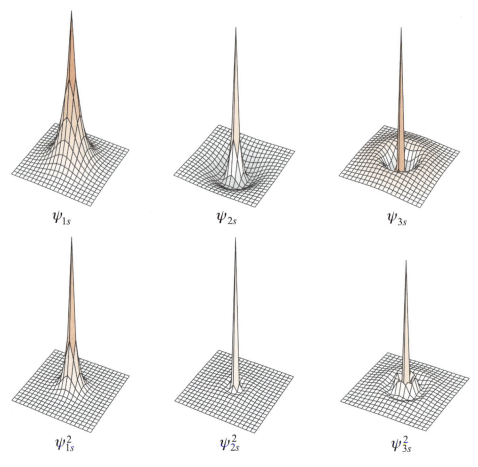

FIGURE 7.3
Surface plots of both $\psi(r)$ and $\psi^2(r)$ for the 1s, 2s, and 3s states of atomic hydrogen. The nucleus lies in the center of the horizontal plane in each case.

We must keep in mind that we are dealing with only the radial parts of the total wave function here. The radial parts are easy to display because they depend on only the one coordinate, r. The angular parts depend on both θ and ϕ and so are somewhat more difficult to display. The $l = 0$ case is easy, however, because when $l = 0$, m_l must equal zero and so we have $Y_0^0(\theta, \phi)$, which according to Table 7.1 is

$$Y_0^0(\theta, \phi) = \frac{1}{\sqrt{4\pi}}$$

$Y_0^0(\theta, \phi)$ is normalized with respect to integration over a spherical surface:

$$\int_0^\pi d\theta \sin\theta \int_0^{2\pi} d\phi \, Y_0^0(\theta, \phi)^* Y_0^0(\theta, \phi) = \frac{1}{4\pi} \int_0^\pi d\theta \sin\theta \int_0^{2\pi} d\phi = 1$$

In this particular case, there is no angular dependence and the wave function is spherically symmetric; in other words, it has the same value for every choice of (θ, ϕ). The complete 1s wave function is

$$\psi_{1s}(r, \theta, \phi) = \psi_{100}(r, \theta, \phi) = R_{10}(r)Y_0^0(\theta, \phi) = (\pi a_0^3)^{-1/2}e^{-r/a_0} \quad (7.21)$$

We have displayed the r, θ, and ϕ dependence on the left side of Equation 7.21, even though the θ and ϕ dependence drops out, to emphasize that $\psi_{1s}(r, \theta, \phi)$ is the complete wave function. For example, the normalization condition is

$$\int_0^\infty dr r^2 \int_0^\pi d\theta \sin\theta \int_0^{2\pi} d\phi \, \psi_{1s}^*(r, \theta, \phi)\psi_{1s}(r, \theta, \phi) = \langle 100 \,|\, 100 \rangle = 1$$

The hydrogen atomic wave functions are called *orbitals*, and, in particular, Equation 7.21 describes the 1s orbital; an electron in the 1s state is called a 1s electron.

The probability that a 1s electron lies between r and $r + dr$ from the nucleus is obtained by integrating $\psi_{1s}^*(r, \theta, \phi)\psi_{1s}(r, \theta, \phi)$ over all values of θ and ϕ according to

$$\text{Prob}(1s) = r^2 dr \int_0^\pi d\theta \sin\theta \int_0^{2\pi} d\phi \, \psi_{1s}^*(r, \theta, \phi)\psi_{1s}(r, \theta, \phi)$$

$$= \frac{4}{a_0^3}r^2 e^{-2r/a_0}dr \quad (7.22)$$

in agreement with Equation 7.20.

We can use Equation 7.22 to calculate the average value of r. For example,

$$\langle r \rangle_{1s} = \langle 100 \,|\, r \,|\, 100 \rangle = \frac{4}{a_0^3}\int_0^\infty r^3 e^{-2r/a_0}dr = \frac{3}{2}a_0 \quad (7.23)$$

Equation 7.22 can be used to determine the most probable distance of a 1s electron from the nucleus.

EXAMPLE 7–3
Show that the most probable value of r (r_{mp}) in a 1s orbital is a_0.

SOLUTION: To determine the most probable value of r, we find the value of r that maximizes the probability density of r or that maximizes

$$f(r) = \frac{4}{a_0^3}r^2 e^{-2r/a_0}$$

If we differentiate $f(r)$ and set the result equal to zero, we find that $r_{mp} = a_0$, the Bohr radius.

The average potential energy in the $1s$ state is given by

$$\langle V(r) \rangle_{1s} = \langle 100 \mid V \mid 100 \rangle = \int_0^{2\pi} d\phi \int_0^{\pi} d\theta \, \sin\theta \int_0^{\infty} dr \, r^2 \psi_{1s}^* \left(-\frac{e^2}{4\pi\epsilon_0 r} \right) \psi_{1s}$$

$$= \frac{e^2}{\pi\epsilon_0 a_0^3} \int_0^{\infty} dr \, r e^{-2r/a_0}$$

$$= -\frac{e^2}{4\pi\epsilon_0 a_0} \tag{7.24}$$

It is interesting to note that $\langle V(r) \rangle = 2\langle E \rangle$. Because $\langle T \rangle + \langle V \rangle = \langle E \rangle$, where $\langle T \rangle$ is the average kinetic energy, we have $\langle V \rangle = -2\langle T \rangle$, or

$$\frac{\langle V \rangle}{\langle T \rangle} = -2 \tag{7.25}$$

for this case. Although we have derived Equation 7.25 only for the $1s$ state of the hydrogen atom, it is generally true for any system in which the potential energy is coulombic. Equation 7.25 is an example of the *virial theorem* and is valid for all atoms and molecules. We shall refer to the virial theorem several times throughout this book. The virial theorem is proved in Problem 5–43.

The next simplest orbital is the $2s$ orbital. A $2s$ orbital is given by

$$\psi_{2s}(r, \theta, \phi) = \psi_{200}(r, \theta, \phi) = R_{20}(r)Y_0^0(\theta, \phi) \tag{7.26}$$

which is also spherically symmetric. In fact, because any s orbital will have the angular factor $Y_0^0(\theta, \phi)$, all s orbitals are spherically symmetric. By referring to Table 7.3, we see that

$$\psi_{2s}(r, \theta, \phi) = \frac{1}{\sqrt{32\pi}} \left(\frac{1}{a_0} \right)^{3/2} \left(2 - \frac{r}{a_0} \right) e^{-r/2a_0} \tag{7.27}$$

Remember that ψ_{2s} is normalized with respect to an integration over r, θ, and ϕ. The average value of r in the $2s$ state of a hydrogen atom is (cf. Problem 7–12)

$$\langle r \rangle_{2s} = \langle 200 \mid r \mid 200 \rangle$$

$$= \int_0^{\infty} dr \, r^3 \int_0^{\pi} d\theta \, \sin\theta \int_0^{2\pi} d\phi \, \psi_{2s}^*(r, \theta, \phi) \psi_{2s}(r, \theta, \phi) = 6a_0 \tag{7.28}$$

showing that a $2s$ electron is on the average a much greater distance from the nucleus than a $1s$ electron. In fact, using the general properties of the radial wave functions, we can show that $\langle r \rangle = \frac{3}{2} a_0 n^2$ for an ns electron.

EXAMPLE 7–4
Show that the virial theorem is valid for the $2s$ state of a hydrogen atom.

SOLUTION: Recall from Equation 7.25 that the virial theorem says that $\langle V \rangle / \langle T \rangle = -2$, or equivalently, that $\langle V(r) \rangle = 2\langle E \rangle$.

$$\langle V(r) \rangle_{2s} = \int_0^{2\pi} d\phi \int_0^{\pi} d\theta \, \sin\theta \int_0^{\infty} dr \, r^2 \psi_{2s}^* \left(-\frac{e^2}{4\pi\epsilon_0 r} \right) \psi_{2s}$$

$$= \int_0^{\infty} dr \, r^2 R_{2s}^*(r) \left(-\frac{e^2}{4\pi\epsilon_0 r} \right) R_{2s}(r)$$

$$= -\frac{e^2}{32\pi\epsilon_0} \left(\frac{1}{a_0} \right)^3 \int_0^{\infty} dr \, r \left(2 - \frac{r}{a_0} \right)^2 e^{-r/a_0}$$

Letting $x = r/a_0$,

$$\langle V(r) \rangle_{2s} = -\frac{e^2}{32\pi\epsilon_0} \frac{1}{a_0} \int_0^{\infty} dx \, (4x - 4x^2 + x^3)e^{-x}$$

$$= -\frac{e^2}{16\pi\epsilon_0 a_0} = 2E_2$$

or equivalently, $\langle V \rangle_{2s} / \langle T \rangle_{2s} = -2$.

Table 7.4 lists values of $\langle r^k \rangle_{n\,l\,m_l} = \langle n\,l\,m_l \mid r^k \mid n\,l\,m_l \rangle$ for $k = 2, 1, -1, -2,$ and -3.

T A B L E 7.4
Values of $\langle r^k \rangle_{n\,l\,m_l} = \langle n\,l\,m_l \mid r^k \mid n\,l\,m_l \rangle$ for a Hydrogen-like Atom or Ion with Nuclear Charge Z for $k = 2, 1, -1, -2,$ and -3

$$\langle r^2 \rangle_{n\,l\,m_l} = \langle n\,l\,m_l \mid r^2 \mid n\,l\,m_l \rangle = \frac{a_0^2 n^4}{Z^2} \left\{ 1 + \frac{3}{2} \left[1 - \frac{l(l+1) - \frac{1}{3}}{n^2} \right] \right\}$$

$$\langle r \rangle_{n\,l\,m_l} = \langle n\,l\,m_l \mid r \mid n\,l\,m_l \rangle = \frac{a_0 n^2}{Z} \left\{ 1 + \frac{1}{2} \left[1 - \frac{l(l+1)}{n^2} \right] \right\}$$

$$\left\langle \frac{1}{r} \right\rangle_{n\,l\,m_l} = \left\langle n\,l\,m_l \mid \frac{1}{r} \mid n\,l\,m_l \right\rangle = \frac{Z}{a_0 n^2}$$

$$\left\langle \frac{1}{r^2} \right\rangle_{n\,l\,m_l} = \left\langle n\,l\,m_l \mid \frac{1}{r^2} \mid n\,l\,m_l \right\rangle = \frac{Z^2}{a_0^2 n^3 (l + \frac{1}{2})}$$

$$\left\langle \frac{1}{r^3} \right\rangle_{n\,l\,m_l} = \left\langle n\,l\,m_l \mid \frac{1}{r^3} \mid n\,l\,m_l \right\rangle = \frac{Z^3}{a_0^3 n^3 l (l + \frac{1}{2})(l + 1)}$$

Source: Pauling, L., Wilson, E. B. *Introduction to Quantum Chemistry.* McGraw-Hill: New York, 1935.

7.3 There Are Three p Orbitals for Each Value of the Principal Quantum Number, $n \geq 2$

When $l \neq 0$, the hydrogen atomic wave functions are not spherically symmetric; they depend on θ and ϕ. In this section, we will concentrate on the angular parts of the hydrogen wave functions. Let's first consider states with $l = 1$, or p orbitals. Because $m_l = 0$ or ± 1 when $l = 1$, there are three p orbitals for each value of n. The angular part of the p orbitals is given by the three spherical harmonics $Y_1^0(\theta, \phi)$ and $Y_1^{\pm 1}(\theta, \phi)$. The simplest of these spherical harmonics is

$$Y_1^0(\theta, \phi) = \left(\frac{3}{4\pi}\right)^{1/2} \cos \theta \tag{7.29}$$

which is readily shown to be normalized, because

$$\frac{3}{4\pi} \int_0^\pi d\theta \sin \theta \int_0^{2\pi} d\phi \cos^2 \theta = \frac{3}{2} \int_0^\pi \sin \theta \cos^2 \theta d\theta = \frac{3}{2} \int_{-1}^1 x^2 dx = 1$$

In the last step, we let $\cos \theta = x$.

A common way to present the angular functions is as three-dimensional figures. Figure 7.4 is the familiar tangent-sphere picture of a p orbital often presented in general chemistry texts. Although the tangent-sphere picture represents the shape of the angular part of p orbitals, it is *not* a faithful representation of the shape of a p_z orbital because the radial functions are not included.

Figure 7.5 shows surface plots of both ψ_{2p_z} and $\psi_{2p_z}^2$ for atomic hydrogen. Note that the positive and negative lobes in ψ_{2p_z} in Figure 7.4 appear as a peak and a depression in Figure 7.5.

Because a complete wave function generally depends on three coordinates, wave functions are difficult to display clearly. One useful and instructive way, however, is the following: The quantity $\psi^* \psi \, d\tau$ is the probability that the electron is located within the

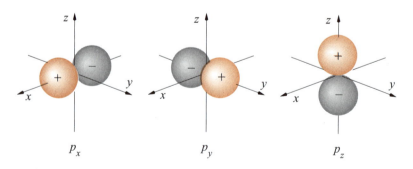

FIGURE 7.4
Three-dimensional polar plots of the angular part of the real representation of the hydrogen atomic wave functions for $l = 1$ (see Equation 7.31).

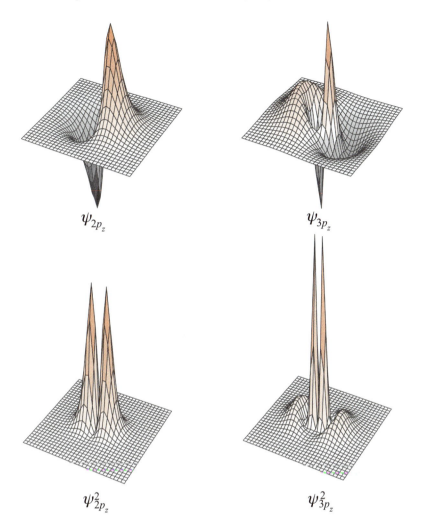

ψ_{2p_z} ψ_{3p_z}

$\psi_{2p_z}^2$ $\psi_{3p_z}^2$

FIGURE 7.5
Surface plots of ψ_{2p_z} and $\psi_{2p_z}^2$ for atomic hydrogen. The nucleus lies in the center of the horizontal plane in each case.

volume element $d\tau$. Thus, we can divide space into little volume elements and compute the average or some representative value of $\psi^*\psi$ within each volume element and then represent the value of $\psi^*\psi$ by the density of dots in a picture. Figure 7.6 shows such plots for several orbitals.

An alternate way to represent complete wave functions is as contour maps. Figure 7.7a shows a contour map for a $1s$ orbital. In each case, the nine contours shown enclose the 10%, 20%, . . . , 90% probability of finding the electron within each contour. Note that the contour maps appear as cross sections of the plots in Figure 7.5.

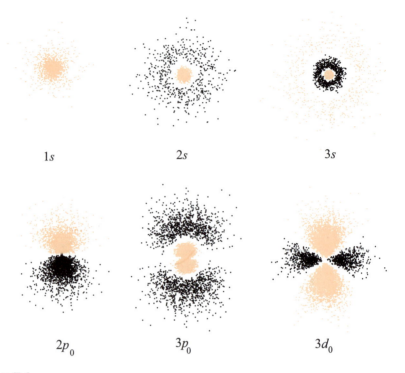

FIGURE 7.6
Probability density plots in a planar cross section of some hydrogen atomic orbitals. The density of the dots is proportional to the probability of finding the electron in that region.

It is interesting to compare the depictions of the $2p_0$ and $3p_0$ orbitals in Figures 7.4 and 7.7. The expressions for these orbitals are

$$\psi_{2p_0}(r, \theta, \phi) = R_{21}(r)Y_1^0(\theta, \phi)$$

and

$$\psi_{3p_0}(r, \theta, \phi) = R_{31}(r)Y_1^0(\theta, \phi)$$

Both orbitals have the same angular part, which is represented in Figure 7.4. The radial functions have $n - l - 1$ nodes, however, and so $R_{21}(r)$ has no nodes and $R_{31}(r)$ has one. The difference in the shapes of the $2p_0$ and $3p_0$ orbitals in Figures 7.4 and 7.7 is due to the node in $R_{31}(r)$. This example illustrates the inadequacy of the "tangent-sphere" representation of p orbitals.

The angular functions with $m_l \neq 0$ are more difficult to represent pictorially because they not only depend on ϕ in addition to θ, but are complex as well. In particular, the $l = 1$ states with $m_l \neq 0$ are

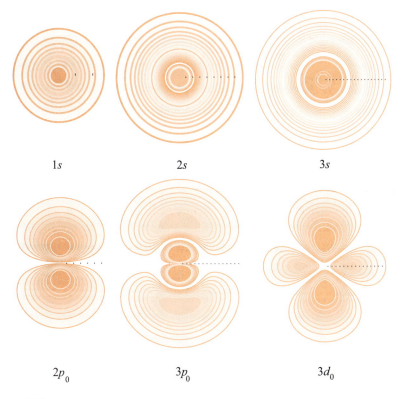

FIGURE 7.7
Probability contour plots for some hydrogen atomic orbitals. The nine contours shown in each case enclose the 10%, 20%, ..., 90% probability of finding the electron within each contour. The scale of the figure is indicated by hash marks: one mark corresponds to one Bohr radius a_0. Note that the different orbitals are presented on different scales (see Figure 7.6). The contour values were taken from Gerhold, G., McMurchie, L., Tye, T., Percentage Contour Maps of Electron Densities in Atoms. *Am. J. Phys.*, **40**, 998 (1972).

$$Y_1^{+1}(\theta, \phi) = -\left(\frac{3}{8\pi}\right)^{1/2} \sin\theta \, e^{+i\phi}$$

$$Y_1^{-1}(\theta, \phi) = \left(\frac{3}{8\pi}\right)^{1/2} \sin\theta \, e^{-i\phi}$$

(7.30)

The probability densities associated with $Y_1^{+1}(\theta, \phi)$ and $Y_1^{-1}(\theta, \phi)$ are the same because

$$|Y_1^{+1}(\theta, \phi)|^2 = \frac{3}{8\pi} \sin^2\theta$$

and

$$|Y_1^{-1}(\theta, \phi)|^2 = \frac{3}{8\pi} \sin^2\theta$$

Because $Y_1^{+1}(\theta, \phi)$ and $Y_1^{-1}(\theta, \phi)$ correspond to the same energy, we know from Section 4.2 that any linear combination of Y_1^{+1} and Y_1^{-1} is also an energy eigenfunction with the same energy. It is customary to use the combinations

$$p_x = \frac{1}{\sqrt{2}}(Y_1^{-1} - Y_1^{+1}) = \left(\frac{3}{4\pi}\right)^{1/2} \sin\theta \cos\phi$$

$$p_y = \frac{i}{\sqrt{2}}(Y_1^{-1} + Y_1^{+1}) = \left(\frac{3}{4\pi}\right)^{1/2} \sin\theta \sin\phi$$

(7.31)

"Tangent-sphere" plots of p_x and p_y are shown in Figure 7.4. They have the same shape as the p_z function except that they are directed along the x and y axes. The three functions p_x, p_y, and p_z are often used as the angular part of hydrogen atomic wave functions because they are real and have easily visualized directional properties.

For the $l = 2$ case, $m_l = 0, \pm1$, and ±2, and so there are five d orbitals. For $m_l = \pm1$ and ±2, we take linear combinations as we did above for the p functions. The customary linear combinations are (Problem 7–23)

$$d_{z^2} = Y_2^0 = \left(\frac{5}{16\pi}\right)^{1/2}(3\cos^2\theta - 1)$$

$$d_{xz} = \frac{1}{\sqrt{2}}(Y_2^{-1} - Y_2^{+1}) = \left(\frac{15}{4\pi}\right)^{1/2} \sin\theta \cos\theta \cos\phi$$

$$d_{yz} = \frac{i}{\sqrt{2}}(Y_2^{-1} + Y_2^{+1}) = \left(\frac{15}{4\pi}\right)^{1/2} \sin\theta \cos\theta \sin\phi$$

(7.32)

$$d_{x^2-y^2} = \frac{1}{\sqrt{2}}(Y_2^{+2} + Y_2^{-2}) = \left(\frac{15}{16\pi}\right)^{1/2} \sin^2\theta \cos 2\phi$$

$$d_{xy} = \frac{1}{\sqrt{2}i}(Y_2^{+2} - Y_2^{-2}) = \left(\frac{15}{16\pi}\right)^{1/2} \sin^2\theta \sin 2\phi$$

The angular parts of the five d orbitals are shown in Figure 7.8. Note that the last four orbitals given in Equations 7.32 differ only in their orientation. Figure 7.8 suggests the rationale of the notation of the d orbitals: d_{z^2} lies along the z axis, $d_{x^2-y^2}$ lies along the x and y axes, d_{xy} lies in the x–y plane, d_{xz} lies in the x–z plane, and d_{yz} lies in the y–z plane. Figures 7.4 and 7.8 illustrate a nice pictorial interpretation of the magnetic quantum number, m_l. Note that $|m_l|$, the magnitude of m_l, is equal to the number of nodal planes that contain the z axis in Figures 7.4 and 7.8. For example, p_z, with $m_l = 0$, has no nodal plane containing the z axis, but p_x and p_y, with $|m_l| = 1$, each has one. Similarly, d_{z^2}, with $m_l = 0$, has none, d_{xz} and d_{yz}, with $|m_l| = 1$, have one, and $d_{x^2-y^2}$ and d_{xy}, with $|m_l| = 2$, have two.

There is no fundamental reason to choose linear combinations of spherical harmonics such that the angular wave functions are real, but most chemists use the five

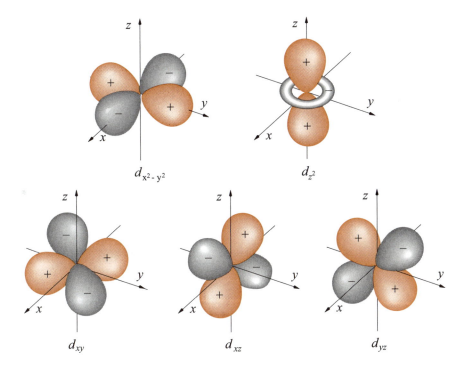

FIGURE 7.8
Three-dimensional plots of the angular part of the real representation of the hydrogen atomic wave functions for $l = 2$. Such plots show the directional character of these orbitals but are not good representations of the shape of these orbitals because the radial functions are not included.

d orbitals given by Equations 7.32 because the functions in Equations 7.32 are real and have convenient directional properties. The real representations of the hydrogen atomic wave functions are given in Table 7.5. The functions in Table 7.5 are the linear combinations of the complex wave functions in Table 7.3. Both sets are equivalent, but chemists normally use the real functions in Table 7.5. We will see in later chapters that molecular wave functions can be built out of atomic orbitals, and if the atomic orbitals have a definite directional character, we can use chemical intuition to decide which are the more important atomic orbitals to use to describe molecular orbitals.

7.4 The Energy Levels of a Hydrogen Atom Are Split by a Magnetic Field

In this section, we shall discuss a hydrogen atom in an external magnetic field. We shall see that the magnetic field causes the energy levels to be split into sublevels, which leads to a splitting of the spectral lines in hydrogen. This splitting is called the *Zeeman effect*. Before discussing the Zeeman effect, however, we shall review some facts and equations concerning magnetic dipoles and magnetic fields.

TABLE 7.5

The Complete Hydrogen-like Atomic Wave Functions Expressed as Real Functions for $n = 1$, 2, and 3 [a]

$$\psi_{1s} = \frac{1}{\sqrt{\pi}} \left(\frac{Z}{a_0}\right)^{3/2} e^{-\rho}$$

$$\psi_{2s} = \frac{1}{4\sqrt{2\pi}} \left(\frac{Z}{a_0}\right)^{3/2} (2 - \rho)e^{-\rho/2}$$

$$\psi_{2p_z} = \frac{1}{4\sqrt{2\pi}} \left(\frac{Z}{a_0}\right)^{3/2} \rho e^{-\rho/2} \cos\theta$$

$$\psi_{2p_x} = \frac{1}{4\sqrt{2\pi}} \left(\frac{Z}{a_0}\right)^{3/2} \rho e^{-\rho/2} \sin\theta \cos\phi$$

$$\psi_{2p_y} = \frac{1}{4\sqrt{2\pi}} \left(\frac{Z}{a_0}\right)^{3/2} \rho e^{-\rho/2} \sin\theta \sin\phi$$

$$\psi_{3s} = \frac{1}{81\sqrt{3\pi}} \left(\frac{Z}{a_0}\right)^{3/2} (27 - 18\rho + 2\rho^2)e^{-\rho/3}$$

$$\psi_{3p_z} = \frac{\sqrt{2}}{81\sqrt{\pi}} \left(\frac{Z}{a_0}\right)^{3/2} \rho(6 - \rho)e^{-\rho/3} \cos\theta$$

$$\psi_{3p_x} = \frac{\sqrt{2}}{81\sqrt{\pi}} \left(\frac{Z}{a_0}\right)^{3/2} \rho(6 - \rho)e^{-\rho/3} \sin\theta \cos\phi$$

$$\psi_{3p_y} = \frac{\sqrt{2}}{81\sqrt{\pi}} \left(\frac{Z}{a_0}\right)^{3/2} \rho(6 - \rho)e^{-\rho/3} \sin\theta \sin\phi$$

$$\psi_{3d_{z^2}} = \frac{1}{81\sqrt{6\pi}} \left(\frac{Z}{a_0}\right)^{3/2} \rho^2 e^{-\rho/3}(3\cos^2\theta - 1)$$

$$\psi_{3d_{xz}} = \frac{\sqrt{2}}{81\sqrt{\pi}} \left(\frac{Z}{a_0}\right)^{3/2} \rho^2 e^{-\rho/3} \sin\theta \cos\theta \cos\phi$$

$$\psi_{3d_{yz}} = \frac{\sqrt{2}}{81\sqrt{\pi}} \left(\frac{Z}{a_0}\right)^{3/2} \rho^2 e^{-\rho/3} \sin\theta \cos\theta \sin\phi$$

$$\psi_{3d_{x^2-y^2}} = \frac{1}{81\sqrt{2\pi}} \left(\frac{Z}{a_0}\right)^{3/2} \rho^2 e^{-\rho/3} \sin^2\theta \cos 2\phi$$

$$\psi_{3d_{xy}} = \frac{1}{81\sqrt{2\pi}} \left(\frac{Z}{a_0}\right)^{3/2} \rho^2 e^{-\rho/3} \sin^2\theta \sin 2\phi$$

a. The quantity Z is the nuclear charge, and $\rho = Zr/a_0$, where a_0 is the Bohr radius.

The motion of an electric charge around a closed loop produces a magnetic dipole m whose magnitude is given by

$$m = iA \tag{7.33}$$

where i is the current in amperes (coulombs per second) and A is the area of the loop in square meters. If we consider a circular loop for simplicity, then

$$i = \frac{qv}{2\pi r} \tag{7.34}$$

where v is the velocity of the charge q and r is the radius of the circle. Substituting Equation 7.34 and $A = \pi r^2$ into Equation 7.33 gives

$$m = \frac{qrv}{2} \tag{7.35}$$

More generally, if the orbit is not circular, then Equation 7.35 becomes

$$\mathbf{m} = \frac{q(\mathbf{r} \times \mathbf{v})}{2} \tag{7.36}$$

Note that Equation 7.36 reduces to Equation 7.35 for the case of a circular orbit (Problem 7–32). We can express $\mathbf{m}$ in terms of angular momentum by using the fact that $\mathbf{L} = \mathbf{r} \times \mathbf{p}$ and $\mathbf{p} = m\,\mathbf{v}$, so that Equation 7.36 becomes

$$\mathbf{m} = \frac{q}{2m}\,\mathbf{L} \tag{7.37}$$

Note that $\mathbf{m}$ and $\mathbf{L}$ are perpendicular to the plane of the motion. For an electron, $q = -|e|$ and Equation 7.37 becomes

$$\mathbf{m} = -\frac{|e|}{2m_e}\,\mathbf{L} \tag{7.38}$$

where m_e is the mass of the electron.

A magnetic dipole will interact with a magnetic field, and the potential energy of a magnetic dipole in a magnetic field is given by

$$V = -\mathbf{m} \cdot \mathbf{B} \tag{7.39}$$

where $\mathbf{B}$ is the strength of the magnetic field. The quantity $\mathbf{B}$ is defined through the equation

$$\mathbf{F} = q\,(\mathbf{v} \times \mathbf{B}) \tag{7.40}$$

where $\mathbf{F}$ is the force acting upon a charge q moving with a velocity $\mathbf{v}$ in a magnetic field of strength $\mathbf{B}$. The SI units of magnetic field strength are *tesla* (T). From Equation 7.40, we see that one tesla is equal to one newton/ampere·meter. If, as usual, we take the magnetic field to be in the z direction, then Equation 7.39 becomes

$$V = -m_z B_z \tag{7.41}$$

Using Equation 7.38 for m_z, we have

$$V = \frac{|e|\,B_z}{2m_e}L_z \tag{7.42}$$

If we replace L_z by its operator equivalent $\hat{L}_z$, then Equation 7.42 gives the part of the hydrogen atom Hamiltonian operator that accounts for the external magnetic field. Thus, the Hamiltonian operator for a hydrogen atom in an external magnetic field is

$$\hat{H} = \hat{H}_0 + \frac{|e| B_z}{2m_e} \hat{L}_z \tag{7.43}$$

where $\hat{H}_0$ is the Hamiltonian operator in the absence of a magnetic field. The corresponding Schrödinger equation is

$$\hat{H}_0 \psi + \frac{|e| B_z}{2m_e} \hat{L}_z \psi = E \psi \tag{7.44}$$

The hydrogen atomic wave functions are eigenfunctions of *both* $\hat{H}_0$ and $\hat{L}_z$, and so they are also eigenfunctions of $\hat{H}$ in Equation 7.43. In particular, we have

$$\hat{H}_0 \psi_{nlm_l}(r, \theta, \phi) = -\frac{m_e e^4}{8\epsilon_0^2 h^2 n^2} \psi_{nlm_l}(r, \theta, \phi)$$

and

$$\hat{L}_z \psi_{nlm_l}(r, \theta, \phi) = m_l \hbar \psi_{nlm_l}(r, \theta, \phi)$$

Therefore, the energy levels of a hydrogen atom in a magnetic field are

$$E = -\frac{m_e e^4}{8\epsilon_0^2 h^2 n^2} + \beta_B m_l B_z \qquad \begin{cases} n = 1, 2, 3, \ldots \\ m_l = 0, \pm 1, \pm 2, \ldots, \pm l \end{cases} \tag{7.45}$$

where β_B, defined as

$$\beta_B = \frac{|e| \hbar}{2m_e} \tag{7.46}$$

is called a *Bohr magneton*. Numerically, a Bohr magneton has the value (Problem 7–34)

$$\beta_B = 9.2740 \times 10^{-24} \text{ J·T}^{-1} \tag{7.47}$$

According to Equation 7.45, a state with given values of n and l is split into $2l + 1$ levels by an external magnetic field. For example, Figure 7.9 shows the results for the $1s$ and $2p$ states of atomic hydrogen. The $1s$ state is not split ($2l + 1 = 1$), but the $2p$ state is split into three levels ($2l + 1 = 3$). Figure 7.9 suggests that the $2p$ to $1s$ transition in atomic hydrogen will be split into three distinct transitions instead of just one.

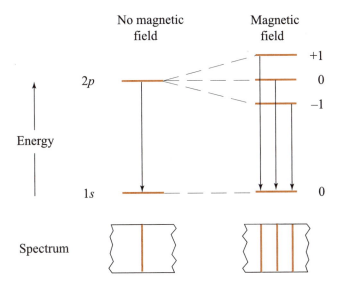

FIGURE 7.9
The splitting of the $2p$ state of a hydrogen atom in an external magnetic field. The $2p$ state is split into three closely spaced levels. In an external magnetic field, the $2p$ to $1s$ transition is split into three distinct transition frequencies.

EXAMPLE 7–5
Calculate the magnitude of the splitting of the $2p$ level of atomic hydrogen for a magnetic field strength of 1.00 tesla and compare the result to the difference in energy between the $1s$ and $2p$ states.

SOLUTION: Equation 7.45 shows that the splitting is given by

$$\Delta E = \beta_B \, m_l B_z$$

$$= (9.274 \times 10^{-24} \text{ J·T}^{-1})(1.00 \text{ T}) \, m_l$$

$$= (9.274 \times 10^{-24} \text{ J}) \, m_l \qquad m_l = 0, \pm 1$$

The energy difference between the unperturbed levels $1s$ and $2p$ is

$$E_{2p} - E_{1s} = -\frac{m_e e^4}{8\epsilon_0^2 \hbar^2} \left(\frac{1}{4} - 1 \right) = 1.635 \times 10^{-18} \text{ J}$$

which shows numerically that the splitting is very small compared to the difference in energy between the $1s$ and $2p$ levels.

The results of Example 7-5 suggest that the three transitions shown in Figure 7.9 lie very close together (Problem 7–35), and we say that the $2p \rightarrow 1s$ transition becomes a *triplet* in the presence of an external magnetic field.

It turns out that this is incorrect. The $n = 2$ to $n = 1$ transition is split into ten lines instead of three in the presence of an external magnetic field. In fact, we haven't mentioned it, but the $n = 2$ to $n = 1$ transition in the absence of an external magnetic field is split into two closely spaced lines. The frequencies of these two lines are $82\,258.921$ cm^{-1} and $82\,259.287$ cm^{-1}; the difference between them is only 0.366 cm^{-1}, but this is well within experimental error. It appears that something is missing. The "something" that is missing is electron spin, which we shall address in the next section.

7.5 An Electron Has an Intrinsic Spin Angular Momentum

As early as 1921, the American physicist Arthur H. Compton, who was studying the scattering of X rays from crystal surfaces, was led to conclude that "the electron itself, spinning like a tiny gyroscope, is probably the ultimate magnetic particle." In 1922, two German physicists, Otto Stern and Walther Gerlach, passed a beam of silver atoms (recall from general chemistry that a silver atom has a $4d^{10}5s^1$ outer electron configuration, and so has a single outer electron) through an inhomogeneous magnetic field in order to split the beam into its $2l + 1$ space-quantized components (Figure 7.10). A homogeneous magnetic field will orient magnetic dipoles but not exert a translational force. An inhomogeneous magnetic field, however, will exert a translational force (Problem 7–36) and hence spatially separate magnetic dipoles that are oriented differently. Classically, a beam of magnetic dipoles will orient themselves through a continuous angle and so will become spread out in a continuous manner. Quantum-mechanically, however, a state with a given value of l will be restricted to

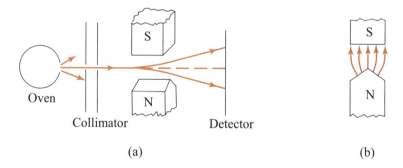

FIGURE 7.10
(a) A schematic diagram of the Stern–Gerlach experiment. (b) A cross-sectional view of the pole pieces of the magnet depicting the inhomogeneous magnetic field that they produce.

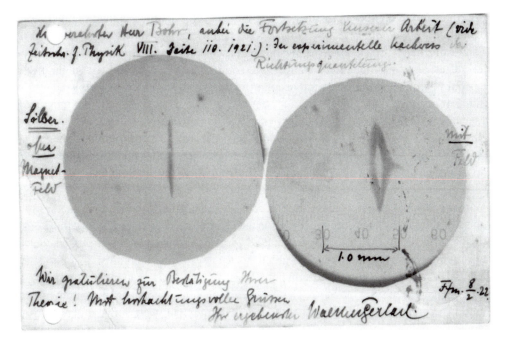

FIGURE 7.11
A postcard from Walther Gerlach to Niels Bohr, dated February 8, 1922. The left side shows the pattern of the beam of silver atoms without a magnetic field, and the right side shows the pattern with the inhomogeneous magnetic field. Reproduced courtesy of Niels Bohr Archive, Copenhagen, Denmark.

$2l + 1$ discrete orientations, and so such a system will be split into $2l + 1$ components by an inhomogeneous magnetic field. Stern and Gerlach found the quite unexpected result that a beam of silver atoms splits into only two parts (Figure 7.11). Note that this corresponds to $2l + 1 = 2$, or to $l = 1/2$. Up to now we have admitted only integer values of l.

Another similar observation is the splitting that occurs in atomic spectra. For example, under high resolution it was observed that the $n = 2$ to $n = 1$ transition in atomic hydrogen is split into two closely spaced lines, called a *doublet*. These observations cannot be explained using the ideas and equations that we have developed up to now, and although there were indeed ingenious theories for all these observations, the later contributions of a number of people made these explanations more and more tenuous. In addition, why all the electrons in the ground state of an atom do not occur in the innermost shell, which is the shell of lowest energy, was not understood. Niels Bohr had done a great deal of work on the periodic system of the elements, and this was always an underlying nagging question.

In 1925, Wolfgang Pauli showed that all these observations could be explained with the postulate that an electron can exist in two distinct states. Pauli introduced a fourth quantum number in a rather ad hoc manner. This fourth quantum number, now called the *spin quantum number* m_s, is restricted to the two values $+\frac{1}{2}$ and $-\frac{1}{2}$. It is interesting that Pauli did not give any interpretation to this fourth quantum number. The existence of a fourth quantum number was somewhat of a mystery because the three spatial coordinates of an electron account for n, l, and m_l, but what is this quantum number due to?

It was finally two young Dutch physicists, George Uhlenbeck and Samuel Goudsmit, in 1925, who showed that the two intrinsic states of an electron could be identified with two angular momenta, or spin, states. We have seen earlier that the orbital motion of electrons leads to an associated magnetic moment.

We are going to simply graft the concept of spin onto the quantum theory and onto the postulates that we have developed earlier. This may appear to be a somewhat unsatisfactory way to proceed, but it turns out to be quite satisfactory for our purposes. In the early 1930s, the British physicist Paul Dirac developed a relativistic extension of quantum mechanics, and one of its greatest successes is that spin arose in a perfectly natural way. We shall introduce spin here, however, in an ad hoc manner.

We have seen from the Stern–Gerlach experiment that a beam of silver atoms splits into two components, implying that the magnetic moment of its single outer electron and its associated angular momentum is described by an angular-momentum quantum number, $l = \frac{1}{2}$. Just as we have the eigenvalue equations for $\hat{L}^2$ and $\hat{L}_z$,

$$\hat{L}^2 Y_l^{m_l}(\theta, \phi) = \hbar^2 l(l+1) Y_l^{m_l}(\theta, \phi) \tag{7.48}$$

and

$$\hat{L}_z Y_l^{m_l}(\theta, \phi) = m_l \hbar Y_l^{m_l}(\theta, \phi) \tag{7.49}$$

we define the spin operators $\hat{S}^2$ and $\hat{S}_z$ and their eigenfunctions α and β by the equations

$$\hat{S}^2 \alpha = \hbar^2 s(s+1)\alpha \qquad \hat{S}^2 \beta = \hbar^2 s(s+1)\beta \qquad s = \frac{1}{2}$$

$$\hat{S}_z \alpha = \frac{1}{2}\hbar\alpha \qquad \hat{S}_z \beta = -\frac{1}{2}\hbar\beta \tag{7.50}$$

In a sense $\alpha = Y_{1/2}^{1/2}$ and $\beta = Y_{1/2}^{-1/2}$, but this is a strictly *formal* association and α and β, and even $\hat{S}^2$ and $\hat{S}_z$ for that matter, do not have to be specified any further.

Just as we can write that the value of the orbital angular momentum of an electron in a hydrogen atom is given by

$$L = \hbar\sqrt{l(l+1)} \tag{7.51}$$

we can say that the spin angular momentum of an electron is

$$S = \hbar\sqrt{s(s+1)} \tag{7.52}$$

Unlike l, which can vary from 0 to ∞, s can have only the value $s = \frac{1}{2}$. Note that because s is not allowed to assume large values, the spin angular momentum can never assume classical behavior. Spin is strictly a nonclassical concept. The functions α and β in Equation 7.50 are called *spin eigenfunctions*. We assume that α and β are orthonormal, which we write *formally* as

$$\int \alpha^* \alpha \, d\sigma = \int \beta^* \beta \, d\sigma = 1$$

$$\int \alpha^* \beta \, d\sigma = \int \alpha \beta^* \, d\sigma = 0$$

(7.53)

where σ is called the *spin variable*. The spin variable has no classical analog. We emphasize that Equations 7.53 are *formal*. The spin variable σ is *not* a continuous variable, having only two z components $\pm \hbar/2$. Some authors write Equations 7.53 as summations over $\sigma = \pm 1/2$, but they are better written in the bracket notation:

$$\langle \alpha \mid \alpha \rangle = \langle \beta \mid \beta \rangle = 1$$

$$\langle \alpha \mid \beta \rangle = \langle \beta \mid \alpha \rangle = 0$$

(7.54)

We can also express α and β in a different notation to emphasize their (spin) angular-momentum nature. For orbital angular momentum, we have

$$Y_l^{m_l}(\theta, \phi) = \mid lm_l \rangle$$

Using this notation, Equations 7.48 and 7.49 become

$$\hat{L}^2 \mid lm_l \rangle = \hbar^2 l(l + 1) \mid lm_l \rangle \qquad l = 0, 1, 2, \ldots$$

(7.55)

and

$$\hat{L}_z \mid lm_l \rangle = \hbar m_l \mid lm_l \rangle \qquad m_l = 0, \pm 1, \ldots, \pm l$$

(7.56)

For spin states, $l \to s = \frac{1}{2}$ and $m_l \to m_s = \pm \hbar/2$, and so we make the correspondence

$$\alpha = \left| \frac{1}{2} \, \frac{1}{2} \right\rangle \qquad \beta = \left| \frac{1}{2} \, -\frac{1}{2} \right\rangle$$

(7.57)

Using this notation, Equations 7.50 become

$$\hat{S}^2 \mid s \, m_s \rangle = \hbar^2 s(s + 1) \mid s \, m_s \rangle \qquad s = \frac{1}{2}$$

(7.58)

and

$$\hat{S}_z \mid s \, m_s \rangle = \hbar m_s \mid s \, m_s \rangle \qquad m_s = \pm \frac{1}{2}$$

(7.59)

and Equation 7.54 becomes

$$\langle s\, m_s \mid s\, m'_s \rangle = \delta_{m_s, m'_s} \tag{7.60}$$

where δ_{m_s, m'_s} is the Kronecker delta.

We must now include the spin functions α and β with the spatial functions. To a first approximation, the spatial and spin parts of the wave function are independent and so we write

$$\Psi_{n,l,m_l,m_s}(r, \theta, \phi, \sigma) = \psi_{n,l,m_l}(r, \theta, \phi)\alpha(\sigma) \qquad \text{or} \qquad \psi_{n,l,m_l}(r, \theta, \phi)\beta(\sigma) \tag{7.61}$$

In the bracket notation, we have $\mid n\, l\, m_l\, m_s \rangle$. Thus, four quantum numbers are required to specify the state of an electron in a hydrogen atom.

The complete one-electron wave function $\mid n\, l\, m_l\, m_s \rangle$ is called a *spin orbital*. The first two spin orbitals of a hydrogen atom are

$$\Psi_{100\frac{1}{2}}(\mathbf{r}, \sigma) = \left| 100\frac{1}{2} \right\rangle = \left(\frac{Z^3}{\pi a_0^3} \right)^{1/2} e^{-Zr/a_0}\alpha$$

and

$$\Psi_{100-\frac{1}{2}}(\mathbf{r}, \sigma) = \left| 100\ -\frac{1}{2} \right\rangle = \left(\frac{Z^3}{\pi a_0^3} \right)^{1/2} e^{-Zr/a_0}\beta$$

where $\mathbf{r} = (r, \theta, \phi)$ and $d\mathbf{r} = r^2 \sin\theta \, dr d\theta d\phi$. It follows that each of the spin orbitals is normalized because we can write

$$\int \Psi^*_{100\frac{1}{2}}(\mathbf{r}, \sigma)\Psi_{100\frac{1}{2}}(\mathbf{r}, \sigma)\, d\mathbf{r} d\sigma = \langle 100\tfrac{1}{2} \mid 100\tfrac{1}{2} \rangle$$

$$= \frac{Z^3}{\pi a_0^3} \int_0^\infty e^{-2Zr/a_0}\, 4\pi r^2\, dr\, \langle \alpha \mid \alpha \rangle = 1$$

EXAMPLE 7–6
Show that the two spin orbitals $\Psi_{100\frac{1}{2}}(\mathbf{r}, \sigma)$ and $\Psi_{100-\frac{1}{2}}(\mathbf{r}, \sigma)$ are orthogonal.

SOLUTION:

$$\int \Psi^*_{100\frac{1}{2}}(\mathbf{r}, \sigma)\Psi_{100-\frac{1}{2}}(\mathbf{r}, \sigma)\, d\mathbf{r} d\sigma = \frac{Z^3}{\pi a_0^3} \int_0^\infty e^{-2Zr/a_0}\, 4\pi r^2\, dr\, \langle \alpha \mid \beta \rangle = 0$$

because $\langle \alpha \mid \beta \rangle = 0$. We can express the orthonormality of $\Psi_{100\frac{1}{2}}$ and $\Psi_{100-\frac{1}{2}}$ as one equation by writing

$$\langle 100\, m_s \mid 100\, m'_s \rangle = \delta_{m_s, m'_s}$$

Note that even though the "100" part in these wave functions is normalized, the two wave functions are orthogonal due to the spin parts.

The Hamiltonian operator given in Equation 7.2 does not contain spin (it is strictly electrostatic), and so the corresponding energies are independent of the spin state. Thus, the wave functions $| n \, l \, m_l \tfrac{1}{2} \rangle$ and $| n \, l \, m_l \, -\tfrac{1}{2} \rangle$ have the same energy if Equation 7.2 is used for the Hamiltonian operator.

7.6 Spin-Orbit Interaction Affects the Energies of a Hydrogen Atom

We saw in Section 7.4 that an electron has a magnetic moment that is proportional to its orbital angular momentum, or that

$$\mathbf{m}_L = -\frac{|e|}{2m_e} \mathbf{L} \tag{7.62}$$

We have subscripted $\mathbf{m}$ with an L here to emphasize that this magnetic moment is due to the orbital angular momentum of the electron. Given that the eigenvalues of $\hat{L}^2$ are $\hbar l(l+1)$, we can write the magnitude of $\mathbf{m}_L$ as

$$m_L = -\frac{|e|\hbar}{2m_e} [l(l+1)]^{1/2} = -\beta_B [l(l+1)]^{1/2} \tag{7.63}$$

where β_B is the Bohr magneton. There is also a magnetic moment associated with the spin of the electron. It's given by an equation similar to Equation 7.62, but in order to obtain agreement with a number of experimental observations, we must modify Equation 7.62 by introducing a factor g and write

$$\mathbf{m}_S = -\frac{g|e|}{2m_e} \mathbf{S} \tag{7.64}$$

and

$$m_S = -g\beta_B [s(s+1)]^{1/2} \tag{7.65}$$

where $g = 2$. For several years, g was called the "anomalous" spin factor, but it appeared as a natural result when Dirac formulated his relativistic extension of quantum mechanics.

According to Equation 7.64, then, we have

$$m_{S_z} = -\frac{g|e|S_z}{2m_e} = -\frac{g|e|\hbar m_s}{2m_e}$$

$$= -g\,\beta_B\,m_s = \pm\beta_B \tag{7.66}$$

In the Stern–Gerlach experiment, the two projections of the beam occurred at positions consistent with magnetic moments $\pm\beta_B$, just as Equation 7.66 predicts.

As we pointed out above, the Hamiltonian operator in Equation 7.2 is strictly electrostatic, and does not contain spin. This Hamiltonian operator yields the energies $E_n = -m_e e^4 / 8\epsilon_0^2 \hbar^2 n^2$, which give a very good description of the hydrogen atomic spectrum. Nevertheless, there are discrepancies. For example, a close examination of the $n = 2$ to $n = 1$ transition shows that it consists of a doublet, which is one of many observations that led Uhlenbeck and Goudsmit to introduce electron spin into quantum mechanics. It was soon recognized that it was necessary to take into account the fact that the magnetic moment due to orbital angular momentum and the magnetic moment due to spin angular momentum interact with each other, just as two magnets interact with each other, and the energy of this interaction must be included in the Hamiltonian operator. When this is done, Equation 7.2 becomes

$$\hat{H} = -\frac{\hbar^2}{2m_e}\nabla^2 - \frac{e^2}{4\pi\epsilon_0 r} + \xi(r)\,\hat{\mathbf{L}} \cdot \hat{\mathbf{S}} \tag{7.67}$$

where $\xi(r)$ is a function whose form is not necessary here. (See, however, Problem 7–43.) The added term is called the *spin-orbit interaction* term.

Realize that spin-orbit interaction is a small effect in a hydrogen atom, and that the resultant energies are going to differ only slightly from the electrostatic energies $E_n = -m_e e^4 / 8\epsilon_0^2 \hbar^2 n^2$. We are not able to solve the Schrödinger equation exactly with the spin-orbit interaction term included, but because the effect of the term is small, we expect that the energies $E_n = -m_e e^4 / 8\epsilon_0^2 \hbar^2 n^2$ will be altered only slightly by its inclusion. We can use a technique called *perturbation theory* to calculate how much the electrostatic energies are altered (perturbed). This will not be necessary for our purposes here, but we'll learn how to calculate the effect of perturbations in Chapter 8.

There is one consequence of spin-orbit interaction that we must discuss here, however. When the spin-orbit interaction is taken into account, $\hat{L}^2$ and $\hat{S}^2$ no longer commute with $\hat{H}$ and so $\mathbf{L}$ and $\mathbf{S}$ are no longer conserved; only the total angular momentum,

$$\mathbf{J} = \mathbf{L} + \mathbf{S} \tag{7.68}$$

is conserved. In analogy with Equations 7.55 and 7.56 (and Equations 7.58 and 7.59), we have

$$\hat{J}^2\,|\,jm_j\rangle = \hbar^2 j(j+1)\,|\,jm_j\rangle \qquad \text{and} \qquad \hat{J}_z\,|\,jm_j\rangle = \hbar m_j\,|\,jm_j\rangle \tag{7.69}$$

If spin-orbit interaction is considered, then the energy of a hydrogen atom depends upon n and j. We know that the allowed values of l are 0, 1, 2, ... and the only allowed value of s is 1/2. What are the allowed values of j? This type of question arises frequently in quantum mechanics; we have two angular momenta and we wish to determine the allowed values of their sum. This general problem is called *addition of angular momenta*. The general theory of the addition of angular momenta is a little bit involved, but it's fairly easy to determine the allowed values of j in this case, however, because $s = 1/2$ and m_s can assume only two values, $\pm 1/2$.

There are $2l + 1$ possible orbital angular-momentum states for a fixed value of l and $2s + 1 = 2$ possible spin angular-momentum states, giving $2(2l + 1)$ as the total number of states. Let's consider the case $l = 0$ first. There are only two possible states, those with $m_j = m_l + m_s = 0 \pm 1/2 = \pm 1/2$. Since $m_j = \pm 1/2$, then $j = 1/2$. (See Table 7.6.)

Now let's consider the case for $l = 1$, for which there is a total of $2(2 \times 1 + 1) = 6$ states. The largest possible value of $m_j = m_l + m_s$ is $m_j = 1 + 1/2 = 3/2$. In that case, $j = 3/2$ with

$$m_j = \frac{3}{2}, \frac{1}{2}, -\frac{1}{2}, -\frac{3}{2} \qquad \left(j = \frac{3}{2} \right)$$

This accounts for four of the six states. There is only one way to obtain $m_j = 3/2$, but there are two ways to obtain $m_j = 1/2$—namely, $m_j = 0 + 1/2$ and $m_j = 1 - 1/2$. One of these is accounted for above as part of the $j = 3/2$ state, but there is one left over. This state, with $m_j = 1/2$, must belong to the $j = 1/2$ state. Because a $j = 1/2$ state has two values of m_j ($\pm 1/2$), we have

$$m_j = \frac{1}{2}, -\frac{1}{2} \qquad \left(j = \frac{1}{2} \right)$$

This accounts for all six states, and so we see that $j = l + s = l + \frac{1}{2}$ and $j = l - s = l - \frac{1}{2}$. (See Table 7.6.)

We need to consider only one more value of l to see the pattern. There are ten possible states when $l = 2$ ($2 \times 2 + 1$). The largest possible value of m_j is $m_j = 2 + 1/2 = 5/2$. Thus, we have $j = 5/2$ with

$$m_j = \frac{5}{2}, \frac{3}{2}, \frac{1}{2}, -\frac{1}{2}, -\frac{3}{2}, -\frac{5}{2} \qquad \left(j = \frac{5}{2} \right)$$

This accounts for six of the ten possible states. There is only one way to obtain $m_j = 5/2$, but there are two ways to obtain $m_j = 3/2$—namely, $m_j = 1 + 1/2$ and $m_j = 2 - 1/2$. One of these is accounted for by the $j = 5/2$ state above, but the other is not. It must belong to the $j = 3/2$ state, with

$$m_j = \frac{3}{2}, \frac{1}{2}, -\frac{1}{2}, -\frac{3}{2} \qquad \left(j = \frac{3}{2} \right)$$

This accounts for all ten states, and we see that $j = l + s = 5/2$ and $j = l + s - 1 = 3/2$. (See Table 7.6.)

TABLE 7.6
The Allowed Values of the Total Angular Momentum, j, for Various Values of l When the Orbital Angular Momentum, l, and the Spin Angular Momentum, $s = 1/2$, of a Single Electron Are Added [a]

$l = 0$	2 states	
	$m_j = \dfrac{1}{2}, -\dfrac{1}{2}$	$j = \dfrac{1}{2}$
$l = 1$	6 states	
	$m_j = \dfrac{3}{2}, \dfrac{1}{2}, -\dfrac{1}{2}, -\dfrac{3}{2}$	$j = \dfrac{3}{2}$
	$m_j = \dfrac{1}{2}, -\dfrac{1}{2}$	$j = \dfrac{1}{2}$
$l = 2$	10 states	
	$m_j = \dfrac{5}{2}, \dfrac{3}{2}, \dfrac{1}{2}, -\dfrac{1}{2}, -\dfrac{3}{2}, -\dfrac{5}{2}$	$j = \dfrac{5}{2}$
	$m_j = \dfrac{3}{2}, \dfrac{1}{2}, -\dfrac{1}{2}, -\dfrac{3}{2}$	$j = \dfrac{3}{2}$
$l = 3$	14 states	
	$m_j = \dfrac{7}{2}, \dfrac{5}{2}, \dfrac{3}{2}, \dfrac{1}{2}, -\dfrac{1}{2}, -\dfrac{3}{2}, -\dfrac{5}{2}, -\dfrac{7}{2}$	$j = \dfrac{7}{2}$
	$m_j = \dfrac{5}{2}, \dfrac{3}{2}, \dfrac{1}{2}, -\dfrac{1}{2}, -\dfrac{3}{2}, -\dfrac{5}{2}$	$j = \dfrac{5}{2}$

a. The total number of states for a given value of l is $2(2l + 1)$.

EXAMPLE 7–7
Repeat the above analysis for $l = 3$.

SOLUTION: There are 14 possible states, $2 (2 \times 3 + 1)$. The largest possible value of m_j is $3 + 1/2 = 7/2$. Thus, we have $j = 7/2$ with

$$m_j = \frac{7}{2}, \frac{5}{2}, \frac{3}{2}, \frac{1}{2}, -\frac{1}{2}, -\frac{3}{2}, -\frac{5}{2}, -\frac{7}{2} \qquad \left(j = \frac{7}{2} \right)$$

which accounts for 8 of the 14 possible states. There is only one way to obtain $m_j = 7/2$, but there are two ways to obtain $m_j = 5/2$—namely, $m_j = 2 + 1/2$ and $m_j = 3 - 1/2$. One of these is accounted for by the $j = 7/2$ state above, but the other is not. It must belong to the $j = 5/2$ state, with

$$m_j = \frac{5}{2}, \frac{3}{2}, \frac{1}{2}, -\frac{1}{2}, -\frac{3}{2}, -\frac{5}{2} \qquad \left(j = \frac{5}{2} \right)$$

These two values of j account for all 14 possible states.

Table 7.6 summarizes these results. Notice that we obtain $j = l + s$ and $l - s$, except when $l = 0$, where we obtain only $j = l + s$. We can summarize all these results in one equation by writing

$$j = l + s \quad \text{and} \quad |l - s| \qquad (7.70)$$

When $l = 0$, we have $j = 1/2$; when $l = 1$, we have $j = 3/2$ and $1/2$; and when $l = 2$, we have $j = 5/2$ and $3/2$, and so on.

7.7 The Electronic Energy Levels of a Hydrogen Atom Are Described by Term Symbols

The electronic energies of a hydrogen atom depend only upon the principal quantum number n if we do not include spin-orbit interaction, but they depend upon n and j if we do. We designate the electronic states of a hydrogen atom by an *atomic term symbol*, which is expressed by the occupied orbital followed by the symbol ^2l_j. The left superscript here represents the multiplicity (degeneracy) of the spin, which is $2s + 1 = 2$. In writing a term symbol for atomic hydrogen, we make the correspondence

$$l = \quad 0 \quad 1 \quad 2 \quad 3 \quad 4$$
$$ \quad S \quad P \quad D \quad F \quad G \quad \text{etc.}$$

Therefore, the ground electronic state, for which $l = 0$ and $j = 1/2$, is designated by $1s\ ^2S_{1/2}$.

Table 7.7 lists the term symbols and the energies for the first few electronic states of atomic hydrogen. Except for a small difference between the various $^2S_{1/2}$ and $^2P_{1/2}$ states, which is due to a subtle quantum-electrodynamic effect, the energies for each value of n depend only on j, due to spin-orbit coupling. Table 7.7 is a screen shot of part of the actual output from a website (*http://physics.nist.gov/PhysRefData/ASD/levels_form.html*) maintained by the National Institutes of Standards and Technology that lists the energy levels for essentially all the atoms and their ions. The notation H I in this case denotes a neutral hydrogen atom. We shall visit this website for other atoms in Chapter 9, where we shall see that multielectron atoms are also described by atomic term symbols $^{2S+1}L_J$, where S is the total spin angular momentum, L is the total orbital angular momentum, and J is the total angular momentum. In the case of atomic hydrogen, $S = 1/2$, $L = l$, and $J = j = l \pm 1/2$.

Let's use Table 7.7 to take a closer look at the atomic hydrogen spectrum. In particular, let's look at the Lyman series, which is the series of lines that arise from transitions from states with $n \geq 2$ to the $n = 1$ state. (See Figure 1.10.) As we did in

TABLE 7.7
A Screen Shot of the First Few Electronic States of Atomic Hydrogen

H I   167 Levels Found (Page 1 of 10)

Data on Lande factors and level compositions are not available for this ion

Configuration	Term	J	Level (cm^{-1})
1s	^{2}S	$1/2$	0
2p	^{2}P$^\circ$	$1/2$	82 258.9206
		$3/2$	82 259.2865
2s	^{2}S	$1/2$	82 258.9559
3p	^{2}P$^\circ$	$1/2$	97 492.2130
		$3/2$	97 492.3214
3s	^{2}S	$1/2$	97 492.2235
3d	^{2}D	$3/2$	97 492.3212
		$5/2$	97 492.3574
4p	^{2}P$^\circ$	$1/2$	102 823.8505
		$3/2$	102 823.8962
4s	^{2}S	$1/2$	102 823.8549
4d	^{2}D	$3/2$	102 823.8961
		$5/2$	102 823.9114
4f	^{2}F$^\circ$	$5/2$	102 823.9113
		$7/2$	102 823.9190
5p	^{2}P$^\circ$	$1/2$	105 291.6306
		$3/2$	105 291.6540

Source: http://physics.nist.gov/PhysRefData/ASD/levels_form.html

Chapter 1, we can use the Rydberg formula to calculate the frequencies of the lines in the Lyman series:

$$\tilde{\nu} = 109\,677.58 \left(1 - \frac{1}{n^2} \right) \text{cm}^{-1} \qquad n = 2, 3, \ldots \qquad (7.71)$$

If we express our results in terms of wave numbers, we obtain the following:

Transition	Frequency/cm^{-1}
$2 \rightarrow 1$	82 258.19
$3 \rightarrow 1$	97 491.18
$4 \rightarrow 1$	102 822.73
$5 \rightarrow 1$	105 290.48

If we use Table 7.7, we see that there are three states for $n = 2$. Not all these states can make a transition to the ground state because of selection rules. Recall that selection rules are restrictions that govern the possible, or *allowed*, transitions from one state to another. In the case of atomic spectra, the selection rules are

$$\Delta L = \pm 1$$
$$\Delta S = 0 \qquad (7.72)$$
$$\Delta J = 0, \pm 1$$

except that a transition from a state with $J = 0$ to another state with $J = 0$ is not allowed (*forbidden*). The selection rules given by Equations 7.72 have been deduced experimentally and corroborated theoretically (see Problem 7–46). (The rule $\Delta L = \pm 1$ follows from the principle of conservation of angular momentum because a photon has a spin angular momentum of $\hbar$.)

The selection rules given in Equations 7.72 tell us that $^2P \rightarrow \,^2S$ transitions are allowed, but that $^2S \rightarrow \,^2S$ transitions are not allowed because $\Delta L = 0$ and that $^2S \rightarrow \,^2D$ transitions and $^2F \rightarrow \,^2P$ transitions are not allowed because $\Delta L = \pm 2$, respectively, in these transitions. Thus, if we look closely at the Lyman series of atomic hydrogen, we see that the allowed transitions into the ground state are

$$np \,^2P_{1/2} \rightarrow 1s \,^2S_{1/2} \begin{pmatrix} \Delta L = 1 \\ \Delta S = 0 \\ \Delta J = 0 \end{pmatrix}$$

or

$$np \,^2P_{3/2} \rightarrow 1s \,^2S_{1/2} \begin{pmatrix} \Delta L = 1 \\ \Delta S = 0 \\ \Delta J = -1 \end{pmatrix}$$

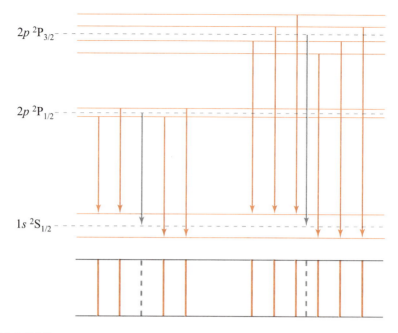

FIGURE 7.13
The Zeeman effect for the $1s$ and $2p$ levels of atomic hydrogen in a weak external magnetic field, showing the allowed transitions. A schematic diagram of the resulting spectrum is shown at the bottom. The dashed lines show the fine structure that is present in the absence of an external magnetic field.

$$
g(j, l) = \begin{cases} 2\left(\dfrac{l+1}{2l+1}\right) & j = l + \tfrac{1}{2} \\[2ex] \dfrac{2l}{2l+1} & j = l - \tfrac{1}{2} \end{cases}
$$

This quantity $g(j, l)$ is known as the *Landé g factor*. Note that $g = 2$ for the $1s\,^2S_{1/2}$ states, 2/3 for the $2p\,^2P_{1/2}$ states, and 4/3 for the $2p\,^2P_{3/2}$ states.

Figure 7.13 shows the splitting of the $1s\,^2S_{1/2}$, $2p\,^2P_{1/2}$, and $2p\,^2P_{3/2}$ states in an external magnetic field. The allowed transitions ($\Delta m_j = 0, \pm 1$) are indicated, along with a schematic sketch of the resulting spectrum. Note that there are ten lines in the $2p \rightarrow 1s$ transition.

EXAMPLE 7–9
Using the data in Table 7.7, calculate the frequency (in cm^{-1}) of the extreme right-hand transition in Figure 7.13 in an external magnetic field of 1.00 tesla.

SOLUTION: The values of l, j, and m_j in the upper state ($2p\,^2P_{3/2}$) are 1, 3/2, and 1/2, respectively. Therefore, $g = 4/3$ in Equation 7.75 and we have

$$\Delta E_{upper} = \frac{4}{3}\beta_B \left(\frac{1}{2}\right)(1.00\ T) = \frac{2(9.274 \times 10^{-24}\ J)}{3}$$

$$= \left(\frac{2}{3}\right)(9.2740 \times 10^{-24}\ J)(5.0341 \times 10^{22}\ cm^{-1} \cdot J^{-1})$$

$$= 0.3112\ cm^{-1}$$

The energy of the upper state is, using the data in Table 7.7,

$$E_{upper} = (82\ 259.2865 + 0.3112)\ cm^{-1} = 82\ 260.5977\ cm^{-1}$$

For the lower state ($1s\ ^2S_{1/2}$), the values of l, j, and m_j are 0, 1/2, and $-1/2$, respectively. Therefore, $g = 2$ in Equation 7.75 and we have

$$E_{lower} = 0\ cm^{-1} - 2\beta_B \left(\frac{1}{2}\right)(1.00\ T) = -0.4669\ cm^{-1}$$

The frequency of the transition, then, is

$$\Delta E = E_{upper} - E_{lower} = 82\ 261.0646\ cm^{-1}$$

The frequency that we would calculate on the basis of a purely electrostatic Hamiltonian operator is

$$\Delta E = (109\ 677.58\ cm^{-1})\left(1 - \frac{1}{4}\right) = 82\ 258.19\ cm^{-1}$$

7.9 The Schrödinger Equation for a Helium Atom Cannot Be Solved Exactly

The next system to study is the helium atom, whose Schrödinger equation is

$$\left(-\frac{\hbar^2}{2M}\nabla^2 - \frac{\hbar^2}{2m_e}\nabla_1^2 - \frac{\hbar^2}{2m_e}\nabla_2^2\right)\psi(\mathbf{R}, \mathbf{r}_1, \mathbf{r}_2) + \left(-\frac{2e^2}{4\pi\epsilon_0|\mathbf{R} - \mathbf{r}_1|}\right.$$

$$\left.-\frac{2e^2}{4\pi\epsilon_0|\mathbf{R} - \mathbf{r}_2|} + \frac{e^2}{4\pi\epsilon_0|\mathbf{r}_1 - \mathbf{r}_2|}\right)\psi(\mathbf{R}, \mathbf{r}_1, \mathbf{r}_2) = E\psi(\mathbf{R}, \mathbf{r}_1, \mathbf{r}_2)$$

$$(7.76)$$

In this equation, $\mathbf{R}$ is the position of the helium nucleus, and $\mathbf{r}_1$ and $\mathbf{r}_2$ are the positions of the two electrons; M is the mass of the nucleus, and m_e is the electronic mass; ∇^2 is the Laplacian operator with respect to the position of the nucleus, and ∇_1^2 and ∇_2^2 are the Laplacian operators with respect to the positions of the electronic coordinates. Realize that this is a three-body problem and *not* a two-body problem, and so the separation into center-of-mass and relative coordinates is much more complicated than it is for hydrogen. Because $M \gg m_e$, however, regarding the nucleus as fixed relative to the motion of the electrons is still an excellent approximation. Under this approximation, we can fix the nucleus at the origin of a spherical coordinate system and write the Schrödinger equation as

$$-\frac{\hbar^2}{2m_e}(\nabla_1^2 + \nabla_2^2)\psi(\mathbf{r}_1, \mathbf{r}_2) - \frac{2e^2}{4\pi\epsilon_0}\left(\frac{1}{r_1} + \frac{1}{r_2}\right)\psi(\mathbf{r}_1, \mathbf{r}_2) + \frac{e^2}{4\pi\epsilon_0|\mathbf{r}_1 - \mathbf{r}_2|}\psi(\mathbf{r}_1, \mathbf{r}_2)$$

$$= E\psi(\mathbf{r}_1, \mathbf{r}_2) \tag{7.77}$$

Even this simplified equation cannot be solved exactly.

The $e^2/4\pi\epsilon_0|\mathbf{r}_1 - \mathbf{r}_2|$ term is called the *interelectronic repulsion* term and is directly responsible for the difficulty associated with Equation 7.77. If this term were not there, the total Hamiltonian operator in Equation 7.77 would be the sum of the Hamiltonian operators of two hydrogen atoms. According to Equations 3.61 through 3.64, the total energy would be the sum of the energies of the two individual hydrogen atoms, and the wave function would be a product of two hydrogen atomic orbitals.

Because we cannot solve Equation 7.77 exactly, we must resort to some approximation method. Fortunately, two quite different approximation methods that can yield extremely good results have found wide use in quantum chemistry. These are called the *variational method* and *perturbation theory* and are presented in the next chapter.

Problems

7–1. Show that both $\hbar^2\nabla^2/2m_e$ and $e^2/4\pi\epsilon_0 r$ have the units of energy (joules).

7–2. Show that the substitution $R(r)Y_l^{m_l}(\theta, \phi)$ into Equation 7.7 yields Equation 7.10.

7–3. Show that the radial functions given in Table 7.2 are normalized.

7–4. Referring to Table 7.2, show that $R_{10}(r)$ and $R_{20}(r)$ are orthonormal.

7–5. Referring to Table 7.3, show that the first few hydrogen atomic wave functions are orthonormal.

7–6. Show explicitly that

$$\hat{H}\psi = -\frac{m_e e^4}{8\epsilon_0^2 h^2}\psi$$

for the ground state of a hydrogen atom.

7–7. Show explicitly that

$$\hat{H}\psi = -\frac{m_e e^4}{32\epsilon_0^2 h^2}\psi$$

for a $2p_0$ state of a hydrogen atom.

7–8. Show that all of the following expressions for the ground state of a hydrogen atom are equivalent:

$$E_0 = -\frac{\hbar^2}{2m_e a_0^2} = -\frac{e^2}{8\pi\epsilon_0 a_0} = -\frac{m_e e^4}{32\pi^2\epsilon_0^2\hbar^2} = -\frac{m_e e^4}{8\epsilon_0^2 h^2}$$

7–9. Calculate the probability that a hydrogen $1s$ electron will be found within a distance $2a_0$ from the nucleus.

7–10. Calculate the radius of the sphere that encloses a 50% probability of finding a hydrogen $1s$ electron. Repeat the calculation for a 90% probability.

7–11. Many problems involving the calculation of average values for the hydrogen atom require doing integrals of the form

$$I_n = \int_0^\infty r^n e^{-\beta r}\,dr$$

This integral can be evaluated readily by starting with the elementary integral

$$I_0(\beta) = \int_0^\infty e^{-\beta r}\,dr = \frac{1}{\beta}$$

Show that the derivatives of $I(\beta)$ are

$$\frac{dI_0}{d\beta} = -\int_0^\infty r e^{-\beta r}\,dr$$

$$\frac{d^2 I_0}{d\beta^2} = \int_0^\infty r^2 e^{-\beta r}\,dr$$

and so on. Using the fact that $I_0(\beta) = 1/\beta$, show that the values of these two integrals are $-1/\beta^2$ and $2/\beta^3$, respectively. Show that, in general,

$$\frac{d^n I_0}{d\beta^n} = (-1)^n \int_0^\infty r^n e^{-\beta r}\,dr = (-1)^n \frac{n!}{\beta^{n+1}}$$

and that

$$I_n = \frac{n!}{\beta^{n+1}}$$

7–12. Prove that the average value of r in the $1s$ and $2s$ states for a hydrogen-like atom is $3a_0/2Z$ and $6a_0/Z$, respectively. (Compare your result to Table 7.4.)

7–13. Prove that $\langle V \rangle = 2\langle E \rangle$ and, consequently, that $\langle V \rangle/\langle T \rangle = -2$, for a $2p_0$ electron.

7–14. By evaluating the appropriate integrals, compute $\langle r \rangle$ in the $2s$, $2p$, and $3s$ states of the hydrogen atom; compare your results with Table 7.4 and the general formula

$$\langle r_{nl} \rangle = \frac{a_0}{2Z}[3n^2 - l(l+1)]$$

7–15. By evaluating the appropriate integrals, compute $\langle 1/r \rangle$ in the $2s$, $2p$, and $3s$ states of the hydrogen atom; compare your results with Table 7.4 and the general formula

$$\left\langle \frac{1}{r} \right\rangle_{n,l} = \frac{Z}{a_0 n^2}$$

7–16. Use the results in Table 7.4 to show that $1/\langle r \rangle \neq \langle 1/r \rangle$.

7–17. Show that the two maxima in the plot of $r^2 R_{20}^2(r)$ against r occur at $(3 \pm \sqrt{5})a_0$. (See Figure 7.2.)

7–18. Calculate the value of $\langle r \rangle$ for the $n = 2$, $l = 1$ state and the $n = 2$, $l = 0$ state of the hydrogen atom. Are you surprised by the answers? Explain.

7–19. The average value of r for a hydrogen-like atom can be evaluated in general and is given by (see Table 7.4)

$$\langle r \rangle_{nl} = \frac{n^2 a_0}{Z} \left\{ 1 + \frac{1}{2} \left[1 - \frac{l(l+1)}{n^2} \right] \right\}$$

Verify this formula explicitly for the ψ_{211} orbital.

7–20. The average value of r^2 for a hydrogen-like atom is given in Table 7.4. Verify the entry explicitly for the ψ_{210} orbital.

7–21. The average values of $1/r$, $1/r^2$, and $1/r^3$ are given in Table 7.4. Verify these entries explicitly for the ψ_{210} orbital.

7–22. In Chapter 4, we learned that if ψ_1 and ψ_2 are solutions of the Schrödinger equation that have the same energy E_n, then $c_1\psi_1 + c_2\psi_2$ is also a solution with that energy. Let $\psi_1 = \psi_{210}$ and $\psi_2 = \psi_{211}$ (see Table 7.3). What is the energy corresponding to $\psi = c_1\psi_1 + c_2\psi_2$, where $c_1^2 + c_2^2 = 1$? What does this result tell you about the uniqueness of the three p orbitals, p_x, p_y, and p_z?

7–23. Verify the linear combinations given in Equations 7.32.

7–24. Show that the total probability density of the $2p$ orbitals is spherically symmetric by evaluating $\displaystyle\sum_{m_l=-1}^{1} \psi_{21m_l}^2$. (Use the wave functions in Table 7.3.)

7–25. Show that the total probability density of the $3d$ orbitals is spherically symmetric by evaluating $\displaystyle\sum_{m_l=-2}^{2} \psi_{32m_l}^2$. (Use the wave functions in Table 7.3.)

7–26. Show that the sum of the probability densities for the $n = 3$ states of a hydrogen atom is spherically symmetric. Do you expect this to be true for all values of n? Explain.

7–27. The designations of the d orbitals can be rationalized in the following way. Equation 7.32 shows that d_{xz} goes as $\sin\theta \cos\theta \cos\phi$. Using the relation between cartesian and spherical coordinates, show that $\sin\theta \cos\theta \cos\phi$ is proportional to xz. Similarly, show that $\sin\theta \cos\theta \sin\phi$ (d_{yz}) is proportional to yz, that $\sin^2\theta \cos 2\phi$ ($d_{x^2-y^2}$) is proportional to $x^2 - y^2$, and that $\sin^2\theta \sin 2\phi$ (d_{xy}) is proportional to xy.

7–28. What is the degeneracy of each of the hydrogen atomic energy levels (neglecting spin-orbit interactions)?

7–29. Set up the Hamiltonian operator for the system of an electron interacting with a fixed nucleus of atomic number Z. The simplest such system is singly ionized helium, where $Z = 2$. We will call this a hydrogen-like system. Observe that the only difference between this Hamiltonian operator and the hydrogen Hamiltonian operator is the correspondence that e^2 for the hydrogen atom becomes Ze^2 for the hydrogen-like ion. Consequently, show that the energy becomes (cf. Equation 7.11)

$$E_n = -\frac{m_e Z^2 e^4}{8\epsilon_0^2 h^2 n^2} \qquad n = 1, 2, \ldots$$

Furthermore, now show that the solutions to the radial equation, Equation 7.10, are

$$R_{nl}(r) = -\left\{ \frac{(n-l-1)!}{2n[(n+l)!]^3} \right\}^{1/2} \left(\frac{2Z}{na_0} \right)^{l+3/2} r^l e^{-Zr/na_0} L_{n+l}^{2l+1} \left(\frac{2Zr}{na_0} \right)$$

Show that the $1s$ orbital for this system is

$$\psi_{1s} = \frac{1}{\sqrt{\pi}} \left(\frac{Z}{a_0} \right)^{3/2} e^{-Zr/a_0}$$

and show that it is normalized. Show that

$$\langle r \rangle = \frac{3 a_0}{2Z} \qquad \text{and} \qquad r_{mp} = \frac{a_0}{Z}$$

Last, compare the ionization energy of a hydrogen atom and that of a singly ionized helium atom. Express your answer in kilojoules per mole.

7–30. What is the ratio of the ground-state energy of atomic hydrogen to that of atomic deuterium?

7–31. The virial theorem is proved in general in Problem 5–43. Show that if $V(x, y, z)$ is a coulombic potential

$$V(x, y, z) = \frac{Ze^2}{4\pi \epsilon_0 (x^2 + y^2 + z^2)^{1/2}}$$

then

$$\langle V \rangle = -2\langle \hat{T} \rangle = 2\langle E \rangle \qquad (1)$$

where

$$\langle E \rangle = \langle \hat{T} \rangle + \langle V \rangle$$

We proved that this result is valid for a $1s$ electron in Section 7.2 and for a $2s$ electron in Example 7–4. Although we proved equation 1 only for the case of one electron in the field of one nucleus, equation 1 is valid for many-electron atoms and molecules. The proof is a straightforward extension of the proof developed in this problem.

7–32. Show that Equation 7.36 reduces to Equation 7.35 in the case of a circular orbit.

7–33. Show that V in Equation 7.42 has units of energy.

7–34. Verify the value of a Bohr magneton, given in Equation 7.47.

7–35. Superconducting magnets have magnetic field strengths of the order of 15 T. Calculate the magnitude of the splitting shown in Figure 7.9 for a magnetic field of 15 T. Compare your result with the energy difference between the unperturbed $1s$ and $2p$ levels. Show that the three distinct transitions shown in Figure 7.9 lie very close together.

7–36. Show that the force acting upon a magnetic dipole moment in a magnetic field B_z that varies in the z direction is given by $F_z = m_z \partial B_z / \partial z$.

7–37. Explain why the spin functions α and β can be represented by $| \frac{1}{2} \frac{1}{2} \rangle$ and $| \frac{1}{2} -\frac{1}{2} \rangle$ in Dirac notation.

7–38. Townsend (see end-of-chapter references or Problem 7–43) gives the following formula for the shift of a hydrogen atom energy level due to spin-orbit coupling:

$$\Delta E_{so} = \frac{m_e c^2 Z^4 \alpha^4}{4n^3 (l + \frac{1}{2}) l (l + 1)} \times \begin{cases} l & j = l + \frac{1}{2} \\ -(l+1) & j = l - \frac{1}{2} \end{cases}$$

The only new quantity in this expression is α, which is called the *fine structure constant*, and is given by $e^2 / 4\pi \epsilon_0 \hbar c$. Use this formula for ΔE_{so} to show that the difference in energy between two states with the same value of n and l is

$$\text{diff} = \frac{m_e c^2 Z^4 \alpha^4}{2n^3 l (l+1)} = \frac{5.8437 Z^4 \text{ cm}^{-1}}{n^3 l (l+1)}$$

Calculate the difference in energies between the $2p\ ^2P_{1/2}$ and $2p\ ^2P_{3/2}$ states of a hydrogen atom and compare your results to what you obtain from Table 7.7. Do the same for the $3p\ ^2P_{1/2}$ and $3p\ ^2P_{3/2}$ states and the $4p\ ^2P_{1/2}$ and $4p\ ^2P_{3/2}$ states.

7–39. Use the equation of the previous problem to calculate the difference in energy between the $3d\ ^2D_{3/2}$ and $3d\ ^2D_{5/3}$ states and the $4d\ ^2D_{3/2}$ and $4d\ ^2D_{5/2}$ states of a hydrogen atom. How about for the $4f\ ^2F_{5/2}$ and $4f\ ^2F_{7/2}$ states?

7–40. Use the formula in Problem 7–38 to determine the value of the fine structure constant. Take its reciprocal.

7–41. Repeat Problem 7–38 for a singly ionized helium atom. Go to *http://physics.nist.gov/PhysRefData.ASD/levels_form.html* to see the energy levels of He+, and compare your results to the experimental values.

7–42. Repeat Problem 7–39 for a singly ionized helium atom. Go to *http://physics.nist.gov/PhysRefData.ASD/levels_form.html* to see the energy levels of He+, and compare your results to the experimental values.

7–43. This problem develops a heuristic derivation of the formula for ΔE_{so}, the shift in the hydrogen atom energy level due to spin-orbit interaction (the first formula in Problem 7–38). We start with an explicit formula for the Hamiltonian operator for spin-orbit interaction (see Townsend, end-of-chapter references):

$$\hat{H}_{so} = \frac{Ze^2}{2m_e^2 c^2 \kappa_0 r^3} \hat{\mathbf{L}} \cdot \hat{\mathbf{S}}$$

We want to take the average of this quantity. Show that $\hat{H}_{so}$ can be written as

$$\hat{H}_{so} = \frac{Ze^2}{4m_e^2 c^2 r^3} (\hat{J}^2 - \hat{L}^2 - \hat{S}^2)$$

(Remember that $\mathbf{J} = \hat{\mathbf{L}} + \hat{\mathbf{S}}$.) Although the inclusion of $\hat{H}_{so}$ into the hydrogen atom Hamiltonian operator alters the wave functions, if the effect of $\hat{H}_{so}$ is small (as we are assuming), then we can use the hydrogen atomic wave functions as a good approximation. The hydrogen atomic wave functions are eigenfunctions of $\hat{L}^2$, $\hat{L}_z$, $\hat{S}^2$, and $\hat{S}_z$, but it's possible to build linear combinations of them that are eigenfunctions of $\hat{J}^2$, $\hat{J}_z$, $\hat{L}^2$, and $\hat{S}^2$ (denote them by $| n, J, M_J, L, S \rangle$), in which case J, M_J, L, and S are good quantum numbers. Use these wave functions to argue that

$$\langle n, J, M_J, L, S \mid \hat{H}_{so} \mid n, J, M_J, L, S \rangle$$

$$= \frac{Ze^2}{4m_e^2 c^2 \kappa_0} \left\langle \frac{1}{r^3} \right\rangle_{n,l} \langle J, M_J, L, S \mid \hat{J}^2 - \hat{L}^2 - \hat{S}^2 \mid J, M_J, L, S \rangle$$

Now use the fact that (see Table 7.4)

$$\left\langle \frac{1}{r^3} \right\rangle_{n,l} = \frac{Z^3}{a_0^3 n^3 l (l+1)(l+\frac{1}{2})}$$

and that

$$(\hat{J}^2 - \hat{L}^2 - \hat{S}^2) \mid J, M_J, L, S \rangle = \hbar^2 \left[J(J+1) - L(L+1) - S(S+1) \right] \mid J, M_J, L, S \rangle$$

to obtain

$$E_{so} = \frac{Z^4 e^2 \hbar^2 [J(J+1) - L(L+1) - S(S+1)]}{4m_e c^2 \kappa_0 a_0^3 n^3 l (l+1)(l+\frac{1}{2})}$$

Finally, show that this result becomes

$$\Delta E_{so} = \frac{m_e c^2 Z^4 \alpha^4}{4 n^3 l (l+1)(l+\frac{1}{2})} \times \begin{cases} l & j = l + \frac{1}{2} \\ -(l+1) & j = l - \frac{1}{2} \end{cases}$$

This is not a rigorous derivation, but it does give an idea of the "flavor" of the derivation.

7–44. Extend Table 7.6 to the case $l = 4$. How many states are there?

7–45. Can you deduce a general formula for the total number of states for each value of l in Table 7.6?

7–46. We'll see in Section 8.6 that the selection rule for electronic transitions in a hydrogen atom depends upon whether the integral

$$I = \langle n, l, m_l | \mathbf{r} | n', l', m_l' \rangle$$

is zero or nonzero. If $I = 0$, then the transition is forbidden; otherwise, it may occur. Show that $I = 0$ unless $\Delta l = \pm 1$ and $\Delta m_l = 0$. Do you find any restriction Δn?

7–47. Using the data in Table 7.7, calculate the frequency (in cm^{-1}) of the extreme left-hand transition in Figure 7.13 in an external magnetic field of 1.00 tesla. Compare your result to the frequency that you would obtain neglecting spin-orbit interaction.

7–48. Using the data in Table 7.7, calculate the frequency (in cm^{-1}) of the second-from-right transition in Figure 7.13 in an external magnetic field of 1.00 tesla. Compare your result to the frequency that you would obtain neglecting spin-orbit interaction.

References

Pauling, L., Wilson, E. B., Jr. *Introduction to Quantum Mechanics*. Dover Publications: Mineola, NY, 1985.

Pilar, F. *Elementary Quantum Chemistry*, 2nd ed. Dover Publications: Mineola, NY, 2001.

Eisenberg, R., Resnick, R. *Quantum Physics of Atoms, Molecules, Solids, Nuclei, and Particles*. Wiley & Sons: New York, 1974.

Cox, P. A. *Introduction to Quantum Theory and Atomic Structure*. Oxford University Press: New York, 1996.

Slater, J. C. *Quantum Theory of Atomic Structure*. McGraw-Hill: New York, 1960.

Woodgate, G. K. *Elementary Atomic Structure*, 2nd ed. Oxford University Press: New York, 1983.

Townsend, J. S. *A Modern Approach to Quantum Mechanics*. University Science Books: Sausalito, CA, 2000.

Matrices

Many physical operations such as magnification, rotation, and reflection through a plane can be represented mathematically by quantities called matrices. A matrix is a two-dimensional array that obeys a certain set of rules called *matrix algebra*. Even if matrices are entirely new to you, they are so convenient that learning some of their simpler properties is worthwhile. Furthermore, a great deal of quantum mechanics or modern quantum chemistry uses matrix algebra frequently.

Consider the lower of the two vectors shown in Figure G.1. The x and y components of the vector are given by $x_1 = r \cos \alpha$ and $y_1 = r \sin \alpha$, where r is the length of $\mathbf{r}_1$. Now let's rotate the vector counterclockwise through an angle θ, so that $x_2 = r \cos(\alpha + \theta)$ and $y_2 = r \sin(\alpha + \theta)$ (see Figure G.1). Using trigonometric formulas, we can write

$$x_2 = r \cos(\alpha + \theta) = r \cos \alpha \cos \theta - r \sin \alpha \sin \theta$$

$$y_2 = r \sin(\alpha + \theta) = r \cos \alpha \sin \theta + r \sin \alpha \cos \theta$$

or

$$
\begin{aligned}
x_2 &= x_1 \cos \theta - y_1 \sin \theta \\
y_2 &= x_1 \sin \theta + y_1 \cos \theta
\end{aligned}
\tag{G.1}
$$

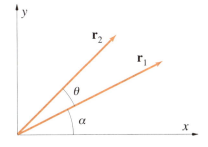

FIGURE G.1
An illustration of the rotation of a vector $\mathbf{r}$ through an angle θ.

Thus, we see that

$$C = BA = \begin{pmatrix} b_{11} & b_{12} \\ b_{21} & b_{22} \end{pmatrix} \begin{pmatrix} a_{11} & a_{12} \\ a_{21} & a_{22} \end{pmatrix} = \begin{pmatrix} b_{11}a_{11} + b_{12}a_{21} & b_{11}a_{12} + b_{12}a_{22} \\ b_{21}a_{11} + b_{22}a_{21} & b_{21}a_{12} + b_{22}a_{22} \end{pmatrix} \quad \text{(G.11)}$$

This result may look complicated, but it has a nice pattern that we will illustrate two ways. Mathematically, the ijth element of C is given by the formula

$$c_{ij} = \sum_k b_{ik}a_{kj} \quad \text{(G.12)}$$

Notice that we sum over the middle index. For example,

$$c_{11} = \sum_k b_{1k}a_{k1} = b_{11}a_{11} + b_{12}a_{21}$$

as in Equation G.11. A more pictorial way is to notice that any element in C can be obtained by multiplying elements in any row in B by the corresponding elements in any column in A, adding them, and then placing them in C where the row and column intersect. In terms of vectors, we take the dot product of the row of B and the column of A and place the result at their intersection. For example, c_{11} is obtained by multiplying the elements of row 1 of B with the elements of column 1 of A, or by the scheme

$$\rightarrow \begin{pmatrix} b_{11} & b_{12} \\ b_{21} & b_{22} \end{pmatrix} \begin{pmatrix} a_{11} & a_{12} \\ a_{21} & a_{22} \end{pmatrix} = \begin{pmatrix} b_{11}a_{11} + b_{12}a_{21} & \cdot \\ \cdot & \cdot \end{pmatrix}$$

and c_{12} by

$$\rightarrow \begin{pmatrix} b_{11} & b_{12} \\ b_{21} & b_{22} \end{pmatrix} \begin{pmatrix} a_{11} & a_{12} \\ a_{21} & a_{22} \end{pmatrix} = \begin{pmatrix} \cdot & b_{11}a_{12} + b_{12}a_{22} \\ \cdot & \cdot \end{pmatrix}$$

EXAMPLE G–2
Find $C = BA$ if

$$B = \begin{pmatrix} 1 & 2 & 1 \\ 3 & 0 & -1 \\ -1 & -1 & 2 \end{pmatrix} \quad \text{and} \quad A = \begin{pmatrix} -3 & 0 & -1 \\ 1 & 4 & 0 \\ 1 & 1 & 1 \end{pmatrix}$$

SOLUTION:

$$C = \begin{pmatrix} 1 & 2 & 1 \\ 3 & 0 & -1 \\ -1 & -1 & 2 \end{pmatrix} \begin{pmatrix} -3 & 0 & -1 \\ 1 & 4 & 0 \\ 1 & 1 & 1 \end{pmatrix}$$

$$= \begin{pmatrix} -3+2+1 & 0+8+1 & -1+0+1 \\ -9+0-1 & 0+0-1 & -3+0-1 \\ 3-1+2 & 0-4+2 & 1+0+2 \end{pmatrix}$$

$$= \begin{pmatrix} 0 & 9 & 0 \\ -10 & -1 & -4 \\ 4 & -2 & 3 \end{pmatrix}$$

EXAMPLE G–3

The matrix R given by Equation G.2 represents a rotation through the angle θ. Show that R^2 represents a rotation through an angle 2θ. In other words, show that R^2 represents two sequential applications of R.

SOLUTION:

$$R^2 = \begin{pmatrix} \cos\theta & -\sin\theta \\ \sin\theta & \cos\theta \end{pmatrix} \begin{pmatrix} \cos\theta & -\sin\theta \\ \sin\theta & \cos\theta \end{pmatrix}$$

$$= \begin{pmatrix} \cos^2\theta - \sin^2\theta & -2\sin\theta\cos\theta \\ 2\sin\theta\cos\theta & \cos^2\theta - \sin^2\theta \end{pmatrix}$$

Using standard trigonometric identities, we get

$$R^2 = \begin{pmatrix} \cos 2\theta & -\sin 2\theta \\ \sin 2\theta & \cos 2\theta \end{pmatrix}$$

which represents rotation through an angle 2θ.

Matrices do not have to be square to be multiplied together, but either Equation G.11 or the pictorial method illustrated above suggests that the number of columns of B must be equal to the number of rows of A. When this is so, A and B are said to be *compatible*. For example, Equations G.4 can be written in matrix form as

$$\begin{pmatrix} x_2 \\ y_2 \end{pmatrix} = \begin{pmatrix} a_{11} & a_{12} \\ a_{21} & a_{22} \end{pmatrix} \begin{pmatrix} x_1 \\ y_1 \end{pmatrix} \tag{G.13}$$

An important aspect of matrix multiplication is that BA does not necessarily equal AB. For example, if

$$\mathsf{A} = \begin{pmatrix} 0 & 2 \\ 1 & 0 \end{pmatrix} \quad \text{and} \quad \mathsf{B} = \begin{pmatrix} 3 & 0 \\ 0 & -1 \end{pmatrix}$$

then

$$\mathsf{AB} = \begin{pmatrix} 0 & 2 \\ 1 & 0 \end{pmatrix} \begin{pmatrix} 3 & 0 \\ 0 & -1 \end{pmatrix} = \begin{pmatrix} 0 & -2 \\ 3 & 0 \end{pmatrix}$$

and

$$\mathsf{BA} = \begin{pmatrix} 3 & 0 \\ 0 & -1 \end{pmatrix} \begin{pmatrix} 0 & 2 \\ 1 & 0 \end{pmatrix} = \begin{pmatrix} 0 & 6 \\ -1 & 0 \end{pmatrix}$$

and so $\mathsf{AB} \neq \mathsf{BA}$. If it does happen that $\mathsf{AB} = \mathsf{BA}$, then A and B are said to *commute*.

EXAMPLE G–4

Do the matrices A and B commute if

$$\mathsf{A} = \begin{pmatrix} 2 & 1 \\ 0 & 1 \end{pmatrix} \quad \text{and} \quad \mathsf{B} = \begin{pmatrix} 1 & 1 \\ 0 & 1 \end{pmatrix}$$

SOLUTION:

$$\mathsf{AB} = \begin{pmatrix} 2 & 3 \\ 0 & 1 \end{pmatrix}$$

and

$$\mathsf{BA} = \begin{pmatrix} 2 & 2 \\ 0 & 1 \end{pmatrix}$$

so they do not commute.

Another property of matrix multiplication that differs from ordinary scalar multiplication is that the equation

$$\mathsf{AB} = \mathsf{O}$$

where O is the zero matrix (all elements equal to zero) does not imply that A or B necessarily is a zero matrix. For example,

$$\begin{pmatrix} 1 & 1 \\ 2 & 2 \end{pmatrix} \begin{pmatrix} -1 & 1 \\ 1 & -1 \end{pmatrix} = \begin{pmatrix} 0 & 0 \\ 0 & 0 \end{pmatrix}$$

A linear transformation that leaves (x_1, y_1) unaltered is called the identity transformation, and the corresponding matrix is called the *identity matrix* or the *unit matrix*. All the elements of the unit matrix are equal to zero, except those along the diagonal, which equal one:

$$I = \begin{pmatrix} 1 & 0 & 0 & \cdots & 0 \\ 0 & 1 & 0 & \cdots & 0 \\ 0 & 0 & 1 & \cdots & 0 \\ \vdots & \vdots & \vdots & \ddots & \vdots \\ 0 & 0 & 0 & \cdots & 1 \end{pmatrix}$$

The elements of I are δ_{ij}, the Kronecker delta, which equals one when $i = j$ and zero when $i \neq j$. The unit matrix has the property that

$$IA = AI \tag{G.14}$$

The unit matrix is an example of a *diagonal matrix*. The only nonzero elements of a diagonal matrix are along its diagonal. Diagonal matrices are necessarily square matrices.

If $BA = AB = I$, then B is said to be the *inverse* of A, and is denoted by A^{-1}. Thus, A^{-1} has the property that

$$AA^{-1} = A^{-1}A = I \tag{G.15}$$

If A represents some transformation, then A^{-1} undoes that transformation and restores the original state. There are recipes for finding the inverse of a matrix, but we won't need them (see Problem G–9, however). Nevertheless, it should be clear on physical grounds that the inverse of R in Equation G.2 is

$$R^{-1} = R(-\theta) = \begin{pmatrix} \cos\theta & \sin\theta \\ -\sin\theta & \cos\theta \end{pmatrix} \tag{G.16}$$

which is obtained from R by replacing θ by $-\theta$. In other words, if $R(\theta)$ represents a rotation through an angle θ, then $R^{-1} = R(-\theta)$ and represents the reverse rotation. It is easy to show that R and R^{-1} satsify Equation G.15. Using Equations G.2 and G.16, we have

$$R^{-1}R = \begin{pmatrix} \cos\theta & \sin\theta \\ -\sin\theta & \cos\theta \end{pmatrix} \begin{pmatrix} \cos\theta & -\sin\theta \\ \sin\theta & \cos\theta \end{pmatrix}$$
$$= \begin{pmatrix} \cos^2\theta + \sin^2\theta & 0 \\ 0 & \cos^2\theta + \sin^2\theta \end{pmatrix}$$
$$= \begin{pmatrix} 1 & 0 \\ 0 & 1 \end{pmatrix}$$

and

$$RR^{-1} = \begin{pmatrix} \cos\theta & -\sin\theta \\ \sin\theta & \cos\theta \end{pmatrix} \begin{pmatrix} \cos\theta & \sin\theta \\ -\sin\theta & \cos\theta \end{pmatrix}$$

$$= \begin{pmatrix} \cos^2\theta + \sin^2\theta & 0 \\ 0 & \cos^2\theta + \sin^2\theta \end{pmatrix}$$

$$= \begin{pmatrix} 1 & 0 \\ 0 & 1 \end{pmatrix}$$

Most of the matrices that occur in quantum mechanics are *orthogonal*. The characteristic property of an orthogonal matrix is that both its columns and its rows form a set of orthogonal vectors. For example,

$$A = \frac{1}{9} \begin{pmatrix} 1 & 8 & -4 \\ 4 & -4 & -7 \\ 8 & 1 & 4 \end{pmatrix} \tag{G.17}$$

is an orthogonal matrix. The vectors that make up the rows, for example, are normalized,

$$\frac{1}{81}[1^2 + 8^2 + (-4)^2] = 1$$

$$\frac{1}{81}[4^2 + (-4)^2 + (-7)^2] = 1$$

and

$$\frac{1}{81}(8^2 + 1^2 + 4^2) = 1$$

and the dot products of its rows are

$$1 \times 4 + 8 \times (-4) + (-4)(-7) = 0$$
$$1 \times 8 + 8 \times 1 + (-4)(4) = 0$$
$$4 \times 8 + (-4)(1) + (-7)(4) = 0$$

The same is true for its columns.

It is very easy to find the inverse of an orthogonal matrix. First, we define the transpose of A, A^T, as the matrix that we obtain from A by interchanging rows and columns. The transpose of Equation G.17 is

$$A^T = \begin{pmatrix} 1 & 4 & 8 \\ 8 & -4 & 1 \\ -4 & -7 & 4 \end{pmatrix} \tag{G.18}$$

We obtain A^T from A by simply flipping it about its diagonal. In terms of the elements of A, we have $a_{ij}^T = a_{ji}$. If $A^T = A$, then the matrix is said to be *symmetric*. In terms of the elements of A, we have $a_{ij} = a_{ji}$. Most matrices in quantum mechanics are symmetric. It turns out that the inverse of an orthogonal matrix is equal to its transpose, or

$$A^{-1} = A^T \qquad \text{(orthogonal matrix)} \qquad \text{(G.19)}$$

EXAMPLE G–5

Show that A^T given by Equation G.18 is equal to A^{-1}.

SOLUTION:

$$A^T A = \frac{1}{81} \begin{pmatrix} 1 & 4 & 8 \\ 8 & -4 & 1 \\ -4 & -7 & 4 \end{pmatrix} \begin{pmatrix} 1 & 8 & -4 \\ 4 & -4 & -7 \\ 8 & 1 & 4 \end{pmatrix} = \begin{pmatrix} 1 & 0 & 0 \\ 0 & 1 & 0 \\ 0 & 0 & 1 \end{pmatrix}$$

Notice that the products of the rows of A^T into the columns of A are exactly the same as those of our original definition of an orthogonal matrix. Thus, $A^T = A^{-1}$ because both the rows and the columns of A form sets of orthogonal vectors. Orthogonal matrices correspond to rotations in space. When an orthogonal matrix A operates on a vector **v**, it simply rotates the vector.

We can associate a determinant with a square matrix by writing

$$\det A = |A| = \begin{vmatrix} a_{11} & a_{12} & \cdots & a_{1n} \\ a_{21} & a_{22} & \cdots & a_{2n} \\ \vdots & \vdots & \ddots & \vdots \\ a_{n1} & a_{n2} & \cdots & a_{nn} \end{vmatrix}$$

Thus, the determinant of R is

$$\begin{vmatrix} \cos\theta & -\sin\theta \\ \sin\theta & \cos\theta \end{vmatrix} = \cos^2\theta + \sin^2\theta = 1$$

and $\det R^{-1} = 1$ also. If $\det A = 0$, then A is said to be a *singular matrix*. Singular matrices do not have inverses.

A quantity that frequently arises in group theory is the sum of the diagonal elements of a matrix, called the *trace* of the matrix. Thus, the trace of the matrix

$$B = \begin{pmatrix} 1/2 & 0 & 1 \\ 0 & 2 & 1 \\ 1 & 1 & 1/2 \end{pmatrix}$$

is 3, which we write as Tr $B = 3$ (Problem G–15).

Problems

G–1. Given the two matrices

$$A = \begin{pmatrix} 1 & 0 & -1 \\ -1 & 2 & 0 \\ 0 & 1 & 1 \end{pmatrix} \quad \text{and} \quad B = \begin{pmatrix} -1 & 1 & 0 \\ 3 & 0 & 2 \\ 1 & 1 & 1 \end{pmatrix}$$

form the matrices $C = 2A - 3B$ and $D = 6B - A$.

G–2. Given the three matrices

$$A = \frac{1}{2} \begin{pmatrix} 0 & 1 \\ 1 & 0 \end{pmatrix} \qquad B = \frac{1}{2} \begin{pmatrix} 0 & -i \\ i & 0 \end{pmatrix} \qquad C = \frac{1}{2} \begin{pmatrix} 1 & 0 \\ 0 & -1 \end{pmatrix}$$

show that $A^2 + B^2 + C^2 = \frac{3}{4}I$, where I is a unit matrix. Also show that

$$AB - BA = iC$$

$$BC - CB = iA$$

$$CA - AC = iB$$

G–3. Given the matrices

$$A = \frac{1}{\sqrt{2}} \begin{pmatrix} 0 & 1 & 0 \\ 1 & 0 & 1 \\ 0 & 1 & 0 \end{pmatrix} \qquad B = \frac{1}{\sqrt{2}} \begin{pmatrix} 0 & -i & 0 \\ i & 0 & -i \\ 0 & i & 0 \end{pmatrix} \qquad C = \begin{pmatrix} 1 & 0 & 0 \\ 0 & 0 & 0 \\ 0 & 0 & -1 \end{pmatrix}$$

show that

$$AB - BA = iC$$

$$BC - CB = iA$$

$$CA - AC = iB$$

and

$$A^2 + B^2 + C^2 = 2I$$

where I is a unit matrix.

G–4. Do you see any similarity between the results of Problems G–2 and G–3 and the commutation relations involving the components of angular momentum?

G–5. A three-dimensional rotation about the z axis can be represented by the matrix

$$R = \begin{pmatrix} \cos\theta & -\sin\theta & 0 \\ \sin\theta & \cos\theta & 0 \\ 0 & 0 & 1 \end{pmatrix}$$

Show that

$$\det R = |R| = 1$$

Also show that

$$R^{-1} = R(-\theta) = \begin{pmatrix} \cos\theta & \sin\theta & 0 \\ -\sin\theta & \cos\theta & 0 \\ 0 & 0 & 1 \end{pmatrix}$$

G–6. Show that the matrix R in Problem G–5 is orthogonal.

G–7. Given the matrices

$$C_3 = \begin{pmatrix} -\frac{1}{2} & -\frac{\sqrt{3}}{2} \\ \frac{\sqrt{3}}{2} & -\frac{1}{2} \end{pmatrix} \qquad \sigma_v = \begin{pmatrix} 1 & 0 \\ 0 & -1 \end{pmatrix}$$

$$\sigma_v' = \begin{pmatrix} -\frac{1}{2} & \frac{\sqrt{3}}{2} \\ \frac{\sqrt{3}}{2} & \frac{1}{2} \end{pmatrix} \qquad \sigma_v'' = \begin{pmatrix} -\frac{1}{2} & -\frac{\sqrt{3}}{2} \\ -\frac{\sqrt{3}}{2} & \frac{1}{2} \end{pmatrix}$$

show that

$$\sigma_v C_3 = \sigma_v'' \qquad C_3\sigma_v = \sigma_v'$$

$$\sigma_v''\sigma_v' = C_3 \qquad C_3\sigma_v'' = \sigma_v$$

Calculate the determinant associated with each matrix. Calculate the trace of each matrix.

G–8. Which of the matrices in Problem G–7 are orthogonal?

G–9. The inverse of a matrix A can be found by using the following procedure:

1. Replace each element of A by its cofactor in the corresponding determinant (see MathChapter F for a definition of a cofactor).
2. Take the transpose of the matrix obtained in step 1.
3. Divide each element of the matrix obtained in step 2 by the determinant of A.

For example, if

$$A = \begin{pmatrix} 1 & 2 \\ 3 & 4 \end{pmatrix}$$

then det $A = -2$ and

$$A^{-1} = -\frac{1}{2}\begin{pmatrix} 4 & -2 \\ -3 & 1 \end{pmatrix}$$

Show that $AA^{-1} = A^{-1}A = I$. Use the above procedure to find the inverse of

$$A = \begin{pmatrix} \frac{1}{2} & \frac{1}{\sqrt{2}} \\ \frac{1}{\sqrt{2}} & 0 \end{pmatrix} \qquad \text{and} \qquad B = \begin{pmatrix} 0 & 2 & 3 \\ 1 & 1 & 1 \\ 2 & 0 & 1 \end{pmatrix}$$

G–10. Recall that a singular matrix is one whose determinant is equal to zero. Referring to the procedure in Problem G–9, do you see why a singular matrix has no inverse?

G–11. Consider the matrices A and S,

$$A = \begin{pmatrix} 1 & 0 & 1 \\ 0 & 1 & 0 \\ 1 & 0 & 1 \end{pmatrix} \qquad S = \begin{pmatrix} \frac{1}{\sqrt{2}} & 0 & \frac{1}{\sqrt{2}} \\ 0 & 1 & 0 \\ \frac{1}{\sqrt{2}} & 0 & -\frac{1}{\sqrt{2}} \end{pmatrix}$$

First, show that S is orthogonal. Then evaluate the matrix $D = S^{-1}AS = S^{T}AS$. What form does D have?

G–12. A matrix whose elements satisfy the relation $a_{ij} = a_{ji}^{*}$ is called *Hermitian*. You can think of a Hermitian matrix as a symmetric matrix in a complex space. Show that the eigenvalues of a Hermitian matrix are real. (Note the similarity between a Hermitian operator and a Hermitian matrix.) *Hint*: Start with $Hx_i = \lambda_i x_i$ and $H^{*}x_j^{*} = \lambda_j^{*}x_j^{*}$ and multiply the first equation from the left by x_j and the second from the left by x_i^{*} and then use the Hermitian property of H.

G–13. Show that $(AB)^{T} = B^{T}A^{T}$.

G–14. Show that $(AB)^{-1} = B^{-1}A^{-1}$.

G–15. Show that $\text{Tr } AB = \text{Tr } BA$.

G–16. Consider the simultaneous algebraic equations

$$x + y = 3$$
$$4x - 3y = 5$$

Show that this pair of equations can be written in the matrix form

$$Ax = c \tag{1}$$

where

$$x = \begin{pmatrix} x \\ y \end{pmatrix} \qquad c = \begin{pmatrix} 3 \\ 5 \end{pmatrix} \qquad \text{and} \qquad A = \begin{pmatrix} 1 & 1 \\ 4 & -3 \end{pmatrix}$$

Now multiply equation 1 from the left by A^{-1} to obtain

$$x = A^{-1}c \qquad (2)$$

Now show that

$$A^{-1} = -\frac{1}{7}\begin{pmatrix} -3 & -1 \\ -4 & 1 \end{pmatrix}$$

and that

$$x = -\frac{1}{7}\begin{pmatrix} -3 & -1 \\ -4 & 1 \end{pmatrix}\begin{pmatrix} 3 \\ 5 \end{pmatrix} = \begin{pmatrix} 2 \\ 1 \end{pmatrix}$$

or that $x = 2$ and $y = 1$. Do you see how this procedure generalizes to any number of simultaneous equations?

G–17. Solve the following simultaneous algebraic equations by the matrix inverse method developed in Problem G–16:

$$x + y - z = 1$$
$$2x - 2y + z = 6$$
$$x + 3z = 0$$

First, show that

$$A^{-1} = \frac{1}{13}\begin{pmatrix} 6 & 3 & 1 \\ 5 & -4 & 3 \\ -2 & -1 & 4 \end{pmatrix}$$

and evaluate $x = A^{-1}c$.

Douglas Hartree and Vladimir Fock formulated an approximate method for calculating atomic (and molecular) properties in the 1930s that is still used today.

Douglas Hartree was born on March 27, 1897, in Cambridge, England, and died in 1958. After receiving his Ph.D. in applied mathematics from the University of Cambridge in 1926, he spent the years 1929 to 1937 as the chair of applied mathematics and 1937 to 1946 as professor of theoretical physics at the University of Manchester. From 1946 until his death, he was Plummer Professor of Mathematical Physics at the University of Cambridge. Hartree pioneered the use of computers in research in the United Kingdom. He developed powerful methods of numerical analysis, which he applied to problems in atomic structure, ballistics, atmospheric physics, and hydrodynamics. Hartree was also an accomplished pianist and drummer.

Vladimir Fock (also Fok) was born on December 22, 1898, in Petrograd (later Leningrad and now St. Petersburg), Russia, and died there in 1974. After graduating from Petrograd University in 1922, he spent the years 1924 to 1936 at the Leningrad Institute of Physics and Technology. From 1936 to 1953 he was at the Institute of Physics, USSR Academy of Science; then he returned to Leningrad University, where he remained until his death. Fock generalized the equations of Hartree to include the fact that electronic wave functions must be antisymmetric under the interchange of any two electrons (the Pauli exclusion principle). Fock's research was in quantum electrodynamics, general relativity, and solid-state physics. He was almost completely deaf as an adult.

CHAPTER 8

Approximation Methods

We ended the previous chapter by saying that the Schrödinger equation cannot be solved exactly for any atom or molecule more complicated than a hydrogen atom. At first thought, this statement would appear to certainly deprive quantum mechanics of any interest to chemists, but, fortunately, approximation methods can be used to solve the Schrödinger equation to almost any desired accuracy. In this chapter, we will present the two most widely used of these methods, the variational method and perturbation theory. We will present the basic equations of the variational method and perturbation theory and then apply them to a variety of simple problems.

8.1 The Variational Method Provides an Upper Bound to the Ground-State Energy of a System

We will first illustrate the *variational method*. Consider the ground state of some arbitrary system. The ground-state wave function ψ_0 and energy E_0 satisfy the Schrödinger equation

$$\hat{H}\psi_0 = E_0\psi_0 \tag{8.1}$$

Multiply Equation 8.1 from the left by ψ_0^* and integrate over all space to obtain

$$E_0 = \frac{\int \psi_0^* \hat{H}\psi_0 \, d\tau}{\int \psi_0^* \psi_0 \, d\tau} = \frac{\langle \psi_0 \mid \hat{H} \mid \psi_0 \rangle}{\langle \psi_0 \mid \psi_0 \rangle} \tag{8.2}$$

where $d\tau$ represents the appropriate volume element. We have not set the denominator equal to unity in Equation 8.2 to allow for the possibility that ψ_0 is not normalized beforehand. A beautiful theorem says that if we substitute any other function ϕ for ψ_0

into Equation 8.2 and calculate the corresponding energy according to

$$E_\phi = \frac{\int \phi^* \hat{H} \phi \, d\tau}{\int \phi^* \phi \, d\tau} = \frac{\langle \phi \mid \hat{H} \mid \phi \rangle}{\langle \phi \mid \phi \rangle} \tag{8.3}$$

then E_ϕ will be greater than the ground-state energy E_0. In an equation, we have the *variational principle*

$$E_\phi \geq E_0 \tag{8.4}$$

where the equality holds only if $\phi = \psi_0$, the exact wave function. We will not prove the variational principle here (although it is fairly easy), but Problem 8–1 takes you through the proof step by step.

The variational principle says that we can calculate an upper bound to E_0 by using any trial function we wish. The closer ϕ is to ψ_0 in some sense, the closer E_ϕ will be to E_0. We can choose a trial function ϕ such that it depends upon some arbitrary parameters, $\alpha, \beta, \gamma, \ldots$, called *variational parameters*. The energy also will depend upon these variational parameters, and Equation 8.4 will read

$$E_\phi(\alpha, \beta, \gamma, \ldots) \geq E_0 \tag{8.5}$$

Now we can minimize E_ϕ with respect to each of the variational parameters and thus approach the exact ground-state energy E_0.

As a specific example, consider the ground state of a hydrogen atom. Although we know from Chapter 7 that we can solve this problem exactly, let's assume that we cannot and use the variational method to determine an approximate solution. We will then compare our variational result to the exact result. Because $l = 0$ in the ground state, the Hamiltonian operator is (cf. Equation 7.10)

$$\hat{H} = -\frac{\hbar^2}{2m_e r^2} \frac{d}{dr} \left(r^2 \frac{d}{dr} \right) - \frac{e^2}{4\pi \epsilon_0 r} \tag{8.6}$$

Even if we did not know the exact solution, we would expect that the wave function decays to zero with increasing r. Consequently, as a *trial function*, we will try a Gaussian function of the form $\phi(r) = e^{-\alpha r^2}$, where α is a variational parameter. By a straightforward calculation, we can show that (cf. Problem 8–2)

$$4\pi \int_0^\infty \phi^*(r) \hat{H} \phi(r) r^2 \, dr = \frac{3\hbar^2 \pi^{3/2}}{4\sqrt{2} m_e \alpha^{1/2}} - \frac{e^2}{4\epsilon_0 \alpha}$$

and that

$$4\pi \int_0^\infty \phi^*(r) \phi(r) r^2 \, dr = \left(\frac{\pi}{2\alpha} \right)^{3/2}$$

Therefore, from Equation 8.3,

$$E(\alpha) = \frac{3\hbar^2\alpha}{2m_e} - \frac{e^2\alpha^{1/2}}{2^{1/2}\epsilon_0\pi^{3/2}} \tag{8.7}$$

We now minimize $E(\alpha)$ with respect to α by differentiating $E(\alpha)$ with respect to α and setting the result equal to zero. We solve the equation

$$\frac{dE(\alpha)}{d\alpha} = \frac{3\hbar^2}{2m_e} - \frac{e^2}{(2\pi)^{3/2}\epsilon_0\alpha^{1/2}} = 0$$

for α to give

$$\alpha = \frac{m_e^2 e^4}{18\pi^3\epsilon_0^2\hbar^4} \tag{8.8}$$

as the value of α that minimizes $E(\alpha)$. Substituting Equation 8.8 back into Equation 8.7, we find that

$$E_{min} = -\frac{4}{3\pi}\left(\frac{m_e e^4}{16\pi^2\epsilon_0^2\hbar^2}\right) = -0.424\left(\frac{m_e e^4}{16\pi^2\epsilon_0^2\hbar^2}\right) \tag{8.9}$$

compared with the exact value (Equation 7.11)

$$E_0 = -\frac{1}{2}\left(\frac{m_e e^4}{16\pi^2\epsilon_0^2\hbar^2}\right) = -0.500\left(\frac{m_e e^4}{16\pi^2\epsilon_0^2\hbar^2}\right) \tag{8.10}$$

Note that $E_{min} > E_0$, as the variational theorem assures us.

The normalized trial function is given by $\phi(r) = (2\alpha/\pi)^{3/4}e^{-\alpha r^2}$, where α is given by Equation 8.8, and the exact ground-state wave function (the hydrogen 1s orbital) is given by $(\pi a_0^3)^{-1/2}e^{-r/a_0}$, where $a_0 = 4\pi\epsilon_0\hbar^2/m_e e^2$ is the Bohr radius. We can compare these two functions by first expressing α in terms of a_0, which comes out to be

$$\alpha = \frac{m_e^2 e^4}{18\pi^3\epsilon_0^2\hbar^4} = \frac{16}{18\pi}\cdot\frac{m_e^2 e^4}{16\pi^2\epsilon_0^2\hbar^4} = \frac{8}{9\pi}\cdot\frac{1}{a_0^2}$$

Thus, we can write the trial function as

$$\phi(r) = \frac{8}{3^{3/2}\pi}\left(\frac{1}{\pi a_0^3}\right)^{1/2}e^{-(8/9\pi)r^2/a_0^2}$$

This result is compared with ψ_{1s} in Figure 8.1.

Our variational calculation for the ground-state energy of a hydrogen atom is within 80% of the exact result. This result was obtained using a trial function with only one

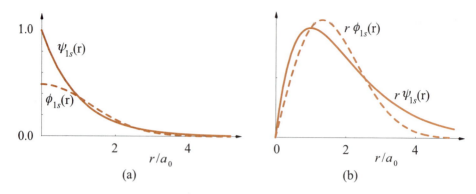

FIGURE 8.1
(a) A comparison of the optimized (normalized) Gaussian trial wave function $\phi_{1s}(r) = (2\alpha/\pi)^{3/4}e^{-\alpha r^2}$, where α is given by Equation 8.8 (dashed line), and the exact ground-state hydrogen wave function, $\psi_{1s}(r) = (1/\pi a_0^3)^{1/2}e^{-r/a_0}$, where $a_0 - 4\pi\epsilon_0\hbar^2/m_e e^2$ is the Bohr radius (solid line). Both functions are plotted against the reduced distance, r/a_0, and the vertical axis is expressed in units of $1/(\pi a_0^3)^{1/2}$. (b) Similar plots for $r\phi_{1s}(r)$ and $r\psi_{1s}(r)$.

variational parameter. We can obtain progressively better results by using more flexible trial functions, containing more parameters. In fact, we will see an example of such a progression (Table 8.1) that approaches the exact energy in Section 8.3.

EXAMPLE 8–1
Consider the ground state of a harmonic oscillator. Even if we had not solved this problem exactly previously, we might expect that the ground-state wave function would be symmetric about $x = 0$ because the potential energy is. As a trial function, then, try

$$\phi = \cos \lambda x \qquad -\frac{\pi}{2\lambda} < x < \frac{\pi}{2\lambda}$$

where λ is a variational parameter. Use this trial function, which is plotted in Figure 8.2, to calculate the ground-state energy of a harmonic oscillator.

SOLUTION: The Hamiltonian operator for a harmonic oscillator is

$$\hat{H} = -\frac{\hbar^2}{2\mu}\frac{d^2}{dx^2} + \frac{k}{2}x^2$$

and so the numerator in Equation 8.3 is

$$\text{numerator} = \int_{-\pi/2\lambda}^{\pi/2\lambda} \cos \lambda x \left(-\frac{\hbar^2}{2\mu}\frac{d^2}{dx^2} + \frac{k}{2}x^2\right)\cos \lambda x \, dx$$

$$= \frac{\pi\hbar^2\lambda}{4\mu} + \left(\frac{\pi^3}{48} - \frac{\pi}{8}\right)\frac{k}{\lambda^3}$$

and the denominator in Equation 8.3 is

$$\text{denominator} = \int_{-\pi/2\lambda}^{\pi/2\lambda} \cos^2 \lambda x \, dx = \frac{\pi}{2\lambda}$$

Equation 8.3 gives

$$E(\lambda) = \frac{\hbar^2 \lambda^2}{2\mu} + \left(\frac{\pi^2}{24} - \frac{1}{4} \right) \frac{k}{\lambda^2}$$

Now, minimizing $E(\lambda)$ with respect to λ gives

$$\lambda = \left[\left(\frac{\pi^2}{12} - \frac{1}{2} \right) \frac{k\mu}{\hbar^2} \right]^{1/4}$$

Substituting this value into $E(\lambda)$ gives

$$E_{\min} = \left(\frac{\pi^2}{3} - 2 \right)^{1/2} \frac{1}{2} \hbar \left(\frac{k}{\mu} \right)^{1/2} = (1.14) \frac{1}{2} \hbar \omega$$

compared to the exact value

$$E_0 = \frac{1}{2} \hbar \omega$$

Again, considering the simplicity of the trial function, we find quite good agreement with the exact result. Figure 8.2 compares the resulting variational wave function and the exact wave function for this case.

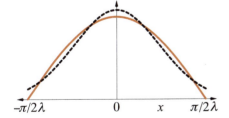

FIGURE 8.2
A comparison of the exact ground-state wave function (dotted curve) of a harmonic oscillator and a trial function of the form $\cos \lambda x$ with $-\pi/2\lambda < x < \pi/2\lambda$ (solid curve), used in Example 8–1, where λ is a variational parameter.

As a final example, we will use the variational method to estimate the ground-state energy of a helium atom. We saw at the end of Chapter 7 that the Hamiltonian operator for a helium atom is

$$\hat{H} = -\frac{\hbar^2}{2m_e} (\nabla_1^2 + \nabla_2^2) - \frac{2e^2}{4\pi \epsilon_0} \left(\frac{1}{r_1} + \frac{1}{r_2} \right) + \frac{e^2}{4\pi \epsilon_0} \frac{1}{r_{12}} \qquad (8.11)$$

The Schrödinger equation cannot be solved exactly for this system because of the term involving r_{12}. Equation 8.11 can be written in the form

$$\hat{H} = \hat{H}_{\mathrm{H}}(1) + \hat{H}_{\mathrm{H}}(2) + \frac{e^2}{4\pi\epsilon_0} \frac{1}{r_{12}} \tag{8.12}$$

where

$$\hat{H}_{\mathrm{H}}(j) = -\frac{\hbar^2}{2m_{\mathrm{e}}} \nabla_j^2 - \frac{2e^2}{4\pi\epsilon_0} \frac{1}{r_j} \qquad j = 1 \text{ and } 2 \tag{8.13}$$

is the Hamiltonian operator for a single electron around a helium nucleus. Thus, $\hat{H}_{\mathrm{H}}(1)$ and $\hat{H}_{\mathrm{H}}(2)$ satisfy the equation

$$\hat{H}_{\mathrm{H}}(j)\psi_{\mathrm{H}}(r_j, \theta_j, \phi_j) = E_j\psi_{\mathrm{H}}(r_j, \theta_j, \phi_j) \qquad j = 1 \text{ or } 2 \tag{8.14}$$

where $\psi_{\mathrm{H}}(r_j, \theta_j, \phi_j)$ is a hydrogen-like wave function with $Z = 2$ and where the E_j are given by (Problem 7–29)

$$E_j = -\frac{Z^2 m_{\mathrm{e}} e^4}{32\pi^2 \epsilon_0^2 \hbar^2 n_j^2} \qquad j = 1 \text{ or } 2 \tag{8.15}$$

with $Z = 2$. If we ignore the interelectronic repulsion term $(e^2/4\pi\epsilon_0 r_{12})$, then the Hamiltonian operator is separable and the ground-state wave function would be (Section 3.9)

$$\phi_0(\mathbf{r}_1, \mathbf{r}_2) = \psi_{1s}(\mathbf{r}_1)\psi_{1s}(\mathbf{r}_2) \tag{8.16}$$

where (Table 7.3)

$$\psi_{1s}(\mathbf{r}_j) = \left(\frac{Z^3}{\pi a_0}\right)^{1/2} e^{-Zr_j/a_0} \qquad j = 1 \text{ or } 2 \tag{8.17}$$

where $a_0 = 4\pi\epsilon_0\hbar^2/m_{\mathrm{e}}e^2$. We can use Equations 8.16 and 8.17 as a trial function using Z as a variational constant. Thus, we must evaluate

$$E(Z) = \int \phi_0(\mathbf{r}_1, \mathbf{r}_2)\hat{H}\phi_0(\mathbf{r}_1, \mathbf{r}_2) \, d\mathbf{r}_1 d\mathbf{r}_2 \tag{8.18}$$

with $\hat{H}$ given by Equation 8.12. The integral is a bit lengthy, albeit straightforward, to evaluate and is carried out step by step in Problem 8–18. The result is

$$E(Z) = \frac{m_{\mathrm{e}}e^4}{16\pi^2\epsilon_0^2\hbar^2} \left(Z^2 - \frac{27}{8}Z\right) \tag{8.19}$$

Equation 8.19 suggests that it is convenient to express E in units of $m_e e^4/16\pi^2\epsilon_0^2\hbar^2$, and so we can write it as

$$E(Z) = Z^2 - \frac{27}{8}Z \qquad (8.20)$$

If we minimize $E(Z)$ with respect to Z, we find that $Z_{min} = 27/16$. We substitute this result back into Equation 8.20 to obtain

$$E_{min} = -\left(\frac{27}{16}\right)^2 = -2.8477 \qquad (8.21)$$

compared with the most accurate calculated result of -2.9037 (in units of $m_e e^4/16\pi^2\epsilon_0^2\hbar^2$), which is in good agreement with the experimental result of -2.9033. (See Problem 8–19.) Thus, we achieve a fairly good result, considering the simplicity of the trial function.

The value of Z that minimizes E can be interpreted as an *effective nuclear charge*. The fact that Z comes out to be less than 2 reflects the fact that each electron partially screens the nucleus from the other, so that the net effective nuclear charge is reduced from 2 to 27/16.

8.2 A Trial Function That Depends Linearly on the Variational Parameters Leads to a Secular Determinant

As another example of the variational method, consider a particle in a one-dimensional box. Even without prior knowledge of the exact ground-state wave function, we should expect it to be symmetric about $x = a/2$ and to go to zero at the walls. One of the simplest functions with these properties is $x^n(a - x)^n$, where n is an integer. Consequently, let's estimate E_0 by using

$$\phi = c_1 x(a - x) + c_2 x^2(a - x)^2 \qquad (8.22)$$

as a trial function, where c_1 and c_2 are to be determined variationally—that is, where c_1 and c_2 are the variational parameters. If ϕ in Equation 8.22 is used as a trial function, we find after quite a lengthy but straightforward calculation that

$$E_{min} = 0.125\,002\frac{h^2}{ma^2} \qquad (8.23)$$

compared with

$$E_{exact} = \frac{h^2}{8ma^2} = 0.125\,000\frac{h^2}{ma^2}$$

So we see that using a trial function with more than one parameter can produce impressive results. The price we pay is a correspondingly more lengthy calculation. Fortunately, there is a systematic way to handle a trial function such as Equation 8.22. Note that Equation 8.22 is a linear combination of functions. Such a trial function can be written generally as

$$\phi = \sum_{n=1}^{N} c_n f_n \tag{8.24}$$

where the c_n are variational parameters and the f_n are arbitrary known functions (that at least satisfy the boundary conditions). We will use such a trial function often in later chapters. For simplicity, we will assume that $N = 2$ in Equation 8.22 and that the c_n and f_n are real. We relax these restrictions in Problem 8–24.

Consider

$$\phi = c_1 f_1 + c_2 f_2$$

Then,

$$\int \phi \hat{H} \phi \, d\tau = \int (c_1 f_1 + c_2 f_2) \hat{H}(c_1 f_1 + c_2 f_2) d\tau$$

$$= c_1^2 \int f_1 \hat{H} f_1 d\tau + c_1 c_2 \int f_1 \hat{H} f_2 d\tau + c_1 c_2 \int f_2 \hat{H} f_1 d\tau + c_2^2 \int f_2 \hat{H} f_2 d\tau$$

$$= c_1^2 H_{11} + c_1 c_2 H_{12} + c_1 c_2 H_{21} + c_2^2 H_{22} \tag{8.25}$$

where the H_{ij} are given by

$$H_{ij} = \int f_i \hat{H} f_j \, d\tau = \langle i \mid \hat{H} \mid j \rangle \tag{8.26}$$

Because $\hat{H}$ is Hermitian, the H_{ij} are symmetric; in other words, $H_{ij} = H_{ji}$. Using this result, Equation 8.25 becomes

$$\int \phi \hat{H} \phi \, d\tau = c_1^2 H_{11} + 2c_1 c_2 H_{12} + c_2^2 H_{22} \tag{8.27}$$

Similarly, we have

$$\int \phi^2 d\tau = c_1^2 S_{11} + 2c_1 c_2 S_{12} + c_2^2 S_{22} \tag{8.28}$$

where

$$S_{ij} = S_{ji} = \int f_i f_j \, d\tau = \langle i \mid j \rangle \tag{8.29}$$

The quantities H_{ij} and S_{ij} in Equations 8.26 and 8.29 are called *matrix elements*. By substituting Equations 8.27 and 8.28 into Equation 8.3, we find

$$E(c_1, c_2) = \frac{c_1^2 H_{11} + 2c_1 c_2 H_{12} + c_2^2 H_{22}}{c_1^2 S_{11} + 2c_1 c_2 S_{12} + c_2^2 S_{22}} \tag{8.30}$$

where we emphasize here that E is a function of the variational parameters c_1 and c_2.

Before differentiating $E(c_1, c_2)$ in Equation 8.30 with respect to c_1 and c_2, it is convenient to write it in the form

$$E(c_1, c_2)(c_1^2 S_{11} + 2c_1 c_2 S_{12} + c_2^2 S_{22}) = c_1^2 H_{11} + 2c_1 c_2 H_{12} + c_2^2 H_{22} \tag{8.31}$$

If we differentiate Equation 8.31 with respect to c_1, we find that

$$(2c_1 S_{11} + 2c_2 S_{12})E + \frac{\partial E}{\partial c_1}(c_1^2 S_{11} + 2c_1 c_2 S_{12} + c_2^2 S_{22}) = 2c_1 H_{11} + 2c_2 H_{12} \tag{8.32}$$

Because we are minimizing E with respect to c_1, $\partial E / \partial c_1 = 0$ and so Equation 8.32 becomes

$$c_1(H_{11} - E S_{11}) + c_2(H_{12} - E S_{12}) = 0 \tag{8.33}$$

Similarly, by differentiating $E(c_1, c_2)$ with respect to c_2 instead of c_1, we find

$$c_1(H_{12} - E S_{12}) + c_2(H_{22} - E S_{22}) = 0 \tag{8.34}$$

Equations 8.33 and 8.34 constitute a pair of linear algebraic equations for c_1 and c_2. There is a nontrivial solution—that is, a solution that is not simply $c_1 = c_2 = 0$, if and only if the determinant of the coefficients vanishes (MathChapter F), or if and only if

$$\begin{vmatrix} H_{11} - E S_{11} & H_{12} - E S_{12} \\ H_{12} - E S_{12} & H_{22} - E S_{22} \end{vmatrix} = 0 \tag{8.35}$$

Thus, we obtain a secular determinant and a secular equation (MathChapter F). The quadratic secular equation gives two values for E, and we take the smaller of the two as our variational approximation for the ground-state energy.

To illustrate the use of Equation 8.35, let's go back to solving the problem of a particle in a one-dimensional box variationally using Equation 8.22 as a trial function. For convenience, we will set $a = 1$. In this case,

$$f_1 = x(1 - x) \qquad \text{and} \qquad f_2 = x^2(1 - x)^2 \tag{8.36}$$

and the matrix elements (see Equations 8.26 and 8.29) are (see Problem 8–20)

$$H_{11} = \frac{\hbar^2}{6m} \qquad\qquad S_{11} = \frac{1}{30}$$

$$H_{12} = H_{21} = \frac{\hbar^2}{30m} \qquad S_{12} = S_{21} = \frac{1}{140} \qquad\qquad (8.37)$$

$$H_{22} = \frac{\hbar^2}{105m} \qquad\qquad S_{22} = \frac{1}{630}$$

EXAMPLE 8–2

Show explicitly that $H_{12} = H_{21}$ in Equation 8.37.

SOLUTION: Using the Hamiltonian operator of a particle in a box, we have

$$H_{12} = \int_0^1 f_1 \hat{H} f_2 \, dx = \langle 1 | \hat{H} | 2 \rangle$$

$$= \int_0^1 x(1-x) \left[-\frac{\hbar^2}{2m} \frac{d^2}{dx^2} x^2 (1-x)^2 \right] dx$$

$$= -\frac{\hbar^2}{2m} \int_0^1 x(1-x)(2 - 12x + 12x^2) \, dx$$

$$= -\frac{\hbar^2}{2m} \left(-\frac{1}{15} \right) = \frac{\hbar^2}{30m}$$

Similarly,

$$H_{21} = \int_0^1 f_2 \hat{H} f_1 \, dx = \langle 2 | \hat{H} | 1 \rangle$$

$$= \int_0^1 x^2 (1-x)^2 \left[-\frac{\hbar^2}{2m} \frac{d^2}{dx^2} x(1-x) \right] dx$$

$$= -\frac{\hbar^2}{2m} \int_0^1 x^2 (1-x)^2 (-2) dx$$

$$= -\frac{\hbar^2}{2m} \left(-\frac{1}{15} \right) = \frac{\hbar^2}{30m}$$

Substituting the values of the matrix elements H_{ij} and S_{ij} given by Equations 8.37 into the secular determinant (Equation 8.35) gives

$$\begin{vmatrix} \dfrac{1}{6} - \dfrac{\varepsilon}{30} & \dfrac{1}{30} - \dfrac{\varepsilon}{140} \\[2ex] \dfrac{1}{30} - \dfrac{\varepsilon}{140} & \dfrac{1}{105} - \dfrac{\varepsilon}{630} \end{vmatrix} = 0$$

where $\varepsilon = Em/\hbar^2$. The corresponding secular equation is

$$\varepsilon^2 - 56\varepsilon + 252 = 0$$

whose roots are

$$\varepsilon = \frac{56 \pm \sqrt{2128}}{2} = 51.065 \quad \text{and} \quad 4.93487$$

We choose the smaller root and obtain

$$E_{min} = 4.93487\frac{\hbar^2}{m} = 0.125\,002\frac{\hbar^2}{m}$$

compared with (recall that $a = 1$)

$$E_{exact} = \frac{h^2}{8m} = 0.125\,000\frac{h^2}{m}$$

The excellent agreement here is better than should be expected normally for such a simple trial function. Note that $E_{min} > E_{exact}$, as it must be.

EXAMPLE 8–3
Determine the normalized trial function for our variational treatment of a particle in a box.

SOLUTION: To determine the normalized trial function, we must determine c_1 and c_2 in Equation 8.22. These quantities are given by Equations 8.33 and 8.34, which are the two algebraic equations that lead to the secular determinant (Equation 8.35). These two equations are not independent of each other, so we will use the first to calculate the ratio c_2/c_1:

$$\frac{c_2}{c_1} = -\frac{H_{11} - ES_{11}}{H_{12} - ES_{12}} = -\frac{\dfrac{\hbar^2}{6m} - \left(4.93487\dfrac{\hbar^2}{m}\right)\dfrac{1}{30}}{\dfrac{\hbar^2}{30m} - \left(4.93487\dfrac{\hbar^2}{m}\right)\dfrac{1}{140}} = 1.13342$$

or $c_2 = 1.13342\,c_1$. So far, then, we have

$$\phi(x) = c_1[\,x(1 - x) + 1.13342\,x^2(1 - x)^2\,]$$

We now determine c_1 by requiring $\phi(x)$ to be normalized.

$$\int_0^1 \phi^2(x)\,dx = c_1^2 \int_0^1 [x^2(1-x)^2 + 2.26684\,x^3(1-x)^3 + 1.28464\,x^4(1-x)^4]\,dx = 1$$

Instead of expanding out each integral, it is more convenient to use (*CRC Standard Mathematical Tables and Formulae*)

$$\int_0^1 x^m(1-x)^n dx = \frac{m!\,n!}{(m+n+1)!}$$

in which case

$$\int_0^1 \phi^2(x)\,dx = c_1^2 \left(\frac{2!\,2!}{5!} + 2.26684\frac{3!\,3!}{7!} + 1.28464\frac{4!\,4!}{9!}\right) = 0.051\,5642\,c_1^2 = 1$$

giving us $c_1 = 4.40378$. Thus, the normalized trial function is

$$\phi(x) = 4.40378\,x(1-x) + 4.99133\,x^2(1-x)^2$$

Figure 8.3 compares $\phi(x)$ with the exact ground-state particle-in-a-box wave function (with $a = 1$), $\psi_1(x) = 2^{1/2}\sin \pi x$.

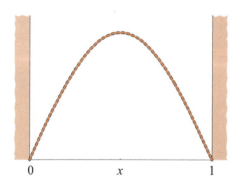

FIGURE 8.3
A comparison of the optimized and normalized trial function determined in Example 8–3 (black dotted curve) with the exact ground-state particle-in-a-box wave function, $\psi_1(x) = 2^{1/2}\sin \pi x$ (solid curve). The two curves are essentially the same.

You may wonder about the physical meaning of the other root to Equation 8.35. It turns out that it is an upper bound to the energy of the first excited state of a particle in a box. The value we calculated above is $1.2935h^2/m$, compared with the exact value of $4h^2/8ma^2$, or $0.5000h^2/m$. Thus, we see that although the second root is an upper bound to E_2, it is a fairly crude one. Although there are methods to give better upper bounds to excited-state energies, we will restrict ourselves to a determination of only the ground-state energy.

If we use a linear combination of N functions as in Equation 8.24, instead of using a linear combination of two functions as we have done so far, then we obtain N simultaneous linear algebraic equations for the c_j's:

$$c_1(H_{11} - ES_{11}) + c_2(H_{21} - ES_{21}) + \cdots + c_N(H_{1N} - ES_{1N}) = 0$$

$$c_1(H_{12} - ES_{12}) + c_2(H_{22} - ES_{22}) + \cdots + c_N(H_{2N} - ES_{2N}) = 0$$

$$\vdots \qquad\qquad \vdots \qquad\qquad \vdots$$

$$c_1(H_{1N} - ES_{1N}) + c_2(H_{2N} - ES_{2N}) + \cdots + c_N(H_{NN} - ES_{NN}) = 0$$

We can express these equations compactly by using the matrix notation

$$\mathsf{H}\,\mathsf{c} = E\,\mathsf{S}\,\mathsf{c} \tag{8.38}$$

where H is an $N \times N$ matrix with matrix elements H_{ij}, S is an $N \times N$ matrix with matrix elements S_{ij}, and c is an $N \times 1$ column matrix whose elements are c_j. To have a nontrivial solution to this set of homogeneous equations, we must have

$$
\begin{vmatrix}
H_{11} - ES_{11} & H_{12} - ES_{12} & \cdots & H_{1N} - ES_{1N} \\
H_{12} - ES_{12} & H_{22} - ES_{22} & \cdots & H_{2N} - ES_{2N} \\
\vdots & \vdots & \ddots & \vdots \\
H_{1N} - ES_{1N} & H_{2N} - ES_{2N} & \cdots & H_{NN} - ES_{NN}
\end{vmatrix} = 0
$$

or

$$|\mathsf{H} - E\mathsf{S}| = 0 \tag{8.39}$$

in matrix notation. In writing these equations, we have used the fact that $\hat{H}$ is a Hermitian operator, so $H_{ij} = H_{ji}$. The secular equation associated with this secular determinant is an Nth-order polynomial in E. We choose the smallest root of the Nth-order secular equation as an approximation to the ground-state energy. The determination of the smallest root must usually be done numerically for values of N larger than two. This is actually a standard numerical problem, and a number of packaged computer programs do this.

Once the smallest root of Equation 8.39 has been determined, we can substitute it back into Equations 8.38 to determine the c_j's. As in Example 8–3, only $N - 1$ of these equations are independent, and so we can use them to determine only the ratios $c_2/c_1, c_3/c_1, \ldots, c_N/c_1$, for example. We can then determine c_1 by requiring that the trial function ϕ be normalized, as we did in Example 8–3.

EXAMPLE 8–4
Use the variational method to calculate the ground-state energy of a particle in a box with a potential given by

$$
V(x) = \begin{cases}
\infty & x < 0 \\
\dfrac{V_0 x}{a} & 0 < x < a \\
\infty & x > a
\end{cases}
$$

Because the potential due to gravity is $V(x) = mgx$, where x is the height above the earth's surface, this problem is sometimes referred to as a particle in a gravitational well. The boundary conditions are the same as that for a particle in a box: $\phi(0) = \phi(a) = 0$. This suggests that we might use a linear combination of the first two particle-in-a-box wave functions as the trial function.

SOLUTION: For simplicity, we use

$$\phi(x) = c_1 \left(\frac{2}{a}\right)^{1/2} \sin \frac{\pi x}{a} + c_2 \left(\frac{2}{a}\right)^{1/2} \sin \frac{2\pi x}{a}$$

The matrix elements are

$$H_{11} = \left\langle \psi_1^{(0)} \left| \hat{H}^{(0)} + \frac{V_0 x}{a} \right| \psi_1^{(0)} \right\rangle$$

$$= \frac{h^2}{8ma^2} + \frac{2V_0}{a^2} \int_0^a dx\, x \sin^2 \frac{\pi x}{a} = \frac{h^2}{8ma^2} + \frac{V_0}{2}$$

$$H_{22} = \frac{4h^2}{8ma^2} + \frac{2V_0}{a^2} \int_0^a dx\, x \sin^2 \frac{2\pi x}{a} = \frac{4h^2}{8ma^2} + \frac{V_0}{2}$$

$$H_{12} = H_{21} = \frac{2V_0}{a^2} \int_0^a dx\, x \sin \frac{\pi x}{a} \sin \frac{2\pi x}{a} = -\frac{16V_0}{9\pi^2}$$

$$S_{11} = S_{22} = 1 \qquad \text{and} \qquad S_{12} = 0$$

The secular determinant is

$$\begin{vmatrix} \dfrac{h^2}{8ma^2} + \dfrac{V_0}{2} - E & -\dfrac{16V_0}{9\pi^2} \\[3mm] -\dfrac{16V_0}{9\pi^2} & \dfrac{4h^2}{8ma^2} + \dfrac{V_0}{2} - E \end{vmatrix} = 0$$

Let $\varepsilon = 8ma^2 E/h^2$ and $v_0 = 8ma^2 V_0/h^2$ to write this secular determinant as

$$\begin{vmatrix} 1 + \dfrac{v_0}{2} - \varepsilon & -\dfrac{16v_0}{9\pi^2} \\[3mm] -\dfrac{16v_0}{9\pi^2} & 4 + \dfrac{v_0}{2} - \varepsilon \end{vmatrix} = 0$$

which leads to the quadratic equation

$$\varepsilon^2 - (5 + v_0)\varepsilon + \left(1 + \frac{v_0}{2}\right)\left(4 + \frac{v_0}{2}\right) - \frac{256v_0^2}{81\pi^4} = 0$$

or

$$\varepsilon = \frac{5 + v_0}{2} \pm \frac{1}{2}\left[9 + \left(\frac{32v_0}{9\pi^2}\right)^2\right]^{1/2}$$

The negative sign gives the smaller value of ε.

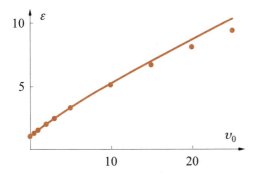

FIGURE 8.4
A plot of the energy of a particle in a gravitational well calculated variationally (solid curve) (Example 8–4) and the exact energy (dots) against v_0, the magnitude of the potential at $x = a$.

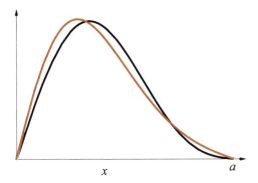

FIGURE 8.5
The variational wave function for $v_0 = 10$ determined from Example 8–4 (orange) and the exact wave function (black) plotted against x.

This problem can actually be solved exactly in terms of Bessel functions [see Langhoff, P. W., Schrödinger Particle in a Gravitational Well, *Am. J. Phys.*, **39**, 954, (1971)], and Figure 8.4 compares the variational energy ε and the exact energy by plotting both against v_0. Note that the variational energy lies above the exact energy, as you should expect. Problem 8–25 has you determine the variational wave function for the case $v_0 = 10$. Figure 8.5 compares this result to the exact wave function.

Problems 8–26, 8–27, and 8–29 use different trial functions to solve this problem.

8.3 Trial Functions Can Be Linear Combinations of Functions That Also Contain Variational Parameters

It is a fairly common practice to use a trial function of the form

$$\phi = \sum_{j=1}^{N} c_j f_j$$

TABLE 8.1
The Ground-State Energy of a Hydrogen
Atom Using a Trial Function of the Form
$\phi = \sum_{j=1}^{N} c_j e^{-\alpha_j r^2}$, Where the c_j and the α_j
Are Treated as Variational Parameters

N	$E_{min}/(m_e e^4/16\pi^2\epsilon_0^2\hbar^2)$
1	$-0.424\,413$
2	$-0.485\,813$
3	$-0.496\,967$
4	$-0.499\,276$
5	$-0.499\,76$
6	$-0.499\,88$
8	$-0.499\,92$
16	$-0.499\,98$

where the f_j themselves contain variational parameters. An example of such a trial function for the hydrogen atom is

$$\phi = \sum_{j=1}^{N} c_j e^{-\alpha_j r^2}$$

where the c_j's and the α_j's are treated as variational parameters. We have seen in Section 8.1 that the use of one term gives an energy $-0.424(m_e e^4/16\pi^2\epsilon_0^2\hbar^2)$ compared with the exact value of $-0.500(m_e e^4/16\pi^2\epsilon_0^2\hbar^2)$. Table 8.1 shows the results for taking more terms. We can see that the exact value is approached as N increases. Realize in this case, however, that we do not obtain a simple secular determinant, because ϕ is linear only in the c_j but not in the α_j. The minimization of E with respect to the c_j and α_j is fairly complicated, involving $2N$ parameters, and must be done numerically. Fortunately, a number of readily available computer programs can be used to do this (Problem 8–30).

8.4 Perturbation Theory Expresses the Solution to One Problem in Terms of Another Problem That Has Been Solved Previously

Suppose we wish to solve the Schrödinger equation

$$\hat{H}\psi = E\psi$$

for some particular system, but we are unable to find an exact solution as we have done for a harmonic oscillator, a rigid rotator, and a hydrogen atom in previous chapters.

It turns out that most systems cannot be solved exactly; some specific examples are a helium atom, an anharmonic oscillator, and a nonrigid rotator.

We saw at the end of Chapter 7 that the Hamiltonian operator for a helium atom is

$$\hat{H} = -\frac{\hbar^2}{2m_e}(\nabla_1^2 + \nabla_2^2) - \frac{2e^2}{4\pi\epsilon_0}\left(\frac{1}{r_1} + \frac{1}{r_2}\right) + \frac{e^2}{4\pi\epsilon_0}\frac{1}{r_{12}} \tag{8.40}$$

Equation 8.40 can be written in the form

$$\hat{H} = \hat{H}_H(1) + \hat{H}_H(2) + \frac{e^2}{4\pi\epsilon_0}\frac{1}{r_{12}} \tag{8.41}$$

where

$$\hat{H}_H(j) = -\frac{\hbar^2}{2m_e}\nabla_j^2 - \frac{2e^2}{4\pi\epsilon_0}\frac{1}{r_j} \qquad j = 1 \text{ and } 2 \tag{8.42}$$

is the Hamiltonian operator for a single electron around a helium nucleus. Thus, $\hat{H}_H(1)$ and $\hat{H}_H(2)$ satisfy the equation

$$\hat{H}_H(j)\psi_H(r_j, \theta_j, \phi_j) = E_j\psi_H(r_j, \theta_j, \phi_j) \qquad j = 1 \text{ and } 2 \tag{8.43}$$

where $\psi_H(r_j, \theta_j, \phi_j)$ is a hydrogen-like wave function with $Z = 2$ (Table 7.5) and where the E_j are given by (Problem 7–29)

$$E_j = -\frac{Z^2 m_e e^4}{8\epsilon_0 h^2 n_j^2} \qquad j = 1 \text{ and } 2 \tag{8.44}$$

with $Z = 2$. Notice that if it were not for the interelectronic repulsion term $e^2/4\pi\epsilon_0 r_{12}$ in Equation 8.41, the Hamiltonian operator for a helium atom would be separable and the helium atomic wave functions would be products of hydrogen-like wave functions (Section 3.9).

Another example of a problem that could be solved readily if it were not for additional terms in the Hamiltonian operator is an anharmonic oscillator. Recall that the harmonic-oscillator potential arises naturally as the first term in a Taylor expansion of a general potential about the equilibrium nuclear separation. Consider an anharmonic oscillator whose potential energy is given by

$$V(x) = \frac{1}{2}kx^2 + \frac{1}{6}\gamma_3 x^3 + \frac{1}{24}\gamma_4 x^4 \tag{8.45}$$

The Hamiltonian operator is

$$\hat{H} = -\frac{\hbar^2}{2\mu}\frac{d^2}{dx^2} + \frac{1}{2}kx^2 + \frac{1}{6}\gamma_3 x^3 + \frac{1}{24}\gamma_4 x^4 \tag{8.46}$$

We substitute Equations 8.55 and 8.56 into Equation 8.51 to obtain

$$(\hat{H}^{(0)} + \hat{H}^{(1)})(\psi_n^{(0)} + \lambda\psi_n^{(1)} + \lambda^2\psi_n^{(2)} + \cdots) = (E_n^{(0)} + \lambda E_n^{(1)} + \lambda^2 E_n^{(2)} + \cdots)$$
$$\times (\psi_n^{(0)} + \lambda\psi_n^{(1)} + \lambda^2\psi_n^{(2)} + \cdots)$$

Each side of this equation is an expansion in λ, which can be written as (Problem 8–34)

$$(\hat{H}^{(0)}\psi_n^{(0)} - E_n^{(0)}\psi_n^{(0)}) + (\hat{H}^{(0)}\psi_n^{(1)} + \hat{H}^{(1)}\psi_n^{(0)} - E_n^{(0)}\psi_n^{(1)} - E_n^{(1)}\psi_n^{(0)})\lambda$$
$$+ (\hat{H}^{(0)}\psi_n^{(2)} + \hat{H}^{(1)}\psi_n^{(1)} - E_n^{(0)}\psi_n^{(2)} - E_n^{(1)}\psi_n^{(1)} - E_n^{(2)}\psi_n^{(0)})\lambda^2 + O(\lambda^3) = 0 \quad (8.57)$$

where $O(\lambda^3)$ means terms of order λ^3 and higher. Notice that both terms in the first set of parentheses, the coefficient of λ^0, are of zero order; all four terms in the second set of parentheses, the coefficient of λ^1, are of first order; and so on.

Because λ is an arbitrary parameter, the coefficients of each power of λ must equal zero separately for Equation 8.57 to hold. The terms in the first set of parentheses, the coefficient of λ^0, cancel because of Equation 8.53. Let's look at the coefficient of λ^1:

$$\hat{H}^{(0)}\psi_n^{(1)} + \hat{H}^{(1)}\psi_n^{(0)} = E_n^{(0)}\psi_n^{(1)} + E_n^{(1)}\psi_n^{(0)} \quad (8.58)$$

Equation 8.58 can be simplified considerably by multiplying both sides from the left by $\psi_n^{(0)*}$ and integrating over all space. By doing this and then rearranging slightly, we get

$$\int \psi_n^{(0)*}(\hat{H}^{(0)} - E_n^{(0)})\psi_n^{(1)} \, d\tau + \int \psi_n^{(0)*}\hat{H}^{(1)}\psi_n^{(0)} \, d\tau = E_n^{(1)} \int \psi_n^{(0)*}\psi_n^{(0)} \, d\tau \quad (8.59)$$

It is convenient (and economical) at this stage to use the bracket notation and write Equation 8.59 as

$$\langle\psi_n^{(0)}|\hat{H}^{(0)} - E_n^{(0)}|\psi_n^{(1)}\rangle + \langle\psi_n^{(0)}|\hat{H}^{(1)}|\psi_n^{(0)}\rangle = E_n^{(1)}\langle\psi_n^{(0)}|\psi_n^{(0)}\rangle \quad (8.60)$$

The integral in the last term in Equation 8.60 is unity because we take $\psi_n^{(0)}$ to be normalized. More importantly, however, the first term on the left side is equal to zero. To see this, remember that $\hat{H}^{(0)} - E^{(0)}$ is Hermitian, and so we have

$$\langle\psi_n^{(0)}|\hat{H}^{(0)} - E_n^{(0)}|\psi_n^{(1)}\rangle = \langle(\hat{H}^{(0)} - E_n^{(0)})\psi_n^{(0)}|\psi_n^{(1)}\rangle = 0$$

because of Equation 8.53. Thus, Equation 8.60 becomes

$$E_n^{(1)} = \langle\psi_n^{(0)}|\hat{H}^{(1)}|\psi_n^{(0)}\rangle = \int d\tau \, \psi_n^{(0)*}\hat{H}^{(1)}\psi_n^{(0)} \quad (8.61)$$

Equation 8.61 gives $E_n^{(1)}$, the *first-order correction* to $E_n^{(0)}$. Through first order, then, the energy is given by

$$E_n = E_n^{(0)} + E_n^{(1)} = E_n^{(0)} + \langle\psi_n^{(0)}|\hat{H}^{(1)}|\psi_n^{(0)}\rangle \quad \text{(first order)} \quad (8.62)$$

Let's use Equation 8.62 to calculate the energy through first order for the potential,

$$V(x) = \begin{cases} \infty & x < 0 \\ \dfrac{V_0 x}{a} & 0 < x < a \\ \infty & x > a \end{cases} \tag{8.63}$$

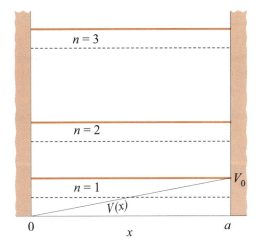

FIGURE 8.6
The potential of a particle in a gravitational well. The unperturbed energies (dashed lines) and the first-order energies (solid lines) are shown. The perturbation potential is $V_0 x/a$, which represents the gravitational potential.

This potential represents a particle in a gravitational well (see Figure 8.6). We discussed this system variationally in Example 8–4. In attempting to solve a problem by perturbation theory, the first and most important step is to formulate the problem into an unperturbed part and a perturbation, or, in other words, to recognize what the unperturbed problem might be. In this case, the unperturbed problem is a particle in a box, and so

$$\psi_n^{(0)}(x) = \left(\frac{2}{a}\right)^{1/2} \sin \frac{n\pi x}{a} \qquad n = 1, 2, 3, \ldots$$
$$E_n^{(0)} = \frac{n^2 h^2}{8ma^2} \qquad n = 1, 2, 3, \ldots \tag{8.64}$$

and

$$\hat{H}^{(1)} = \frac{V_0 x}{a} \qquad 0 < x < a \tag{8.65}$$

Using Equation 8.61,

$$E_n^{(1)} = \left\langle \psi_n^{(0)} \left| \frac{V_0 x}{a} \right| \psi_n^{(0)} \right\rangle = \frac{2V_0}{a^2} \int_0^a dx \, x \sin^2 \frac{n\pi x}{a}$$
$$= \frac{V_0}{2} \qquad n = 1, 2, \ldots \tag{8.66}$$

The ground-state energy through first order is

$$E_1 = E_1^{(0)} + E_1^{(1)} = \frac{h^2}{8ma^2} + \frac{V_0}{2} \tag{8.67}$$

It is convenient to introduce units such that $\varepsilon = 8m E a^2 / h^2 a^2$ and $v_0 = 8m V_0 a^2 / h^2$, in which case Equation 8.67 becomes

$$\varepsilon_1 = 1 + \frac{v_0}{2} \tag{8.68}$$

Note that Equation 8.68 is the same as the result obtained for the ground state in Example 8–4 through first order in v_0 (Problem 8–34). Figure 8.6 shows the potential well, the unperturbed energies, and their first-order corrections.

EXAMPLE 8–5

Calculate the first-order correction to the ground-state energy of an anharmonic oscillator whose potential is

$$V(x) = \frac{1}{2}kx^2 + \frac{1}{6}\gamma_3 x^3 + \frac{1}{24}\gamma_4 x^4$$

SOLUTION: In this case, the unperturbed system is a harmonic oscillator and the perturbation is

$$\hat{H}^{(1)} = \frac{1}{6}\gamma_3 x^3 + \frac{1}{24}\gamma_4 x^4$$

The ground-state wave function of a harmonic oscillator is

$$\psi_0(x) = \left(\frac{\alpha}{\pi}\right)^{1/4} e^{-\alpha x^2/2}$$

where $\alpha = (k\mu/\hbar^2)^{1/2}$. The first-order correction to the ground-state energy is given by

$$E_0^{(1)} = \left(\frac{\alpha}{\pi}\right)^{1/2} \int_{-\infty}^{\infty} \left(\frac{1}{6}\gamma_3 x^3 + \frac{1}{24}\gamma_4 x^4\right) e^{-\alpha x^2} dx$$

The integral involving $\gamma_3 x^3/6$ here vanishes because the integrand is an odd function of x. The remaining integral is

$$E_0^{(1)} = \left\langle 0 \left| \frac{1}{24}\gamma_4 x^4 \right| 0 \right\rangle = \frac{\gamma_4}{12} \left(\frac{\alpha}{\pi}\right)^{1/2} \int_0^{\infty} x^4 e^{-\alpha x^2} dx$$

The integral here can be found in tables (see also Problem 5–23) and is equal to $3\pi^{1/2}/8\alpha^{5/2}$, and so

$$E_n^{(1)} = \frac{\gamma_4}{32\alpha^2} = \frac{\hbar^2\gamma_4}{32k\mu}$$

The total ground-state energy is

$$E_0 = \frac{h\nu}{2} + \frac{\hbar^2\gamma_4}{32k\mu} + \text{higher-order terms}$$

Problem 8–35 has you calculate $E_n^{(1)}$ for $n = 1$.

We can apply perturbation theory to a helium atom whose Hamiltonian operator is given by Equations 8.41 and 8.42. For simplicity, we will consider only the ground-state energy. If we consider the interelectronic repulsion term, $e^2/4\pi\epsilon_0 r_{12}$, to be the perturbation, then the unperturbed wave functions and energies are the hydrogen-like quantities given by

$$\hat{H}^{(0)} = \hat{H}_H(1) + \hat{H}_H(2)$$

$$\psi^{(0)} = \psi_{1s}(r_1, \theta_1, \phi_1)\psi_{1s}(r_2, \theta_2, \phi_2) \tag{8.69}$$

$$E^{(0)} = -\frac{Z^2 m_e e^4}{32\pi^2\epsilon_0^2\hbar^2 n_1^2} - \frac{Z^2 m_e e^4}{32\pi^2\epsilon_0^2\hbar^2 n_2^2}$$

and

$$\hat{H}^{(1)} = \frac{e^2}{4\pi\epsilon_0 r_{12}}$$

with $Z = 2$. Using Equation 8.61, we have

$$E^{(1)} = \iint d\mathbf{r}_1 d\mathbf{r}_2 \psi_{1s}(\mathbf{r}_1)\psi_{1s}(\mathbf{r}_2)\frac{e^2}{4\pi\epsilon_0 r_{12}}\psi_{1s}(\mathbf{r}_1)\psi_{1s}(\mathbf{r}_2) \tag{8.70}$$

where $\psi_{1s}(\mathbf{r}_j)$ is given by

$$\psi_{1s}(\mathbf{r}_j) = \left(\frac{Z^3}{\pi a_0}\right)^{1/2} e^{-Zr_j/a_0} \qquad j = 1 \text{ or } 2 \tag{8.71}$$

The evaluation of the integral in Equation 8.70 is a little lengthy, but Problem 8–39 (see also Problem 8–40) carries it out step by step. The final result is

$$E^{(1)} = \frac{5Z}{8}\left(\frac{m_e e^4}{16\pi^2\epsilon_0^2\hbar^2}\right) \tag{8.72}$$

or $E^{(1)} = 5Z/8$ in units of $m_e e^4/16\pi^2\epsilon_0^2\hbar^2$. If we add this to $E^{(0)}$, with $n_1 = n_2 = 1$,

Assume now that initially the system is in state 1. We let the perturbation begin at $t = 0$ and assume that $\Psi(t)$ is a linear combination of $\Psi_1(t)$ and $\Psi_2(t)$ with coefficients that depend upon time. Thus, we write

$$\Psi(t) = a_1(t)\Psi_1(t) + a_2(t)\Psi_2(t) \tag{8.81}$$

where $a_1(t)$ and $a_2(t)$ are to be determined. Recall from Chapter 4 that for such a linear combination, $a_i^* a_i$ is the probability that the molecule is in state i. We substitute Equation 8.81 into Equation 8.77 to obtain

$$a_1(t)\hat{H}^{(0)}\Psi_1 + a_2(t)\hat{H}^{(0)}\Psi_2 + a_1(t)\hat{H}^{(1)}\Psi_1 + a_2(t)\hat{H}^{(1)}\Psi_2$$
$$= a_1(t)i\hbar\frac{\partial\Psi_1}{dt} + a_2(t)i\hbar\frac{\partial\Psi_2}{dt} + i\hbar\Psi_1\frac{da_1}{dt} + i\hbar\Psi_2\frac{da_2}{dt} \tag{8.82}$$

Using the result given in Example 8–6, we can cancel the first two terms on both sides of Equation 8.82 to obtain

$$a_1(t)\hat{H}^{(1)}\Psi_1 + a_2(t)\hat{H}^{(1)}\Psi_2 = i\hbar\Psi_1\frac{da_1}{dt} + i\hbar\Psi_2\frac{da_2}{dt} \tag{8.83}$$

We now multiply Equation 8.83 by ψ_2^* and integrate over the spatial coordinates to get

$$a_1(t)\int \psi_2^*\hat{H}^{(1)}\Psi_1 d\tau + a_2(t)\int \psi_2^*\hat{H}^{(1)}\Psi_2 d\tau$$
$$= i\hbar\frac{da_1}{dt}\int \psi_2^*\Psi_1 d\tau + i\hbar\frac{da_2}{dt}\int \psi_2^*\Psi_2 d\tau \tag{8.84}$$

The first integral on the right side vanishes because $\Psi_1 = \psi_1 e^{-iE_1 t/\hbar}$ (Equation 8.80) and because ψ_2 and ψ_1 are orthogonal. Similarly, the second integral on the right side is equal to $i\hbar e^{-iE_2 t/\hbar} da_2/dt$ because $\Psi_2 = \psi_2 e^{-iE_2 t/\hbar}$ and ψ_2 is normalized. Solving Equation 8.84 for $i\hbar da_2/dt$ gives

$$i\hbar\frac{da_2}{dt} = a_1(t)e^{iE_2 t/\hbar}\int \psi_2^*\hat{H}^{(1)}\Psi_1 d\tau + a_2(t)e^{iE_2 t/\hbar}\int \psi_2^*\hat{H}^{(1)}\Psi_2 d\tau$$

Using Equation 8.80 for Ψ_1 and Ψ_2 finally gives

$$i\hbar\frac{da_2}{dt} = a_1(t)\exp\left[\frac{-i(E_1 - E_2)t}{\hbar}\right]\int \psi_2^*\hat{H}^{(1)}\psi_1 d\tau + a_2(t)\int \psi_2^*\hat{H}^{(1)}\psi_2 d\tau \tag{8.85}$$

The system is initially in state 1, and so

$$a_1(0) = 1 \quad \text{and} \quad a_2(0) = 0 \tag{8.86}$$

Because $\hat{H}^{(1)}$ is considered a small perturbation, there are not enough transitions out of state 1 to cause a_1 and a_2 to differ appreciably from their initial values. Thus, as an approximation, we may replace $a_1(t)$ and $a_2(t)$ on the right side of Equations 8.85 by their initial values $[a_1(0) = 1, a_2(0) = 0]$ to get

$$i\hbar \frac{da_2}{dt} = \exp\left[\frac{-i(E_1 - E_2)t}{\hbar}\right] \int \psi_2^* \hat{H}^{(1)} \psi_1 d\tau \qquad (8.87)$$

For convenience only, we will take the electric field to be in the z direction, in which case we can write

$$\hat{H}^{(1)} = -\mu_z E_{0z} \cos 2\pi \nu t = -\mu_z E_{0z} \cos \omega t$$

$$= -\frac{\mu_z E_{0z}}{2}(e^{i\omega t} + e^{-i\omega t}) \qquad (8.88)$$

where μ_z is the z component of the molecular dipole moment and E_{0z} is the magnitude of the electric field along the z axis. We substitute this expression for $\hat{H}^{(1)}$ into Equation 8.87 and obtain

$$\frac{da_2}{dt} = \frac{(\mu_z)_{12} E_{0z}}{2i\hbar}\left\{\exp\left[\frac{i(E_2 - E_1 + \hbar\omega)t}{\hbar}\right] + \exp\left[\frac{i(E_2 - E_1 - \hbar\omega)t}{\hbar}\right]\right\} \qquad (8.89)$$

where we have defined

$$(\mu_z)_{12} = \int \psi_2^* \mu_z \psi_1 d\tau \qquad (8.90)$$

The quantity $(\mu_z)_{12}$ is the z component of the *transition dipole moment* between states 1 and 2. Note that if $(\mu_z)_{12} = 0$, then $da_2/dt = 0$ and there will be no transitions out of state 1 into state 2. The dipole transition moment is what underlies selection rules. We have used this result in Sections 5.12 and 6.7. Transitions occur only between states for which the transition moment is nonzero.

Before leaving this discussion, let's integrate Equation 8.89 between 0 and t to obtain

$$a_2(t) = \frac{(\mu_z)_{12} E_{0z}}{2}\left\{\frac{1 - \exp[i(E_2 - E_1 + \hbar\omega)t/\hbar]}{E_2 - E_1 + \hbar\omega} + \frac{1 - \exp[i(E_2 - E_1 - \hbar\omega)t/\hbar]}{E_2 - E_1 - \hbar\omega}\right\}$$

$$(8.91)$$

Because we have taken $E_2 > E_1$, the so-called *resonance denominators* in Equation 8.91 cause the second term in this equation to become much larger than the first term and to be of major importance in determining $a_2(t)$ when

$$E_2 - E_1 \approx \hbar\omega = h\nu \qquad (8.92)$$

Thus, we obtain in a natural way the Bohr frequency condition we have used repeatedly. When a system makes a transition from one state to another, it absorbs (or emits) a photon whose energy is equal to the difference in the energies of the two states.

The probability of absorption or the intensity of absorption is proportional to the probability of observing the molecules to be in state 2, which is given by $a_2^*(t)a_2(t)$. Using only the second term in Equation 8.91, we obtain (Problem 8–43)

$$P_{1 \to 2}(\omega, t) = a_2^*(t)a_2(t) = \frac{(\mu_z)_{12}^2 E_{0z}^2 \sin^2[(E_2 - E_1 - \hbar\omega)t/2\hbar]}{(E_2 - E_1 - \hbar\omega)^2} \qquad (8.93)$$

Equation 8.93 is plotted in Figure 8.7. Note that the plot indicates strong absorption when $\hbar\omega = h\nu \approx E_2 - E_1$.

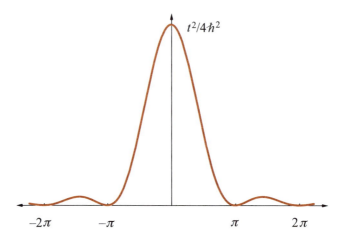

FIGURE 8.7
The function $F(\omega) = \sin^2[(E_2 - E_1 - \hbar\omega)t/2\hbar]/(E_2 - E_1 - \hbar\omega)^2$, which represents the probability of making a $1 \to 2$ transition at a frequency ω in the time interval 0 to t, plotted against $(E_2 - E_1 - \hbar\omega)t/2\hbar$ for $t = 1$. Note that this function peaks when $E_2 - E_1 = \hbar\omega = h\nu$.

Equation 8.93 is not applicable under normal conditions because the irradiating source consists of at least a narrow band of frequencies, and so Equation 8.93 must be averaged over this band. If we let $g(\omega)$ be the frequency distribution of the irradiating source, then $P_{1 \to 2}(\omega, t)$ becomes

$$P_{1 \to 2}(t) = (\mu_z)_{12}^2 E_{0z}^2 \int_{\text{band}} \frac{\sin^2[(E_2 - E_1 - \hbar\omega)t/2\hbar]}{(E_2 - E_1 - \hbar\omega)^2} g(\omega)\, d\omega \qquad (8.94)$$

Figure 8.7 shows that $\sin^2[(E_2 - E_1 - \hbar\omega)t/2\hbar]/(E_2 - E_1 - \hbar\omega)^2$ is strongly peaked around $\omega_{12} = (E_2 - E_1)/\hbar$, and so if $g(\omega)$ does not vary too strongly around ω_{12}, then to a good approximation, we may take $g(\omega_{12})$ out from under the integral sign and write Equation 8.94 as

$$P_{1 \to 2}(t) = (\mu_z)_{12}^2 E_{0z}^2 g(\omega_{12}) \int_{\text{band}} \frac{\sin^2[(E_2 - E_1 - \hbar\omega)t/2\hbar]}{(E_2 - E_1 - \hbar\omega)^2}\, d\omega \qquad (8.95)$$

Furthermore, because the integrand is peaked around $\omega = \omega_{12}$, we can write the integration limits as $-\infty$ to ∞ and write

$$P_{1 \to 2}(t) = (\mu_z)_{12}^2 E_{0z}^2 g(\omega_{12}) \int_{-\infty}^{\infty} \frac{\sin^2[(E_2 - E_1 - \hbar\omega)t/2\hbar]}{(E_2 - E_1 - \hbar\omega)^2} \, d\omega \qquad (8.96)$$

Using the fact that

$$\int_{-\infty}^{\infty} \frac{\sin^2 x}{x^2} \, dx = \pi$$

Equation 8.96 becomes

$$P_{1 \to 2}(t) = \frac{\pi}{2} \left[\frac{(\mu_z)_{12} E_{0z}}{\hbar} \right]^2 t g(\omega_{12}) \qquad (8.97)$$

The spectroscopic absorption coefficient is the rate at which transitions occur, and so equals the time derivative of Equation 8.97, or

$$W_{1 \to 2} = \frac{\pi}{2} \left[\frac{(\mu_z)_{12} E_{0z}}{\hbar} \right]^2 g(\omega_{12}) \qquad (8.98)$$

This formula simply says that there must be radiation at the frequency $\omega_{12} = (E_2 - E_1)/\hbar$ for a transition to occur, which is just a formal statement of the Bohr frequency condition. Equation 8.98 is a form of what is called *Fermi's golden rule*.

Before we finish this chapter, let's go back to Equation 8.93. We said above that Equation 8.93 is not applicable under normal conditions. This is so not only because the irradiation source is not perfectly monochromatic, but also because collisions interfere with the absorption process. There are certain experiments, however, in which $P_{1 \to 2}(\omega, t)$ can be observed directly. An example of such an experiment is the following. A gas is irradiated by a very short pulse from an infrared laser that populates a specific excited rotational state (call it state 1). Meanwhile, the gas is irradiated by microwave radiation that induces a transition from state 1 to another rotational state (call it state 2). The interaction between the molecules in state 1 and the microwave field acts as the time-dependent perturbation in Equation 8.88. After a short time t (of the order of nanoseconds), state 2 is interrogated by a pulse from a visible laser that dissociates the molecules in state 2, and the dissociation products are observed by fluorescence. The intensity of the fluorescence is directly proportional to the population of state 2, or to $|a_2(t)|^2$ in Equation 8.93. The time t in Equation 8.93 is the time between the two laser pulses, $E_2 - E_1$ is the difference in energy between the two rotational states (1 and 2), and ω is the frequency of the microwave radiation. Figure 8.8 shows the result of such an experiment, where the solid line is the experimental result and the dashed line is calculated according to Equation 8.93, modified somewhat for the experimental conditions.

8–10. Consider a three-dimensional, spherically symmetric, isotropic harmonic oscillator with $V(r) = kr^2/2$. Using a trial function $e^{-\alpha r^2}$ with α as a variational parameter, calculate the ground-state energy. Do the same using $e^{-\alpha r}$. The Hamiltonian operator is

$$\hat{H} = -\frac{\hbar^2}{2\mu r^2}\frac{d}{dr}\left(r^2\frac{d}{dr}\right) + \frac{k}{2}r^2$$

Compare these results with the exact ground-state energy, $E = \frac{3}{2}h\nu$. Why is one of these so much better than the other?

8–11. Use a trial function of the form $e^{-\alpha x^2/2}$ to calculate the ground-state energy of a quartic oscillator, whose potential is $V(x) = cx^4$.

8–12. Use the variational method to calculate the ground-state energy of a particle constrained to move within the region $0 \leq x \leq a$ in a potential given by

$$V(x) = \begin{cases} V_0 x & 0 \leq x \leq \dfrac{a}{2} \\[2mm] V_0(a - x) & \dfrac{a}{2} \leq x \leq a \end{cases}$$

As a trial function, use a linear combination of the first two particle-in-a-box wave functions:

$$\phi(x) = c_1\left(\frac{2}{a}\right)^{1/2}\sin\frac{\pi x}{a} + c_2\left(\frac{2}{a}\right)^{1/2}\sin\frac{2\pi x}{a}$$

8–13. Consider a particle of mass m in the potential energy field described by

$$V(x) = \begin{cases} V_0 & x < -a \\ 0 & -a < x < a \\ V_0 & x > a \end{cases}$$

(See also the figure in Problem 4–55.) This problem describes a particle in a finite well. If $V_0 \to \infty$, then we have a particle in a box. Using $\phi(x) = l^2 - x^2$ for $-l < x < l$ and $\phi(x) = 0$ otherwise as a trial function with l as a variational parameter, calculate the ground-state energy of this system for $\alpha = 2m V_0 a^2/\hbar^2 = 4$ and 12. The exact ground-state energies are $0.530\hbar^2/ma^2$ and $0.736\hbar^2/ma^2$, respectively (see Problem 4–55).

8–14. Repeat the calculation in the previous problem for a trial function $\phi(x) = \cos\lambda x$ for $-\pi/2\lambda < x < \pi/2\lambda$ and $\phi(x) = 0$ otherwise. Use λ as a variational parameter.

8–15. Consider a particle that is confined to a sphere of radius a. The Hamiltonian operator for this system is (see Equation 7.10)

$$\hat{H} = -\frac{\hbar^2}{2mr^2}\frac{d}{dr}\left(r^2\frac{d}{dr}\right) + \frac{\hbar^2 l(l + 1)}{2mr^2} \qquad 0 < r \leq a$$

In the ground state, $l = 0$ and so

$$\hat{H} = -\frac{\hbar^2}{2mr^2}\frac{d}{dr}\left(r^2\frac{d}{dr}\right) \qquad 0 < r \leq a$$

As in the case of a particle in a rectangular box, $\phi(a) = 0$. Use $\phi(r) = a - r$ to calculate an upper bound to the ground-state energy of this system. There is no variational parameter in this case, but the calculated energy is still an upper bound to the ground-state energy. The exact ground-state energy is $\pi^2\hbar^2/2ma^2$ (see Problem 8–17).

8–16. Repeat the calculation in Problem 8–15 using $\phi(r) = (a - r)^2$ as a trial function. Compare your result to the one obtained in the previous problem. The exact (normalized) wave function is given in the next problem. Compare plots of $(1 - r)$ and $(1 - r)^2$ (after normalizing them) and the exact wave functions.

8–17. In this problem, we will solve the Schrödinger equation for the ground-state wave function and energy of a particle confined to a sphere of radius a. The Schrödinger equation is given by Equation 7.10 with $l = 0$ (ground state) and without the $e^2/4\pi\epsilon_0 r$ term:

$$-\frac{\hbar^2}{2mr^2}\frac{d}{dr}\left(r^2\frac{d\psi}{dr}\right) = E\psi$$

Substitute $u = r\psi$ into this equation to get

$$\frac{d^2u}{dr^2} + \frac{2mE}{\hbar^2}u = 0$$

The general solution to this equation is

$$u(r) = A\cos\alpha r + B\sin\alpha r$$

or

$$\psi(r) = \frac{A\cos\alpha r}{r} + \frac{B\sin\alpha r}{r}$$

where $\alpha = (2mE/\hbar^2)^{1/2}$. Which of these terms is finite at $r = 0$? Now use the fact that $\psi(a) = 0$ to prove that

$$\alpha a = \pi$$

for the ground state, or that the ground-state energy is

$$E = \frac{\pi^2\hbar^2}{2ma^2}$$

Show that the normalized ground-state wave function is

$$\psi(r) = (2\pi a)^{-1/2}\frac{\sin\pi r/a}{r}$$

8–18. This problem fills in the steps of the variational treatment of a helium atom. We use a trial function of the form

$$\phi(\mathbf{r}_1, \mathbf{r}_2) = \frac{Z^3}{a_0^3\pi}e^{-Z(r_1+r_2)/a_0}$$

with Z as an adjustable parameter. The Hamiltonian operator of a helium atom is

$$\hat{H} = -\frac{\hbar^2}{2m_e}\nabla_1^2 - \frac{\hbar^2}{2m_e}\nabla_2^2 - \frac{2e^2}{4\pi\epsilon_0 r_1} - \frac{2e^2}{4\pi\epsilon_0 r_2} + \frac{e^2}{4\pi\epsilon_0 r_{12}}$$

We now evaluate

$$E(Z) = \int d\mathbf{r}_1 d\mathbf{r}_2\, \phi^* \hat{H} \phi$$

The evaluation of this integral is greatly simplified if you recall that

$$\psi(r_j) = (Z^3/a_0^3\pi)^{1/2} e^{-Zr_j/a_0}$$

is an eigenfunction of a hydrogen-like Hamiltonian operator, one for which the nucleus has a charge Z. Show that the helium atom Hamiltonian operator can be written as

$$\hat{H} = -\frac{\hbar^2}{2m_e}\nabla_1^2 - \frac{Ze^2}{4\pi\epsilon_0 r_1} - \frac{\hbar^2}{2m_e}\nabla_2^2 - \frac{Ze^2}{4\pi\epsilon_0 r_2} + \frac{(Z-2)e^2}{4\pi\epsilon_0 r_1} + \frac{(Z-2)e^2}{4\pi\epsilon_0 r_2} + \frac{e^2}{4\pi\epsilon_0 r_{12}}$$

where

$$\left(-\frac{\hbar^2}{2m_e}\nabla^2 - \frac{Ze^2}{4\pi\epsilon_0 r}\right)\left(\frac{Z^3}{a_0^3\pi}\right)^{1/2} e^{-Zr/a_0} = -\frac{Z^2 e^2}{8\pi\epsilon_0 a_0}\left(\frac{Z^3}{a_0^3\pi}\right)^{1/2} e^{-Zr/a_0}$$

Show that

$$E(Z) = \frac{Z^6}{a_0^6\pi^2} \iint d\mathbf{r}_1 d\mathbf{r}_2\, e^{-Z(r_1+r_2)/a_0} \left[-\frac{Z^2 e^2}{8\pi\epsilon_0 a_0} - \frac{Z^2 e^2}{8\pi\epsilon_0 a_0} + \frac{(Z-2)e^2}{4\pi\epsilon_0 r_1}\right.$$

$$\left. + \frac{(Z-2)e^2}{4\pi\epsilon_0 r_2} + \frac{e^2}{4\pi\epsilon_0 r_{12}}\right] e^{-Z(r_1+r_2)/a_0}$$

The last integral is evaluated in Problem 8–39 or 8–40 and the others are elementary. Show that $E(Z)$, in units of $(m_e e^4/16\pi^2\epsilon_0^2\hbar^2)$, is given by

$$E(Z) = -Z^2 + 2(Z-2)\frac{Z^3}{\pi}\int d\mathbf{r}\frac{e^{-2Zr}}{r} + \frac{5}{8}Z$$

$$= -Z^2 + 2(Z-2)Z + \frac{5}{8}Z$$

$$= Z^2 - \frac{27}{8}Z$$

Now minimize E with respect to Z and show that

$$E = -\left(\frac{27}{16}\right)^2 = -2.8477$$

in units of $m_e e^4/16\pi^2\epsilon_0^2\hbar^2$. Interpret the value of Z that minimizes E.

8–19. Use the spectral data for He and He$^+$ from the website *http://physics.nist.gov/PhysRef Data/ASD/levels_form.html* to determine the experimental ground-state energy of a helium atom.

8–20. Verify all the matrix elements in Equation 8.37.

8–21. Consider a system subject to the potential

$$V(x) = \frac{k}{2}x^2 + \frac{\gamma_3}{6}x^3 + \frac{\gamma_4}{24}x^4$$

Calculate the ground-state energy of this system using a trial function of the form

$$\phi = c_1\psi_0(x) + c_2\psi_2(x)$$

where $\psi_0(x)$ and $\psi_2(x)$ are the harmonic-oscillator wave functions. Why did we not include $\psi_1(x)$?

8–22. It is quite common to assume a trial function of the form

$$\phi = c_1\phi_1 + c_2\phi_2 + \cdots + c_n\phi_n$$

where the variational parameters and the ϕ_n may be complex. Using the simple, special case

$$\phi = c_1\phi_1 + c_2\phi_2$$

show that the variational method leads to

$$E_\phi = \frac{c_1^* c_1 H_{11} + c_1^* c_2 H_{12} + c_1 c_2^* H_{21} + c_2^* c_2 H_{22}}{c_1^* c_1 S_{11} + c_1^* c_2 S_{12} + c_1 c_2^* S_{21} + c_2^* c_2 S_{22}}$$

where

$$H_{ij} = \int \phi_i^* \hat{H}_j \phi_j d\tau = H_{ji}^*$$

and

$$S_{ij} = \int \phi_i^* \phi_j d\tau = S_{ji}^*$$

because $\hat{H}$ is a Hermitian operator. Now write the above equation for E_ϕ as

$$c_1^* c_1 H_{11} + c_1^* c_2 H_{12} + c_1 c_2^* H_{21} + c_2^* c_2 H_{22}$$
$$= E_\phi(c_1^* c_1 S_{11} + c_1^* c_2 S_{12} + c_1 c_2^* S_{21} + c_2^* c_2 S_{22})$$

and show that if we set

$$\frac{\partial E_\phi}{\partial c_1^*} = 0 \qquad \text{and} \qquad \frac{\partial E_\phi}{\partial c_2^*} = 0$$

8–31. In this problem, we shall calculate the polarizability of a hydrogen atom using the variational method. The polarizability of an atom is a measure of the distortion of the electronic distribution of the atom when it is placed in an external electric field. When an atom is placed in an external electric field, the field induces a dipole moment in the atom. It is a good approximation to say that the magnitude of the induced dipole moment is proportional to the strength of the electric field. In an equation, we have

$$\mu = \alpha \mathcal{E} \tag{1}$$

where μ is the magnitude of the induced dipole moment, $\mathcal{E}$ is the strength of the electric field, and α is a proportionality constant called the *polarizability*. The value of α depends upon the particular atom.

The energy required to induce a dipole moment is given by

$$E = -\int_0^{\mathcal{E}} \mu \, d\mathcal{E}' = -\int_0^{\mathcal{E}} \alpha \mathcal{E}' \, d\mathcal{E}' = -\frac{\alpha \mathcal{E}^2}{2} \tag{2}$$

Equation 2 is the energy associated with a polarizable atom in an electric field.

Consider now a hydrogen atom for simplicity. In a hydrogen atom, there is an instantaneous dipole moment pointing from the electron to the nucleus. This instantaneous dipole moment is given by $-e\mathbf{r}$ and interacts with an external electric field according to

$$E = -\boldsymbol{\mu} \cdot \boldsymbol{\mathcal{E}} = e\mathbf{r} \cdot \boldsymbol{\mathcal{E}} \tag{3}$$

If $\mathcal{E}$ is taken to be in the z direction, then equation 3 introduces a perturbation term to the Hamiltonian operator of the hydrogen atom that is of the form

$$\hat{H}^{(1)} = e\mathcal{E}_z r \cos\theta$$

and so the complete Hamiltonian operator is

$$\hat{H} = \hat{H}^{(0)} + e\mathcal{E}_z r \cos\theta \tag{4}$$

We can solve this problem using perturbation theory, but we shall use the variational method here to calculate the ground-state energy of a hydrogen atom in an external electric field.

Problem 8–24 shows that it is convenient to write a trial function as a linear combination of orthonormal functions. In particular, in this case it is convenient to choose the orthonormal functions to be the eigenfunctions of the unperturbed system. Because the field induces a dipole in the z direction, let's take

$$\phi = c_1 \psi_{1s} + c_2 \psi_{2p_z} \tag{5}$$

as our trial function. Using the hydrogen atomic wave functions given in Table 7.5, show that

$$H_{11} = -\frac{e^2}{2\,\kappa_0\,a_0}$$

$$H_{22} = -\frac{e^2}{8\,\kappa_0\,a_0} \tag{6}$$

$$H_{12} = \frac{8}{\sqrt{2}}\left(\frac{2}{3}\right)^5 e\mathcal{E}a_0$$

where $\kappa_0 = 4\pi\epsilon_0$. Show that the two roots of the corresponding secular equation are

$$E = -\frac{5\,e^2}{16\,\kappa_0\,a_0} \pm \frac{3\,e^2}{16\,\kappa_0\,a_0}\left(1 + \frac{2^{23}\,\mathcal{E}_z^2\kappa_0^2a_0^4}{3^{12}\,e^2}\right)^{1/2} \tag{7}$$

Now use the expansion of $(1+x)^{1/2}$ given in Equation D.14 to obtain

$$E = -\frac{e^2}{2\,\kappa_0\,a_0} - 2.96\,\kappa_0\,a_0^3\frac{\mathcal{E}_z^2}{2} + \cdots \tag{8}$$

Compare this result to the macroscopic equation (equation 2) to show that the polarizability of a hydrogen atom is

$$\alpha = 2.96\,\kappa_0\,a_0^3 \tag{9}$$

The exact value for the hydrogen atom is $9\,\kappa_0\,a_0^3/2$. Although the numerical value is in error by 35%, we do see that the polarizability is proportional to a_0^3 (i.e., to a measure of the volume of the atom). This is a general result and can be used to estimate polarizabilities. Why is there no linear term in $\mathcal{E}_z$ in the above equation for E? What do you think a first-order perturbation calculation of the 1s state would give?

8–32. It is instructive to redo the calculation of the polarizability of a hydrogen atom in the previous problem using a trial function of the form

$$\phi = c_1\psi_{1s} + c_2\psi_{3p_z}$$

This trial function has the same symmetry as equation 5 in the previous problem, but it involves the ψ_{3p_z} orbital instead of the ψ_{2p_z}. Show that in this case

$$E = -\frac{5\,e^2}{18\,\kappa_0\,a_0} \pm \frac{1}{2}\left(\frac{16}{81}\frac{e^4}{\kappa_0^2\,a_0^2} + \frac{3^6}{2^{11}}e^2\mathcal{E}_z^2a_0^2\right)^{1/2}$$

or

$$E = -\frac{e^2}{2\,\kappa_0\,a_0} - 0.400\,\kappa_0\,a_0^3\frac{\mathcal{E}_z^2}{2}$$

for a polarizability, $\alpha = 0.400\,\kappa_0\,a_0^3$. Note that, in this case, the energy is quite a bit higher than that in equation 8 of the previous problem, and in fact it is not very far from the 1s energy, $-e^2/2 \kappa_0 a_0$. This result suggests that the $3p_z$ orbital somehow does not play much of a role in the trial function, particularly compared to the trial function involving the $1s$ and $2p_z$ orbitals. These two calculations of the polarizability of a hydrogen atom illustrate a general principle that we discuss in Problem 8–28.

8–33. Verify the expansion in Equation 8.57.

8–34. Use the series expansion of $(1 + x)^{1/2}$ (Equation D.14) to show that the variational result of Example 8–4 can be written as

$$\varepsilon = 1 + \frac{v_0}{2} - \frac{256}{243\pi^4} v_0^2 + O(v_0^3)$$

8–35. Calculate the first-order correction to the first excited state of an anharmonic oscillator whose potential is given in Example 8–5.

8–36. Calculate the first-order correction to the energy of a particle constrained to move within the region $0 \leq x \leq a$ in the potential

$$V(x) = \begin{cases} V_0 x & 0 \leq x \leq \dfrac{a}{2} \\ V_0(a - x) & \dfrac{a}{2} \leq x \leq a \end{cases}$$

where V_0 is a constant.

8–37. Use first-order perturbation theory to calculate the first-order correction to the ground-state energy of a quartic oscillator whose potential energy is

$$V(x) = cx^4$$

In this case, use a harmonic oscillator as the unperturbed system. What is the perturbing potential?

8–38. In Example 5–2, we introduced the Morse potential

$$V(x) = D(1 - e^{-\beta x})^2$$

as a description of the internuclear potential energy of a diatomic molecule. First expand the Morse potential in a power series about x. (*Hint:* Use the expansion $e^x = 1 + x + \frac{x^2}{2} + \frac{x^3}{6} + \cdots$.) What is the Hamiltonian operator for the Morse potential? Show that the Hamiltonian operator can be written in the form

$$\hat{H} = -\frac{\hbar^2}{2\mu} \frac{d^2}{dx^2} + ax^2 + bx^3 + cx^4 + \cdots \tag{1}$$

How are the constants a, b, and c related to the constants D and β? What part of the Hamiltonian operator would you associate with $\hat{H}^{(0)}$, and what are the functions $\psi_n^{(0)}$ and energies $E_n^{(0)}$? Use perturbation theory to evaluate the first-order corrections to the energy of the first three states that arise from the cubic and quartic terms.

8–39. In applying first-order perturbation theory to a helium atom, we must evaluate the integral (Equation 8.70)

$$E^{(1)} = \frac{e^2}{4\pi\epsilon_0} \iint d\mathbf{r}_1 d\mathbf{r}_2 \psi_{1s}^*(\mathbf{r}_1) \psi_{1s}^*(\mathbf{r}_2) \frac{1}{r_{12}} \psi_{1s}(\mathbf{r}_1) \psi_{1s}(\mathbf{r}_2)$$

where

$$\psi_{1s}(\mathbf{r}_j) = \left(\frac{Z^3}{a_0^3 \pi}\right)^{1/2} e^{-Zr_j/a_0}$$

and $Z = 2$ for a helium atom. This same integral occurs in a variational treatment of a helium atom, where in that case the value of Z is left arbitrary. This problem proves that

$$E^{(1)} = \frac{5Z}{8} \left(\frac{m_e e^4}{16\pi^2 \epsilon_0^2 \hbar^2}\right)$$

Let $\mathbf{r}_1$ and $\mathbf{r}_2$ be the radius vectors of electrons 1 and 2, respectively, and let θ be the angle between these two vectors. Now this is generally *not* the θ of spherical coordinates, but if we choose one of the radius vectors, say $\mathbf{r}_1$, to be the z axis, then the two θ's are the same. Using the law of cosines,

$$r_{12} = (r_1^2 + r_2^2 - 2r_1 r_2 \cos\theta)^{1/2}$$

show that $E^{(1)}$ becomes

$$E^{(1)} = \frac{e^2}{4\pi\epsilon_0} \frac{Z^6}{a_0^6 \pi^2} \int_0^\infty dr_1 e^{-Zr_1/a_0} 4\pi r_1^2 \int_0^\infty dr_2 e^{-Zr_2/a_0} r_2^2$$

$$\times \int_0^{2\pi} d\phi \int_0^\pi \frac{d\theta \sin\theta}{(r_1^2 + r_2^2 - 2r_1 r_2 \cos\theta)^{1/2}}$$

Letting $x = \cos\theta$, show that the integrand over θ is

$$\int_0^\pi \frac{d\theta \sin\theta}{(r_1^2 + r_2^2 - 2r_1 r_2 \cos\theta)^{1/2}} = \int_{-1}^1 \frac{dx}{(r_1^2 + r_2^2 - 2r_1 r_2 x)^{1/2}}$$

$$= \frac{2}{r_1} \quad r_1 > r_2$$

$$= \frac{2}{r_2} \quad r_1 < r_2$$

Substituting this result into $E^{(1)}$, show that

$$E^{(1)} = \frac{e^2}{4\pi\epsilon_0}\frac{16Z^6}{a_0^6}\int_0^\infty dr_1 e^{-2Zr_1/a_0}r_1^2 \left(\frac{1}{r_1}\int_0^{r_1} dr_2 e^{-2Zr_2/a_0}r_2^2\right.$$

$$\left. + \int_{r_1}^\infty dr_2 e^{-2Zr_2/a_0}r_2\right)$$

$$= \frac{e^2}{4\pi\epsilon_0}\frac{4Z^3}{a_0^3}\int_0^\infty dr_1 e^{-2Zr_1/a_0}r_1^2\left[\frac{1}{r_1} - e^{-2Zr_1/a_0}\left(\frac{Z}{a_0}+\frac{1}{r_1}\right)\right]$$

$$= \frac{5}{8}Z\left(\frac{e^2}{4\pi\epsilon_0 a_0}\right) = \frac{5}{8}Z\left(\frac{m_e e^4}{16\pi^2\epsilon_0^2\hbar^2}\right)$$

Show that the energy through first order is

$$E^{(0)} + E^{(1)} = \left(-Z^2 + \frac{5}{8}Z\right)\left(\frac{m_e e^4}{16\pi^2\epsilon_0^2\hbar^2}\right) = -\frac{11}{4}\left(\frac{m_e e^4}{16\pi^2\epsilon_0^2\hbar^2}\right)$$

$$= -2.75\left(\frac{m_e e^4}{16\pi^2\epsilon_0^2\hbar^2}\right)$$

compared with the exact result, $E_{\text{exact}} = -2.9037(m_e e^4/16\pi^2\epsilon_0^2\hbar^2)$.

8–40. In the previous problem we evaluated the integral that occurs in the first-order perturbation theory treatment of a helium atom (see Equation 8.70). In this problem we will evaluate the integral by another method, one that uses an expansion for $1/r_{12}$ that is useful in many applications. We can write $1/r_{12}$ as an expansion in terms of spherical harmonics,

$$\frac{1}{r_{12}} = \frac{1}{|\mathbf{r}_1 - \mathbf{r}_2|} = \sum_{l=0}^\infty \sum_{m=-l}^{+l} \frac{4\pi}{2l+1}\frac{r_<^l}{r_>^{l+1}}Y_l^m(\theta_1,\phi_1)Y_l^{m*}(\theta_2,\phi_2)$$

where θ_i and ϕ_i are the angles that describe $\mathbf{r}_i$ in a spherical coordinate system and $r_<$ and $r_>$ are, respectively, the smaller and larger values of r_1 and r_2. In other words, if $r_1 < r_2$, then $r_< = r_1$ and $r_> = r_2$. Substitute $\psi_{1s}(r_i) = (Z^3/a_0^3\pi)^{1/2}e^{-Zr_i/a_0}$, and the above expansion for $1/r_{12}$ into Equation 8.70, integrate over the angles, and show that all the terms except for the $l = 0$, $m = 0$ term vanish. Show that

$$E^{(1)} = \frac{e^2}{4\pi\epsilon_0}\frac{16Z^6}{a_0^6}\int_0^\infty dr_1 r_1^2 e^{-2Zr_1/a_0}\int_0^\infty dr_2 r_2^2 \frac{e^{-2Zr_2/a_0}}{r_>}$$

Now show that

$$E^{(1)} = \frac{e^2}{4\pi\epsilon_0} \frac{16Z^6}{a_0^6} \int_0^\infty dr_1 r_1 e^{-2Zr_1/a_0} \int_0^{r_1} dr_2 r_2^2 e^{-2Zr_2/a_0}$$

$$+ \frac{e^2}{4\pi\epsilon_0} \frac{16Z^6}{a_0^6} \int_0^\infty dr_1 r_1^2 e^{-2Zr_1/a_0} \int_{r_1}^\infty dr_2 r_2 e^{-2Zr_2/a_0}$$

$$= -\frac{e^2}{4\pi\epsilon_0} \frac{4Z^6}{a_0^6} \int_0^\infty dr_1 r_1 e^{-2Zr_1/a_0} \left[e^{-2Zr_1/a_0} \left(\frac{2Z^2 r_1^2}{a_0^2} + \frac{2Zr_1}{a_0} + 1 \right) - 1 \right]$$

$$+ \frac{e^2}{4\pi\epsilon_0} \frac{4Z^6}{a_0^6} \int_0^\infty dr_1 r_1^2 e^{-2Zr_1/a_0} \left[e^{-2Zr_1/a_0} \left(\frac{2Zr_1}{a_0} + 1 \right) \right]$$

$$= -\frac{e^2}{4\pi\epsilon_0} \frac{4Z^6}{a_0^6} \int_0^\infty dr_1 e^{-4Zr_1/a_0} \left[\frac{r_1^2 a_0^2}{Z^2} + \frac{r_1 a_0^3}{Z^3} \right]$$

$$+ \frac{e^2}{4\pi\epsilon_0} \frac{4Z^3}{a_0^3} \int_0^\infty dr_1 r_1 e^{-2Zr_1/a_0}$$

$$= \frac{5}{8} Z \left(\frac{e^2}{4\pi\epsilon_0 a_0} \right)$$

as in Problem 8–39.

8–41. Consider a molecule with a dipole moment μ in an electric field $\mathcal{E}$. We picture the dipole moment as a positive charge and a negative charge of magnitude q separated by a vector $\mathbf{l}$.

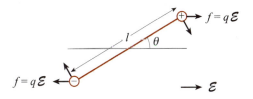

The field $\mathcal{E}$ causes the dipole to rotate into a direction parallel to $\mathcal{E}$. Therefore, work is required to rotate the dipole to an angle θ to $\mathcal{E}$. The force causing the molecule to rotate is actually a torque (torque is the angular analog of force) and is given by $l/2$ times the force perpendicular to $\mathbf{l}$ at each end of the vector $\mathbf{l}$. Show that this torque is equal to $\mu\mathcal{E} \sin\theta$ and that the energy required to rotate the dipole from some initial angle θ_0 to some arbitrary angle θ is

$$V = \int_{\theta_0}^\theta \mu\mathcal{E} \sin\theta' d\theta'$$

Given that θ_0 is customarily taken to be $\pi/2$, show that

$$V = -\mu\mathcal{E} \cos\theta = -\boldsymbol{\mu} \cdot \boldsymbol{\mathcal{E}}$$

8–42. Derive Equation 8.93 using the second term in Equation 8.91 (cf. Problem A–6).

8–43. In this problem we shall derive the formulas of perturbation theory to second order in the perturbation. First, substitute Equation 8.61 into Equation 8.58 to obtain an equation in which the only unknown quantity is $\psi_n^{(1)}$. A standard way to solve the equation for $\psi_n^{(1)}$ is to expand the unknown $\psi_n^{(1)}$ in terms of the eigenfunctions of the unperturbed problem. Show that if we substitute

$$\psi_n^{(1)} = \sum_j a_{nj} \psi_j^{(0)} \tag{1}$$

into Equation 8.58, multiply by $\psi_k^{(0)*}$, and integrate, then we obtain

$$\sum_j a_{nj} \langle \psi_k^{(0)} | \hat{H}^{(0)} - E_n^{(0)} | \psi_j^{(0)} \rangle = E_n^{(1)} \langle \psi_k^{(0)} | \psi_n^{(0)} \rangle - \langle \psi_k^{(0)} | \hat{H}^{(1)} | \psi_n^{(0)} \rangle \tag{2}$$

There are two cases to consider here, $k = n$ and $k \neq n$. Show that when $k = n$, we obtain Equation 8.61 again and that when $k \neq n$, we obtain

$$a_{nk} = \frac{\langle \psi_k^{(0)} | \hat{H}^{(1)} | \psi_n^{(0)} \rangle}{E_n^{(0)} - E_k^{(0)}} = \frac{H_{kn}^{(1)}}{E_n^{(0)} - E_k^{(0)}} \tag{3}$$

where

$$H_{kn}^{(1)} = \langle \psi_k^{(0)} | \hat{H}^{(1)} | \psi_n^{(0)} \rangle = \int \psi_k^{(0)*} \hat{H}^{(1)} \psi_n^{(0)} \, d\tau \tag{4}$$

Thus, we have determined all the a's in equation 1 except for a_{nn}. We can determine a_{nn} by requiring that ψ_n in Equation 8.55 be normalized through first order, or through terms linear in λ. Show that this requirement is equivalent to requiring that $\psi_n^{(0)}$ be orthogonal to $\psi_n^{(1)}$ and that it gives $a_{nn} = 0$. The complete wave function to first order, then, is

$$\psi_n = \psi_n^{(0)} + \lambda \sum_{j \neq n} \frac{H_{jn}^{(1)} \psi_j^{(0)}}{E_n^{(0)} - E_j^{(0)}} = \psi_n^{(0)} + \lambda \psi_n^{(1)} \tag{5}$$

which defines $\psi_n^{(1)}$. Now that we have $\psi_n^{(1)}$, we can determine the second-order energy by setting the λ^2 term in Equation 8.57 equal to zero:

$$H^{(0)} \psi_n^{(2)} + \hat{H}^{(1)} \psi_n^{(1)} - E_n^{(0)} \psi_n^{(2)} - E_n^{(1)} \psi_n^{(1)} - E_n^{(2)} \psi_n^{(0)} = 0 \tag{6}$$

As with $\psi_n^{(1)}$, we write $\psi_n^{(2)}$ as

$$\psi_n^{(2)} = \sum_s b_{ns} \psi_s^{(0)} \tag{7}$$

Substitute this expression into equation 6, multiply from the left by $\psi_m^{(0)*}$, and integrate to obtain

$$b_{nm} E_m^{(0)} + \sum_{j \neq n} a_{nj} H_{mj}^{(1)} = b_{nm} E_n^{(0)} + a_{nm} E_n^{(1)} + \delta_{nm} E_n^{(2)} \tag{8}$$

Let $n = m$ to get (remember that $a_{nn} = 0$)

$$E_n^{(2)} = \sum_{j \neq n} \frac{H_{nj}^{(1)} H_{jn}^{(1)}}{E_n^{(0)} - E_j^{(0)}} = \sum_{j \neq n} \frac{|H_{nj}^{(1)}|}{E_n^{(0)} - E_j^{(0)}} \qquad (9)$$

Thus, the energy through second order is

$$E_n = E_n^{(0)} + E_n^{(1)} + E_n^{(2)}$$

where $E_n^{(1)}$ is given by Equation 8.61 and $E_n^{(2)}$ is given by equation 9.

8–44. Derive the equation for E_n through second order by starting with

$$E = \frac{\langle \psi_n \mid \hat{H} \mid \psi_n \rangle}{\langle \psi_n \mid \psi_n \rangle}$$

8–45. Problem 8–43 shows that $\psi_n^{(1)}$ has the form $\psi_n^{(1)} = \sum_{j \neq n} a_{nj} \psi_j^{(0)}$. Given that $\psi_n^{(2)}$ has a similar form, $\psi_n^{(2)} = \sum_{j \neq n} b_{nj} \psi_j^{(0)}$, show that a knowledge of the wave function through first order determines the energy through third order.

8–46. In this problem we'll calculate the ground-state energy of a particle in a gravitational well (Example 8–4 and Equation 8.68) through second order in perturbation theory using the results of Problem 8–43. From equation 9 of Problem 8–43, we see that the second-order correction to the ground-state ($n = 1$) energy is given by

$$E_1^{(2)} = \sum_{j \neq 1} \frac{[H_{1j}^{(1)}]^2}{E_1^{(0)} - E_j^{(0)}}$$

where

$$H_{1j}^{(1)} = \left\langle \psi_1^{(0)} \left| \frac{V_0 x}{a} \right| \psi_j^{(0)} \right\rangle$$

and $\psi_k^{(0)} = (2/a)^{1/2} \sin k\pi x/a$. Show that

$$H_{1j}^{(1)} = \frac{2V_0}{a^2} \int_0^a dx \, x \sin \frac{\pi x}{a} \sin \frac{j\pi x}{a} = \begin{cases} -\dfrac{8 j V_0}{\pi^2 (1 - j^2)^2} & j \text{ even} \\ 0 & j \text{ odd} \end{cases}$$

for $j \geq 2$. Now show that

$$E_1^{(2)} = \frac{64 V_0^2}{\pi^4} \frac{8ma^2}{h^2} \sum_{\substack{j \geq 2 \\ (j \text{ even})}}^{\infty} \frac{j^2}{(j^2 - 1)^5}$$

$$= \frac{64 v_0^2}{\pi^4} \frac{h^2}{8ma^2} \sum_{\substack{j \geq 2 \\ (j \text{ even})}}^{\infty} \frac{j^2}{(j^2 - 1)^5}$$

where $v_0 = 8m V_0 a^2 / h^2$.

Show that the energy through second order is

$$\varepsilon = \varepsilon_1^{(0)} + \varepsilon_1^{(1)} + \varepsilon_1^{(2)} = 1 + \frac{v_0}{2} + v_0^2 \left[\frac{64}{\pi^4} \sum_{\substack{j \geq 2 \\ (j \text{ even})}}^{\infty} \frac{j^2}{(j^2 - 1)^5} \right]$$

where $\varepsilon = 8mEa^2/h^2$.

The summation here converges very rapidly; two terms give 0.01648, which is accurate to four significant figures. Therefore,

$$\varepsilon = 1 + \frac{v_0}{2} + 0.01083 \, v_0^3 + O(v_0^3)$$

Compare this result to that obtained in Problem 8–34. Comment on the comparison. The following figure compares this result to the exact energy as a function of v_0. Note that the two sets of values agree for small values of v_0, but diverge as v_0 increases.

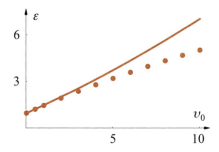

References

Pauling, L., Wilson, E. B., Jr. *Introduction to Quantum Mechanics*. Dover Publications: Mineola, NY, 1985.

Eyring, H., Walter, J., Kimball, G. E. *Quantum Chemistry*. Wiley & Sons: New York, 1944.

Shankar, R. *Principles of Quantum Mechanics*. Springer: New York, 2005.

Townsend, J. S. *A Modern Approach to Quantum Mechanics*. University Science Books: Sausalito, CA, 2000.

Matrix Eigenvalue Problems

The Schrödinger equation

$$\hat{H}\psi = E\psi \tag{H.1}$$

is an eigenvalue problem; ψ is the eigenfunction and E is the corresponding eigenvalue. We've seen in MathChapter G that operators can be represented by matrices, and so the matrix equation

$$\textsf{Ac} = \lambda \textsf{c} \tag{H.2}$$

which is analogous to Equation H.1, is called a *matrix eigenvalue problem*, where $\textsf{c}$ is an *eigenvector* of the matrix $\textsf{A}$ and λ is the corresponding eigenvalue. Equations H.1 and H.2 suggest that there is a strong relationship between the Schrödinger equation and a matrix eigenvalue problem. We have seen this relationship in Chapter 8, but we didn't develop it there. In fact, quantum mechanics can be presented entirely in terms of matrices instead of differential equations as we have done in this book. It's traditional for quantum chemistry to be presented in terms of differential equations because chemistry students are presumably more comfortable or familiar with differential equations than with matrices; but, in fact, matrix algebra is much easier than differential equations, and most research in molecular quantum mechanics is couched in terms of matrices and matrix eigenvalue problems.

To see explicitly the relation between the Schrödinger equation and a matrix eigenvalue problem, we expand the (unknown) eigenfunction ψ in Equation H.1 in terms of some convenient set of (real and normalized) functions ϕ_i:

$$\psi = \sum_{i=1}^{N} c_i \phi_i \tag{H.3}$$

As N gets larger and larger, we expect Equation H.3 to become more and more exact if we choose the ϕ_i well. The unknown nature of ψ is now represented by the set of

427

unknown coefficients $\{c_i\}$. We substitute Equation H.3 into Equation H.1, multiply by ϕ_j^*, and then integrate over all the coordinates to obtain the set of algebraic equations

$$
\begin{aligned}
H_{11}c_1 + H_{12}c_2 + \cdots + H_{1N}c_N &= E(c_1 + c_2 S_{12} + \cdots + c_N S_{1N}) \\
H_{21}c_1 + H_{22}c_2 + \cdots + H_{2N}c_N &= E(c_1 S_{21} + c_2 + \cdots + c_N S_{1N}) \\
\vdots \qquad \vdots \qquad\qquad \vdots \;\; &= \qquad \vdots \\
H_{N1}c_1 + H_{N2}c_2 + \cdots + H_{NN}c_N &= E(c_1 S_{N1} + c_2 S_{N2} + \cdots + c_N)
\end{aligned}
\tag{H.4}
$$

where the

$$
H_{ij} = \int d\tau\, \phi_i \hat{H} \phi_j \tag{H.5}
$$

and the

$$
S_{ij} = \int d\tau\, \phi_i \phi_j
$$

are called *matrix elements*. We have used the fact that the ϕ_i are normalized ($S_{ii} = 1$) in writing Equation H.4. We can write Equation H.4 as a matrix eigenvalue problem

$$
\mathsf{H}\mathsf{c} = E\mathsf{S}\mathsf{c} \tag{H.6}
$$

Equation H.6 is equivalent to Equation 8.38. This type of equation appears often in quantum chemistry, and will appear repeatedly in later chapters. Equation H.6 becomes the same as Equation H.2 (with $\mathsf{A} = \mathsf{S}^{-1}\mathsf{H}$) if we multiply Equation H.6 from the left by S^{-1}. Thus, we see that the Schrödinger equation can be expressed as a matrix eigenvalue problem.

Let's look at Equation H.2 more closely. Equation H.2 represents the system of homogeneous linear equations

$$
\begin{aligned}
(a_{11} - \lambda)c_1 + a_{12}c_2 + \cdots + a_{1N}c_N &= 0 \\
a_{21}c_1 + (a_{22} - \lambda)c_2 + \cdots + a_{2N}c_N &= 0 \\
\vdots \qquad \vdots \qquad\qquad \vdots \qquad \vdots \\
a_{N1}c_1 + a_{N2}c_2 + \cdots + (a_{NN} - \lambda)c_N &= 0
\end{aligned}
\tag{H.7}
$$

As we have seen a number of times before, the determinant of the c_j's must be equal to zero in order to have a nontrivial solution; in other words, a solution where not all the $c_j = 0$. Thus, we write

$$
\det(\mathsf{A} - \lambda \mathsf{I}) = 0 \tag{H.8}
$$

which leads to the secular equation, which is an Nth-degree polynomial equation in λ. The solution to this equation gives us N eigenvalues in Equation H.2. Associated with each eigenvalue is an eigenvector. We obtain each eigenvector by substituting one of the values of λ into Equation H.4 and then solving for the c_j's. We did this repeatedly in Chapter 8.

EXAMPLE H–1

Find the eigenvalues and eigenvectors of

$$\mathbf{A} = \begin{pmatrix} a & 1 \\ 1 & a \end{pmatrix}$$

where a is a constant.

SOLUTION: The determinant of $\mathbf{A} - \lambda\mathbf{I}$ is given by

$$\det(\mathbf{A} - \lambda\mathbf{I}) = \begin{vmatrix} a - \lambda & 1 \\ 1 & a - \lambda \end{vmatrix} = (a - \lambda)^2 - 1 = 0$$

and so the eigenvalues are given by the solution to $(a - \lambda)^2 - 1 = 0$, or $\lambda = a \pm 1$. The equations for the eigenvectors are (see Equation H.7)

$$(a - \lambda)c_1 + c_2 = 0$$

$$c_1 + (a - \lambda)c_2 = 0$$

If we substitute $\lambda = a + 1$ into these equations, we obtain

$$-c_1 + c_2 = 0$$

$$c_1 - c_2 = 0$$

or $c_1 = c_2$. Thus, the eigenvector is (c_1, c_1), where c_1 is an arbitrary constant. We can fix the value of c_1 by requiring that the eigenvector be normalized, in which case we have

$$\mathbf{c}_1 = \begin{pmatrix} 1/\sqrt{2} \\ 1/\sqrt{2} \end{pmatrix}$$

The other normalized eigenvector is given by

$$\mathbf{c}_2 = \begin{pmatrix} 1/\sqrt{2} \\ -1/\sqrt{2} \end{pmatrix}$$

It's an easy exercise to verify that $\mathbf{A}\mathbf{c}_1 = \lambda_1\mathbf{c}_1$ and that $\mathbf{A}\mathbf{c}_2 = \lambda_2\mathbf{c}_2$.

In Example H–1, we solved a 2×2 eigenvalue problem. The algebra was simple because we had to solve only a quadratic equation to find the two eigenvalues. The algebra increases drastically as we go on to problems of dimension greater than two; even a

3×3 system leads to a cubic equation for λ, which is usually quite tedious to solve, and the tedium grows rapidly with the size of the matrix. There are a number of user-friendly mathematical computer programs available nowadays that can easily handle very large matrices. Three such programs are *MathCad, Maple,* and *Mathematica,* each of which can perform algebraic manipulations as well as do numerical calculations. At least one of these programs is available in most chemistry departments, and you should learn how to use one of these programs. Any of these programs, as well as others, can solve for all the eigenvalues and corresponding eigenvectors of a sizable matrix in seconds.

Note that the two eigenvectors in Example H–1 are orthonormal because

$$c_1 \cdot c_1 = \left(\frac{1}{\sqrt{2}} \frac{1}{\sqrt{2}} + \frac{1}{\sqrt{2}} \frac{1}{\sqrt{2}} \right) = 1$$

$$c_2 \cdot c_2 = \left[\frac{1}{\sqrt{2}} \frac{1}{\sqrt{2}} + \left(-\frac{1}{\sqrt{2}} \right) \left(-\frac{1}{\sqrt{2}} \right) \right] = 1$$

and

$$c_1 \cdot c_2 = \left[\frac{1}{\sqrt{2}} \frac{1}{\sqrt{2}} + \frac{1}{\sqrt{2}} \left(-\frac{1}{\sqrt{2}} \right) \right] = 0$$

This is generally true for eigenvectors of distinct eigenvalues of a symmetric matrix; in other words, one for which $A = A^T$. This result is completely analogous to the fact that the nondegenerate eigenfunctions of Hermitian operators are orthonormal (see Section 4.6). Recall that a definition of a Hermitian operator $\hat{A}$ is

$$\int d\tau \, \psi_i^* \hat{A} \psi_j = \int d\tau \, (\hat{A}\psi_i)^* \psi_j \tag{H.9}$$

If we let $A_{ij} = \int d\tau \, \psi_i^* \hat{A} \psi_j$, then Equation H.9 says that

$$A_{ij} = A_{ji}^* \qquad \text{(Hermitian matrix)} \tag{H.10}$$

A symmetrical matrix would have $A_{ij} = A_{ji}$. Equation H.10 is the extension of the definition of a symmetric matrix to a complex space, where the elements of the matrices may be complex. If A satisfies Equation H.10, it is said to be a *Hermitian matrix.* All matrices in quantum mechanics must be Hermitian because the eigenvalues of a Hermitian matrix are real, just as the eigenvalues of a Hermitian operator are real (see Section 4.6).

EXAMPLE H–2
Show that the eigenvectors of a Hermitian matrix are real and that the eigenvectors corresponding to distinct eigenvalues are orthogonal.

SOLUTION: Start with $Ac_j = \lambda_j c_j$. Multiply both sides from the left by c_i^* to obtain $c_i^* Ac_j = \lambda_j c_i^* c_j$, which we write in the notation

$$A_{ij} = \lambda_j c_i^* c_j \tag{H.11}$$

Now multiply $A^*c_i^* = \lambda_i^* c_i^*$ from the left by c_j to obtain $c_j A^* c_i^* = \lambda_i^* c_j c_i^*$, which we write as

$$A_{ji}^* = \lambda_i^* c_j c_i^* \tag{H.12}$$

But A is Hermitian, so $A_{ij} = A_{ji}^*$. Furthermore, $c_j c_i^* = c_i^* c_j$ because the dot product of two vectors is commutative. Comparing Equations H.11 and H.12 gives

$$(\lambda_i^* - \lambda_j) c_i^* c_j = 0 \tag{H.13}$$

If $i = j$, $c_i^* c_j \geq 0$, and so $\lambda_j = \lambda_j^*$, which says that the eigenvalues are real. If $i \neq j$, then $\lambda_i \neq \lambda_j$ if there is no degeneracy, and so $c_i^* c_j = 0$, which says that c_i and c_j are orthogonal.

Let's go back to Equation H.2, which we will write in the form

$$Ac_k = \lambda_k c_k \qquad k = 1, 2, \ldots, N \tag{H.14}$$

There are N eigenvalues λ_k and N corresponding eigenvectors, c_k. Now let's normalize the c_k and form a matrix

$$S = (c_1, c_2, \ldots, c_N) \tag{H.15}$$

where the notation means that the columns of S are the (normalized) eigenvectors of A. Because the columns of S consist of the eigenvectors of A, and because these eigenvectors form an orthonormal set if A is symmetric (which it usually is), S is an orthogonal matrix. In other words, $S^{-1} = S^T$. Furthermore, the matrix S has a remarkable property that we can see by operating on S with A to obtain (Problem H–6)

$$AS = (Ac_1, Ac_2, \ldots, Ac_N)$$

$$= (\lambda_1 c_1, \lambda_2 c_2, \ldots, \lambda_N c_N)$$

$$= SD \tag{H.16}$$

where

$$D = \begin{pmatrix} \lambda_1 & 0 & 0 & \cdots & 0 \\ 0 & \lambda_2 & 0 & \cdots & 0 \\ \vdots & \vdots & \vdots & \ddots & \vdots \\ 0 & 0 & 0 & \cdots & \lambda_N \end{pmatrix} \tag{H.17}$$

is a diagonal matrix whose elements are the eigenvalues of A.
  If we multiply Equation H.16 from the left by S^{-1}, then we obtain

$$D = S^{-1}AS = S^T AS \tag{H.18}$$

because S is orthogonal. Equation H.18 is called a *similarity transformation*. We say that the matrix A has been *diagonalized* by the similarity transformation in Equation H.18. Diagonalizing a matrix A is *completely equivalent* to solving the eigenvalue problem in Equation H.2, or, because Equations H.1 and H.2 are equivalent, diagonalizing the Hamiltonian matrix is completely equivalent to solving the Schrödinger equation. Physically, A and D represent the same operation (such as a rotation or a reflection through a plane). Their different forms result from the fact that D is expressed in an optimum, or natural, coordinate system. Because of the central importance of matrix diagonalization in quantum mechanics, there are many sophisticated and efficient algorithms for matrix diagonalization in the numerical analysis literature.

EXAMPLE H–3
Diagonalize the matrix A in Example H–1.

SOLUTION: The matrix S is given by

$$S = \begin{pmatrix} \frac{1}{\sqrt{2}} & \frac{1}{\sqrt{2}} \\ \frac{1}{\sqrt{2}} & -\frac{1}{\sqrt{2}} \end{pmatrix}$$

The inverse of S is

$$S^{-1} = \begin{pmatrix} \frac{1}{\sqrt{2}} & \frac{1}{\sqrt{2}} \\ \frac{1}{\sqrt{2}} & -\frac{1}{\sqrt{2}} \end{pmatrix}$$

Using Equation H.18, we have

$$S^{-1}AS = \begin{pmatrix} \frac{1}{\sqrt{2}} & \frac{1}{\sqrt{2}} \\ \frac{1}{\sqrt{2}} & -\frac{1}{\sqrt{2}} \end{pmatrix} \begin{pmatrix} a & 1 \\ 1 & a \end{pmatrix} \begin{pmatrix} \frac{1}{\sqrt{2}} & \frac{1}{\sqrt{2}} \\ \frac{1}{\sqrt{2}} & -\frac{1}{\sqrt{2}} \end{pmatrix}$$

$$= \begin{pmatrix} a+1 & 0 \\ 0 & a-1 \end{pmatrix} = D$$

Notice that the elements of D are the eigenvalues of A. Notice also that the trace of A is equal to the trace of D, which equals $\lambda_1 + \lambda_2$ (Problem H–12).

Problems

H–1. Determine the eigenvalues and eigenvectors of $A = \begin{pmatrix} 1 & 1 \\ 1 & 1 \end{pmatrix}$.

H–2. Determine the eigenvalues and eigenvectors of $A = \begin{pmatrix} 1 & -2 \\ -2 & 1 \end{pmatrix}$.

H–3. Determine the eigenvalues and eigenvectors of $A = \begin{pmatrix} 1 & 0 & 1 \\ 0 & 1 & 0 \\ 1 & 0 & 0 \end{pmatrix}$.

H–4. Determine the eigenvalues and eigenvectors of $A = \begin{pmatrix} 1 & 0 & -1 \\ 0 & 1 & 0 \\ -1 & 0 & 1 \end{pmatrix}$.

H–5. Show that the matrix $A = \begin{pmatrix} 1 & i & 1-i \\ -i & 0 & -1+i \\ 1+i & -1-i & 3 \end{pmatrix}$ is Hermitian.

H–6. Verify that $(\lambda_1 c_1, \lambda_2 c_2, \ldots, \lambda_N c_N) = SD$ in Equation H.16.

H–7. The three eigenvectors of A in Problem H–4 are $c_1(-1, 0, 1)$, $c_2(0, 1, 0)$, and $c_3(1, 0, 1)$, where c_1, c_2, and c_3 are arbitrary. Choose them so that the three eigenvectors are normalized. Now form the matrix S whose columns consist of the three normalized eigenvectors. Find the inverse of S and then show explicitly that $S^{-1} = S^T$, or that S is indeed orthogonal.

H–8. Diagonalize the matrix in Problem H–1.

H–9. Diagonalize the matrix in Problem H–2.

H–10. Diagonalize the matrix in Problem H–3.

H–11. Diagonalize the matrix in Problem H–4.

H–12. Show that $\text{Tr } D = \text{Tr } S^{-1}AS$.

H–13. Programs such as *MathCad* and *Mathematica* can find the eigenvalues and corresponding eigenvectors of large matrices in seconds. Use one of these programs to find the eigenvalues and corresponding eigenvectors of

$$A = \begin{pmatrix} a & 1 & 0 & 0 & 0 & 1 \\ 1 & a & 1 & 0 & 0 & 0 \\ 0 & 1 & a & 1 & 0 & 0 \\ 0 & 0 & 1 & a & 1 & 0 \\ 0 & 0 & 0 & 1 & a & 1 \\ 1 & 0 & 0 & 0 & 1 & a \end{pmatrix}$$

Charlotte E. Moore was born in Ercildoun, Pennsylvania, on September 24, 1898, and died in 1990. After graduating from Swarthmore College in 1920, she worked at the Princeton University Observatory and the Mt. Wilson Observatory on stellar spectra and the determination of the sun's chemical composition. She earned a Ph.D. in astronomy in 1931 from the University of California at Berkeley on a Lick Fellowship. After receiving her Ph.D., she returned to Princeton, and in 1945 joined the spectroscopy section at the National Bureau of Standards (now the National Institute of Standards and Technology) until her retirement in 1968. Moore was placed in charge of an Atomic Energy Level Program, a program whose mission was to produce a current and more complete compilation of spectral data and atomic energy levels. She not only compiled published data but critically analyzed the data for each spectrum. When the data were insufficient or of dubious quality, she persuaded competent spectroscopists to carry out new observations and analysis. The result of her effort, *Atomic Energy Levels* (1949–1958), is a classic work that provides data for 485 atomic species in a uniform, clear format with standardized notation. In 1949, Moore was elected as an Associate of the Royal Astronomical Society, the first woman to receive this honor, breaking a 129-year tradition. In 1937, she married a fellow astronomer, Bancroft Sitterly, but always published under her maiden name.

CHAPTER 9

Many-Electron Atoms

We concluded Chapter 7 with an introduction to the helium atom. We showed there that if we considered the nucleus to be fixed at the origin, then the Schrödinger equation has the form

$$\left[\hat{H}_{\mathrm{H}}(1) + \hat{H}_{\mathrm{H}}(2) + \frac{e^2}{4\pi \epsilon_0 r_{12}} \right] \psi(\mathbf{r}_1, \mathbf{r}_2) = E \psi(\mathbf{r}_1, \mathbf{r}_2) \tag{9.1}$$

where $\hat{H}_{\mathrm{H}}(j)$ is the hydrogen-like Hamiltonian operator of electron j (Equation 7.77). If it were not for the presence of the interelectronic repulsion term, Equation 9.1 would be immediately solvable. Its eigenfunctions would be products of hydrogen-like wave functions and its eigenvalues would be sums of the hydrogen-like energies of the two electrons (see Section 3.9). Helium is our first multielectron system, and although the helium atom may seem to be of minimal interest to chemists, we will discuss it in detail in this chapter because the solution of the helium atom illustrates the techniques used for more complex atoms. Then, after discussing electron spin and the Pauli exclusion principle, we will discuss the Hartree–Fock theory of many-electron atoms. Finally, we will discuss the term symbols of atoms and ions and how they are used to label electronic states. This chapter illustrates the powerful utility of quantum mechanics in analyzing the electronic properties of atoms.

9.1 Atomic and Molecular Calculations Are Expressed in Atomic Units

We will apply both perturbation theory and the variational method to a helium atom, but before doing so, we will introduce a system of units, called *atomic units*, that is widely used in atomic and molecular calculations to simplify the equations. Natural units of mass and charge on an atomic or molecular scale are the mass of an electron and the magnitude of the charge on an electron (the charge on a proton). Equation 1.22 suggests that a natural unit of angular momentum on an atomic or molecular scale is $\hbar$.

TABLE 9.1
Atomic Units and Their SI Equivalents

Property	Atomic unit	SI equivalent
Mass	Mass of an electron, m_e	9.1094×10^{-31} kg
Charge	Charge on a proton, e	1.6022×10^{-19} C
Angular momentum	Planck constant divided by 2π, $\hbar$	1.0546×10^{-34} J·s
Length	Bohr radius, $a_0 = \dfrac{4\pi\epsilon_0\hbar^2}{m_e e^2}$	5.2918×10^{-11} m
Energy	$\dfrac{m_e e^4}{16\pi^2\epsilon_0^2\hbar^2} = \dfrac{e^2}{4\pi\epsilon_0 a_0} = E_h$	4.3597×10^{-18} J
Permittivity	$\kappa_0 = 4\pi\epsilon_0$	1.1127×10^{-10} C²·J⁻¹·m⁻¹

A natural unit of length on an atomic scale is the Bohr radius (Equation 1.24),

$$a_0 = \frac{4\pi\epsilon_0\hbar^2}{m_e e^4} \tag{9.2}$$

and we saw repeatedly in Chapter 7 that a natural unit of energy is

$$E = \frac{m_e e^4}{16\pi^2\epsilon_0^2\hbar^2} \tag{9.3}$$

It is convenient in atomic and molecular calculations to use units that are natural on that scale. The units that we will adopt for atomic and molecular calculations are given in Table 9.1. This set of units is called *atomic units*. The atomic unit of energy is called a *hartree* and is denoted by E_h. Note that in atomic units the ground-state energy of a hydrogen atom is $-\frac{1}{2} E_h$ (cf. Equation 7.11).

EXAMPLE 9–1
The unit of energy in atomic units is given by

$$1\,E_h = \frac{m_e e^4}{16\pi^2\epsilon_0^2\hbar^2}$$

Express $1\,E_h$ in units of joules (J), kilojoules per mole (kJ·mol⁻¹), wave numbers (cm⁻¹), and electron volts (eV).

SOLUTION: To find $1\,E_h$ expressed in joules, we substitute the SI values of m_e, e, $4\pi\epsilon_0$, and $\hbar$ into the above equation. Using these values from Table 9.1, we find

$$1\,E_h = \frac{(9.1094 \times 10^{-31}\,\text{kg})(1.6022 \times 10^{-19}\,\text{C})^4}{(1.1127 \times 10^{-10}\,\text{C}^2 \cdot \text{J}^{-1} \cdot \text{m}^{-1})^2 (1.0546 \times 10^{-34}\,\text{J} \cdot \text{s})^2}$$

$$= 4.3597 \times 10^{-18}\,\text{J}$$

If we multiply this result by the Avogadro constant, we obtain

$$1\,E_h = 2625.5\,\text{kJ} \cdot \text{mol}^{-1}$$

To express $1\,E_h$ in wave numbers (cm^{-1}), we use the fact that $1\,E_h = 4.3597 \times 10^{-18}\,\text{J}$ along with the equation

$$\tilde{v} = \frac{1}{\lambda} = \frac{hv}{hc} = \frac{E}{ch} = \frac{4.3597 \times 10^{-18}\,\text{J}}{(2.9979 \times 10^8\,\text{m} \cdot \text{s}^{-1})(6.6261 \times 10^{-34}\,\text{J} \cdot \text{s})}$$

$$= 2.1947 \times 10^7\,\text{m}^{-1} = 2.1947 \times 10^5\,\text{cm}^{-1}$$

so that we can write

$$1\,E_h = 2.1947 \times 10^5\,\text{cm}^{-1}$$

Last, to express $1\,E_h$ in terms of electron volts, we use the conversion factor

$$1\,\text{eV} = 1.6022 \times 10^{-19}\,\text{J}$$

Using the value of $1\,E_h$ in joules obtained previously, we have

$$1\,E_h = (4.3597 \times 10^{-18}\,\text{J})\left(\frac{1\,\text{eV}}{1.6022 \times 10^{-19}\,\text{J}}\right)$$

$$= 27.211\,\text{eV}$$

The use of atomic units greatly simplifies most of the equations we will use in atomic and molecular calculations. For example, the Hamiltonian operator of a helium atom

$$\hat{H} = -\frac{\hbar^2}{2m_e}\nabla_1^2 - \frac{\hbar^2}{2m_e}\nabla_2^2 - \frac{2e^2}{4\pi\epsilon_0 r_1} - \frac{2e^2}{4\pi\epsilon_0 r_2} + \frac{e^2}{4\pi\epsilon_0 r_{12}} \tag{9.4}$$

becomes simply

$$\hat{H} = -\frac{1}{2}\nabla_1^2 - \frac{1}{2}\nabla_2^2 - \frac{2}{r_1} - \frac{2}{r_2} + \frac{1}{r_{12}} \tag{9.5}$$

in atomic units (Problem 9–6). An important aspect of the use of atomic units in atomic and molecular calculations is that the calculated energies are independent of the values

TABLE 9.2
Ground-State Energy of a Helium Atom [a]

Method	Energy/E_h	Ionization energy/E_h	Ionization energy/kJ·mol^{-1}
Perturbation calculations			
Complete neglect of the interelectronic repulsion term	−4.0000	2.000	5250
First-order perturbation theory	−2.7500	0.7500	1969
Second-order perturbation theory	−2.9077	0.9077	2383
Thirteenth-order perturbation theory [b]	−2.903 724 33	0.903 724 33	2373
Variational calculations			
$(1s)^2$ with $\zeta = 1.6875$	−2.8477	0.8477	2226
Eckart, Equation 9.13 [c]	−2.8757	0.8757	2299
Hartree–Fock [d]	−2.861 68	0.8617	2262
Hylleraas, [e] 10 parameters	−2.903 63	0.903 63	2372
Pekeris, [f] 1078 parameters	−2.903 724 375	0.903 724 375	2373
Experimental value	−2.9033	0.9033	2373

a. These are nonrelativistic, fixed-nucleus approximation energies. Corrections for nuclear motion and relativistic corrections can be estimated to be $10^{-4}E_h$.
b. Scheer, C. W., Knight, R. E. Two-Electron Atoms III. A Sixth-Order Perturbation Study of the $1\,^1S$ Ground State. *Rev. Mod. Phys.,* **35**, 426 (1963).
c. Eckart, C. E. The Theory and Calculation of Screening Constants. *Phys. Rev.,* **36**, 878 (1930).
d. Clementi, E., Roetti, C. Roothaan–Hartree–Fock atomic wavefunctions: Basis functions and their coefficients for ground and certain excited states of neutral and ionized atoms, $Z \leq 54$. *At. Data Nucl. Data Tables,* **14 (3–4)**, 177 (1974).
e. Hylleraas, E. A. Neue Berechnung der Energie des Heliums in Grundzustande, sowie des tiefsten Terms von Ortho-Helium. *Z. Physik,* **54**, 347 (1929).
f. Pekeris, C. L. $1\,^1S$ and $2\,^3S$ States of Helium. *Phys. Rev.,* **115**, 1216 (1959).

of physical constants such as the electron mass, the Planck constant, etc. As the values of physical constants are further refined by advances in experimental methodology, the energies calculated using atomic units will not be affected by these refinements. For example, we will see in the next section that the most accurate calculation of the ground-state energy of a helium atom gives −2.903 724 375 E_h (Table 9.2), which took months of computer time at the time the calculation was done. Because atomic units were used, this value will never have to be redetermined.

9.2 Both Perturbation Theory and the Variational Method Can Yield Good Results for a Helium Atom

We applied first-order perturbation theory to a helium atom in Section 8.5 and found that the first-order correction to the energy is $5Z/8\, E_h$ and that the energy of a helium atom through first order is given by

$$E = -Z^2 + \frac{5}{8}Z = -\frac{11}{4}E_h = -2.750\, E_h \tag{9.6}$$

or $-7219\,\text{kJ}\cdot\text{mol}^{-1}$. The experimental value of the energy is $-2.9033\, E_h$, or $-7621\,\text{kJ}\cdot\text{mol}^{-1}$, and so we see that first-order perturbation theory gives a result that is approximately 5% in error. Scheer and Knight (see Table 9.2) calculated the energy through many orders of perturbation theory and found that

$$E = -Z^2 + \frac{5}{8}Z - 0.157\,66 + \frac{0.008\,70}{Z} + \frac{0.000\,889}{Z^2} + \cdots \tag{9.7}$$

Equation 9.7 yields a value of $-2.9037\, E_h$, in good agreement with the experimental value of $-2.9033\, E_h$.

We also used the variational method to calculate the ground-state energy of a helium atom in Section 8.1. We used a (normalized) trial function of the form $\psi(\mathbf{r}_1, \mathbf{r}_2) = \psi_{1s}(\mathbf{r}_1)\psi_{1s}(\mathbf{r}_2)$ with $\psi_{1s}(\mathbf{r}) = (\zeta^3/\pi)e^{-\zeta r}$, where ζ is a variational parameter. In other words, we used

$$\psi(\mathbf{r}_1, \mathbf{r}_2) = \frac{\zeta^3}{\pi}e^{-\zeta(r_1+r_2)} \tag{9.8}$$

Thus, we had to evaluate

$$E = \iint d\mathbf{r}_1 d\mathbf{r}_2 \psi(\mathbf{r}_1, \mathbf{r}_2)\hat{H}\psi(\mathbf{r}_1, \mathbf{r}_2) \tag{9.9}$$

In Section 8.1, we found that E comes out to be

$$E(\zeta) = \zeta^2 - \frac{27}{8}\zeta \tag{9.10}$$

Minimizing E with respect to ζ gives $\zeta_{\min} = 27/16$ and

$$E_{\min} = -\left(\frac{27}{16}\right)^2 E_h = -2.847\,66\, E_h \tag{9.11}$$

compared to the first-order perturbation theory result of $-2.7500\, E_h$ and the higher-order result of $-2.9037\, E_h$. [Problems 9–7 through 9–10 explain why Equation 9.10

consists of a term that is quadratic in ζ and a term that is linear in ζ, and then develop the relation between minimizing $E(\zeta)$ with respect to ζ and the virial theorem.]

The agreement we have found between first-order perturbation theory or our simple variational approximation and the experimental value of the energy may appear to be quite good, but let's examine this agreement more closely. The ionization energy (IE) of a helium atom is given by

$$IE = E_{He^+} - E_{He}$$

The energy of He^+ is $-2E_h$ (Problem 9–2), so we have

$$IE = -2 + \frac{11}{4} = 0.7500\, E_h$$

$$= 1969\ \text{kJ·mol}^{-1} \quad \text{(first-order perturbation theory)}$$

or

$$IE = -2 + \left(\frac{27}{16}\right)^2 = 0.8477\, E_h$$

$$= 2226\ \text{kJ·mol}^{-1} \quad \text{(our variational result)}$$

whereas the experimental value of the ionization energy is $0.9033\, E_h$, or $2373\ \text{kJ·mol}^{-1}$. Even our variational result, with its 6% discrepancy with the experimental total energy, is not too satisfactory if you realize that an error of $0.0056\, E_h$ is equivalent to about $150\ \text{kJ·mol}^{-1}$, which is the same order of magnitude as the strength of a chemical bond.

Example 9–2 illustrates another shortcoming of the simple variational trial function that we have been using.

EXAMPLE 9–2
In Problem 9–11, we show that the generalization of Equation 9.10 for a helium-like two-electron atom or ion of nuclear charge Z is

$$E(\zeta) = -\zeta^2 + 2\zeta(\zeta - Z) + \frac{5}{8}\zeta \tag{9.12}$$

(Note that Equation 9.12 reduces to Equation 9.10 when $Z = 2$.) Use this result to calculate the ionization energy of a hydride ion, and interpret your result.

SOLUTION: To find the energy of a hydride ion, we let $Z = 1$ in Equation 9.12 to obtain

$$E(\zeta) = \zeta^2 - \frac{11}{8}\zeta$$

Minimize this with respect to ζ to obtain $\zeta_{min} = 11/16$ and $E = -(11/16)^2 E_h = -0.472\,66\,E_h$. The ionization energy is given by

$$IE = E_H - E_{H^-} = -\frac{1}{2} + \left(\frac{11}{16}\right)^2 = -0.0273\,E_h$$

This negative result for IE implies that the hydride ion is not stable with respect to a separated hydrogen atom and an electron (at rest); in other words, it will ionize spontaneously. This is not the case experimentally and, in fact, the ionization energy of a hydride ion is $0.0275\,E_h$. Clearly, we must be able to do better than this.

One approach is the following. The trial function in Equation 9.8 assumes that both electrons have the same effective nuclear charge. This may be so in some average sense, but it will not be true at all times because there will be instants of time where one electron is far from the nucleus and the other close to it, and so the effective nuclear charges will not be the same. We can account for this by using two variational parameters, ζ_1 and ζ_2, in Equation 9.8. If we were to just write a single product trial function, however, we would be distinguishing one electron from the other by the labels 1 and 2. (We shall have more to say about this in Section 9.4.) In order to treat the two electrons on an equal footing, we write $\psi(\mathbf{r}_1, \mathbf{r}_2)$ as

$$\psi(\mathbf{r}_1, \mathbf{r}_2) = N(e^{-\zeta_1 r_1}e^{-\zeta_2 r_2} + e^{-\zeta_2 r_1}e^{-\zeta_1 r_2}) \tag{9.13}$$

where N is a normalization constant.

When we use Equation 9.13 in Equation 9.9 to calculate $E(\zeta_1, \zeta_2)$, the integral is similar to the one for Equation 9.8, only a bit messier. Nevertheless, when $E(\zeta_1, \zeta_2)$ is minimized with respect to ζ_1 and ζ_2, we find that $E = -2.875\,66\,E_h$ (Problem 9–12), which is a significant improvement over our simple variational treatment. The ionization energy comes out to be $0.8757\,E_h$, compared to $0.8477\,E_h$ using Equation 9.8. (The accepted value is $0.9037\,E_h$.) Furthermore, the ionization energy of H^- comes out to be $+0.0133\,E_h$, which is positive but still about a factor of 2 less than the accepted value.

Although the trial function given by Equation 9.13 leads to a significant improvement over the trial function given by Equation 9.8, we can do better yet. Because a suitable trial function may be almost any convenient function (that satisfies the boundary conditions), we are not restricted to using 1s hydrogen-like orbitals as we have done up to now. For example, we could use a linear combination of a 1s orbital and a 2s orbital and write

$$\psi(\mathbf{r}_1, \mathbf{r}_2) = \phi(\mathbf{r}_1)\phi(\mathbf{r}_2) \tag{9.14}$$

where $\phi(r) = N[c_1 e^{-\zeta r} + c_2(2 - r)e^{-\zeta r/2}]$ and N is a normalization constant. This trial function has three variational parameters, and will yield a better energy than using just a 1s orbital. There is no reason (other than computational) not to go on and include more hydrogen-like orbitals. Such a procedure has been very well developed and used extensively over the years. There is an important modification, however. Rather than using hydrogen-like orbitals, it is customary to use a set of functions that were introduced by the American physicist John Slater in the 1930s. These functions, which are called *Slater orbitals*, are of the form

$$S_{nlm_l}(r, \theta, \phi) = N_{nl} r^{n-1} e^{-\zeta r} Y_l^{m_l}(\theta, \phi) \tag{9.15}$$

where $N_{nl} = (2\zeta)^{n+\frac{1}{2}}/[(2n)!]^{1/2}$ is a normalization constant (Problem 9–13) and the $Y_l^{m_l}(\theta, \phi)$ are the spherical harmonics (Section 6.6 and Table 6.5). The parameter ζ (zeta) is taken to be arbitrary and is not necessarily equal to Z/n as in the hydrogen-like orbitals. Note that the radial parts of Slater orbitals do not have nodes like hydrogen atomic orbitals do (Problem 9–14).

EXAMPLE 9–3

Show that $S_{nlm_l}(r, \theta, \phi)$ is not orthogonal to $S_{n'lm_l}(r, \theta, \phi)$.

SOLUTION: We must show that

$$I = \int_0^\infty dr\, r^2 \int_0^\pi d\theta \sin\theta \int_0^{2\pi} d\phi\, S_{nlm_l}^*(r, \theta, \phi) S_{n'lm_l}(r, \theta, \phi)$$

$$\propto \int_0^\infty dr\, r^{n+n'} e^{-2\zeta r} \int_0^\pi d\theta \sin\theta \int_0^{2\pi} d\phi\, Y_l^{m_l}(\theta, \phi)^* Y_l^{m_l}(\theta, \phi)$$

$$\neq 0$$

The integral over θ and ϕ equals one because of the orthonormality of the spherical harmonics, leaving

$$I = \int_0^\infty r^{n+n'} e^{-2\zeta r} dr$$

But this integral cannot equal zero because the integrand is always positive.

It has become a standard procedure in quantum chemistry to use a trial function of the form of Equation 9.14, where $\phi(r)$ is a linear combination of Slater orbitals. As we include more and more Slater orbitals, we reach a limit that is both practical and theoretical. In this limit, $E = -2.8617\, E_h$ and the ionization energy is $0.8617\, E_h$, compared with the best variational values, $-2.9037\, E_h$ and $0.9037\, E_h$, respectively. This limiting value is the best value of the energy that can be obtained using a trial function of the form of a product of one-electron wave functions (Equation 9.14). This limit is called the *Hartree–Fock limit*, and we will discuss it more fully in the next section. Note that the concept of electron orbitals is preserved in the *Hartree–Fock*

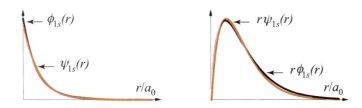

FIGURE 9.1
A comparison of the orbital that we obtained from a simple variational calculation [in other words, $\psi_{1s}(\zeta = 27/16)$] and the Hartree–Fock orbital given by Equation 9.16.

approximation because Equation 9.14 represents the trial function as a product of one-electron functions, or *orbitals*. For example, we shall see in the next section that the $1s$ orbital of a helium atom is well approximated by the two-term expression (Problem 9–15)

$$\phi_{1s}(r) = 0.843\,785\,S_{1s}(\zeta = 1.453\,63) + 0.180\,687\,S_{1s}(\zeta = 2.910\,93) \quad (9.16)$$

where $S_{1s}(\zeta)$ is a Slater $1s$ orbital (which happens to be the same as a hydrogen $1s$ orbital). Because Equation 9.16 is a sum of two terms with different values of ζ, $\phi_{1s}(r)$ given by Equation 9.16 is called a *double-zeta orbital*. Figure 9.1 compares Equation 9.16 to the simple one-parameter variational trial function $(\zeta^3/\pi)^{1/2}e^{-\zeta r}$ with $\zeta = 27/16$. The energy associated with Equations 9.14 and 9.16 is $-2.8617\,E_h$. We shall see that this is the best energy that we can achieve using a trial function of the form of Equation 9.14 (in other words, using orbitals).

If we do not restrict the trial function to be a product of single-electron orbitals, then we can go on and obtain essentially the exact energy. It has been found to be efficacious to include terms containing the interelectronic distance r_{12} explicitly in the trial function. This was first done by Hylleraas in 1930, who used a trial function of the (unnormalized) form

$$\psi(r_1, r_2, r_{12}) = e^{-\zeta r_1}e^{-\zeta r_2}(1 + cr_{12}) \quad (9.17)$$

Using ζ and c as variational parameters, Hylleraas obtained a value of $E = -2.8913\,E_h$, within less than 0.5% of the exact value. Hylleraas used a mechanical calculator to carry out his now classic calculations, but nowadays we can use a computer to easily carry out this procedure to a large number of terms to yield an energy that is essentially exact. The most extensive such calculation was carried out in 1959 by Pekeris, who obtained $E = -2.903\,724\,375\,E_h$ using a trial function containing 1078 parameters.

Although these calculations do show that we can obtain essentially exact energies by using the variational method with r_{12} in the trial function explicitly, these calculations are quite difficult computationally and do not readily lend themselves to large atoms and molecules. Furthermore, we have abandoned the orbital concept altogether. The orbital concept has been of great use to chemists, so the scheme nowadays is to find

hydrogen-like wave functions as specific examples, the first two spin orbitals of a hydrogen-like atom are

$$\Psi_{100\frac{1}{2}} = \psi_{1s}(r)\alpha = \left(\frac{Z^3}{\pi}\right)^{1/2} e^{-Zr}\alpha$$

(9.35)

$$\Psi_{100-\frac{1}{2}} = \psi_{1s}(r)\beta = \left(\frac{Z^3}{\pi}\right)^{1/2} e^{-Zr}\beta$$

We showed in Section 7.5 that $\Psi_{100\frac{1}{2}}$ and $\Psi_{100-\frac{1}{2}}$ are orthonormal:

$$\langle \Psi_{100\frac{1}{2}} \mid \Psi_{100\frac{1}{2}} \rangle = \langle \psi_{1s} \mid \psi_{1s} \rangle \langle \alpha \mid \alpha \rangle = 1$$

$$\langle \Psi_{100-\frac{1}{2}} \mid \Psi_{100-\frac{1}{2}} \rangle = \langle \psi_{1s} \mid \psi_{1s} \rangle \langle \beta \mid \beta \rangle = 1$$

and

$$\langle \Psi_{100\frac{1}{2}} \mid \Psi_{100-\frac{1}{2}} \rangle = \langle \psi_{1s} \mid \psi_{1s} \rangle \langle \alpha \mid \beta \rangle = 0$$

because the spatial part is normalized and $\langle \alpha \mid \alpha \rangle = \langle \beta \mid \beta \rangle = 1$ and $\langle \alpha \mid \beta \rangle = \langle \beta \mid \alpha \rangle = 0$.

You probably remember from general chemistry that no two electrons in an atom can have the same values of all four quantum numbers, n, l, m_l, and m_s. This restriction is called the *Pauli exclusion principle*. There is another, more fundamental statement of the exclusion principle that restricts the form of a multielectron wave function. We will present the Pauli exclusion principle as another postulate of quantum mechanics, but before doing so we must introduce the idea of an *antisymmetric wave function*. Let's go back to a helium atom and write

$$\psi(1, 2) = 1s\alpha(1)1s\beta(2)$$

(9.36)

where $1s\alpha$ and $1s\beta$ are shorthand notation for $\Psi_{100\frac{1}{2}}$ and $\Psi_{100-\frac{1}{2}}$, respectively, and where the arguments 1 and 2 denote all four coordinates (x, y, z, and σ) of electrons 1 and 2, respectively. Note that Equation 9.36 corresponds to a product of the two wave functions given by Equation 9.35. Because no known experiment can distinguish one electron from another, we say that electrons are indistinguishable and, therefore, cannot be labeled. Thus, the wave function

$$\psi(2, 1) = 1s\alpha(2)1s\beta(1)$$

(9.37)

is equally as good as Equation 9.36. Mathematically, indistinguishability requires that we take linear combinations involving all possible labelings of the electrons. For a two-electron atom, we take the linear combinations of Equations 9.36 and 9.37 and write

$$\Psi_1(1, 2) = \psi(1, 2) + \psi(2, 1) = 1s\alpha(1)1s\beta(2) + 1s\alpha(2)1s\beta(1)$$

(9.38)

and

$$\Psi_2(1, 2) = \psi(1, 2) - \psi(2, 1) = 1s\alpha(1)\,1s\beta(2) - 1s\alpha(2)\,1s\beta(1) \qquad (9.39)$$

Both Ψ_1 and Ψ_2 describe states in which there are two indistinguishable electrons; one electron is in the spin orbital $1s\alpha$ and the other is in $1s\beta$. Neither wave function specifies which electron is in each spin orbital, nor should they because the electrons are indistinguishable.

Both of the wave functions Ψ_1 and Ψ_2 appear to be acceptable wave functions for the ground state of a helium atom, but it turns out experimentally that we must use the wave function Ψ_2. Note that Ψ_2 has the property that it changes sign when the two electrons are interchanged because

$$\Psi_2(2, 1) = \psi(2, 1) - \psi(1, 2) = -\Psi_2(1, 2) \qquad (9.40)$$

We say that $\Psi_2(1, 2)$ is *antisymmetric* under the interchange of the two electrons. The observation that the ground state of a helium atom is described by only Ψ_2 is but one example of the Pauli exclusion principle:

Postulate 6
All electronic wave functions must be antisymmetric under the interchange of any two electrons.

In Section 9.5, we will show that Postulate 6 implies the more familiar statement of the Pauli exclusion principle, that no two electrons in an atom can have the same values of the four quantum numbers, n, l, m_l, and m_s.

EXAMPLE 9–4
The wave function $\Psi_2(1, 2)$ given by Equation 9.39 is not normalized as it stands. Determine the normalization constant of $\Psi_2(1, 2)$ given that the "$1s$" parts are normalized.

SOLUTION: We want to find the constant c such that

$$I = c^2 \langle \Psi_2(1, 2) \,|\, \Psi_2(1, 2) \rangle = 1$$

First notice that $\Psi_2(1, 2)$ can be factored into the product of a spatial part and a spin part:

$$\Psi_2(1, 2) = 1s(1)\,1s(2)[\alpha(1)\beta(2) - \alpha(2)\beta(1)]$$

$$= 1s(\mathbf{r}_1)\,1s(\mathbf{r}_2)[\alpha(\sigma_1)\beta(\sigma_2) - \alpha(\sigma_2)\beta(\sigma_1)] \qquad (9.41)$$

The normalization integral becomes the product of three integrals:

$$I = c^2 \langle 1s(1) \,|\, 1s(1) \rangle \langle 1s(2) \,|\, 1s(2) \rangle \langle \alpha(1)\beta(1) - \alpha(2)\beta(1) \,|\, \alpha(1)\beta(2) - \alpha(2)\beta(1) \rangle$$

The spatial integrals are equal to 1 because we have taken the $1s$ orbitals to be normalized. Now let's look at the spin integrals. When the two terms in the integrand of the spin integral are multiplied, we get four integrals. One of them is

$$\iint \alpha^*(\sigma_1)\beta^*(\sigma_2)\alpha(\sigma_1)\beta(\sigma_2)d\sigma_1 d\sigma_2 = \langle \alpha(1)\beta(2) \mid \alpha(1)\beta(2) \rangle$$

$$= \langle \alpha(1) \mid \alpha(1) \rangle \langle \beta(2) \mid \beta(2) \rangle = 1$$

where once again we point out that integrating over σ_1 and σ_2 is purely symbolic; σ_1 and σ_2 are discrete variables. Another is

$$\langle \alpha(1)\beta(2) \mid \alpha(2)\beta(1) \rangle = \langle \alpha(1) \mid \beta(1) \rangle \langle \beta(2) \mid \alpha(2) \rangle = 0$$

The other two are equal to 1 and 0, and so

$$I = c^2 \langle \Psi_2(1, 2) \mid \Psi_2(1, 2) \rangle = 2c^2 = 1$$

or $c = 1/\sqrt{2}$.

9.5 Antisymmetric Wave Functions Can Be Represented by Slater Determinants

Now that we have introduced spin and have seen that we must use antisymmetric wave functions, we must ask why we could ignore the spin part of the wave function when we treated a helium atom in Section 9.2. The reason is that Ψ_2 can be factored into a spatial part and a spin part, as we saw in Equation 9.41 in Example 9–4. In Section 9.2, we used only the spatial part of Ψ_2, and the spatial part is just a product of two Slater $1s$ orbitals. If we use Ψ_2 in Equation 9.39 to calculate the ground-state energy of a helium atom, then we obtain

$$E = \frac{\langle \Psi_2(1, 2) \mid \hat{H} \mid \Psi_2(1, 2) \rangle}{\langle \Psi_2(1, 2) \mid \Psi_2(1, 2) \rangle} \tag{9.42}$$

The numerator in Equation 9.42 is

$$\int 1s^*(\mathbf{r}_1) 1s^*(\mathbf{r}_2)[\alpha^*(\sigma_1)\beta^*(\sigma_2) - \alpha^*(\sigma_2)\beta^*(\sigma_1)]$$

$$\times \hat{H} 1s(\mathbf{r}_1) 1s(\mathbf{r}_2)[\alpha(\sigma_1)\beta(\sigma_2) - \alpha(\sigma_2)\beta(\sigma_1)] \, d\mathbf{r}_1 d\mathbf{r}_2 d\sigma_1 d\sigma_2 \tag{9.43}$$

Because the Hamiltonian operator does not contain any spin operators, it does not affect the spin functions and so we can factor the integral in Equation 9.43 to give

$$\int 1s^*(\mathbf{r}_1) 1s^*(\mathbf{r}_2) \hat{H} 1s(\mathbf{r}_1) 1s(\mathbf{r}_2) d\mathbf{r}_1 d\mathbf{r}_2$$

$$\times \int [\alpha^*(\sigma_1)\beta^*(\sigma_2) - \alpha^*(\sigma_2)\beta^*(\sigma_1)][\alpha(\sigma_1)\beta(\sigma_2) - \alpha(\sigma_2)\beta(\sigma_1)] d\sigma_1 d\sigma_2 \qquad (9.44)$$

We showed in Example 9–4 that the total spin integral is equal to 2. The spin integral in the denominator in Equation 9.42 is also equal to 2 (they are the same) and so Equation 9.42 becomes

$$E = \frac{\int \psi^*(\mathbf{r}_1, \mathbf{r}_2) \hat{H} \psi(\mathbf{r}_1, \mathbf{r}_2) d\mathbf{r}_1 d\mathbf{r}_2}{\int \psi^*(\mathbf{r}_1, \mathbf{r}_2) \psi(\mathbf{r}_1, \mathbf{r}_2) d\mathbf{r}_1 d\mathbf{r}_2} \qquad (9.45)$$

where $\psi(\mathbf{r}_1, \mathbf{r}_2)$ is just the spatial part of $\Psi_2(1, 2)$. Equation 9.45 is equivalent to Equation 9.42. It is important to realize that a factorization into a spatial part and a spin part does *not* occur in general.

It is fairly easy to write the antisymmetric two-electron wave function by inspection, but what if we have a set of N spin orbitals and we need to construct an antisymmetric N-electron wave function? In the early 1930s, Slater introduced the use of determinants (MathChapter F) to construct antisymmetric wave functions. If we use Equation 9.41 as an example, then we see that we can write Ψ (we will drop the subscript 2) in the form

$$\Psi(1, 2) = \begin{vmatrix} 1s\alpha(1) & 1s\beta(1) \\ 1s\alpha(2) & 1s\beta(2) \end{vmatrix} \qquad (9.46)$$

We obtain Equation 9.41 upon expanding this determinant. The wave function $\Psi(1, 2)$ given by Equation 9.46 is called a *determinantal wave function*.

Two properties of determinants are of particular importance to us. The first is that the value of a determinant changes sign when we interchange any two rows or any two columns of the determinant. The second is that a determinant is equal to zero if any two rows or any two columns are the same (MathChapter F).

Notice that when we interchange the two electrons in the determinantal wave function $\Psi(1, 2)$ (Equation 9.46), we interchange the two rows and so change the sign of $\Psi(1, 2)$. Furthermore, if we place both electrons in the same spin orbital, say, the $1s\alpha$ spin orbital, then $\Psi(1, 2)$ becomes

$$\Psi(1, 2) = \begin{vmatrix} 1s\alpha(1) & 1s\alpha(1) \\ 1s\alpha(2) & 1s\alpha(2) \end{vmatrix} = 0$$

This determinant is equal to zero because the two columns are the same. Thus, we see that the determinantal representation of wave functions automatically satisfies the Pauli exclusion principle. Determinantal wave functions are always antisymmetric and vanish when any two electrons have the same four quantum numbers—that is, when both electrons occupy the same spin orbital.

We need to consider one more factor before our discussion of determinantal wave functions is complete. Recall from Example 9–4 that the normalization constant for $\Psi(1, 2)$ given by Equation 9.46 is $1/\sqrt{2}$. Therefore,

$$\Psi(1, 2) = \frac{1}{\sqrt{2}} \begin{vmatrix} 1s\alpha(1) & 1s\beta(1) \\ 1s\alpha(2) & 1s\beta(2) \end{vmatrix} \tag{9.47}$$

is a *normalized* two-electron determinantal wave function. The factor of $1/\sqrt{2}$ assures that $\Psi(1, 2)$ is normalized.

We have developed the determinantal representation of wave functions using a two-electron system as an example. To generalize this development for an N-electron system, we use an $N \times N$ determinant. Furthermore, one can show (Problem 9–27) that the normalization constant is $\sqrt{N!}$, and so we have the normalized N-electron determinantal wave function

$$\Psi(1, 2, \ldots, N) = \frac{1}{\sqrt{N!}} \begin{vmatrix} u_1(1) & u_2(1) & \cdots & u_N(1) \\ u_1(2) & u_2(2) & \cdots & u_N(2) \\ \vdots & \vdots & \ddots & \vdots \\ u_1(N) & u_2(N) & \cdots & u_N(N) \end{vmatrix} \tag{9.48}$$

where the u's in Equation 9.48 are spin orbitals. Notice that $\Psi(1, 2, \ldots, N)$ changes sign whenever two electrons (rows) are interchanged and vanishes if any two electrons occupy the same spin orbital (two identical columns).

Let's consider a lithium atom. We cannot put all three electrons into $1s$ orbitals because two columns in the determinantal wave function would be the same. Thus, an appropriate wave function for a lithium atom is

$$\Psi(1, 2, 3) = \frac{1}{\sqrt{3!}} \begin{vmatrix} 1s\alpha(1) & 1s\beta(1) & 2s\alpha(1) \\ 1s\alpha(2) & 1s\beta(2) & 2s\alpha(2) \\ 1s\alpha(3) & 1s\beta(3) & 2s\alpha(3) \end{vmatrix}$$

EXAMPLE 9–5
Write down the determinantal wave function of a beryllium atom.

SOLUTION: The ground-state electron configuration of a beryllium atom is $1s^2 2s^2$. Therefore,

$$\Psi(1, 2, 3, 4) = \frac{1}{\sqrt{4!}} \begin{vmatrix} 1s\alpha(1) & 1s\beta(1) & 2s\alpha(1) & 2s\beta(1) \\ 1s\alpha(2) & 1s\beta(2) & 2s\alpha(2) & 2s\beta(2) \\ 1s\alpha(3) & 1s\beta(3) & 2s\alpha(3) & 2s\beta(3) \\ 1s\alpha(4) & 1s\beta(4) & 2s\alpha(4) & 2s\beta(4) \end{vmatrix}$$

The standard method for determining the optimal form of the spatial part of the spin orbitals in a determinantal wave function like this one is the Hartree–Fock self-consistent field (SCF) method, which we discuss in the next section.

9.6 The Hartree–Fock Method Uses Antisymmetric Wave Functions

In Section 9.3, we discussed the Hartree–Fock method for a helium atom. The Hartree–Fock equation for this system is given by Equation 9.21, where $\hat{H}_1^{\text{eff}}$ is given by Equation 9.20. A helium atom is a special case because the Slater determinant factors into a spatial part and a spin part, and so we were able to use Equation 9.18 as the helium atomic wave function. This factorization into a spatial part and a spin part does not occur for atoms with more than two electrons, and so we must start with a complete Slater determinant such as Equation 9.48. The application of the Hartree–Fock method to atoms that contain three or more electrons introduces new terms that occur because of the determinantal nature of the wave functions. For simplicity, we shall consider only closed-shell systems consisting of $2N$ electrons, in which the wave functions are represented by N doubly occupied spatial orbitals. In such cases, the atomic wave function is given by one Slater determinant.

The Hamiltonian operator for a $2N$-electron atom is

$$
\hat{H} = -\frac{1}{2}\sum_{j=1}^{2N}\nabla_j^2 - \sum_{j=1}^{2N}\frac{Z}{r_j} + \sum_{j=1}^{2N}\sum_{j>i}\frac{1}{r_{ij}}
$$

$$
= \sum_{j=1}^{2N}\hat{h}_j + \sum_{j=1}^{2N}\sum_{j>i}\frac{1}{r_{ij}} \tag{9.49}
$$

where

$$
\hat{h}_j = -\frac{1}{2}\nabla_j^2 - \frac{Z}{r_j} \tag{9.50}
$$

and the wave function is

$$
\Psi(1, 2, \ldots, 2N)
$$

$$
= \frac{1}{\sqrt{(2N)!}}
\begin{vmatrix}
\psi_1\alpha(1) & \psi_1\beta(1) & \cdots & \psi_N\alpha(1) & \psi_N\beta(1) \\
\psi_1\alpha(2) & \psi_1\beta(2) & \cdots & \psi_N\alpha(2) & \psi_N\beta(2) \\
\vdots & \vdots & \ddots & \vdots & \vdots \\
\psi_1\alpha(2N) & \psi_1\beta(2N) & \cdots & \psi_N\alpha(2N) & \psi_N\beta(2N)
\end{vmatrix} \tag{9.51}
$$

and $\hat{K}_j(\mathbf{r}_1)$, called the *exchange operator*, is given by

$$\hat{K}_j(\mathbf{r}_1)\psi_i(\mathbf{r}_1) = \psi_j(\mathbf{r}_1) \int d\mathbf{r}_2\, \psi_j^*(\mathbf{r}_2)\frac{1}{r_{12}}\psi_i(\mathbf{r}_2) \tag{9.62}$$

Note that the summation in the definition of the Fock operator $\hat{F}(\mathbf{r})$ goes from 1 to N because there are N spatial orbitals. Note also that the Coulomb operator operating on $\psi_i(\mathbf{r}_1)$ gives a function of $\mathbf{r}_1$ times $\psi_i(\mathbf{r}_1)$. The exchange operator acts differently, however; the function that it acts upon ends up under the integral sign.

We can obtain an expression for the energy of the ith molecular orbital by multiplying Equation 9.58 from the left by $\psi_i^*(\mathbf{r}_1)$ and integrating over $\mathbf{r}_1$ to obtain

$$\varepsilon_i = \int d\mathbf{r}_1\, \psi_i^*(\mathbf{r}_1)\hat{F}(\mathbf{r}_1)\psi_i(\mathbf{r}_1) \tag{9.63}$$

Using the above definition of the Fock operator, Equation 9.63 becomes

$$\varepsilon_i = I_i + \sum_{j=1}^{N}(2J_{ij} - K_{ij}) \tag{9.64}$$

where I_i, J_{ij}, and K_{ij} are given by Equations 9.54, 9.55, and 9.56, respectively. If we compare this result to Equation 9.53, we see that

$$E = \sum_{i=1}^{N}(I_i + \varepsilon_i) \tag{9.65}$$

Note that E is *not* simply the sum of the Hartree–Fock orbital energies.

Equations 9.58 through 9.62 are called the *Hartree–Fock equations*. As we have seen already, the Hartree–Fock equations must be solved by a self-consistent procedure, where we use an initial guess of the $\psi_i(\mathbf{r})$ to calculate $F(\mathbf{r})$ and then use this result for $F(\mathbf{r})$ to calculate a new set of $\psi_i(\mathbf{r})$. This iterative process is continued until a self-consistent set of orbitals is obtained.

We carry out a simple Hartree–Fock calculation explicitly for a helium atom in the appendix to this chapter, where we express the helium $1s$ atomic orbital as a linear combination of two Slater orbitals. This procedure was developed in the 1950s, just as computers were becoming generally available, by Clemens Roothaan of the University of Chicago. He expressed the atomic orbitals, ψ_i, as linear combinations of functions, $\phi_v(\mathbf{r})$, $v = 1, \ldots, K$,

$$\psi = \sum_{v=1}^{K} c_v\phi_v \tag{9.66}$$

which were usually, but not necessarily, taken to be Slater orbitals. The set of functions $\{\phi_v\}$ is called a *basis set*, and the functions themselves are called *basis functions*. As the basis set becomes larger and larger, the expansion, Equation 9.66, leads to more and more accurate representations of the atomic orbitals. Eventually a limit is reached where the orbitals no longer improve. This limit is called the *Hartree–Fock limit*, and gives the best atomic orbitals. Don't forget, however, that even these optimum atomic orbitals constitute an approximation to the true atomic wave function because the Hartree–Fock approximation assumes that each electron experiences an average potential of all the other electrons. In other words, the motion of the electrons is uncorrelated. The true wave function cannot be written in terms of a single Slater determinant.

If we substitute Equation 9.66 into Equation 9.58, we obtain

$$\hat{F}(\mathbf{r}_1) \left[\sum_{\nu=1}^{K} c_\nu \phi_\nu(\mathbf{r}_1) \right] = \varepsilon \sum_{\nu=1}^{K} c_\nu \phi_\nu(\mathbf{r}_1) \tag{9.67}$$

Now multiply both sides from the left by $\phi_\mu^*(\mathbf{r}_1)$ and integrate over $\mathbf{r}_1$ to obtain

$$\sum_\nu c_\nu \int d\mathbf{r}_1 \, \phi_\mu^*(\mathbf{r}_1) \hat{F}(\mathbf{r}_1) \phi_\nu(\mathbf{r}_1) = \varepsilon \sum_\nu c_\nu \int d\mathbf{r}_1 \, \phi_\mu^*(\mathbf{r}_1) \phi_\nu(\mathbf{r}_1)$$

We now define the overlap matrix S with matrix elements

$$S_{\mu\nu} = \int d\mathbf{r}_1 \, \phi_\mu^*(\mathbf{r}_1) \phi_\nu(\mathbf{r}_1) \tag{9.68}$$

and the Fock matrix F with matrix elements

$$F_{\mu\nu} = \int d\mathbf{r}_1 \, \phi_\mu^*(\mathbf{r}_1) \hat{F}(\mathbf{r}_1) \phi_\nu(\mathbf{r}_1) \tag{9.69}$$

Both of these matrices are $K \times K$ Hermitian matrices; they are real and symmetric if the basis set is chosen to be a set of real functions, which is usually the case.

With these definitions, the Hartree–Fock equation, Equation 9.58, becomes

$$\sum_\nu F_{\mu\nu} c_\nu = \varepsilon \sum_\nu S_{\mu\nu} c_\nu \qquad \mu = 1, 2, \ldots, K \tag{9.70}$$

Equations 9.70 are just a set of simultaneous equations for the c_ν and are called the *Hartree–Fock–Roothaan equations*. They can be written more compactly by writing them in matrix notation:

$$F c = \varepsilon S c \tag{9.71}$$

In Equation 9.71, F and S are $K \times K$ matrices and c is a $K \times 1$ column vector. The equation leads to a $K \times K$ secular determinant, which gives us K orbital energies, ε, and K eigenvectors, c, which gives us K atomic orbitals. Equation 9.71 looks similar to Equation 8.38, but there is a difference. The F matrix in Equation 9.71 depends upon the c_ν, so Equations 9.70 and 9.71 are not linear algebraic equations. Equation 9.71 must be solved self-consistently. The calculation done in the appendix is the solution of Equation 9.71 for the case $K = 2$ (cf. Equation 9.32).

9.7 Hartree–Fock–Roothaan Atomic Wave Functions Are Available On-Line

The *Hartree–Fock–Roothaan procedure* is widely used for the calculation of atomic and molecular orbitals. An enormous amount of work was done along these lines in the 1960s, particularly by Enrico Clementi and others at the University of Chicago. For a helium atom, for example, they found the five-term orbital

$$\phi_{1s}(r) = 0.768\,38\,S_{1s}(\zeta = 1.417\,14) + 0.223\,46\,S_{1s}(\zeta = 2.376\,82)$$

$$+ 0.004\,082\,S_{1s}(\zeta = 4.396\,28) - 0.009\,94\,S_{1s}(\zeta = 0.526\,99)$$

$$+ 0.002\,30\,S_{1s}(\zeta = 7.842\,52)$$

which leads to an energy $-2.861\,680\,E_{\rm h}$. The time-consuming (and costly) part of these calculations is the determination of the exponents because this involves a nonlinear optimization. Much of the value of this work was not just to determine the Hartree–Fock atomic orbitals, but to provide a set of "best" exponents to use in molecular calculations. Consequently, only the linear coefficients in the atomic orbitals need be optimized in molecular calculations. Although the resulting molecular energies and corresponding molecular orbitals are not the *very* best that can be obtained (since the exponents in the orbitals are fixed), they turn out to be almost indistinguishable from molecular energies and molecular orbitals obtained from a complete (and much more demanding) optimization. We shall discuss this more fully in Chapter 12.

Atomic orbitals calculated by the Hartree–Fock–Roothaan method have been re-calculated by Carlos Bunge and others and are available on-line at *http://www.ccl.net/cca/data/atomic-RHF-wavefunctions/tables*. Figure 9.3 shows a screen capture of the first few lines of the website, giving the full citations and also the results for a helium atom.

Let's look just at the SCF atomic orbital, which is given in the last set of lines in the figure. The first column gives the type of Slater orbital, the second column gives the orbital exponent of the Slater orbital, and the third column gives the linear coefficient. Thus, the SCF atomic orbital for a helium atom is given by

$$\psi_{1s} = 1.347\,900\,S_{1s}(\zeta = 1.4595) - 0.001\,613\,S_{3s}(\zeta = 5.3244)$$

$$- 0.100\,506\,S_{2s}(\zeta = 2.6298) - 0.270\,779\,S_{2s}(\zeta = 1.7504) \qquad (9.72)$$

where

$$S_{ns}(\zeta) = \frac{(2\zeta)^{n+\frac{1}{2}}}{[4\pi(2n)!]^{1/2}} r^{n-1} e^{-\zeta r} \qquad (9.73)$$

```
            ROOTHAAN-HARTREE-FOCK GROUND STATE ATOMIC WAVE FUNCTIONS

     Slater-type orbital expansions and expectation values for Z=2-54

     Reference: C.F.Bunge, J.A.Barrientos and A.V.Bunge,
                Atomic Data and Nuclear Data Tables 53,113-162(1993)

     Complementary reference:
                 C.F.Bunge, J.A.Barrientos, A.V.Bunge and J.A.Cogordan,
                 Phys. Rev. A46,3691-3696(1992).


  HELIUM, Z=2      1s(2)   1S

  TOTAL ENERGY         KINETIC ENERGY        POTENTIAL ENERGY        VIRIAL RATIO
  -2.861679993          2.861681613           -5.723361606           -1.999999434

  RHOat0 = 22.593709    Kato cusp = 1.999972

                     1s
  ORB.ENERGY     -0.917955
  <R>             0.927272
  <R**2>          1.184820
  <1/R>           1.687283
  <1/R**2>        5.995503

  1S    1.4595    1.347900
  3S    5.3244   -0.001613
  2S    2.6298   -0.100506
  2S    1.7504   -0.270779
```

FIGURE 9.3
A screen shot of the first few lines of the website *http://www.ccl.net/cca/data/atomic-RHF-wavefunctions/tables*, showing the Hartree–Fock–Roothaan ground-state atomic wave function of a helium atom.

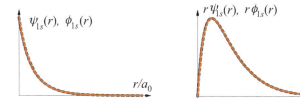

FIGURE 9.4
A comparison of the SCF wave functions of a helium atom given by Equation 9.72 (solid line) with the one calculated in the appendix (dashed line).

Problem 9–32 has you show that ψ_{1s} is normalized. Figure 9.4 compares this SCF wave function to the one that we calculate in the appendix. The two wave functions are essentially indistinguishable.

Figure 9.5 shows the SCF results for a lithium atom. In this case, there are two orbitals, 1s and 2s, to consider. The first column of the last set of lines gives the type

Using these values, we have

$$\psi_{2s} = -0.016\,378\ S_{1s}(\zeta = 5.7531) - 0.155\,066\ S_{1s}(\zeta = 3.7156)$$
$$+ 0.000\,426\ S_{3s}(\zeta = 9.9670) - 0.059\,234\ S_{3s}(\zeta = 3.7128)$$
$$- 0.031\,925\ S_{2s}(\zeta = 4.4661) + 0.387\,968\ S_{2s}(\zeta = 1.2919)$$
$$+ 0.685\,674\ S_{2s}(\zeta = 0.8555)$$

According to Koopmans's approximation, ε_i in Equation 9.58 is the ionization energy of an electron from the ith orbital. Table 9.3 compares some ionization energies of neon and argon obtained by using Koopmans's approximation with those obtained by subtracting the Hartree–Fock energy of the neutral atom from that of the ion. You can see that Koopmans's approximation gives results that are almost as good as the direct calculation. Figure 9.7 shows the ionization energies of the elements hydrogen through xenon plotted against atomic number. Both ionization energies obtained by Koopmans's approximation and experimental data are shown in the figure. This plot clearly shows the shell and subshell structure that students first learn in general chemistry. Given that there are no adjustable parameters involved in the calculated values in Figure 9.7, the agreement with experimental data is remarkable.

Note that the order of the energies of the various subshells is in general agreement with observation for neutral atoms. In particular, the energies of the $2s$ and $2p$ orbitals are not the same as they are for the hydrogen atom. The degeneracy of the $2s$ and

T A B L E 9.3
Ionization Energies of Neon and Argon Obtained from Neutral Atom Orbital Energies (Koopmans's Approximation) and by Subtracting the Hartree–Fock Energy of the Neutral Atom from the Hartree–Fock Energy of the Appropriate State of the Positive Ion

		Ionization energies/MJ·mol^{-1}		
Electron removed	Resulting orbital occupancy	Koopmans's approximation	Direct Hartree–Fock calculation	Experimental
Neon				
$1s$	$1s2s^22p^6$	86.0	83.80	83.96
$2s$	$1s^22s2p^6$	5.07	4.76	4.68
$2p$	$1s^22s^22p^5$	2.23	1.92	2.08
Argon				
$1s$	$1s2s^22p^63s^23p^6$	311.4	308.25	309.32
$2s$	$1s^22s2p^63s^23p^6$	32.35	31.33	
$2p$	$1s^22s^22p^53s^23p^6$	25.13	24.01	23.97
$3s$	$1s^22s^22p^63s3p^6$	3.35	3.20	2.82
$3p$	$1s^22s^22p^63s^23p^5$	1.55	1.43	1.52

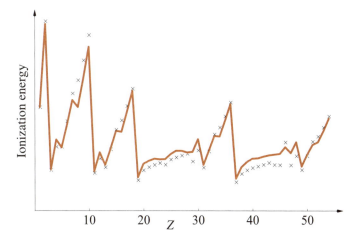

FIGURE 9.7
The ionization energies of neutral atoms of hydrogen through xenon plotted versus atomic number. The solid line connects the experimental data, and the crosses are calculated according to Koopmans's approximation. (From *http://www.ccl.net/cca/data/atomic-RHF-wavefunctions/tables.*)

$2p$ orbitals or, more generally, the fact that the energy depends on only the principal quantum number, is unique to the purely $1/r$ coulombic potential in the hydrogen atom. In a Hartree–Fock calculation, the effective potential $V_j^{\text{eff}}(\mathbf{r}_j)$ is more complicated than $1/r$ (see Figure 9.2), and $V_j^{\text{eff}}(\mathbf{r}_j)$ breaks up the degeneracy found in the hydrogen atom, giving us the familiar ordering of the orbital energies we first learned in general chemistry.

9.8 Correlation Energy Is the Difference Between the Hartree–Fock Energy and the Exact Energy

The Hartree–Fock approximation yields the best energy that can be obtained using a determinantal wave function consisting of spin orbitals. The electrons are assumed to interact only through an average, or effective, potential and we say that the electrons are uncorrelated. In Section 9.3, we defined a *correlation energy* (E_{corr}) by the equation

$$E_{\text{corr}} = E_{\text{exact}} - E_{\text{HF}} \tag{9.75}$$

When using Equation 9.75, both calculations should be based on the same Hamiltonian operator; for example, if spin-orbit interaction terms are included in one, then they should be included in the other.

   We saw in Section 9.3 that the correlation energy for a helium atom is $-0.0420\, E_{\text{h}}$ or $-110\,\text{kJ}\cdot\text{mol}^{-1}$. Although the Hartree–Fock energy is almost 99% of the exact energy, the difference is of the order of the strength of a chemical bond. The correlation energy increases with the number of electrons, so the inclusion of correlation is an important goal in quantum chemistry.

The Eckart wave function given by Equation 9.13 included some correlation. Using two values of zeta acknowledges that the two electrons are not screened to equal extents, and we say that it includes some "in–out" correlation. Its energy is $0.0140\,E_h$ better than Hartree–Fock, and so we see that it accounts for about a third of the correlation energy. The Hylleraas wave function given by Equation 9.17 accounts for about 70% of the correlation energy (Problem 9–38).

Although both of these wave functions include a fair amount of correlation in spite of their relative simplicity, neither one can be extended in a convenient way to more complex atoms and molecules. There is a method, however, that is based upon the Hartree–Fock approximation, but can be used to calculate energies much better than Hartree–Fock energies. To explain this method, we have to use the results of the SCF calculation that we carry out in the appendix, where we use a linear combination of two Slater $1s$ orbitals (Equation 9.32),

$$\psi(r) = c_1 \, S_{1s}(\zeta = 1.453\,63) + c_2 \, S_{1s}(\zeta = 2.910\,93)$$

This led to a 2×2 secular determinant (cf. Equation 9.71 for $K = 2$) from which we found the lower eigenvalue ($\varepsilon = -0.917\,935$) and the corresponding (normalized) orbital to be

$$\psi_1(r) = 0.843\,785 \, S_{1s}(\zeta = 1.453\,63) + 0.180\,687 \, S_{1s}(\zeta = 2.910\,93)$$

by an iterative process. (See the final table in the appendix.) The 2×2 secular determinant yields a second eigenvalue and a corresponding (normalized) second orbital, which comes out to be (Problem 9–39)

$$\psi_2(r) = 1.624\,04 \, S_{1s}(\zeta = 1.453\,63) - 1.821\,22 \, S_{1s}(\zeta = 2.910\,93)$$

This orbital represents an excited-state orbital.

EXAMPLE 9–8
Show that $\psi_1(r)$ and $\psi_2(r)$ are orthogonal. Take
$\langle S_{1s}(\zeta = 1.453\,63) \mid S_{1s}(\zeta = 2.910\,93)\rangle = 0.837\,524$ (Problem 9–40).

SOLUTION:

$$\langle \psi_2 \mid \psi_1 \rangle = (1.624\,04)(0.843\,785)\langle S_{1s}(\zeta = 1.453\,63) \mid S_{1s}(\zeta = 1.453\,63)\rangle$$

$$+ [(1.624\,04)(0.180\,687)$$

$$- (0.843\,785)(1.821\,22)]\langle S_{1s}(\zeta = 1.453\,63) \mid S_{1s}(\zeta = 2.910\,93)\rangle$$

$$- (0.180\,687)(1.821\,22)\langle S_{1s}(\zeta = 2.910\,93) \mid S_{1s}(\zeta = 2.910\,93)\rangle$$

$$= 1.370\,34 - 1.041\,27 - 0.329\,07 = 0$$

The ground state of a helium atom has the electron configuration $\psi_1(1)\psi_1(2)$. (Remember that we do not have to include spin here because we are considering only a two-electron system at this point.) The orbital $\psi_2(r)$ is not occupied in the ground state, and is called a *virtual orbital*. We can form an excited state by promoting an electron from the ground-state orbital ψ_1 to the virtual orbital ψ_2. For example, $\psi_2(1)\psi_2(2)$ represents an electron configuration where both electrons are in the orbital ψ_2. Instead of using just $\psi_1(1)\psi_1(2)$, we could use

$$\Psi_{CI}(1,\, 2) = c_1\psi_1(1)\psi_1(2) + c_2\psi_2(1)\psi_2(2) \tag{9.76}$$

and let c_1 and c_2 be determined variationally. Because $\psi_1(r)$ and $\psi_2(r)$ are orthonormal, this leads to the secular equation

$$\begin{vmatrix} H_{11} - E & H_{12} \\ H_{12} & H_{22} - E \end{vmatrix} = 0$$

where

$$H_{11} = \langle \psi_1(1)\psi_1(2) \mid \hat{H} \mid \psi_1(1)\psi_1(2) \rangle$$

$$H_{12} = \langle \psi_1(1)\psi_1(2) \mid \hat{H} \mid \psi_2(1)\psi_2(2) \rangle$$

$$H_{22} = \langle \psi_2(1)\psi_2(2) \mid \hat{H} \mid \psi_2(1)\psi_2(2) \rangle$$

All the necessary integrals to evaluate H_{11}, H_{12}, and H_{22} are contained in Table 9.7 in the appendix, and we get (Problem 9–41)

$$\begin{vmatrix} -2.861\,67 - E & 0.290\,023 \\ 0.290\,023 & 3.247\,37 - E \end{vmatrix} = 0$$

from which we obtain $E = -2.875\,41\,E_h$ as the lower root. The difference between this value and the Hartree–Fock energy is $\Delta E = -2.875\,41\,E_h + 2.861\,67E_h = -0.013\,74\,E_h$, which is about one-third of the correlation energy. The values of c_1 and c_2 in Equation 9.76 are $c_1 = 0.998\,864$ and $c_2 = -0.047\,644$, so $\Psi_{CI}(1,\, 2)$ comes out to be

$$\Psi_{CI}(1,\, 2) = 0.998\,864\,\psi_1(1)\psi_1(2) - 0.047\,644\,\psi_2(1)\psi_2(2)$$

showing that $\psi_1(1)\psi_1(2)$ is the dominant contribution. After all, the Hartree–Fock approximation does give about 99% of the exact energy.

An advantage of this procedure is that all the necessary integrals had already been calculated in the previous Hartree–Fock calculation. A much more important feature, however, is that it becomes essentially exact in the limit. If we use a Hartree–Fock orbital that is a linear combination of K Slater orbitals, instead of just two as we did in Equation 9.27,

$$\psi(r) = \sum_{j=1}^{K} c_j \phi_j(r) \tag{9.77}$$

then Equation 9.71 becomes a $K \times K$ secular determinant, and we obtain K orbitals, $\psi_1(r)$ through $\psi_K(r)$. The first orbital (the one of lowest energy) is doubly occupied in the ground state of a helium atom, and the remaining $K - 1$ orbitals are virtual orbitals. We can use all these orbitals to form many-electron configurations and write $\Psi_{CI}(1, 2)$ as

$$\Psi_{CI}(1, 2) = \sum_i \sum_j c_{ij} \psi_i(1) \psi_j(2) \tag{9.78}$$

Equation 9.76 above is just a simple version of Equation 9.78. We can then use Equation 9.78 as a variational function to determine the energy and the corresponding wave function.

The method we are describing is called *configuration interaction* (CI). That's why we subscripted Ψ with "CI" in the above equations. As we use larger and larger basis sets and include more and more configurations, we approach the exact result. Configuration interaction is one of several (post-Hartree–Fock) methods used by quantum chemists to go beyond the Hartree–Fock limit. For many-electron systems, the spin does not factor out as it does for a helium atom, and so $\Psi_{CI}(1, 2, \ldots, N)$ involves a summation over electron configurations described by Slater determinants. We shall have more to say about configuration interaction when we discuss molecular calculations.

9.9 A Term Symbol Gives a Detailed Description of an Electron Configuration

In Section 7.7, we discussed term symbols for a hydrogen atom. In this section, we shall discuss term symbols for multielectron atoms. Electron configurations of atoms are ambiguous in the sense that a number of sets of m_l and m_s are consistent with a given electron configuration. For example, consider the ground-state electron configuration of a carbon atom, $1s^2 2s^2 2p^2$. The two $2p$ electrons could be in any of the three $2p$ orbitals ($m_l = 0, \pm 1$) and have any spins consistent with the Pauli exclusion principle. The energies of these different states may differ, and so we require a more detailed designation of the electronic states of atoms. The scheme we will present here is based upon the idea of determining the total orbital angular momentum $\mathbf{L}$ and the total spin angular momentum $\mathbf{S}$ and then adding $\mathbf{L}$ and $\mathbf{S}$ together vectorially to obtain the total angular momentum $\mathbf{J}$. The result of such a calculation, called *Russell–Saunders coupling*, is presented as an *atomic term symbol*, which has the form

$$^{2S+1}L_J$$

In a term symbol, L is the orbital angular-momentum quantum number, S is the spin quantum number, and J is the angular-momentum quantum number. We will see that L will necessarily have values such as 0, 1, 2, ... Similar to assigning the letters s, p, d, and f to the values $l = 0$, 1, 2, and 3 of the orbital angular momentum for the hydrogen atom, we will make the following correspondence:

$$L = \quad 0 \quad 1 \quad 2 \quad 3 \quad 4 \quad 5 \quad \ldots$$
$$ S \quad P \quad D \quad F \quad G \quad H \quad \ldots$$

We will also see that the spin quantum number S will necessarily have values such as $0, \frac{1}{2}, 1, \frac{3}{2}, \ldots$, and so the $2S + 1$ left superscript on a term symbol will have values such as $1, 2, 3, \ldots$. The quantity $2S + 1$ is called the *spin multiplicity*. Thus, ignoring for now the subscript J, term symbols will be of the type

$$^3S \qquad ^2D \qquad ^1P$$

The orbital angular momentum and the spin angular momentum are given by the vector sums

$$\mathbf{L} = \sum_i \mathbf{l}_i \tag{9.79}$$

and

$$\mathbf{S} = \sum_i \mathbf{s}_i \tag{9.80}$$

where the summations are over the electrons in the atom. The z components of $\mathbf{L}$ and $\mathbf{S}$ are given by the scalar sums

$$L_z = \sum_i l_{zi} = \sum_i m_{li} = M_L \tag{9.81}$$

and

$$S_z = \sum_i s_{zi} = \sum_i m_{si} = M_S \tag{9.82}$$

Thus, although the angular momenta add vectorially as in Equations 9.79 and 9.80, the z components add as scalars (Figure 9.8).

Just as the z component of $\mathbf{l}$ can assume the $2l + 1$ values $m_l = l, l - 1, \ldots, 0, \ldots, -l + 1, -l$, the z component of $\mathbf{L}$ can assume the $2L + 1$ values $M_L = L, L - 1, \ldots, 0, \ldots, -L + 1, -L$. Similarly, M_S can take on the $2S + 1$ values $S, S - 1, \ldots, 0, \ldots, -S + 1, -S$. Thus, the spin multiplicity is simply the $2S + 1$ projections that the z component of $\mathbf{S}$ can assume.

Let's consider the electron configuration ns^2 (two electrons in an ns orbital). There is only one possible set of values of $m_{l1}, m_{s1}, m_{l2},$ and m_{s2}:

m_{l1}	m_{s1}	m_{l2}	m_{s2}	M_L	M_S
0	$+\frac{1}{2}$	0	$-\frac{1}{2}$	0	0

The fact that the only value of M_L is $M_L = 0$ implies that $L = 0$. Similarly, the fact that the only value of M_S is $M_S = 0$ implies that $S = 0$. The total angular momentum $\mathbf{J}$

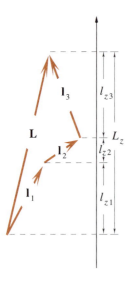

FIGURE 9.8
A schematic illustration of the addition of angular-momentum vectors.

is given by

$$\mathbf{J} = \mathbf{L} + \mathbf{S} \tag{9.83}$$

and the only value of its z component is

$$J_z = L_z + S_z = (M_L + M_S) = M_J = 0 \tag{9.84}$$

which implies that $J = 0$. Consequently, for an ns^2 electron configuration, $L = 0$, $S = 0$, and $J = 0$. The value $L = 0$ is written as S in the term symbol, and so we find that the term symbol corresponding to an ns^2 electron configuration is $^1S_0$ (singlet S zero). Because the two electrons have opposite spins, the total spin angular momentum is zero. Both electrons also occupy an orbital that has no angular momentum, so the total angular momentum must be zero, which is what the $^1S_0$ term indicates.

An np^6 electron configuration also will have a $^1S_0$ term symbol. To understand this, realize that the six electrons in the three np orbitals have the quantum numbers $(n, 1, 1, \pm\frac{1}{2})$, $(n, 1, 0, \pm\frac{1}{2})$, and $(n, 1, -1, \pm\frac{1}{2})$. Therefore, when we add up all the m_{li} and m_{si}, we get $M_L = 0$ and $M_S = 0$, and we have $^1S_0$.

EXAMPLE 9–9
Show that the term symbol corresponding to an nd^{10} electron configuration is $^1S_0$.

SOLUTION: The ten d orbital electrons have the quantum numbers $(n, 2, 2, \pm\frac{1}{2})$, $(n, 2, 1, \pm\frac{1}{2})$, $(n, 2, 0, \pm\frac{1}{2})$, $(n, 2, -1, \pm\frac{1}{2})$, and $(n, 2, -2, \pm\frac{1}{2})$. Therefore, $M_L = 0$ and $M_S = 0$, as for ns^2 and np^6 electron configurations, and the term symbol is $^1S_0$.

Notice that M_L and M_S are necessarily equal to zero for completely filled subshells because for every electron with a negative value of m_{li}, there is another electron with a corresponding positive value to cancel it; the same holds true for the values of m_{si}. Thus, we can ignore the electrons in completely filled subshells when considering other electron configurations. For example, we can ignore the contributions of the $1s^2 2s^2$ orbitals to the $1s^2 2s^2 2p^2$ electron configurations of a carbon atom when we discuss a carbon atom later.

An example of an electron configuration that has a term symbol other than $^1S_0$ is that of a helium atom with the excited-state electron configuration $1s^1 2s^1$. To determine the possible values of m_{l1}, m_{s1}, m_{l2}, and m_{s2}, we set up a table in the following manner: Because m_{l1} and m_{l2} can both have a maximum value of 0, the maximum value of M_L is 0 (see Equation 9.81), and 0 is its only possible value. Similarly, because m_{s1} and m_{s2} can both have values of $\pm 1/2$, M_s can be -1, 0, or 1. We now set up a table with its columns headed by the possible values of M_S and its rows headed by the possible values of M_L, and we then fill in the sets of values of m_{l1}, m_{s1}, m_{l2}, and m_{s2} that are consistent with each value of M_L and M_S.

| | M_S | | |
M_L	1	0	-1
0	$0^+, 0^+$	$0^+, 0^-; 0^-, 0^+$	$0^-, 0^-$

The notation $0^\pm$ means that $m_l = 0$ and $m_s = \pm 1/2$. The possible sets of values of m_{l1}, m_{s1}, m_{l2}, and m_{s2} that are consistent with each value of M_L and M_S are called *microstates*, which are separated by semicolons when there is more than one microstate for a given value of M_L and M_S.

There are four microstates in this table because there are two possible spins $(\pm\frac{1}{2})$ for the electron in the $1s$ orbital and two possible spins for the electron in the $2s$ orbital. Note that we include both $0^+, 0^-$ and $0^-, 0^+$ because the electrons are in *nonequivalent orbitals* (i.e., $1s$ and $2s$). Note that all the values of M_L in the above table are zero, so they all must correspond to $L = 0$. In addition, the largest value of M_S is 1. Consequently, S must equal 1 and the values $M_S = 1$, 0, and -1 correspond to $L = 0$, $S = 1$, corresponding to a ^3S state. This ^3S state accounts for one microstate from each column in the above table. The middle column contains two microstates, but it makes no difference which one we choose. After eliminating one microstate from each column ($0^+, 0^+; 0^-, 0^-$; and either $0^+, 0^-$ or $0^-, 0^+$), we are left with only the entry with $M_L = 0$ and $M_S = 0$ (either $0^-, 0^+$ or $0^+, 0^-$), which implies that $L = 0$ and $S = 0$, corresponding to a ^1S state. These two pairs of $L = 0$, $S = 1$ and $L = 0$, $S = 0$, along with their possible values of M_J, can be summarized as

$$L = 0, \ S = 1 \qquad\qquad L = 0, \ S = 0$$
$$M_L = 0, \ M_S = 1, 0, -1 \qquad M_L = 0, \ M_S = 0$$
$$M_J = M_L + M_S = 1, 0, -1 \quad M_J = M_L + M_S = 0$$

The values of M_J here imply that $J = 1$ for the $L = 0$, $S = 1$ case and that $J = 0$ for the $L = 0$, $S = 0$ case. The two term symbols corresponding to the electron configuration $1s^1 2s^1$ are

$$^3S_1 \quad \text{and} \quad {}^1S_0$$

The $^3S_1$ is called a *triplet S state*. These two term symbols correspond to two different electronic states with different energies. We will see below that the triplet state ($^3S_1$) has a lower energy than the singlet state ($^1S_0$).

9.10 The Allowed Values of J Are $L + S$, $L + S - 1$, ..., $|L - S|$

As a final example of deducing atomic term symbols, we will consider a carbon atom, whose ground-state electron configuration is $1s^2 2s^2 2p^2$. We have shown previously that we do not need to consider completely filled subshells because M_L and M_S are necessarily zero for completely filled subshells. Consequently, we can focus on the electron configuration $2p^2$. Actually, our results will be valid for an np^2 electron configuration where $n = 2, 3, 4, \ldots$, and so we shall consider two equivalent np orbitals. As for the case of $1s^1 2s^1$ above, we will make a table of possible values of m_{l1}, m_{s1}, m_{l2}, and m_{s2}. Before we do this, however, let's see how many entries there will be in the table for np^2. We are going to assign two electrons to two of six possible spin orbitals, $(m_l, m_s) = (1, 1/2), (1,-1/2), (0,1/2), (0,-1/2), (-1,1/2), (-1,-1/2)$. There are six choices for the first spin orbital and five choices for the second, giving a total of $6 \times 5 = 30$ choices. Because the electrons are indistinguishable, however, the order of the two spin orbitals chosen is irrelevant. Thus, we should divide the 30 choices by 2 to give 15 as the number of distinct ways of assigning the two electrons to the six spin orbitals. Generally, the number of distinct ways to assign N electrons to G spin orbitals belonging to the same subshell (*equivalent orbitals*) is given by

$$\frac{G!}{N!(G - N)!} \qquad \text{(equivalent orbitals)} \qquad (9.85)$$

Note that Equation 9.85 gives 15 if $G = 6$ and $N = 2$.

EXAMPLE 9–10
How many distinct ways are there of assigning two electrons to the nd orbitals? In other words, how many sets of m_{li} and m_{si} are there for an nd^2 electron configuration?

SOLUTION: There are five nd orbitals, or 10 nd spin orbitals. Thus, the number of distinct ways of placing two electrons in nd orbitals is

$$\frac{10!}{2!\,8!} = 45$$

To determine the 15 possible sets of m_{l1}, m_{s1}, m_{l2}, and m_{s2} for an np^2 electron configuration, we first determine the possible values of M_L and M_S. Because m_{l1} and m_{l2} can both have a maximum value of 1, the maximum value of M_L is 2 (see Equation 9.81), and so its possible values are 2, 1, 0, -1, and -2. Similarly, because m_{s1} and m_{s2} can each have a maximum value of 1/2, the maximum value of M_S is 1 (see Equation 9.82), and so its possible values are 1, 0, and -1. Using this information, we set up a table with its columns headed by the possible values of M_S and its rows headed by the possible values of M_L, and then fill in the microstates consistent with each value of M_L and M_S, as shown.

		M_S	
M_L	1	0	-1
2	~~$1^+, 1^+$~~	$1^+, 1^-$	~~$1^-, 1^-$~~
1	$0^+, 1^+$	$1^+, 0^-; 1^-, 0^+$	$0^-, 1^-$
0	~~$0^+, 0^+$~~; $1^+, -1^+$	$1^+, -1^-; -1^+, 1^-; 0^+, 0^-$	$1^-, -1^-$; ~~$0^-, 0^-$~~
-1	$0^+, -1^+$	$0^+, -1^-; 0^-, -1^+$	$0^-, -1^-$
-2	~~$-1^+, -1^+$~~	$-1^+, -1^-$	~~$-1^-, -1^-$~~

For example, the notation 1^+ above means that $m_{l1} = 1$ and $m_{s1} = +1/2$, and -1^- means that $m_{l2} = -1$ and $m_{s2} = -1/2$. Unlike the earlier example in which we treated nonequivalent orbitals, we do *not* include both $1^+, 0^-$ and $0^-, 1^+$ in the $M_S = 0$, $M_L = 1$ position, because in this case, the orbitals are equivalent (two np orbitals). Consequently, the two microstates $1^+, 0^-$ and $0^-, 1^+$ are indistinguishable. The six microstates that are crossed out in the above table violate the Pauli exclusion principle. The remaining 15 microstates constitute all the possible microstates for an np^2 electron configuration.

We must now deduce the possible values of L and S from the tabulated values of M_L and M_S. The largest value of M_L is 2, which occurs only with $M_S = 0$. Therefore, there must be a state with $L = 2$ and $S = 0$ (^1D). Because $L = 2$, $M_L = 2, 1, 0, -1$, and -2, and so the ^1D state will account for one microstate in each row of the middle column of the above table. For those rows that contain more than one microstate (the second, third, and fourth rows), it makes no difference which microstate is chosen. We will arbitrarily choose the microstates $1^+, 0^-; 1^+, -1^-;$ and $0^+, -1^-$.

If we eliminate these microstates from the table, we are left with the following table.

		M_S	
M_L	1	0	-1
2			
1	$0^+, 1^+$	$1^-, 0^+$	$0^-, 1^-$
0	$1^+, -1^+$	$-1^+, 1^-; 0^+, 0^-$	$1^-, -1^-$
-1	$0^+, -1^+$	$0^-, -1^+$	$0^-, -1^-$
-2			

The largest value of M_L remaining is $M_L = 1$, implying $L = 1$. There are microstates with $M_L = 1, 0, -1$ associated with $M_S = 1$ $(0^+, 1^+; 1^+, -1^+; 0^+, -1^+)$, with $M_S = 0$ $(1^-, 0^+;$ either $-1^+, 1^-$ or $0^+, 0^-; 0^-, -1^+)$, and with $M_S = -1$ $(0^-, 1^-;$ $1^-, -1^-; 0^-, -1^-)$. Therefore, these nine microstates correspond to $L = 1$ and $S = 1$, or a ^3P (triplet P) state. If we eliminate these nine microstates from the table, then we are left with only one microstate with $M_L = 0$ and $M_S = 0$ at the center of the table, which implies $L = 0$ and $S = 0$ (^1S).

So far, we have found the partially specified term symbols, ^1D, ^3P, and ^1S. To complete the specification of these term symbols, we must determine the possible values of J in each case. Recall that $M_J = M_L + M_S$. For the five entries corresponding to the ^1D state, $M_S = 0$, and so the values of M_J are 2, 1, 0, -1, and -2, which implies that $J = 2$. Thus, the complete term symbol of the ^1D state is $^1D_2$. Note that the degeneracy of this state is 5, or $2J + 1$. The values of M_J for the nine entries for the ^3P state are 2, 1, 1, 0, 0, -1, 0, -1, and -2. We clearly have one set of 2, 1, 0, -1, -2 corresponding to $J = 2$. If we eliminate these five values, then we are left with 1, 0, 0, -1, which corresponds to $J = 1$ and $J = 0$. Thus, the ^3P state has three possible values of J, so the term symbols are $^3P_2$, $^3P_1$, and $^3P_0$. The ^1S state must be $^1S_0$. In summary, then, the electronic states associated with an np^2 configuration are

$$^1D_2, \quad ^3P_0, \quad ^3P_1, \quad ^3P_2, \quad \text{and} \quad ^1S_0$$

The degeneracies of these states are 5, 1, 3, 5, and 1, respectively, which adds up to 15, the number of microstates that we started with. Table 9.4 lists the term symbols that arise from various electron configurations.

T A B L E 9.4
The Possible Term Symbols (Excluding the J Subscript) for Various Electron Configurations for Equivalent Electrons

Electron configuration	Term symbol (excluding the J subscript)
s^1	^2S
p^1	^2P
p^2, p^4	$^1S, ^1D, ^3P$
p^3	$^2P, ^2D, ^4S$
p^1, p^5	^2P
d^1, d^9	^2D
d^2, d^8	$^1S, ^1D, ^1G, ^3P, ^3F$
d^3, d^7	$^2P, ^2D$ (twice), $^2F, ^2G, ^2H, ^4P, ^4F$
d^4, d^6	^1S (twice), ^1D (twice), $^1F, ^1G$ (twice), $^1I, ^3P$ (twice), $^3D, ^3F$ (twice), $^3G, ^3H, ^5D$
d^5	$^2S, ^2P, ^2D$ (thrice), ^2F (twice), ^2G (twice), $^2H, ^2I, ^4P, ^4D, ^4F, ^4G, ^6S$

The values of J for the term symbols in Table 9.4 can be determined in terms of the values of L and S if we recall that

$$\mathbf{J} = \mathbf{L} + \mathbf{S}$$

The largest value that J can have is in the case when both $\mathbf{L}$ and $\mathbf{S}$ are pointing in the same direction, so that $J = L + S$. The smallest value that J can have is when $\mathbf{L}$ and $\mathbf{S}$ are pointing in opposite directions, so that $J = |L - S|$. The values of J lying between $L + S$ and $|L - S|$ are obtained from

$$J = L + S, \quad L + S - 1, \quad L + S - 2, \ldots, \quad |L - S| \tag{9.86}$$

If we apply Equation 9.86 to the ^{3}P term symbol above, then we see that the values of J are given by

$$J = (1 + 1), \quad (1 + 1) - 1, \quad 1 - 1$$

or $J = 2, 1, 0$, as we deduced above.

EXAMPLE 9–11
Use Equation 9.86 to deduce the values of J associated with the term symbols ^{2}S, ^{3}D, and ^{4}F.

SOLUTION: For a ^{2}S state, $L = 0$ and $S = 1/2$. According to Equation 9.86, the only possible value of J is 1/2, and so the term symbol will be $^2S_{1/2}$. For a ^{3}D state, $L = 2$ and $S = 1$. Therefore, the values of J will be 3, 2, and 1, and so the term symbols will be

$$^3D_1, \quad ^3D_2, \quad \text{and} \quad ^3D_3$$

For a ^{4}F state, $L = 3$ and $S = 3/2$. Therefore, the values of J will be 9/2, 7/2, 5/2, and 3/2, and so the term symbols will be

$$^4F_{9/2}, \quad ^4F_{7/2}, \quad ^4F_{5/2}, \quad \text{and} \quad ^4F_{3/2}$$

Example 9–11 shows that the "L and S part" of a term symbol is sufficient to deduce the complete term symbol.

There is a useful consistency test between Equation 9.85 and the term symbols associated with a given electron configuration. A term symbol ^{2S+1}L will have $2S + 1$ entries for each value of M_L in a table of possible values of the m_{li} and m_{si} (see the table of entries for np^2). Because there are $2L + 1$ values of M_L for a given value of L, the total number of entries for each term symbol is $(2S + 1)(2L + 1)$. Applying this result to the np^2 case gives

$$^1S \qquad ^3P \qquad ^1D$$

$$(1 \times 1) + (3 \times 3) + (1 \times 5) = 15$$

in agreement with the value that we calculated from Equation 9.85.

EXAMPLE 9–12

Show that Equation 9.85 and the term symbols for nd^2 given in Table 9.4 are consistent.

SOLUTION: The total number of entries in a table of possible values of $m_{l i}$ and $m_{s i}$ for an nd^2 electron configuration is

$$\frac{G!}{N!\,(G-N)!} = \frac{10!}{2!\,8!} = 45$$

The term symbols given in Table 9.4 are

$$^1S \qquad ^1D \qquad ^1G \qquad ^3P \qquad ^3F$$

$$(1 \times 1) + (1 \times 5) + (1 \times 9) + (3 \times 3) + (3 \times 7) = 45$$

9.11 Hund's Rules Are Used to Determine the Term Symbol of the Ground Electronic State

Each of the states designated by a term symbol corresponds to a determinantal wave function that is an eigenfunction of $\hat{L}^2$ and $\hat{S}^2$, and each state corresponds to a certain energy. Although we could calculate the energy associated with each state, in practice, the various states are ordered according to three empirical rules formulated by the German spectroscopist Friedrich Hund. Hund's rules are as follows:

1. The state with the largest value of S is the most stable (has the lowest energy), and stability decreases with decreasing S.

2. For states with the same value of S, the state with the largest value of L is the most stable.

3. If the states have the same value of L and S, then for a subshell that is less than half-filled, the state with the smallest value of J is the most stable; for a subshell that is more than half-filled, the state with the largest value of J is the most stable.

EXAMPLE 9–13
Use Hund's rules to deduce the lowest energy state of an excited state of a beryllium atom whose electron configuration is $1s^2 2s^1 3s^1$ and the ground state of a carbon atom.

SOLUTION: The term symbols for a $2s^1 3s^1$ configuration are (Problem 9–42)

$$^3S_1 \quad \text{and} \quad ^1S_0$$

According to the first of Hund's rules, the more stable state is the $^3S_1$ state.

The ground-state electron configuration of a carbon atom is p^2. The term symbols for a p^2 configuration are (see Table 9.4)

$$^1S_0, \quad ^3P_0, \quad ^3P_1, \quad ^3P_2, \quad \text{and} \quad ^1D_2$$

According to the first of Hund's rules, the ground state is one of the ^3P states. According to Hund's third rule, the most stable state is the $^3P_0$ state.

9.12 Atomic Term Symbols Are Used to Describe Atomic Spectra

Atomic term symbols are sometimes called spectroscopic term symbols because atomic spectral lines can be assigned to transitions between states that are described by atomic term symbols. The energies of the electronic states of many atoms and ions are well tabulated in United States government publications. The standard hard-copy publication for many years was *Atomic Energy Levels* by Charlotte E. Moore (National Bureau of Standards Circular No. 467, U.S. Government Printing Office, 1949). These tables were usually referred to as "Moore's Tables." Nowadays, however, all these data and updates of these data are available on-line at a website *http://physics.nist.gov/PhysRefData/ASD/* maintained by the National Institute of Standards and Technology (NIST).

Table 9.5 is a screen shot from the NIST website of the first few energy levels of atomic sodium. The ground-state electron configuration $1s^2 2s^2 2p^6 3s$ gives rise to the term $^2S_{1/2}$, which is doubly degenerate, corresponding to $M_J = +1/2$ and $-1/2$. The first excited state arises from the promotion of the outer-shell electron from the $3s$ orbital to the $3p$ orbital, and this gives rise to two states, $^2P_{1/2}$ and $^2P_{3/2}$. The first of these is two-fold degenerate ($M_J = 1/2, -1/2$) and the second is four-fold degenerate ($M_J = 3/2, 1/2, -1/2, -3/2$). Notice in the table that terms with the same values of S and L are separated from each other, and that there is a small separation of energies within each term. The separation within each term is due to the different values of J, and is the result of spin-orbit interaction, which we discussed in Section 7.6 for a hydrogen atom and which we will discuss again in the next section.

The spectrum of atomic sodium is due to transitions from one state to another in Table 9.5. As for many of the systems that we have studied earlier, not all transitions

SOLUTION: The two transitions are

$$3p\ ^2P_{1/2} \rightarrow 3s\ ^2S_{1/2} \qquad \tilde{\nu} = 16\ 956.172\ \text{cm}^{-1}$$

and

$$3p\ ^2P_{3/2} \rightarrow 3s\ ^2S_{1/2} \qquad \tilde{\nu} = 16\ 973.368\ \text{cm}^{-1}$$

The wavelengths are given by $\lambda = 1/\tilde{\nu}$, or

$$\lambda = 5897.557\ \text{Å} \qquad \text{and} \qquad 5891.583\ \text{Å}$$

If we compare these wavelengths with those in Figure 9.9, we see that there is a small discrepancy. This discrepancy is caused by the fact that wavelengths determined experimentally are measured in air, whereas the calculations using Table 9.5 provide wavelengths in a vacuum. We use the index of refraction in air in the region of these wavelengths (1.000 29 at 0°C) to convert from one wavelength to another:

$$\lambda_{\text{vac}} = 1.000\ 29\ \lambda_{\text{air}}$$

If we divide each of the wavelengths obtained above by 1.000 29, then we obtain

$$\lambda_{\text{expt}} = 5895.9\ \text{Å} \qquad \text{and} \qquad 5889.9\ \text{Å}$$

in excellent agreement with Figure 9.9. These wavelengths occur in the yellow region of the spectrum and account for the intense yellow doublet, called the *sodium D line*, which is characteristic of the emission spectrum of sodium atoms.

Figure 9.10 shows the energy-level diagram of a helium atom at a resolution at which the spin-orbit splittings are not significant. The principal feature of the helium energy-level diagram is that it indicates two separate sets of transitions. Notice from the figure that one set of transitions is among singlet states ($S = 0$) and the other is among triplet ($S = 1$) states. No transitions occur between the two sets of states because of the $\Delta S = 0$ selection rule. Thus, the only allowed transitions are between states with the same spin multiplicity. The observed spectrum of helium consists of two overlapping sets of lines. We should point out that the selection rules presented here are useful only for small spin-orbit coupling, and so they apply only to atoms with small atomic numbers. As the atomic number increases, the selection rules break down. For example, mercury has both singlet and triplet states like helium does, but many singlet–triplet state transitions are observed in the atomic spectrum of atomic mercury. In fact, the familiar intense purple line in the spectrum of atomic mercury is a single–triplet transition.

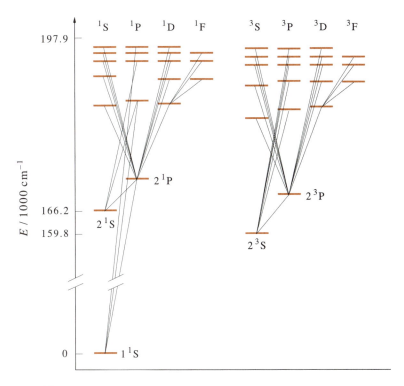

FIGURE 9.10
The energy-level diagram of a helium atom, showing the two separate sets of singlet and triplet states.

9.13 Russell–Saunders Coupling Is Most Useful for Light Atoms

We shall conclude this chapter with a brief discussion of the theory behind atomic term symbols. In addition to the usual kinetic energy and electrostatic terms in the Hamiltonian operator of a many-electron atom, there are a number of magnetic and spin terms. The most important of these magnetic and spin terms is the *spin-orbit interaction* term, which, as we discussed in Section 7.6, represents the interaction of the magnetic moment associated with the spin of an electron with the magnetic field generated by the electric current produced by the electron's own orbital motion. There are other terms such as spin-spin interaction and orbit-orbit interaction, but these are numerically less important. The Hamiltonian operator for a multielectron atom can be written as

$$\hat{H} = -\frac{1}{2} \sum_j \nabla_j^2 - \sum_j \frac{Z}{r_j} + \sum_{i<j} \frac{1}{r_{ij}} + \sum_j \xi(r_j)\, \mathbf{l}_j \cdot \mathbf{s}_j \tag{9.88}$$

where $\mathbf{l}_j$ and $\mathbf{s}_j$ are the individual electronic orbital momenta and spin angular momenta, respectively, and where $\xi(r_j)$ is a scalar function of r_j whose form is not necessary here.

We can abbreviate Equation 9.88 by writing

$$\hat{H} = \hat{H}_0 + \hat{H}_{ee} + \hat{H}_{so} \tag{9.89}$$

where $\hat{H}_0$ represents the first two terms (no interelectronic interactions), $\hat{H}_{ee}$ represents the third term (interelectronic repulsion), and $\hat{H}_{so}$ represents the fourth term (spin-orbit coupling) in Equation 9.88.

Figure 9.11 is a schematic diagram of how the energies of the $1s^2$, $1s2s$, and $1s2p$ electron configurations of a helium atom are split by the $\hat{H}_{ee}$ term and then by the $\hat{H}_{so}$ term. In the absence of the electron-electron repulsion term, $\hat{H}_{ee}$, the electron

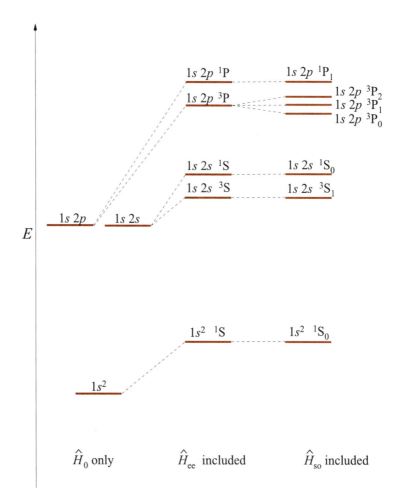

FIGURE 9.11
A schematic diagram of how the energies of the $1s^2$, $1s2s$, and $1s2p$ electron configurations of a helium atom are split by the electron-electron repulsion term and the spin-orbit interaction term. The energies in this figure are not to scale, the splitting due to spin-orbit coupling is much smaller than that due to electron-electron repulsion.

configurations $1s2s$ and $1s2p$ have the same energy. The inclusion of $\hat{H}_{ee}$ splits each of the electron configurations $1s2s$ and $1s2p$ into two terms, as shown in the figure. When spin-orbit interaction is included, the ^{3}P term that results from the $1s2p$ electron configuration is further split into ^{3}P$_2$, ^{3}P$_1$, and ^{3}P$_0$. The energies shown in Figure 9.11 are not to scale; their actual values can be found at *http://physics.nist.gov/PhysRefData/ ASD/levels_form.html*. The splitting due to spin-orbit coupling is much smaller than that due to electron-electron repulsion. The scheme is known as *Russell–Saunders coupling*, and is usually a very good approximation for light atoms.

We can also express the effect of spin-orbit coupling in terms of the multielectron wave functions. Because $\hat{H}_{so}$ is usually small relative to $\hat{H}_{ee}$ for light atoms ($Z < 40$), it can be treated as a perturbation. In that case, the zero-order Hamiltonian operator is

$$\hat{H}^{(0)} = \hat{H}_0 + \hat{H}_{ee} \tag{9.90}$$

Although the individual orbital angular momenta, $\mathbf{l}_j$, and spin angular momenta, $\hat{\mathbf{s}}_i$, commute with $\hat{H}_0$, they do not commute with $\hat{H}^{(0)} = \hat{H}_0 + \hat{H}_{ee}$ and so are not conserved even to zero order. The total orbital angular-momentum operator, $\hat{L}^2$, with $\mathbf{L} = \mathbf{l}_1 + \mathbf{l}_2 + \cdots + \mathbf{l}_N$, and the total spin angular-momentum operator $\hat{S}^2$, with $\mathbf{S} = \mathbf{s}_1 + \mathbf{s}_2 + \cdots + \mathbf{s}_N$, however, do commute with $\hat{H}^{(0)}$, and so are conserved to zero order in the perturbation scheme. Recall from Equation 4.84 that the time dependence of the average of a physical quantity represented by an operator $\hat{A}$ is given by

$$\frac{d\langle A \rangle}{dt} = \frac{d}{dt} \langle \psi(t) \mid \hat{A} \mid \psi(t) \rangle$$

$$= \frac{i}{\hbar} \langle \psi(t) \mid [\hat{H}, \hat{A}] \mid \psi(t) \rangle + \langle \psi(t) \left| \frac{\partial \hat{A}}{\partial t} \right| \psi(t) \rangle$$

Thus, if $\hat{A}$ commutes with $\hat{H}$ (and does not depend explicitly upon time), the time derivative of its average value is zero, and $\langle A \rangle$ is a constant of motion, or, in other words, is conserved. Thus, the resulting quantum numbers L and S are good quantum numbers to zero order because they represent conserved quantities. This is why the term symbols are made up of L and S.

The operators $\hat{L}^2$ and $\hat{S}^2$ do not commute with $\hat{H} = \hat{H}_0 + \hat{H}_{ee} + \hat{H}_{so}$, but because $\hat{H}_{so}$ is small, we say that $\hat{L}$ and $\hat{S}$ *almost* commute with $\hat{H}$ and that the quantum numbers L and S are *almost* good quantum numbers. They would be perfectly good quantum numbers if there were no spin-orbit interaction. This makes sense physically because $\hat{H}^{(0)} = \hat{H}_0 + \hat{H}_{ee}$ contains no term that represents an interaction between the orbital angular momentum and the spin angular momentum, and so each one is conserved separately. Neither one is conserved separately when we include the spin-orbit interaction term, in which case it is the total angular momentum, $\mathbf{J} = \mathbf{L} + \mathbf{S}$, that is conserved. As we said above, however, if the spin-orbit interaction term is small relative to $\hat{H}^{(0)} = \hat{H}_0 + \hat{H}_{ee}$, then $\mathbf{L}$ and $\mathbf{S}$ are almost conserved, and L and S are almost good quantum numbers.

Because $\hat{L}^2$, $\hat{S}^2$, $\hat{L}_z$, and $\hat{S}_z$ commute with $\hat{H}^{(0)}$, the zero-order wave functions are simultaneous eigenfunctions of $\hat{H}$, $\hat{L}^2$, $\hat{L}_z$, $\hat{S}^2$, and $\hat{S}_z$, and so can be labeled by L, M_L, S, and M_S with

$$\hat{H}^{(0)} \mid L, M_L, S, M_S \rangle = E^{(0)} \mid L, M_L, S, M_S \rangle$$

$$\hat{L}^2 \mid L, M_L, S, M_S \rangle = \hbar^2 L(L+1) \mid L, M_L, S, M_S \rangle$$

$$\hat{L}_z \mid L, M_L, S, M_S \rangle = \hbar M_L \mid L, M_L, S, M_S \rangle$$

$$\hat{S}^2 \mid L, M_L, S, M_S \rangle = \hbar^2 S(S+1) \mid L, M_L, S, M_S \rangle$$

$$\hat{S}_z \mid L, M_L, S, M_S \rangle = \hbar M_S \mid L, M_L, S, M_S \rangle$$

As we said above, this scheme is known as Russell–Saunders coupling.

As the atomic number of the atom increases, the spin-orbit term becomes larger than the interelectronic term, and now H_{ee} can be considered to be a small perturbation relative to the other terms in H. In this case L and S are no longer meaningful, and the individual total angular momenta $\mathbf{j}_i = \mathbf{l}_i + \mathbf{s}_i$ become the approximately conserved quantities. One then couples the $\mathbf{j}$'s to obtain the total angular momentum. This scheme is called j–j coupling and is applicable to heavier atoms. In spite of the deterioration of L–S coupling as Z increases, it is still approximately useful, and so the electronic states of even heavy atoms are designated by term symbols of the form $^{2S+1}L_J$.

Appendix: An SCF Calculation of a Helium Atom

We're going to illustrate how an SCF calculation is done, using a helium atom as an example in this appendix. [This calculation is due to Snow, R. L., Bills, J. L., A Simple Illustration of the SCF-LCAO-MO Method. *J. Chem. Educ.*, **52**, 506 (1975).] We'll use a trial function of the form given by Equation 9.18 with $\psi(r_1)$ and $\psi(r_2)$ given by

$$\psi(r) = c_1 \phi_1(r) + c_2 \phi_2(r) \tag{1}$$

where the ϕ_i are the (normalized) Slater orbitals.

$$\phi_i(r) = S_{1s}(\zeta_i) = \left(\frac{\zeta_i^3}{\pi} \right)^{1/2} e^{-\zeta_i r} \tag{2}$$

For simplicity, we'll fix the values of ζ_1 and ζ_2 in the Slater orbitals by $\zeta_1 = 1.453\,63$ and $\zeta_2 = 2.910\,93$ and determine c_1 and c_2 self-consistently. The procedure may seem to be a little complicated, but it is not. Just follow the procedure step by step, finishing with Table 9.8. As we said earlier, in practice, this procedure is carried out computationally, but it's very instructive to follow the procedure by hand at least once.

We take Equation 9.18 as the wave function of a helium atom

$$\Psi(\mathbf{r}_1, \mathbf{r}_2) = \psi(r_1)\psi(r_2) \tag{3}$$

with $\psi(r_i)$ given by equation 1 (Equation 9.32 of the chapter). If we substitute $\Psi(\mathbf{r}_1, \mathbf{r}_2)$ into $E = \langle \Psi \mid \hat{H} \mid \Psi \rangle$, we get (Problem 9–59)

$$E = \frac{2(c_1^2 h_{11} + 2c_1 c_2 h_{12} + c_2^2 h_{22}) + c_1^2 g_{11} + 2c_1 c_2 g_{12} + c_2^2 g_{22}}{c_1^2 + 2c_1 c_2 S_{12} + c_2^2} \tag{4}$$

where

$$h_{ij} = h_{ji} = \int d\mathbf{r}_1\, \phi_i(r_1) \left(-\frac{1}{2}\nabla_1^2 - \frac{Z}{r_1} \right) \phi_j(r_1)$$

$$= \int d\mathbf{r}_1\, \phi_i(r_1)\hat{h}(\mathbf{r}_1)\phi_j(r_1)$$

$$= \langle \phi_i(1) \mid \hat{h}(1) \mid \phi_j(1) \rangle = \langle \phi_i(2) \mid \hat{h}(2) \mid \phi_j(2) \rangle$$

$$= \langle \phi_i \mid \hat{h} \mid \phi_j \rangle \tag{5}$$

$$g_{ij} = g_{ji}$$

$$= c_1^2 \left\langle \phi_i(1)\phi_1(2) \left| \frac{1}{r_{12}} \right| \phi_j(1)\phi_1(2) \right\rangle + c_1 c_2 \left\langle \phi_i(1)\phi_1(2) \left| \frac{1}{r_{12}} \right| \phi_j(1)\phi_2(2) \right\rangle$$

$$+ c_2 c_1 \left\langle \phi_i(1)\phi_2(2) \left| \frac{1}{r_{12}} \right| \phi_j(1)\phi_1(2) \right\rangle + c_2^2 \left\langle \phi_i(1)\phi_2(2) \left| \frac{1}{r_{12}} \right| \phi_j(1)\phi_2(2) \right\rangle \tag{6}$$

and

$$S_{ij} = S_{ji} = \langle \phi_i(1) \mid \phi_j(1) \rangle = \langle \phi_i(2) \mid \phi_j(2) \rangle = \langle \phi_i \mid \phi_j \rangle \tag{7}$$

Let's look at each set of matrix elements in turn. The simplest to evaluate is S_{ij}.

$$S_{ij} = \langle \phi_i \mid \phi_j \rangle = \int_0^\infty \phi_i(r)\phi_j(r)4\pi r^2\, dr$$

$$= \left(\frac{\zeta_i^3 \zeta_j^3}{\pi^2} \right)^{1/2} 4\pi \int_0^\infty dr\, r^2 e^{-(\zeta_i + \zeta_2)r}$$

$$= \frac{8(\zeta_i^3 \zeta_j^3)^{1/2}}{(\zeta_i + \zeta_j)^3} \tag{8}$$

Notice that $S_{11} = S_{22} = 1$. To evaluate h_{ij}, first use the fact that $\phi_i(r)$ is an eigenfunction of a hydrogen-like Hamiltonian operator with charge ζ_i with an eigenvalue $-\zeta_i^2/2$ in atomic units:

$$\left(-\frac{1}{2}\nabla^2 - \frac{\zeta_i}{r}\right)\phi_i(r) = -\frac{\zeta_i^2}{2}\phi_i(r) \tag{9}$$

Now write

$$-\frac{1}{2}\nabla^2 - \frac{Z_i}{r} = -\frac{1}{2}\nabla^2 - \frac{\zeta_i}{r} - \frac{Z - \zeta_i}{r} \tag{10}$$

to obtain

$$\langle\phi_i \mid \hat{h} \mid \phi_j\rangle - -\frac{\zeta_j^2}{2}\langle\phi_i \mid \phi_j\rangle - (Z - \zeta_j)\left\langle\phi_i \left| \frac{1}{r} \right| \phi_j\right\rangle$$

The first integral is just S_{ij} and the second integral is (Problem 9–60)

$$\left\langle\phi_i \left| \frac{1}{r} \right| \phi_j\right\rangle = \int_0^\infty \phi_1(r)\phi_j(r)4\pi r\, dr$$

$$= \frac{4(\zeta_i^3\zeta_2^3)^{1/2}}{(\zeta_i + \zeta_j)^2}$$

Therefore,

$$\langle\phi_i \mid \hat{h} \mid \phi_j\rangle = -\frac{\zeta_j^2}{2}S_{ij} - (Z - \zeta_j)\frac{4(\zeta_i^3\zeta_j^3)^{1/2}}{(\zeta_i + \zeta_j)^2}$$

$$= \frac{4(\zeta_i^3\zeta_j^3)^{1/2}}{(\zeta_i + \zeta_j)^2}\left(-\frac{\zeta_j^2}{\zeta_i + \zeta_j} - Z + \zeta_j\right)$$

$$= \frac{4(\zeta_i^3\zeta_j^3)^{1/2}}{(\zeta_i + \zeta_j)^3}[\zeta_i\zeta_j - Z(\zeta_i + \zeta_j)] \tag{11}$$

The integrals in g_{ij} are more involved because they consist of an integration over the coordinates of both electrons. As i and j take on the values 1 and 2, there are six different integrals in equation 6. The integrals for $i = j = 1$ and $i = j = 2$ are the same as the integral for $E^{(1)}$ in the perturbation treatment of a helium atom (Equation 8.72) and are equal to

$$\left\langle\phi_1(1)\phi_1(2) \left| \frac{1}{r_{12}} \right| \phi_1(1)\phi_1(2)\right\rangle = \frac{5}{8}\zeta_1 \tag{12}$$

$$\left\langle \phi_2(1)\phi_2(2) \left| \frac{1}{r_{12}} \right| \phi_2(1)\phi_2(2) \right\rangle = \frac{5}{8}\zeta_2 \tag{13}$$

The other integrals are listed analytically in Table 9.6, but they are also easy to evaluate numerically for given values of ζ_1 and ζ_2 (Problem 9–61). The numerical values of all the integrals in E in equations 5 and 6 for $\zeta_1 = 1.453\,63$ and $\zeta_2 = 2.910\,93$ are given in Table 9.7. These are the values that we shall use in performing the SCF calculation for a helium atom.

You may have noticed that the integral $\left\langle \phi_2(1)\phi_1(2) \left| \dfrac{1}{r_{12}} \right| \phi_2(1)\phi_1(2) \right\rangle$ (which you get from the first term in equation 6 by letting $i = j = 2$) does not appear in Table 9.6. The reason for this is that there are a number of symmetry relations among the integrals in Table 9.6. For example, we can show that $\left\langle \phi_2(1)\phi_1(2) \left| \dfrac{1}{r_{12}} \right| \phi_2(1)\phi_1(2) \right\rangle$ is equivalent to the third entry; we can relabel the coordinates without changing the value of the

TABLE 9.6
The Analytic Expressions for the Various Integrals in Equations 5 Through 8

$$S_{ij} = \frac{8(\zeta_i^3\zeta_j^3)^{1/2}}{(\zeta_i + \zeta_j)^3}$$

$$h_{ij} = \frac{4(\zeta_i^3\zeta_j^3)^{1/2}}{(\zeta_i + \zeta_j)^3}[\zeta_i\zeta_j - Z(\zeta_i + \zeta_j)]$$

$$\left\langle \phi_1(1)\phi_1(2) \left| \frac{1}{r_{12}} \right| \phi_1(1)\phi_1(2) \right\rangle = \frac{5}{8}\zeta_1$$

$$\left\langle \phi_2(1)\phi_2(2) \left| \frac{1}{r_{12}} \right| \phi_2(1)\phi_2(2) \right\rangle = \frac{5}{8}\zeta_2$$

$$\left\langle \phi_1(1)\phi_2(2) \left| \frac{1}{r_{12}} \right| \phi_1(1)\phi_2(2) \right\rangle = \frac{\zeta_1\zeta_2}{\zeta_1 + \zeta_2} + \frac{\zeta_1^2\zeta_2^2}{(\zeta_1 + \zeta_2)^3}$$

$$\left\langle \phi_1(1)\phi_2(2) \left| \frac{1}{r_{12}} \right| \phi_2(1)\phi_1(2) \right\rangle = \frac{20(\zeta_1\zeta_2)^3}{(\zeta_1 + \zeta_2)^5}$$

$$\left\langle \phi_1(1)\phi_1(2) \left| \frac{1}{r_{12}} \right| \phi_1(1)\phi_2(2) \right\rangle = \frac{8(\zeta_1\zeta_2)^{3/2}\zeta_1(11\zeta_1^2 + 8\zeta_1\zeta_2 + \zeta_2^2)}{(\zeta_1 + \zeta_2)^2(3\zeta_1 + \zeta_2)^3}$$

$$\left\langle \phi_1(1)\phi_2(2) \left| \frac{1}{r_{12}} \right| \phi_2(1)\phi_2(2) \right\rangle = \frac{8(\zeta_1\zeta_2)^{3/2}\zeta_2(\zeta_1^2 + 8\zeta_1\zeta_2 + 11\zeta_2^2)}{(\zeta_1 + \zeta_2)^2(\zeta_1 + 3\zeta_2)^3}$$

Using these values and the values of S_{ij} in Table 9.7, we obtain

$$\begin{vmatrix} -0.874\,370 - \varepsilon & -0.853\,493 - 0.837\,524\,\varepsilon \\ -0.853\,493 - 0.837\,524\,\varepsilon & -0.183\,860 - \varepsilon \end{vmatrix} = 0$$

or

$$0.298\,554\,\varepsilon^2 - 0.371\,414\,\varepsilon - 0.567\,691 = 0$$

The value of the lower root comes out to be $-0.890\,717$.

Using this value in equations 14 gives $c_1/c_2 = 0.107\,496/0.016\,348 = 6.575\,48$. The normalization condition for $\phi(r)$ in Equation 9.26 gives

$$c_1^2 + 2c_1 c_2 S_{12} + c_2^2 = 1$$

or that

$$(6.575\,48)^2 c_1^2 + 2c_2(6.575\,48c_2)S_{12} + c_2^2 = 1$$

or $c_2 = 0.134\,530$ and $c_1 = 0.884\,622$. Finally, we can use equation 4 to calculate E, as shown in Table 9.8. We now repeat the calculation with these values of c_1 and c_2 until the results stabilize, as you can see by comparing successive lines in Table 9.8. Figure 9.12 shows c_1 and c_2 plotted against the number of iterations.

Before we leave this appendix, let's look at the entries in Table 9.8 more carefully. Notice that the total energy is stabilized to its final value by iteration 4, but the values of c_1 and c_2 continue to change for another five iterations, to the precision reported in the

TABLE 9.8
The Successive Values of c_1, c_2, F_{11}, F_{12}, F_{22}, ε, and E for the Iterative Procedure Used to Solve the Hartree–Fock–Roothaan Equations (Equations 14)

Iter-ation	c_1	c_2	F_{11}	F_{12}	F_{22}	ε	E
1	0.500 000	0.500 000	−0.874 370	−0.853 493	−0.183 860	−0.890 717	−2.492 00
2	0.884 622	0.134 530	−0.902 671	−0.923 175	−0.315 450	−0.934 544	−2.857 13
3	0.835 790	0.189 639	−0.886 271	−0.900 502	−0.279 784	−0.914 721	−2.861 50
4	0.845 342	0.178 940	−0.889 460	−0.904 919	−0.286 743	−0.918 563	−2.861 67
5	0.843 481	0.181 028	−0.888 838	−0.904 058	−0.285 386	−0.917 813	−2.861 67
6	0.843 844	0.180 621	−0.888 960	−0.904 226	−0.285 651	−0.917 959	−2.861 67
7	0.843 773	0.180 700	−0.888 936	−0.904 193	−0.285 599	−0.917 931	−2.861 67
8	0.843 787	0.180 685	−0.888 940	−0.904 199	−0.285 609	−0.917 936	−2.861 67
9	0.843 785	0.180 687	−0.888 940	−0.904 199	−0.285 607	−0.917 935	−2.861 67
10	0.843 785	0.180 687	−0.888 940	−0.904 199	−0.285 607	−0.917 935	−2.861 67

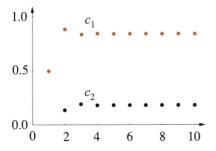

FIGURE 9.12
A plot of c_1 and c_2 in Equation 9.32 plotted against the number of iterations.

table. This shows that the Hartree–Fock energy is stable to first order in perturbations of the wave function (the c_1 and c_2), whereas the wave function is not. This means that a stronger criterion on convergence of the solution is needed if properties other than the energy are going to be calculated from the wave function.

Problems

9–1. Show that the atomic unit of energy can be written as

$$E_h = \frac{\hbar^2}{m_e a_0^2} = \frac{e^2}{4\pi \epsilon_0 a_0} = \frac{m_e e^4}{16\pi^2 \epsilon_0^2 \hbar^2}$$

9–2. Show that the energy of a helium ion in atomic units is $-2\,E_h$.

9–3. Show that the speed of an electron in the first Bohr orbit is $e^2/4\pi\epsilon_0\hbar = 2.188 \times 10^6$ m·s^{-1}. This speed is the unit of speed in atomic units (see Table 9.1).

9–4. Show that the speed of light is equal to 137 in atomic units. (*Hint*: Use the result of the previous problem.)

9–5. Another way to introduce atomic units is to express mass as multiples of m_e, the mass of an electron (instead of kg); charge as multiples of e, the protonic charge (instead of C); angular momentum as multiples of $\hbar$ (instead of in J·s = kg·m^2·s^{-1}); and permittivity as multiples of $4\pi\epsilon_0$ (instead of in C^2·s^2·kg^{-1}·m^{-3}). This conversion can be achieved in all of our equations by letting $m_e = e = \hbar = 4\pi\epsilon_0 = 1$. Show that this procedure is consistent with the definition of atomic units used in the chapter.

9–6. Derive Equation 9.5 from Equation 9.4. Be sure to remember that ∇^2 has units of (distance)$^{-2}$.

9–7. In this problem, we shall show that Equation 9.10 has the form of a sum of a quadratic term in ζ and a linear term in ζ because of the form of the Hamiltonian operator of an atom. Start with $\hat{H}$ for an N-electron atom with nuclear charge Z:

$$\hat{H} = -\frac{1}{2}\sum_{j=1}^{N}\nabla_j^2 - \sum_{j=1}^{N}\frac{Z}{r_j} + \sum_{i<j}\frac{1}{r_{ij}} = \hat{T} + \hat{V}$$

where $\hat{T}$ is the kinetic energy operator and $\hat{V}$ is the potential energy operator. Write out the expressions for $T = \langle \psi \mid \hat{T} \mid \psi \rangle / \langle \psi \mid \psi \rangle$ and $V = \langle \psi \mid \hat{V} \mid \psi \rangle / \langle \psi \mid \psi \rangle$ using a wave function of the form $\psi(x_1, y_1, \ldots, y_N, z_N)$. Now write out the expressions for $T(\zeta) = \langle \psi(\zeta) \mid \hat{T} \mid \psi(\zeta) \rangle / \langle \psi(\zeta) \mid \psi(\zeta) \rangle$ using a wave function of the form $\psi(\zeta x_1, \zeta y_2, \ldots, \zeta y_N, \zeta z_N)$, where all the coordinates are scaled by a factor of ζ. Let $x_1' = \zeta x_1$, $y_1' = \zeta y_1$, and so on to show that

$$T(\zeta) = \zeta^2 T(\zeta = 1)$$

Similarly, write out the expression for $V(\zeta) = \langle \psi(\zeta) \mid \hat{V} \mid \psi(\zeta) \rangle / \langle \psi(\zeta) \mid \psi(\zeta) \rangle$ and then show that

$$V(\zeta) = \zeta V(\zeta = 1)$$

9–8. Show that when we use Equation 9.8 in Equation 9.9, the average kinetic energy and average potential energy come out to be ζ^2 and $-27\zeta/8$, respectively. (Use the results of either Problem 8–43 or 8–44.)

9–9. Show that Equation 9.10 can be written as

$$E(\zeta) = T(\zeta) + V(\zeta) = \zeta^2 T(\zeta = 1) + \zeta V(\zeta = 1)$$

where $T(\zeta = 1)$ and $V(\zeta = 1)$ are the average kinetic energy and the average potential energy, respectively, calculated with Equation 9.8 with $\zeta = 1$ (cf. Problem 9–7).

9–10. Use the results of the previous two problems to show that $T(\zeta = 1)/V(\zeta = 1) = -8/27$ when you use Equation 9.8. What should this ratio equal according to the virial theorem? Now determine the value of ζ such that the ratio $T(\zeta)/V(\zeta)$ satisfies the virial theorem and compare your result to the value of ζ obtained variationally.

9–11. Show that the generalization of Equation 9.10 for a two-electron atom or two-electron ion of nuclear charge Z is

$$E(\zeta) = -\zeta^2 + 2\zeta(\zeta - Z) + \frac{5}{8}\zeta$$

Show that this equation reduces to Equation 9.10 when $Z = 2$.

9–12. Use a program such as *MathCad* or *Mathematica* to evaluate $E(\zeta_1, \zeta_2)$ using Equation 9.13 and then minimize the result with respect to both ζ_1 and ζ_2 to obtain $E = -2.875\,66\,E_h$.

9–13. Show that the normalization constant for the radial part of Slater orbitals is

$$(2\zeta)^{n+\frac{1}{2}}/[(2n)!]^{1/2}$$

9–14. Compare the Slater orbital $S_{200}(r)$ to the hydrogen atomic orbital $\psi_{200}(r)$ (Table 7.3) by plotting them together. Do the same for $S_{300}(r)$ and $\psi_{300}(r)$.

9–15. Show that Equation 9.16 is normalized.

9–16. Compare our simple variational trial function $(\zeta^3/\pi)^{1/2}e^{-\zeta r}$ with $\zeta = 27/16$ with the SCF orbital given by Equation 9.16 by plotting them together (cf. Figure 9.1).

9–17. Substitute Equation 9.5 for $\hat{H}$ into

$$E = \iint d\mathbf{r}_1 d\mathbf{r}_2 \psi^*(\mathbf{r}_1)\psi^*(\mathbf{r}_2)\hat{H}\psi(\mathbf{r}_1)\psi(\mathbf{r}_2)$$

and show that

$$E = I_1 + I_2 + J_{11}$$

where

$$I_j = \int d\mathbf{r}_j \psi^*(\mathbf{r}_j)\left[-\frac{1}{2}\nabla_j^2 - \frac{Z}{r_j}\right]\psi(\mathbf{r}_j)$$

and

$$J_{11} = \iint d\mathbf{r}_1 d\mathbf{r}_2 \psi^*(\mathbf{r}_1)\psi^*(\mathbf{r}_2)\frac{1}{r_{12}}\psi(\mathbf{r}_1)\psi(\mathbf{r}_2)$$

9–18. Why do you think that J_{11} in Equation 9.25 is called a Coulomb integral?

9–19. The normalized variational helium orbital we determined in Chapter 7 is

$$\psi_{1s}(r) = 1.2368\,e^{-27r/16}$$

A two-term Hartree–Fock orbital is given in Equation 9.16,

$$\psi_{1s}(r) = 0.843\,785\,S_{1s}(\zeta = 1.453\,63) + 0.280\,687\,S_{1s}(\zeta = 2.910\,93)$$

and a five-term orbital given by Equation 9.72 is

$$\psi_{1s} = 1.347\,900\,S_{1s}(\zeta = 1.4595) - 0.001\,613\,S_{3s}(\zeta = 5.3244)$$
$$- 0.100\,506\,S_{2s}(\zeta = 2.6298) - 0.270\,779\,S_{2s}(\zeta = 1.7504)$$

Compare these orbitals by plotting them on the same graph.

9–20. Use the SCF orbital given in Equation 9.16 to calculate $V_1^{\text{eff}}(r_1)$ given in Equation 9.19 and compare your result to Figure 9.2.

9–21. Given that $\Psi(1, 2) = 1s\alpha(1)1s\beta(2) - 1s\alpha(2)1s\beta(1)$, prove that

$$\int d\tau_1 d\tau_2 \Psi^*(1, 2)\Psi(1, 2) = 2$$

if the spatial part is normalized.

This final expression for $\rho(x, y, z)$ is the probability density of finding an electron in the volume $dxdydz$ surrounding the point (x, y, z).

Consider the SCF wave function of a helium atom, given by Equation 9.47:

$$\Psi(1, 2) = \frac{1}{\sqrt{2}}[1s\alpha(1)1s\beta(2) - 1s\alpha(2)1s\beta(1)]$$

Now square $\Psi(1, 2)$ and integrate over the spin coordinate of electron 1 and both the spatial and spin coordinates of electron 2 to obtain $1s^2(1)$. Electrons 1 and 2 are indistinguishable, so the probability density of a ground-state helium atom is given by $2\ 1s^2(r)$.

Now let's look at the Slater determinant of a lithium atom (Problem 9–30). Show that if we square $\Psi(1, 2, 3)$ and integrate over the spin coordinate of electron 1 and all the coordinates of electrons 2 and 3, then we obtain

$$\rho(r) = 2\psi_{1s}^2(r) + \psi_{2s}(r)$$

The general result is

$$\rho(\mathbf{r}) = \sum n_j \mid \psi_j(\mathbf{r})\mid^2$$

where the summation runs over all the orbitals and $n_j = 0, 1,$ or 2 is the number of electrons in that orbital.

9–35. Go to the website *http://www.ccl.net/cca/data/atomic-RHF-wavefunctions/tables* and write down the ψ_{1s} orbital for a beryllium atom. Use a program such as *MathCad* or *Mathematica* to show that the orbital is normalized and orthogonal to ψ_{2s} (see also Example 9–7). Using

$$\rho(r) = 2\psi_{1s}^2(r) + 2\psi_{2s}^2(r)$$

for the electron probability density of a beryllium atom, plot $r^2\rho(r)$ against r.

9–36. Go to the website *http://www.ccl.net/cca/data/atomic-RHF-wavefunctions/tables* to verify Koopmans's ionization energies of a neon atom and an argon atom and compare your results to those in Table 9.3.

9–37. Go to the website *http://www.ccl.net/cca/data/atomic-RHF-wavefunctions/tables* to reproduce Figure 9.7.

9–38. Use the entries in Table 9.2 to calculate the percentage of the correlation energy of the Eckart and Hylleraas wave functions for a helium atom.

9–39. Verify that the virtual orbital obtained in the appendix is

$$\psi_2(r) = 1.624\ 04\ S_{1s}(\zeta = 1.453\ 63) - 1.821\ 22\ S_{1s}(\zeta = 2.910\ 93)$$

9–40. Show that

$$\langle S_{1s}(\zeta = 1.453\ 63) \mid S_{1s}(\zeta = 2.910\ 93)\rangle = 0.837\ 524$$

9–41. Verify that the Ψ_{CI} trial function given by Equation 9.76 leads to an energy of $-2.875\,41\,E_h$.

9–42. Determine the term symbols for a $2s^13s^1$ electron configuration.

9–43. Determine the term symbols associated with an np^1 electron configuration. Show that these term symbols are the same as for an np^5 electron configuration.

9–44. Show that the number of sets of magnetic quantum numbers (m_l) and spin quantum numbers (m_s) associated with any term symbol is equal to $(2L+1)(2S+1)$. Apply this result to the np^2 case discussed in Section 9.10, and show that the term symbols ^1S, ^3P, and ^1D account for all the possible sets of magnetic quantum numbers and spin quantum numbers.

9–45. Calculate the number of sets of magnetic quantum numbers (m_l) and spin quantum numbers (m_s) for an nd^8 electron configuration. Show that the term symbols ^1S, ^1D, ^3P, ^3F, and ^1G account for all possible term symbols.

9–46. Determine the term symbols for the electron configuration $nsnp$. Which term symbol corresponds to the lowest energy?

9–47. How many sets of magnetic quantum numbers (m_l) and spin quantum numbers (m_s) are there for an $nsnd$ electron configuration? What are the term symbols? Which term symbol corresponds to the lowest energy?

9–48. The term symbols for an nd^2 electron configuration are ^1S, ^1D, ^1G, ^3P, and ^3F. Calculate the values of J associated with each of these term symbols. Which term symbol represents the ground state?

9–49. The term symbols for an np^3 electron configuration are ^2P, ^2D, and ^4S. Calculate the values of J associated with each of these term symbols. Which term symbol represents the ground state?

9–50. What is the electron configuration of a magnesium atom in its ground state, and what is its ground-state term symbol?

9–51. Given that the electron configuration of a zirconium atom is $[Kr]\,4d^25s^2$, what is the ground-state term symbol for Zr?

9–52. Given that the electron configuration of a palladium atom is $[Kr]\,4d^{10}$, what is the ground-state term symbol for Pd?

9–53. Consider the $1s2p$ electron configuration for helium. What are the states (term symbols) that correspond to this electron configuration? What are the degeneracies of each state if we do not consider spin-orbit coupling? What will happen if you include the effect of spin-orbit coupling?

9–54. Use Table 9.5 to calculate the wavelength of the $4d\,^2D_{3/2} \rightarrow 3p\,^2P_{1/2}$ transition in atomic sodium and compare your result with that given in Figure 9.9. Be sure to use the relation $\lambda_{vac} = 1.000\,29\,\lambda_{air}$ (see Example 9–14).

9–55. The orbital designations s, p, d, and f come from an analysis of the spectrum of atomic sodium. The series of lines due to $ns\,^2S \rightarrow 3p\,^2P$ transitions is called the *sharp* (s) series;

the series due to $np\,^2P \rightarrow 3s\,^2S$ transitions is called the *principal* (*p*) series; the series due to $nd\,^2D \rightarrow 3p\,^2P$ transitions is called the *diffuse* (*d*) series; and the series due to $nf\,^2F \rightarrow 3d\,^2D$ transitions is called the *fundamental* (*f*) series. Identify each of these series in Figure 9.9, and tabulate the wavelengths of the first few lines in each series. Now go to *http://www.ccl.net/cca/data/atomic-RHF-wavefunctions/tables* to calculate the wavelengths of the first few lines in each series and compare your results with those in Figure 9.9. Be sure to use the relation $\lambda_{\text{vac}} = 1.000\,29\,\lambda_{\text{air}}$ (see Example 9–14).

9–56. Use the spectroscopic data for Li, Li$^+$, and Li^{++} in *http://www.physics.nist.gov/ PhysRefData/ASD/levels* to calculate the energy of a lithium atom and compare your result to the Hartree–Fock value in Figure 9.5.

9–57. In this problem, we will derive an explicit expression for $V^{\text{eff}}(r_1)$ given by Equation 9.19 using $\phi(\mathbf{r})$ of the form $(Z^3/\pi)^{1/2}e^{-Zr}$. (We have essentially done this problem in Problem 8–43). Start with

$$V^{\text{eff}}(\mathbf{r}_1) = \frac{Z^3}{\pi} \int d\mathbf{r}_2 \frac{e^{-2Zr_2}}{r_{12}}$$

As in Problem 8–43, we use the law of cosines to write

$$r_{12} = (r_1^2 + r_2^2 - 2r_1r_2 \cos\theta)^{1/2}$$

and so V^{eff} becomes

$$V^{\text{eff}}(r_1) = \frac{Z^3}{\pi} \int_0^\infty dr_2 e^{-2Zr_2}r_2^2 \int_0^{2\pi} d\phi \int_0^\pi \frac{d\theta \sin\theta}{(r_1^2 + r_2^2 - 2r_1r_2 \cos\theta)^{1/2}}$$

Problem 8–43 asks you to show that the integral over θ is equal to $2/r_1$ if $r_1 > r_2$ and is equal to $2/r_2$ if $r_1 < r_2$. Thus, we have

$$V^{\text{eff}}(r_1) = 4Z^3 \left[\frac{1}{r_1} \int_0^{r_1} e^{-Zr_2}r_2^2 dr_2 + \int_{r_1}^\infty e^{-2Zr_2}r_2 dr_2 \right]$$

Now show that

$$V^{\text{eff}}(r_1) = \frac{1}{r_1} - e^{-2Zr_1} \left(Z + \frac{1}{r_1} \right)$$

9–58. Repeat the previous problem using the expansion of $1/r_{12}$ given in Problem 8–44.

9–59. Derive equation 4 of the appendix.

9–60. Evaluate $\left\langle \phi_i \left| \dfrac{1}{r} \right| \phi_j \right\rangle$ for a 1s Slater orbital.

9–61. Use a program such as *MathCad* or *Mathematica* to verify the numerical values in Table 9.7.

References

Eisberg, R., Resnick, R. *Quantum Physics of Atoms, Molecules, Solids, Nuclei, and Particles*. Wiley & Sons: New York, 1974.

Woodgate, G. K. *Elementary Atomic Structure,* 2nd ed. Oxford University Press: New York, 1983.

Slater, J. C. *Quantum Theory of Atomic Structure*. McGraw-Hill: New York, 1960.

Cox, P. A. *Introduction to Quantum Theory and Atomic Structure*. Oxford University Press: New York, 1996.

Herzberg, G. *Atomic Spectra and Atomic Structure*. Dover Publications: Mineola, NY, 1994.

Shore, B. W. *Principles of Atomic Spectra*. Wiley & Sons: New York, 1968.

Snow, R. L., Bills, J. L. A Simple Illustration of the SCF-LCAO-MO Method. *J. Chem. Educ.,* **52**, 506 (1975).

National Institute of Standards and Technology (NIST), *http://physics.nist.gov/PhysRefData/ASD*

Bunge, C. F., Barrientos, J. A., Bunge, A. V. *Atomic Data and Nuclear Data Tables,* **53**, 113 (1993). *http://www.ccl.net/cca/data/atomic-RHF-wavefunctions/tables*

Robert S. Mulliken was born in Newburyport, Massachusetts, on June 25, 1896, and died in Chicago in 1986. He received his Ph.D. in physical chemistry from the University of Chicago in 1921, where his dissertation was on the separation of mercury isotopes by fractional distillation. He then went to Harvard University to continue his study of the behavior of isotopes. Realizing the importance of the new quantum theory to the understanding of atomic and molecular structure, he then spent a year in Europe studying quantum theory. After one year as an assistant professor at Washington Square College in New York City, he spent a year with Friedrich Hund at the University of Göttingen, during which time they developed molecular orbital theory. Upon his return to the United States in 1928, Mulliken accepted a position at the University of Chicago, where he remained until his formal retirement in 1961. He continued working on molecular orbital theory at Chicago, and one of his early important contributions was the introduction of the LCAO approximation. After World War II, Mulliken and his collaborators pioneered the use of computers for calculating and elucidating molecular structure. He was known by all his associates as an unassuming and good-natured man. He was awarded the Nobel Prize in Chemistry in 1966 "for his fundamental work concerning bonds and the electronic structure of molecules by the molecular orbital method."

CHAPTER 10

The Chemical Bond: One- and Two-Electron Molecules

One of the great achievements of quantum mechanics is its ability to describe the chemical bond. Before the development of quantum mechanics, chemists did not understand why two hydrogen atoms come together to form a stable chemical bond. We will see in this chapter that the existence of stable chemical bonds is described by quantum mechanics. Because the molecular ion H_2^+ involves the simplest chemical bond, we will discuss it in detail. We shall use H_2^+ to introduce molecular orbitals, and how they form from linear combinations of atomic orbitals (LCAO-MO). We shall then construct wave functions for H_2 by forming Slater determinants from these molecular orbitals. We shall apply what is called molecular orbital theory to H_2 by using these Slater determinants to calculate the electronic energy of H_2, and to show that molecular orbital theory describes a stable chemical bond in H_2. After discussing the deficiencies of molecular orbital theory, we shall show how to remedy them in a systematic way by a method called configuration interaction, in which the molecular wave function is written as a linear combination of Slater determinants that represent the ground state and various excited states, called configurations. We then develop a procedure to determine the optimum molecular orbitals, which is based on the Hartree–Fock self-consistent field theory that we developed in the previous chapter for atoms. We do this by introducing the Hartree–Fock–Roothaan equations, which give us the optimum molecular orbitals, called self-consistent field linear combinations of atomic orbitals, or SCF-LCAO-MO. The resulting SCF-LCAO-MO can be used to construct Slater determinants representing various configurations, which can then be used to improve the Hartree–Fock–Roothaan wave functions by configuration interaction. Configuration interaction is called a post-Hartree–Fock method because it systematically improves the Hartree–Fock–Roothaan wave functions.

10.1 The Born–Oppenheimer Approximation Simplifies the Schrödinger Equation for Molecules

For simplicity, let's consider the simplest neutral molecule, H_2. The molecular Schrödinger equation for H_2 is

$$\hat{H}_{mol}\psi_{mol}(\mathbf{r}_1, \mathbf{r}_2, \mathbf{R}_A, \mathbf{R}_B) = E_{mol}\psi_{mol}(\mathbf{r}_1, \mathbf{r}_2, \mathbf{R}_A, \mathbf{R}_B) \tag{10.1}$$

where $\hat{H}_{mol}$ is given by

$$\hat{H}_{mol} = -\frac{\hbar^2}{2M}(\nabla_A^2 + \nabla_B^2) - \frac{\hbar^2}{2m_e}(\nabla_1^2 + \nabla_2^2) - \frac{e^2}{4\pi\epsilon_0 r_{1A}}$$

$$-\frac{e^2}{4\pi\epsilon_0 r_{1B}} - \frac{e^2}{4\pi\epsilon_0 r_{2A}} - \frac{e^2}{4\pi\epsilon_0 r_{2B}} + \frac{e^2}{4\pi\epsilon_0 r_{12}} \quad | \quad \frac{e^2}{4\pi\epsilon_0 R} \tag{10.2}$$

where $R = |\mathbf{R}_A - \mathbf{R}_B|$. In Equation 10.2, M is the mass of each hydrogen nucleus, m_e is the mass of an electron, the subscripts A and B refer to the nuclei of the individual atoms, the subscripts 1 and 2 refer to the individual electrons, and the various distances, r_{1A}, r_{1B}, etc., are illustrated in Figure 10.1.

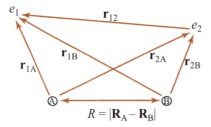

FIGURE 10.1
The definitions of the distances between the nuclei and the electrons involved in the Hamiltonian operator for a hydrogen molecule (Equation 10.2).

The terms involving ∇_A^2 and ∇_B^2 represent the kinetic energy of the nuclei; those involving ∇_1^2 and ∇_2^2 represent the kinetic energy of the electrons; the next four terms represent the interaction of each electron with each nucleus; the term $e^2/4\pi\epsilon_0 r_{12}$ represents the repulsive interaction between the two electrons; and $e^2/4\pi\epsilon_0 R$ represents the repulsion between the two nuclei, which are separated by a distance R.

Because the nuclei are so much more massive than the electrons, they move slowly compared to the electrons, and so the electrons adjust essentially instantaneously to any motion of the nuclei. A systematic treatment of this observation assumes that

$$\psi_{mol}(\mathbf{r}_1, \mathbf{r}_2, \mathbf{R}_A, \mathbf{R}_B) \approx \psi_{el}(\mathbf{r}_1, \mathbf{r}_2, R)\psi_{nucl}(\mathbf{R}_A, \mathbf{R}_B) \tag{10.3}$$

If Equation 10.3 is substituted into Equation 10.1 and terms of order m_e/M are neglected, then two separate equations are obtained. One of these equations represents the motion of the electrons,

$$\hat{H}_{el}\psi_{el}(\mathbf{r}_1, \mathbf{r}_2, R) = E_{el}(R)\psi_{el}(\mathbf{r}_1, \mathbf{r}_2, R) \tag{10.4}$$

where

$$\hat{H}_{el} = -\frac{\hbar}{2m_e}(\nabla_1^2 + \nabla_2^2) - \frac{e^2}{4\pi\varepsilon_0 r_{1A}} - \frac{e^2}{4\pi\varepsilon_0 r_{1B}} - \frac{e^2}{4\pi\varepsilon_0 r_{2A}} - \frac{e^2}{4\pi\varepsilon_0 r_{2B}}$$

$$+ \frac{e^2}{4\pi\varepsilon_0 r_{12}} + \frac{e^2}{4\pi\varepsilon_0 R} \tag{10.5}$$

Because the nuclei are considered to be fixed, the quantity R in Equation 10.5 is treated as a parameter; the energy E_{el} in Equation 10.4 will depend parametrically upon R. As usual, we will express all our equations in atomic units (Section 9.1), and so Equation 10.5 becomes (Problem 10–1)

$$\hat{H}_{el} = -\frac{1}{2}(\nabla_1^2 + \nabla_2^2) - \frac{1}{r_{1A}} - \frac{1}{r_{1B}} - \frac{1}{r_{2A}} - \frac{1}{r_{2B}} + \frac{1}{r_{12}} + \frac{1}{R} \tag{10.6}$$

The other equation that we obtain from Equation 10.3, the one for $\psi_{nucl}(\mathbf{R}_A, \mathbf{R}_B)$, represents the motion of the nuclei and leads to vibrational and rotational motion, which we have already discussed in Chapters 5 and 6.

The approximate separation of a molecular Schrödinger equation such as Equation 10.1 into one for the electronic motion (Equation 10.4) and one for the nuclear motion is called the *Born–Oppenheimer approximation*. The essence of the Born–Oppenheimer approximation is Equation 10.3, which assumes that the complete molecular wave function can be factored into an electronic part and a nuclear part. The practical result of the Born–Oppenheimer approximation is that we can ignore the nuclear kinetic energy terms in the molecular Hamiltonian operator. Although we have discussed the application of the Born–Oppenheimer equation only to the case of H_2, its extension to other molecules is straightforward.

10.2 The Hydrogen Molecular Ion, H_2^+, Is the Prototype Diatomic Molecule

The hydrogen molecular ion, H_2^+, is a stable molecular species with an equilibrium bond length of 105.7 pm (1.997 a_0) and a potential-well depth, D_e, of 0.102 64 E_h (269.5 kJ·mol^{-1}) (Problem 10–3). Figure 10.2 shows the ground-state electronic energy of H_2^+ plotted against the internuclear separation, R. The ground electronic state of H_2^+ dissociates into a ground-state hydrogen atom (whose electronic energy is $-1/2 E_h$) and a proton (no electronic energy), and so the curve in Figure 10.1 goes to $-1/2 E_h$ as $R \to \infty$. Carrington (see references) gives a nice discussion of the experimental properties of H_2^+ in an article titled "The Simplest Molecule."

The hydrogen molecular ion to some extent plays the same role for molecular calculations as the hydrogen atom does for atomic calculations. Just as we build up

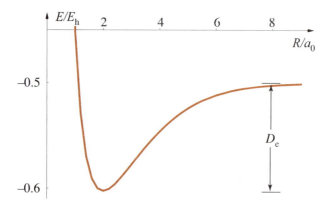

FIGURE 10.2
The ground-state electronic energy of H_2^+ plotted against the internuclear separation R.

atomic orbitals for multielectron atoms by forming Slater determinants of hydrogen-like atomic orbitals, we shall build up molecular orbitals for multielectron molecules by forming Slater determinants of H_2^+-like molecular orbitals. The electronic Schrödinger equation for H_2^+ is

$$\hat{H}\psi_j(r_A, r_B, R) = E_j\psi_j(r_A, r_B, R) \tag{10.7}$$

where the Hamiltonian operator (in atomic units) in the Born–Oppenheimer approximation is

$$\hat{H} = -\frac{1}{2}\nabla^2 - \frac{1}{r_A} - \frac{1}{r_B} + \frac{1}{R} \tag{10.8}$$

where r_A and r_B are the distances of the electron from nucleus A and B, respectively, and R is the internuclear separation, which we treat as a fixed parameter (Figure 10.3).

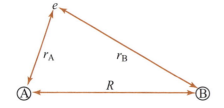

FIGURE 10.3
The parameters r_A, r_B, and R for the H_2^+ molecular ion.

We shall see that the molecular wave function spreads over both nuclei, and so we refer to it as a *molecular orbital*, as opposed to an atomic orbital, which is centered on one nucleus. Like the hydrogen atom, with its one electron, the Schrödinger equation for the hydrogen molecular ion can be solved exactly. There is a coordinate system called

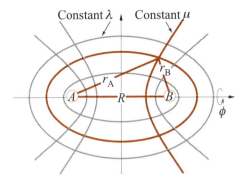

Constant λ Constant μ

FIGURE 10.4

The elliptical coordinate system: lines of constant $\lambda = (r_A + r_B)/R$ map out ellipses in the plane; lines of constant $\mu = (r_A - r_B)/R$ map out hyperbolas; and ϕ is the angle that the plane makes about the internuclear axis.

elliptical coordinates in which Equation 10.7 separates into three ordinary differential equations, just as the Schrödinger equation for the hydrogen atom does. A point in this elliptical coordinate system (Figure 10.4) is given by three coordinates

$$\lambda = \frac{r_A + r_B}{R} \qquad \mu = \frac{r_A - r_B}{R} \qquad (10.9)$$

and the angle ϕ, which is the angle that the (r_A, r_B, R) triangle makes about the internuclear axis. In terms of these coordinates, the solution to Equation 10.7 can be written as

$$\psi(r_A, r_B, R) = F(\lambda)S(\mu)\Phi(\phi) \qquad (10.10)$$

where each factor in Equation 10.10 satisfies an ordinary differential equation. Unfortunately, however, these equations cannot be solved analytically in terms of simple functions, and so the results must be expressed in the form of numerical tables. Figure 10.5 shows the calculated electronic energy plotted against R for the ground state and several excited states of H_2^+. The curve for the ground state goes through a minimum, thus indicating a stable molecular species whose equilibrium internuclear separation is 2.00 a_0 (106 pm). The calculated minimum energy is $-0.602\,64\,E_h$, in excellent agreement with experiment.

It is customary to plot the energy relative to a separated hydrogen atom and a proton so that the energy goes to zero at large values of R. This is shown in Figure 10.6. Note that the only difference between Figures 10.5 and 10.6 is that the zero of energy is shifted by $-1/2\,E_h$, the energy of an isolated hydrogen atom. The minimum in Figure 10.6 is $-0.102\,64\,E_h$ instead of $-0.602\,64\,E_h$.

The states in Figures 10.5 and 10.6 are designated by the following notation. The operator, $\hat{L}_z$, which represents the angular momentum about the internuclear axis, commutes with $\hat{H}$, and so the total wave function must be an eigenfunction of $\hat{L}_z$. We saw in Section 6.6 that $\hat{L}_z = -i\hbar\partial/\partial\phi$ and that its eigenfunctions are $\Phi(\phi) = (2\pi)^{-1/2}e^{im\phi}$ with $m = 0, \pm1, \pm2, \ldots$. Thus, m is a good quantum number and can be used to designate the states of H_2^+. States with $m = 0$ are called σ states, those with

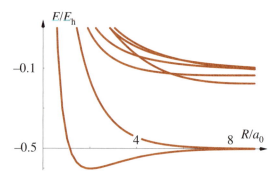

FIGURE 10.5
The electronic energy of H_2^+ plotted against R for the ground state and several excited states. The curve for the ground state goes through a minimum and thus represents a stable molecular species whose energy is the depth of the minimum.

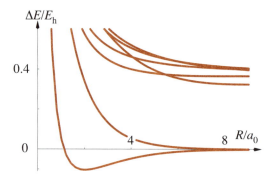

FIGURE 10.6
The electronic energy of H_2^+ relative to a ground-state hydrogen atom and a proton plotted against R for the ground state and several excited states. The energy of the ground state in this case goes to zero for large values of R and the depth of the minimum is $-0.102\,64\,E_h$.

$m = \pm 1$ are called π states, those with $m = \pm 2$ are called δ states, and so on. Figure 10.7 shows sketches of σ states and π states of H_2^+. The σ state wave function is cylindrically symmetric, or has a circular cross section viewed along the internuclear axis. (Note the correspondence between an atomic s orbital and a molecular σ orbital.) The π state wave function has a cross section similar to an atomic p orbital. Note that σ orbitals have no nodal planes and that π orbitals have one nodal plane containing the internuclear axis. It turns out that δ orbitals have two nodal planes containing the internuclear axis, and so on. This result coincides nicely with the angular wave functions of a hydrogen atom where $|m_l|$, the magnitude of magnetic quantum number, is the number of nodal planes in the angular wave function containing the z axis.

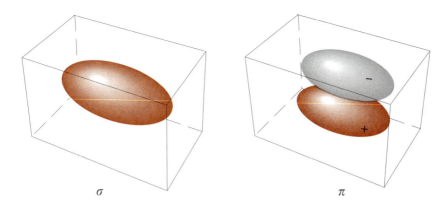

FIGURE 10.7
A sketch of σ states and π states: σ states are cylindrically symmetric when viewed along the internuclear axis; π states have cross sections similar to p orbitals.

In addition to the value of m, the states can be labeled by their behavior under an inversion of the molecule through its center of symmetry—that is, reflecting the positions of all the particles through the point midway between the nuclei. Because of the symmetry of H_2^+, the inversion operator, call it $\hat{P}_{inv}$, commutes with $\hat{H}$. Letting $\hat{P}_{inv}$ operate on $\psi(r_A, r_B, R)$ gives

$$\hat{P}_{inv}\psi(r_A, r_B, R) = c\psi(r_A, r_B, R)$$

A second operation

$$\hat{P}_{inv}^2 \psi(r_A, r_B, R) = c^2\psi(r_A, r_B, R)$$

amounts to the identity operation, and so we see that $c^2 = 1$, or that $c = \pm 1$. Thus, the wave functions of H_2^+ either do not change or change in sign under inversion. If the wave function remains unchanged, then the state is designated by a g (for gerade, the German word for "even"); if it changes in sign, then the state is designated by a u (for ungerade, the German word for "odd"). Figure 10.8 is a scatter plot of the wave functions of the ground state and the first excited state of H_2^+. Both wave functions are cylindrically symmetric and so are σ states. Note that the ground-state wave function is gerade and the first-excited-state wave function is ungerade. The ground state is designated by σ_g and the first excited-state by σ_u. Figure 10.8 gives two designations for each state. The $1\sigma_g$ means that this state is the first σ_g state, in order of increasing energy around the equilibrium bond length for the ground state; the $\sigma_g 1s$ means that this state dissociates to a proton and a hydrogen atom in a $1s$ state.

Even though the Schrödinger equation for H_2^+ can be solved exactly, the solutions are not all that easy to use numerically. Nevertheless, H_2^+ serves as an excellent system to develop approximate methods that can be used for more complicated systems. We can

$$1\sigma_g \quad \sigma_g 1s \qquad\qquad 1\sigma_u \quad \sigma_u 1s$$

FIGURE 10.8
An illustration of the cross section of the wave functions of the electronic ground state and first excited state of H_2^+. Note that the ground state is gerade and that the first excited state is ungerade. The orbital has positive values in the orange regions and negative values in the black regions.

compare the results to assess their efficacy and gain physical insight into the nature of the solutions. In the next section, we shall use H_2^+ to develop molecular orbital theory, which is by far the most widely used approach to the bonding in multielectron molecules.

10.3 Molecular Orbitals Are Constructed from a Linear Combination of Atomic Orbitals

The method we will use to describe the bonding properties of molecules is called *molecular orbital theory*. Molecular orbital theory was developed in the early 1930s by Friedrich Hund and Robert Mulliken and is now the most commonly used method to calculate molecular properties. In molecular orbital theory, we construct molecular wave functions in a manner similar to the way we constructed atomic wave functions in Chapter 9, where we expressed atomic wave functions in terms of determinants involving single-electron wave functions called atomic orbitals. Here we will express molecular wave functions in terms of determinants involving single-electron wave functions called molecular orbitals. The question that arises, then, is how to construct molecular orbitals. In this case, the Schrödinger equation for the one-electron molecular ion H_2^+ is solved approximately using a linear combination of atomic orbitals, and the resulting orbitals are then used to construct determinantal wave functions for more complicated molecules. Although this approach may seem a crude way to proceed (after all, H_2^+ can be solved exactly), it provides good physical insight into the nature of chemical bonds in molecules and yields results in good agreement with experimental observations. Furthermore, this approach can be systematically improved to give any desired degree of accuracy.

The Hamiltonian operator for H_2^+ in the Born–Oppenheimer approximation is given by Equation 10.8. Recall that the variational principle (Chapter 8) says that we can get a good approximation to the energy if we use an appropriate trial function. As a trial function for $\psi_j(r_A, r_B; R)$, we take the linear combination

$$\psi = c_A 1s_A + c_B 1s_B \tag{10.11}$$

where $1s_A$ and $1s_B$ are hydrogen atomic orbitals centered on nuclei A and B, respectively. The molecular orbital given by Equation 10.11 is a *linear* combination of *atomic* orbitals, and is called an *LCAO molecular orbital*. The $1s_A$ and $1s_B$ orbitals, from which the molecular orbital is constructed, are called *basis functions*, and constitute a *basis set*. The basis set in this case is called a *minimal basis set* because two orbitals is the minimum number that can be used to enclose both nuclei. Equation 10.11 consists of a single orbital for each hydrogen atom.

The secular equation associated with Equation 10.11 is

$$\begin{vmatrix} H_{AA} - E & H_{AB} - ES \\ H_{BA} - ES & H_{BB} - E \end{vmatrix} = 0 \tag{10.12}$$

where

$$H_{AA} = \int d\mathbf{r}\, 1s_A \hat{H} 1s_A = \langle 1s_A \mid \hat{H} \mid 1s_A \rangle = \langle 1s_B \mid \hat{H} \mid 1s_B \rangle = H_{BB} \tag{10.13a}$$

$$H_{AB} = \int d\mathbf{r}\, 1s_A \hat{H} 1s_B = \langle 1s_A \mid \hat{H} \mid 1s_B \rangle = H_{BA} \tag{10.13b}$$

and

$$S = \int d\mathbf{r}\, 1s_A 1s_B = \langle 1s_A \mid 1s_B \rangle \tag{10.13c}$$

In Equation 10.12, we used the fact that the $1s$ orbitals are normalized ($\langle 1s_A \mid 1s_A \rangle = \langle 1s_B \mid 1s_B \rangle = 1$) and in Equation 10.13a, we used the fact that $H_{AA} = H_{BB}$ by symmetry. If we write out Equation 10.12, then we obtain

$$(H_{AA} - E)^2 - (H_{AB} - ES)^2 = 0$$

or

$$E_{\pm} = \frac{H_{AA} \pm H_{AB}}{1 \pm S} \tag{10.14}$$

Let's now look at the three integrals in Equations 10.13 in turn. We'll start with S. Note that S involves the product of a $1s$ orbital situated on nucleus A and one situated on nucleus B, where the two nuclei are separated by a distance R, as in Figure 10.9. This product is significant only for regions where the two atomic orbitals have a large overlap (Problem 10–5). Consequently, S is called an *overlap integral*. The extent of overlap which the electron is situated on both nuclei, and so are *exchange integrals*. Exchange integrals arise because we have written the wave function in Equation 10.11 as a linear combination of atomic orbitals situated on both nuclei. This procedure is strictly a quantum-mechanical procedure and so we say that an exchange integral is a quantum-mechanical quantity. Figure 10.12 shows the atomic integrals $J(R) + 1/R$ and the exchange integrals $K(R) + S(R)/R$ plotted against R. This figure nicely illustrates that it is the exchange integrals that lead to the stability of the bond in H_2^+, and that the existence of a stable chemical bond is a quantum-mechanical effect.

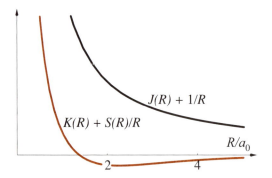

FIGURE 10.12
A plot of the atomic integrals, $J(R) + 1/R$ (black curve), and the exchange integrals, $K(R) + S(R)/R$ (orange curve), in Equation 10.23 plotted against R.

The other energy from Equation 10.23, $\Delta E_-(R)$, represents the first excited state of H_2^+ and is plotted in Figure 10.13. Note that there is no minimum in the curve indicating that this excited state has no bound states. For comparison, we also show the exact energy. The agreement is fairly good.

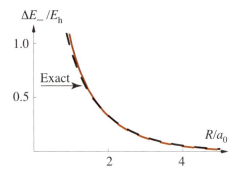

FIGURE 10.13
A comparison of the energy $\Delta E_-(R)$ of the first excited state of H_2^+ calculated from Equation 10.23 with the exact energy.

10.4 The Simplest Molecular Orbital Treatment of H_2^+ Yields a Bonding and an Antibonding Orbital

Let's investigate the molecular orbitals corresponding to $E_{\pm}(R)$ in Equation 10.22.

EXAMPLE 10–2

Determine the molecular orbitals that correspond to the energies E_+ and E_- in Equation 10.22.

SOLUTION: To determine the coefficients c_A and c_B in Equation 10.11, we must use the algebraic equations that lead to the secular determinant in Equation 10.12. These equations are

$$c_A(H_{AA} - E) + c_B(H_{AB} - ES) = 0$$

and

$$c_A(H_{AB} - ES) + c_B(H_{BB} - E) = 0$$

If we substitute E_+ from Equation 10.14 into either of the above equations, say, the first, and recognize that $H_{AA} = H_{BB}$, then we obtain

$$c_A\left(H_{AA} - \frac{H_{AA} + H_{AB}}{1 + S}\right) + c_B\left(H_{AB} - \frac{H_{AA} + H_{AB}}{1 + S}S\right) = 0$$

or

$$c_A\left(\frac{H_{AA}S - H_{AB}}{1 + S}\right) + c_B\left(\frac{H_{AB} - H_{AA}S}{1 + S}\right) = 0$$

or that $c_A = c_B$. Consequently, the molecular orbital corresponding to E_+ is

$$\psi_+ = c_A(1s_A + 1s_B)$$

By requiring ψ_+ to be normalized, we find that

$$c_A^2(1 + 2S + 1) = 1$$

or that $c_A = 1/\sqrt{2(1 + S)}$. Finally, then, we have

$$\psi_+ = \frac{1}{\sqrt{2(1 + S)}}(1s_A + 1s_B) \tag{10.24}$$

To find ψ_-, we substitute E_- into either equation

$$c_A(H_{AA} - E) + c_B(H_{AB} - ES) = 0$$

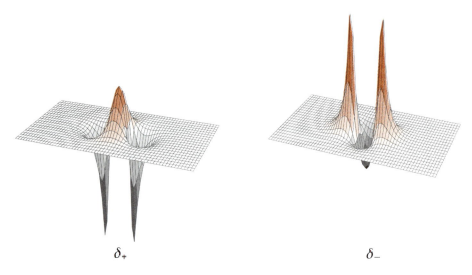

δ_+ δ_-

FIGURE 10.15
Surface plots of δ_+ and δ_-, the difference between the electron density in which the electron is delocalized over the two nuclei and the electron density in which the electron is localized on one of the nuclei.

Let's calculate $\langle \hat{T} \rangle$ and $\langle \hat{V} \rangle$ using the bonding orbital ψ_+ in Equation 10.24. Problem 10–15 has you show that

$$\langle \hat{T} \rangle = \left\langle \psi_+ \left| -\frac{1}{2}\nabla^2 \right| \psi_+ \right\rangle = \frac{\dfrac{1}{2} - \dfrac{S(R)}{2} - K(R)}{1 + S(R)} \tag{10.27}$$

and

$$\langle \hat{V} \rangle = \left\langle \psi_+ \left| -\frac{1}{r_A} \right| \psi_+ \right\rangle + \left\langle \psi_+ \left| -\frac{1}{r_B} \right| \psi_+ \right\rangle + \frac{1}{R}$$
$$= \frac{-1 + J(R) + 2K(R)}{1 + S(R)} + \frac{1}{R} \tag{10.28}$$

EXAMPLE 10–3
Show that $\langle \hat{T} \rangle$ and $\langle \hat{V} \rangle$ given by Equations 10.27 and 10.28 do not satisfy the virial theorem.

SOLUTION: Letting $R = 2.493\, a_0$ in Equations 10.27 and 10.28 (the values of R_{eq} using ψ_+ from Example 10–2) and using the entries in Table 10.1, we have

$$\langle \hat{T} \rangle = 0.3827\, E_h \qquad \text{and} \qquad \langle \hat{V} \rangle = -0.9475\, E_h$$

and

$$\frac{\langle \hat{V} \rangle}{\langle \hat{T} \rangle} = -1.6156 \neq -2$$

Thus, we see that the molecular orbital given by Equation 10.24 does not satisfy the virial theorem. Furthermore, Problem 10–17 has you show that it gives us a skewed picture of the bonding process in diatomic molecules.

We can use a slight modification of our simple molecular orbital, however, to achieve a molecular orbital that does satisfy the virial theorem. Instead of using a linear combination of atomic 1s orbitals, let's use (normalized) Slater 1s orbitals,

$$\phi(\zeta, r) = \left(\frac{\zeta^3}{\pi}\right)^{1/2} e^{-\zeta r} \tag{10.29}$$

and let ζ (zeta) be a variational parameter. Note that we have simply scaled r by a factor ζ. Our molecular orbital now is

$$\psi_+ = c_A \left(\frac{\zeta^3}{\pi}\right)^{1/2} e^{-\zeta r_A} + c_B \left(\frac{\zeta^3}{\pi}\right)^{1/2} e^{-\zeta r_B}$$

where, by symmetry, $c_A^2 = c_B^2$, as in the previous section.

The normalized molecular orbitals are given by

$$\psi_\pm(\zeta, r_A, r_B, R) = \frac{\phi(\zeta, r_A) \pm \phi(\zeta, r_B)}{\sqrt{2(1 \pm S(\zeta, R))}} \tag{10.30}$$

where (Table 10.1 and Problem 10–18)

$$S(\zeta, R) = e^{-\zeta R}\left(1 + \zeta R + \frac{1}{3}\zeta^2 R^2\right)$$

Note that $S(\zeta, R) = S(\zeta R)$, which we can write as $S(w)$, where $w = \zeta R$ as in Table 10.1.

The kinetic energy and the potential energy that we obtain from Equation 10.30 are a simple extension of Equations 10.27 and 10.28 (Problem 10–19):

$$\langle \hat{T}_+ \rangle = \frac{\dfrac{\zeta^2}{2} - \zeta^2 \left[\dfrac{S(w)}{2} + K(w)\right]}{1 + S(w)} = \zeta^2 T_+(w) \tag{10.31}$$

and

$$\langle \hat{V}_+ \rangle = \frac{-\zeta + \zeta J(w) + 2\zeta K(w)}{1 + S(w)} + \frac{\zeta}{w} = \zeta V_+(w) \tag{10.32}$$

Note that these expressions reduce to Equations 10.27 and 10.28 when $\zeta = 1$. The total energy is given by

$$E_+(\zeta, w) = \zeta^2 T_+(w) + \zeta V_+(w) \tag{10.33}$$

where $T_+(w)$ and $V_+(w)$ are defined in Equations 10.31 and 10.32. The fact that $E_+(\zeta, w)$ has the form $\zeta^2 T_+(w) + \zeta V_+(w)$ is a direct result of the form of the Hamiltonian operator (Problem 10–20).

Even though we are discussing the simplest molecular ion, H_2^+, its energy given by Equation 10.33 is a lengthy equation in ζ and $w = \zeta R$, and we're going to see that the corresponding equations for H_2 are even lengthier. Nevertheless, programs such as *MathCad* or *Mathematica* can handle these equations easily. We have reached a point in this book where you should seriously consider using such programs to carry out the calculations. These programs are relatively easy to learn and use, and are indispensable not only for the calculations in this chapter, but for many other applications as well. Furthermore, these programs are capable of carrying out algebraic manipulations and can relieve you of a lot of tedious algebra and allow you to concentrate on the content of equations instead of getting bogged down in algebra.

Equation 10.33 gives the energy as a function of ζ and R because $w = \zeta R$. We can use one of the above programs to minimize $E_+(\zeta, R)$ numerically with respect to ζ and R, in which case we obtain (see also Problem 10–21)

$$E_{min} = -0.586\,51\,E_h \quad \text{at} \quad \zeta = 1.238 \quad \text{and} \quad R_{eq} = 2.003\,a_0$$

This result is about $0.02\,E_h$ (50 kJ·mol^{-1}) lower than that which we obtained for the molecular orbital with $\zeta = 1$, where $E_{min} = -0.564\,83\,E_h$. In addition, the equilibrium bond length is in excellent agreement with the experimental value. Thus, the optimized energy provides a significant improvement. Table 10.2 summarizes the various calculations of the ground-state energy of H_2^+ that we present in this chapter.

EXAMPLE 10–4
Show that the optimized energy of the ground state of H_2^+ satisfies the virial theorem.

SOLUTION: Using the values $\zeta = 1.238$ and $w = (1.238)/(2.003\,a_0) = 2.480\,a_0$ in Equations 10.31 and 10.32 gives

$$\langle \hat{T}_+ \rangle = 0.5865\,E_h \quad \text{and} \quad \langle \hat{V}_+ \rangle = -1.1730\,E_h$$

and so

$$\frac{\langle \hat{V}_+ \rangle}{\langle \hat{T}_+ \rangle} = -2.0000$$

TABLE 10.2
Results of Various Calculations of the Ground-State Electronic Energy of H$_2^+$ [a]

ϕ	E_{min}/E_h	R_{eq}/a_0
$1s\,(\zeta = 1.000)$	−0.564 83	2.49
$1s\,(\zeta = 1.238)$	−0.586 51	2.00
$1s\,(\zeta = 1.000) + a2p_z(\zeta = 1.000)$	−0.565 91	2.00
$1s\,(\zeta = 1.247) + b2p_z(\zeta = 1.247)$	−0.599 07	2.00
$1s\,(\zeta = 1.2458) + c2p_z(\zeta = 1.4224)$	−0.600 36	2.00
$1s\,(\zeta = 1.244) + c_1 2p_z(\zeta = 1.152) + c_2 3d_{z^2}(\zeta = 1.333)$ [b]	−0.6020	2.00
Exact [c]	−0.602 64	2.00

a. The molecular orbitals are of the form $\psi_b = c_A\phi_A + c_B\phi_B$, where ϕ is given in the table.
b. Mulliken, R. S., Ermler, W. C. *Diatomic Molecules*. Academic Press: New York, 1977.
c. Bates, D. R., Ledsham, K., Stewart, A. L. Wave Functions of the Hydrogen Molecular Ion. *Philos. Trans. Roy. Soc. London*, Ser. A. **246**, 215 (1953).

Because $E_+(\zeta, R)$ in Equation 10.33 is a function of both ζ and R through $w = \zeta R$, the value of ζ that minimizes E_+ depends upon R. The result $\zeta = 1.238$ that we showed above is valid only at $R = 2.003\,a_0$. To determine the optimized energy as a function of R, we must determine $\zeta(R)$, the optimum value of ζ as a function of R. We do this by minimizing $E_+(\zeta, R)$ with respect to ζ for various values of R. (See also Problem 10–22.) In this manner, we can determine $\zeta(R)$ numerically, which is plotted in Figure 10.16. Note that $\zeta \to 2$ when $R \to 0$ (the two nuclei have merged into one with a charge of +2) and that $\zeta \to 1$ as $R \to \infty$ (the two nuclei, with charge +1, are widely separated).

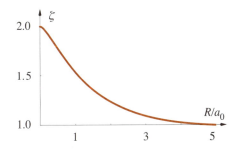

FIGURE 10.16
A plot of the optimum value of ζ (the one that minimizes the energy) plotted against R. Note that $\zeta \to 2$ as $R \to 0$ and that $\zeta \to 1$ as $R \to \infty$.

Now that we know $\zeta(R)$ (numerically), we can use Equation 10.33 to calculate $\Delta E_+(R)$ as a function of R. (Remember that $w = \zeta R$ in Equation 10.33.) Figure 10.17 is a plot of this optimized $\Delta E_+(R)$ against R. Note that the optimized energy is a significant improvement over the energy calculated with $\zeta = 1$.

We can use Equations 10.31 and 10.32 along with $\zeta(R)$ to plot the total energy, the kinetic energy, and the potential energy as a function of R, as shown in Figure 10.18.

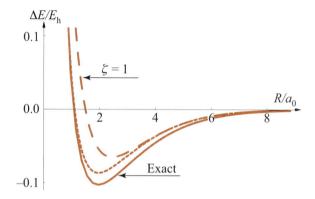

FIGURE 10.17

A plot of the ground-state electronic energy of H_2^+, calculated with $\zeta = 1$ (dashed curve), the optimized energy $[\zeta = \zeta(R)]$ (dotted curve), and the exact energy (solid curve) against R. Note that the optimized energy is a considerable improvement over the energy calculated with $\zeta = 1$.

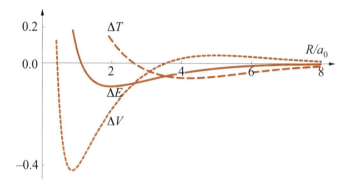

FIGURE 10.18

A plot of the kinetic energy (dashed curve), the potential energy (dotted curve), and the total energy (solid curve) of the optimized minimal basis set calculation of the ground electronic state of H_2^+ plotted against R.

The vertical scales have been adjusted so that all three curves go to zero as $R \rightarrow \infty$ by adding $1/2$ to the total energy, subtracting $1/2$ from the kinetic energy, and adding 1 to the potential energy. Note that as the two nuclei are brought together, the kinetic energy decreases at first, and then rises rather steeply well before the equilibrium internuclear separation is reached. The potential energy, on the other hand, increases at first and then decreases monotonically as R approaches R_{eq} and then rises as $R \rightarrow 0$. The net effect is the curve for the total energy, showing the formation of a stable bond. This behavior is believed to be characteristic of bond formation. A key factor in bond formation is the increase in charge density near the nuclei due to the increased value of $\zeta = 1.238$ at $R = R_{eq}$ compared to $\zeta = 1$ as $R \rightarrow \infty$. This effect lowers the potential energy, and

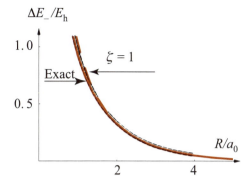

FIGURE 10.19
A comparison of the plot of the exact energy (solid curve) and the $\zeta = 1$ (dashed curve) and optimized energies (black dotted curve) of the first excited state $(\sigma_u^* 1s)$ of H_2^+ versus R.

although it increases the kinetic energy (because the electron is restricted to a smaller region), the net effect is a lowering of the total energy. (The reference to Mulliken and Ermler at the end of the chapter gives a fairly thorough discussion of the formation of chemical bonds.)

Although we shall be primarily interested in ground-state calculations, we show both the $\zeta = 1$ and optimized energies of the $\sigma_u^* 1s$ state of H_2^+ along with the exact energy plotted against R in Figure 10.19. The agreement among all three is quite good.

10.6 The Inclusion of Polarization in a Basis Set Leads to a Considerable Improvement in the Energy

We certainly are not limited to using a linear combination of only two atomic orbitals. After all, it *is* called a minimal basis set. For example, we could use a linear combination of the form

$$\psi = c_1 \, 1s_A + c_2 \, 1s_B + c_3 \, 2s_A + c_4 \, 2s_B$$

where $1s$ and $2s$ are (normalized) Slater-type orbitals

$$1s = \left(\frac{\zeta^3}{\pi} \right)^{1/2} e^{-\zeta r} \qquad \text{and} \qquad 2s = \left(\frac{\zeta^5}{3\pi} \right)^{1/2} r e^{-\zeta r}$$

This will lead to a minimum energy of $-0.586\,51\,E_h$ at $R = 2.00\,a_0$ with $\zeta = 1.24$, which is essentially no improvement over using just a linear combination of two $1s$ orbitals $(-0.586\,51\,E_h)$. The ground-state molecular orbital comes out to be

$$\psi = 0.7071 \, (1s_A + 1s_B) + 0.001\,45 \, (2s_A + 2s_B)$$

so you can see from the relative coefficients in ψ that the $2s$ orbitals contribute very little to ψ. This exercise shows us that the inclusion of more terms in ψ does not necessarily lead to significantly better results. If we use some chemical intuition in choosing which type of atomic orbitals to include, however, we certainly can achieve much better results.

For example, it is clear that the electron distribution in the hydrogen atom does not remain spherical as the two nuclei are brought together. The charge distribution in an isolated hydrogen atom is spherical. As a proton approaches a hydrogen atom, however, the proton attracts the electron and so the electronic charge distribution about the hydrogen atom becomes distorted, or *polarized*. If we let the internuclear axis be the z axis, then we might try a linear combination of a $1s$ and a $2p_z$ orbital to represent the polarized charge distribution and write ψ as

$$\psi = c_1\, 1s_A + c_2\, 1s_B + c_3\, 2p_{zA} + c_4\, 2p_{zB}$$

By symmetry, we expect that $c_1 = c_2$ and $c_3 = c_4$ in the ground-state molecular orbital, so that we can write

$$\psi = c_1(1s_A + a\, 2p_{zA}) + c_1(1s_B + a\, 2p_{zB}) \tag{10.34}$$

where a is a variational parameter. (We can determine c_1 by requiring that ψ be normalized.) Equation 10.34 emphasizes that we are taking a linear combination of two orbitals of the form $\phi = 1s + a\, 2p_z$. Figure 10.20 shows a contour plot of ϕ for $a = 0.14$, which is the value of a that we obtain when we use Equation 10.34 to carry out a variational calculation. Note that ϕ represents a charge distribution that is polarized by the neighboring nucleus. The two atomic orbitals $1s$ and $2p_z$ are said to constitute a *polarized basis set*.

FIGURE 10.20
A surface plot of an orbital of the form $1s + 0.14\, 2p_z$.

The secular equation that arises from Equation 10.34 is a 4×4 determinantal equation, which is easily handled with any number of readily available computer programs. If we take the $1s$ and $2p_z$ orbitals to be the Slater orbitals $S_{100}(r, \zeta)$ and $S_{210}(r, \zeta)$ (Equation 9.15),

$$1s = \left(\frac{\zeta_1^3}{\pi}\right)^{1/2} e^{-\zeta_1 r} \qquad \text{and} \qquad 2p_z = \left(\frac{\zeta_2^5}{\pi}\right)^{1/2} r\cos\theta\, e^{-\zeta_2 r}$$

and let $\zeta_1 = \zeta_2$, then $E_{\min} = -0.599\,07\, E_h$ at $R = 2.00\, a_0$ for $\zeta = 1.247$ and $a = 0.161$. If we vary both ζ_1 and ζ_2 independently, then it turns out that $E_{\min} = -0.600\,36\, E_h$

when $\zeta_1 = 1.2458$, $\zeta_2 = 1.4224$, and $a = 0.1380$. Notice that a polarized basis set can yield a significant improvement in the energy.

If we add some d-orbital character to ψ by including a $3d_{z^2}$ Slater orbital,

$$3d_{z^2} = \left(\frac{\zeta^7}{18\pi} \right)^{1/2} (3\cos^2 - 1) r^2 e^{-\zeta r}$$

in Equation 10.34, then we obtain $E_{min} = -0.6020\,E_h$ at $R = 2.000\,a_0$ with $\zeta_1 = 1.244$, $\zeta_2 = 1.152$, $\zeta_3 = 1.333$, and the coefficients of the $2p_z$ and $3d_{z^2}$ orbitals being 0.2214 and 0.0782, respectively. Recall that the exact energy is $-0.602\,64\,E_h$, so these calculations nicely illustrate that you can achieve very good results by using larger basis sets, particularly if the basis set is chosen judiciously. (See Table 10.2.)

10.7 The Schrödinger Equation for H_2 Cannot Be Solved Analytically

The simplest neutral molecule is H_2, having two electrons. Figure 10.21 shows the experimental ground-state electronic energy of H_2 plotted against the internuclear separation R. The ground electronic state of H_2 dissociates into two ground-state hydrogen atoms, whose electronic energy is $-1\,E_h$. The energy plotted in Figure 10.21 goes to $-1\,E_h$ as $R \to \infty$ and the depth of the well is $-1.1738\,E_h$.

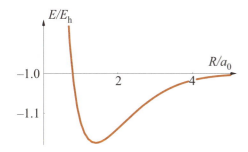

FIGURE 10.21
The ground-state electronic energy of H_2 plotted against the internuclear separation R. The minimum value of $E(R)$ is equal to $-1.174\,E_h$ at $R_{eq} = 1.40\,a_0$.

The electronic Schrödinger equation of H_2 in the Born–Oppenheimer approximation is

$$\hat{H}\psi(\mathbf{r}_1, \mathbf{r}_2, R) = E_{el}\psi(\mathbf{r}_1, \mathbf{r}_2, R)$$

where

$$\hat{H} = -\frac{1}{2}\nabla_1^2 - \frac{1}{2}\nabla_2^2 - \frac{1}{r_{1A}} - \frac{1}{r_{1B}} - \frac{1}{r_{2A}} - \frac{1}{r_{2B}} + \frac{1}{r_{12}} + \frac{1}{R} \qquad (10.35)$$

Unlike for H_2^+, with its one electron, the Schrödinger equation for H_2 cannot be solved exactly. We saw in Chapter 9 that Hylleraas was able to obtain essentially the exact ground-state energy of a helium atom by including the interelectronic distance r_{12}

explicitly in a trial wave function. A similar approach was applied to H_2 with equal success by James and Coolidge as early as 1933. Using a trial function with 13 terms, they obtained an equilibrium bond length of 1.40 a_0 and an energy of $-1.1735 E_h$, in very good agreement with the experimental values. This pioneering work was extended in a classic series of papers by the two Polish physicists Wlodzimierz Kolos and Lutoslav Wolniewicz in the 1960s. They used trial functions containing up to 100 terms, and found a minimum energy of $-1.174\,475\,E_h$ at $R = 1.401 a_0$. This value includes corrections for the Born–Oppenheimer approximation and relativistic effects, and so is essentially exact.

This calculation was published in 1968, at which time the authors pointed out that their calculated dissociation energy of H_2, 36 117.3 cm^{-1}, was in disagreement with the experimental value, 36 113.6 cm^{-1}. In the next year, Herzberg reported new measurements of D_0 and found that D_0 was between 36 116.3 cm^{-1} and 36 118.3 cm^{-1}, in superb agreement with the calculated value. Table 10.3, from Herzberg's paper, compares the theoretical and experimental values of D_0 for H_2, HD, and D_2. Herzberg also gives an observed ionization energy of H_2 of 124 418.4 $\pm$ 0.4 cm^{-1}, compared to the calculated value of 124 417.3 cm^{-1}.

TABLE 10.3
A Comparison of the Theoretical (Kolos and Wolniewicz) and Experimental Values (Herzberg) of D_0 (in Units of cm^{-1}) for H_2, HD, and D_2 [a]

	D_0/cm^{-1}	
	Theoretical value	Experimental value
H_2	36 117.9	36 116.3–36 118.3
HD	36 405.5	36 405.8–36 406.6
D_2	36 748.2	36 748.9

a. *Source:* Herzberg, G., Dissociation Energy and Ionization Energy of Molecular Hydrogen. *Phys. Rev. Lett.,* **23**, 1081 (1969).

So far, we have discussed only the ground state of H_2. Kolos and Wolniewicz have also calculated the electronic energies of several excited states of H_2, which are shown in Figure 10.22. As you can see in the figure, the electronic states of diatomic molecules are denoted by term symbols, just as the electronic states of atoms are. We denote the states by the notation

$$^{2S+1}\Lambda$$

where $\Lambda = |m|$ is the magnitude of the component of the total orbital angular momentum about the internuclear axis. Atoms have a center of symmetry, so their electronic angular momentum is conserved. Thus, the total orbital angular momentum, L, is a good quantum number (neglecting spin-orbit interaction), and we use that fact to denote an atomic term symbol by ^{2S+1}L, where L is the total orbital angu-

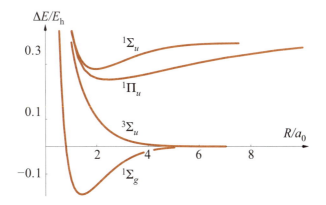

FIGURE 10.22
The energies of several electronic states of H_2 calculated by Kolos and Wolniewicz [Potential-Energy Curves for the $X^1\Sigma_g^+$, $B^3\Sigma_u^+$, and $C^1\Pi_u$ States of the Hydrogen Molecule. *J. Chem. Phys.*, **43**, 2429 (1965), and Potential-Energy Curve for the $B^1\Sigma_u^+$ State of the Hydrogen Molecule. *J. Chem. Phys.*, **45**, 509 (1966).]

lar momentum. A diatomic molecule is symmetric about the internuclear axis, and so its angular momentum about that axis is conserved. Just as we used the notation S, P, D, ... for $L = 0, 1, 2, \ldots$, for atomic term symbols, we use $\Lambda = \Sigma, \Pi, \Delta, \ldots$ for $\Lambda = 0, 1, 2, \ldots$, for diatomic molecules. When $\Lambda = 0$, the orbitals are symmetric about the internuclear axis and so are σ orbitals; when $\Lambda = 1$, the orbitals are π orbitals; similarly, $\Lambda = 2$ for δ orbitals, and so on. Note also that Λ is equal to the number of nodal planes containing the internuclear axis. In addition, for homonuclear diatomic molecules, we add a g or a u right subscript to $^{2S+1}\Lambda$ to indicate whether the electronic wave function does not change sign (g) or does change sign (u) upon inversion through the midpoint of the two nuclei. As we pointed out earlier, the bonding orbital of H_2^+ (Figure 10.8) has gerade symmetry and the antibonding orbital has ungerade symmetry.

EXAMPLE 10–5
Determine the term symbol of the minimal basis set of bonding and antibonding orbitals of H_2^+.

SOLUTION: Both orbitals are σ orbitals, so $\Lambda = 0$, and are denoted by Σ. Furthermore, the total spin is 1/2, so $2S + 1 = 2$. The bonding orbital is gerade, so its term symbol is $^2\Sigma_g$ (doublet-sigma-g). The antibonding state is denoted by $^2\Sigma_u$ (doublet-sigma-u).

Although Kolos and Wolniewicz were able to solve the Schrödinger equation for H_2 essentially exactly using a trial function consisting of many terms, this approach is not practicable for molecules containing more than two electrons because of the terms involving the interelectronic distance r_{12} that they include in their trial functions. We need an approach that we can use for all types of molecules.

10.8 The Ground-State Electron Configuration of H_2 Is $\sigma_g 1s^2$

Using our success that we had for atoms as a guide, where we constructed atomic wave functions in terms of Slater determinants of single-electron orbitals, we'll construct the wave functions for molecules in terms of Slater determinants involving one-electron molecular orbitals. For one-electron molecular orbitals, we'll use H_2^+ molecular orbitals. In the simplest case of the minimal basis set calculation of H_2^+, we obtained two molecular orbitals: σ_b, a bonding orbital, and σ_a, an antibonding orbital. These orbitals, which are a sum and a difference of Slater $1s$ orbitals, can be written as

$$\sigma_b = \frac{(\zeta^3/\pi)^{1/2}e^{-\zeta r_A} + (\zeta^3/\pi)^{1/2}e^{-\zeta r_D}}{\{2[1+S(\zeta R)]\}^{1/2}} \tag{10.36}$$

and

$$\sigma_a = \frac{(\zeta^3/\pi)^{1/2}e^{-\zeta r_A} - (\zeta^3/\pi)^{1/2}e^{-\zeta r_B}}{\{2[1-S(\zeta R)]\}^{1/2}} \tag{10.37}$$

where we let $\zeta \neq 1$ here for generality.

Because σ_b is the molecular orbital corresponding to the ground-state energy of H_2^+, we can describe the ground state of H_2 by placing two electrons with opposite spins in σ_b, just as we place two electrons in a $1s$ atomic orbital to describe the helium atom. The Slater determinant corresponding to this assignment is

$$\psi = \frac{1}{\sqrt{2!}} \begin{vmatrix} \sigma_b\alpha(1) & \sigma_b\beta(1) \\ \sigma_b\alpha(2) & \sigma_b\beta(2) \end{vmatrix}$$

$$= \sigma_b(1)\sigma_b(2) \left\{ \frac{1}{\sqrt{2}}[\alpha(1)\beta(2) - \alpha(2)\beta(1)] \right\} \tag{10.38}$$

Once again, we see the spatial and spin parts of the wave function separate for this two-electron Slater determinant (see Example 9–4). Because the Hamiltonian operator is taken to be independent of spin, we can calculate the energy using only the spatial part of Equation 10.38. Using Equation 10.36 for σ_b, we have a molecular wave function, ψ_{MO}, of the form

$$\psi_{MO} = \frac{1}{2(1+S)}[1s_A(1) + 1s_B(1)][1s_A(2) + 1s_B(2)] \tag{10.39}$$

when $1s$ denotes a Slater $1s$ orbital, $1s = (\zeta^3/\pi)^{1/2}e^{-\zeta r}$. Note that ψ_{MO} is a product of *molecular orbitals*, which in turn are linear combinations of atomic orbitals. This method of constructing molecular wave functions is known as the *LCAO-MO (linear combination of atomic orbitals–molecular orbitals*) method and has been successfully extended and applied to a variety of molecules, as we will see in this and the following chapters.

The energy is given by

$$E_{MO} = \langle \psi_{MO} \mid \hat{H} \mid \psi_{MO} \rangle \tag{10.40}$$

where $\hat{H}$ is given by Equation 10.35. When Equation 10.39 is substituted into Equation 10.40, we obtain a lengthy expression in ζ and $w = \zeta R$. In fact, it has a form similar to that for H_2^+ in Equation 10.33, except that the expressions for $T(w)$ and $V(w)$ are more complicated for H_2. We've placed these results in the appendix at the end of the chapter so that they are readily available to reproduce some of the H_2 results in this section (if you wish).

We emphasize the dependence of E_{MO} on ζ and R by writing Equation 10.40 as

$$E_{MO}(\zeta, R) = \langle \psi_{MO} \mid \hat{H} \mid \psi_{MO} \rangle \tag{10.41}$$

Figure 10.23 shows E_{MO} plotted against R for $\zeta = 1$. The minimum energy comes out to be $-1.0991 E_h$ at $R_{eq} = 1.603 a_0$, compared to the "exact" value of $-1.1744 E_h$ and $R_{eq} = 1.401 a_0$ (Table 10.4). Notice that Figure 10.23 shows something very disconcerting, however; the energy does not go to $-1 E_h$, that of two isolated ground-state hydrogen atoms, as $R \rightarrow \infty$. In other words, this molecular orbital theory wave function gives the wrong dissociation limit! Nor do things get any better if we use ζ as a variational parameter to yield an optimized energy, as shown in Figure 10.23. The values of the minimum energy ($-1.1282 E_h$) and R_{eq} ($1.385 a_0$) improve some, but we still obtain a wrong dissociation limit. We can also see this difficulty if we plot the optimized ζ against R, as in Figure 10.24. As $R \rightarrow 0$, H_2 merges into a helium atom, and Figure 10.24 shows that $\zeta = 27/16 = 1.6875$ as $R \rightarrow 0$, which is the value of ζ when

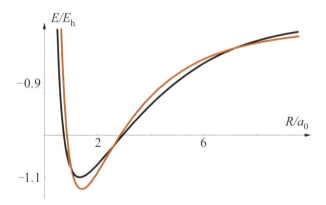

FIGURE 10.23
Both the optimized (orange) and the $\zeta = 1$ (black) molecular orbital energies calculated with Equation 10.41. In neither case does the energy go to the correct limit of $-1 E_h$ as $R \rightarrow \infty$.

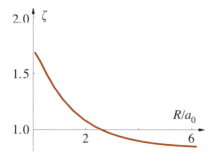

FIGURE 10.24
The optimized value of ζ plotted against R for the molecular orbital given by Equation 10.36. The value of ζ should go to 1, that of an isolated hydrogen atom, as $R \to \infty$, but it does not.

we optimize the energy of a helium atom using Slater orbitals (see Equation 9.11). For large values of R, we have isolated hydrogen atoms, and so ζ should go to 1 as $R \to \infty$, which it does not do. Fortunately, the explanation of this apparent disaster is well understood, as we shall now show.

Let's expand ψ_{MO} in Equation 10.39 to write

$$
\begin{aligned}
\psi_{\text{MO}} &\approx \frac{[1s_A(1) + 1s_B(1)][1s_A(2) + 1s_B(2)]}{2(1+S)} \\[2mm]
&= \frac{1s_A(1)\, 1s_B(2) + 1s_A(2)\, 1s_B(1) + 1s_A(1)\, 1s_A(2) + 1s_B(1)\, 1s_B(2)}{2(1+S)}
\end{aligned}
\tag{10.42}
$$

The first two terms represent two hydrogen atoms, each with an electron. This representation requires two terms because of the indistinguishability of the electrons; just one term, such as $1s_A(1)\, 1s_B(2)$, would not do because $1s_A(2)\, 1s_B(1)$ would be an equally good description. We shall come back to these terms later. The last two terms represent a situation where both electrons are on one atom. We can describe these two terms by the electron-dot formulas,

$$
\text{H}_A\colon \quad \text{H}_B \quad\quad \text{and} \quad\quad \text{H}_A \quad\quad \colon \text{H}_B
$$

or as

$$
\text{H}_A^- \quad\quad \text{H}_B^+ \quad\quad \text{and} \quad\quad \text{H}_A^+ \quad\quad \text{H}_B^-
$$

Thus, we see that the third and fourth terms in Equation 10.42 represent ionic structures for H_2. For large values of R, the molecular orbital wave function given by Equation 10.42 represents a dissociation limit in which H_2 dissociates to an average of two neutral hydrogen atoms and an ion pair, rather than two ground-state hydrogen atoms with an energy of $-1\,E_h$. This is a well-known deficiency of simple molecular orbital theory, and we shall show how to correct for it in the next section.

Because ψ_{MO} given by Equation 10.42 leads to a dissociation limit that consists of an average of two neutral hydrogen atoms and an ion pair, both E_{MO} and ζ go to incorrect limits for large values of R. In fact, $E_{\text{MO}} \to -0.7119\,E_h$ and $\zeta \to 0.843\,75$ as $R \to \infty$, and Examples 10–6 and 10–7 give a nice interpretation of these values.

TABLE 10.4
Results of Various Calculations of the Ground-State Energy of H_2

	Wave function	ζ	E_{min}/E_h	R_{eq}/a_0
MO	Minimal basis set	1.000	−1.0991	1.603
MO	Minimal basis set	1.193	−1.1282	1.385
	Hartree–Fock [a]		−1.1336	1.400
CI	Minimal basis set	1.000	−1.1187	1.668
CI	Minimal basis set	1.194	−1.1479	1.430
CI	Minimal basis set with polarization [b]		−1.1514	1.40
CI	Five terms [b]		−1.1672	1.40
CI	33 terms [c]		−1.1735	1.40
	Trial function with r_{12} 13 terms [d]		−1.1735	1.40
	Trial function with r_{12} with 100 terms [e]		−1.1744	1.401
	Experimental [f]		−1.174	1.401

a. Kolos, W., Roothaan, C. C. J. Accurate Electronic Wave Functions for the H_2 Molecule. *Rev Mod. Phys.,* **32,** 219 (1960).
b. McLean, A. D., Weiss, A., Yoshimine, M. Configuration Interaction in the Hydrogen Molecule. *Rev Mod. Phys.,* **32,** 211 (1960).
c. Hagstrom, S., Shull, H. The Nature of the Two-Electron Chemical Bond: III. Natural Orbitals for H_2. *Rev. Mod. Phys.,* **35,** 624 (1963).
d. James, H. M., Coolidge, A. S. The Ground State of the Hydrogen Molecule. *J. Chem. Phys.,* **1,** 825 (1963).
e. Kolos, W., Wolniewicz, L. Accurate Adiabatic Treatment of the Ground State of the Hydrogen Molecule. *J. Chem. Phys.,* **41,** 3663 (1964); Improved Theoretical Ground-State Energy of the Hydrogen Molecule. *J. Chem. Phys.,* **49,** 404 (1968).
f. Herzberg, G. The Dissociation Energy of the Hydrogen Molecule. *J. Mol. Spectroscopy,* **33,** 147 (1970).

EXAMPLE 10–6
Use the fact that ψ_{MO} leads to a dissociation limit that consists of an average of two neutral hydrogen atoms and an ion pair to show that $\zeta \to 0.84375$ as $R \to \infty$. Use the result of Example 9–2.

SOLUTION: The optimum value of ζ for a ground-state hydrogen atom using a Slater 1s orbital is 1. Example 9–2 shows that the optimum value of ζ for a hydride ion using a Slater 1s orbital is $\zeta = 11/16$. (The H^+ associated with H^- has no electronic energy, and so can be ignored here.) The average of 1 and 11/16 is

$$\zeta = \frac{1}{2}\left(1 + \frac{11}{16}\right) = \frac{27}{32} = 0.84375$$

in agreement with the value in Figure 9.24. (See also Problem 10–46 for a proof of this result.)

EXAMPLE 10–7

According to Example 10–6, the value of ζ in Figure 10.24 goes to 27/32 as $R \to \infty$. Use this result to show that E_{MO} in Figure 10.23 goes to $-0.7119\,E_h$ as $R \to \infty$. Use the result of Example 9–2.

SOLUTION: The second and third entries in Table 10.1 give

$$E_{H_2}(\zeta) = \left\langle \phi_A \left| -\frac{1}{2}\nabla^2 \right| \phi_A \right\rangle + \left\langle \phi_A \left| -\frac{1}{r} \right| \phi_A \right\rangle = \frac{\zeta^2}{2} - \zeta$$

where we used the fact that ϕ_A in Table 10.1 denotes a Slater $1s$ orbital. According to Example 9–2, the corresponding value for the energy of a hydride ion is

$$E_{H^-}(\zeta) = \zeta^2 + 2\zeta(\zeta - 1) + \frac{5}{8}\zeta = \zeta^2 - \frac{11}{8}\zeta$$

(As in Example 10–6, the H^+ has no electronic energy, and so can be ignored here.) Therefore, E_{MO} in Figure 10.24 is the average of $2E_{H_2}(\zeta)$ and $E_{H^-}(\zeta)$, or

$$E_{MO}(\zeta) \to \frac{1}{2}[2E_{H_2}(\zeta) + E_{H^-}(\zeta)] = \frac{1}{2}\left(\zeta^2 - 2\zeta + \zeta^2 - \frac{11}{8}\zeta\right)$$

$$= \zeta^2 - \frac{27}{16}\zeta$$

According to Example 10–6, however, $\zeta = 27/32$, and so

$$E_{MO} \to \left(\frac{27}{32}\right)^2 - 2\left(\frac{27}{32}\right)^2 = -\left(\frac{27}{32}\right)^2 = -0.7119\,E_h$$

in agreement with the value in Figure 10.23.

Before we go on to the next section, let's go back and look at the first two terms in Equation 10.42, which we write as

$$\psi_{VB} = 1s_A(1)\,1s_B(2) + 1s_A(2)\,1s_B(1) \tag{10.43}$$

In 1927, Heitler and London used Equation 10.43 to give the first satisfactory explanation of the stability of a chemical bond. Their method is known as the *valence bond method* (the reason for the VB subscript in Equation 10.43). Although the valence bond method gave the first satisfactory description of a chemical bond, the method has been largely superseded by molecular orbital theory, and we mention it here primarily for historical reasons.

To summarize, the simple minimal basis set molecular orbital for H_2 can be written as

$$\psi_{MO} = \psi_{VB} + \psi_{ionic} \tag{10.44}$$

suggesting that the molecular orbital approach overemphasizes ionic terms, whereas the valence bond method underemphasizes (ignores) them. A correct approach would be to use a linear combination of the form

$$\psi = c_1 \psi_{\text{VB}} + c_2 \psi_{\text{ionic}} \tag{10.45}$$

and let the variational principle decide the relative importance of each contribution. We shall see in the next section that we achieve this result if we use configuration interaction to improve the molecular orbital wave function.

10.9 Configuration Interaction Gives the Correct Dissociation Limit

Consider our minimal basis set LCAO treatment of H_2^+, in which we obtained molecular orbitals

$$\sigma_b = \frac{1s_A + 1s_B}{\sqrt{2(1+S)}} \quad \text{and} \quad \sigma_a = \frac{1s_A - 1s_B}{\sqrt{2(1-S)}} \tag{10.46}$$

In our molecular orbital discussion of H_2, we used

$$\psi_1 = \frac{1}{\sqrt{2}} \begin{vmatrix} \sigma_b \alpha(1) & \sigma_b \beta(1) \\ \sigma_b \alpha(2) & \sigma_b \beta(2) \end{vmatrix} \tag{10.47}$$

or simply its spatial part, $\sigma_b(1)\sigma_b(2)$. However, we can extend our molecular orbital treatment by using the antibonding molecular orbital σ_a as well. We can combine σ_b and σ_a with either of the spin functions α and β to form four possible molecular spin orbitals $\sigma_b \alpha$, $\sigma_b \beta$, $\sigma_a \alpha$, and $\sigma_a \beta$. We can assign the two electrons to these four spin orbitals in $4 \times 3/2 = 6$ different ways. This leads to a total of six Slater determinants,

$$D_1 = \frac{1}{\sqrt{2}} \begin{vmatrix} \sigma_b \alpha(1) & \sigma_b \beta(1) \\ \sigma_b \alpha(2) & \sigma_b \beta(2) \end{vmatrix}$$

$$D_2 = \frac{1}{\sqrt{2}} \begin{vmatrix} \sigma_a \alpha(1) & \sigma_a \beta(1) \\ \sigma_a \alpha(2) & \sigma_a \beta(2) \end{vmatrix}$$

$$D_3 = \frac{1}{\sqrt{2}} \begin{vmatrix} \sigma_b \alpha(1) & \sigma_a \alpha(1) \\ \sigma_b \alpha(2) & \sigma_a \alpha(2) \end{vmatrix}$$

$$D_4 = \frac{1}{\sqrt{2}} \begin{vmatrix} \sigma_b \alpha(1) & \sigma_a \beta(1) \\ \sigma_b \alpha(2) & \sigma_a \beta(2) \end{vmatrix} \tag{10.48}$$

$$D_5 = \frac{1}{\sqrt{2}} \begin{vmatrix} \sigma_b \beta(1) & \sigma_a \alpha(1) \\ \sigma_b \beta(2) & \sigma_a \alpha(2) \end{vmatrix}$$

$$D_6 = \frac{1}{\sqrt{2}} \begin{vmatrix} \sigma_b \beta(1) & \sigma_a \beta(1) \\ \sigma_b \beta(2) & \sigma_a \beta(2) \end{vmatrix}$$

where the $1/\sqrt{2}$ are normalization constants (Problem 10–29).

Because H_2 is a two-electron system, we can work with only the spatial parts of the above determinants and use (Problem 10–30)

$$\psi_1 = \sigma_b(1)\sigma_b(2)$$

$$\psi_2 = \sigma_a(1)\sigma_a(2)$$

$$\psi_3, \psi_4, \psi_5 = \sigma_b(1)\sigma_a(2) - \sigma_a(1)\sigma_b(2) \tag{10.49}$$

$$\psi_6 = \sigma_b(1)\sigma_a(2) + \sigma_a(1)\sigma_b(2)$$

If we take a linear combination of these six wave functions as a variational trial function, then we will get a 6×6 secular determinant, $|H_{ij} - ES_{ij}|$. This 6×6 secular determinant will lead to a sixth-order polynomial equation in E, and the lowest root will be the ground-state energy. It turns out that there is a beautiful theorem that we can use to reduce the above problem to a quadratic equation.

Let $\hat{P}_{AB}$ be an operator that interchanges the nuclei A and B. If $\hat{P}_{AB}$ operates on the wave functions in Equation 10.49, then notice that

$$\hat{P}_{AB}\psi_j = +\psi_j \qquad j = 1 \text{ and } 2$$

$$\hat{P}_{AB}\psi_j = -\psi_j \qquad j = 3, 4, 5, \text{ and } 6 \tag{10.50}$$

The wave functions ψ_1 and ψ_2 have eigenvalue $+1$ and ψ_3 through ψ_6 have eigenvalue -1. The Hamiltonian operator is symmetric in A and B, and so $\hat{P}_{AB}$ commutes with $\hat{H}$, or

$$[\hat{P}_{AB}, \hat{H}] = 0 \tag{10.51}$$

Here is the theorem: Because $\hat{P}_{AB}$ and $\hat{H}$ commute, matrix elements of $\hat{H}$,

$$H_{ij} = \langle \psi_i \mid \hat{H} \mid \psi_j \rangle$$

between states with different eigenvalues of $\hat{P}_{AB}$ equal zero; or, H_{ij} will be nonzero only between states with the same eigenvalue of $\hat{P}_{AB}$. Applying this theorem to the problem at hand gives

$$H_{1j} = 0 \qquad \text{if } j = 3, 4, 5, \text{ or } 6$$

$$H_{2j} = 0 \qquad \text{if } j = 3, 4, 5, \text{ or } 6 \tag{10.52}$$

This theorem and its variations are used extensively in quantum chemistry to simplify secular determinants, which are ubiquitous in molecular calculations. The proof of this theorem and its relatives is short and fairly easy, but is left to Problem 10–34.

EXAMPLE 10–8
Show that

$$I_{AB} = \langle \sigma_b(1)\sigma_b(2) \mid \hat{H} \mid \sigma_b(1)\sigma_a(2) \rangle = 0$$

by interchanging A and B and showing that $I_{BA} = -I_{AB}$, or that $I_{AB} = 0$.

SOLUTION: Interchanging A and B has no effect on σ_b, but $\sigma_a = -\sigma_a$ when A and B are interchanged. Because there is only one factor of σ_a in I_{AB}, we see that $I_{BA} = -I_{AB}$, or that $I_{AB} = 0$. You can verify all the results in Equation 10.52 using this approach.

If we are going to use a linear combination of the six determinantal wave functions in Equation 10.49, then the above theorem tells us that the ground-state wave function ψ_1 has nonzero matrix elements only with ψ_2. We say that ψ_1 mixes only with ψ_2. None of the other wave functions affect the final result, and so we shall use

$$\psi_{CI} = c_1\psi_1 + c_2\psi_2 \tag{10.53}$$

instead of a linear combination of all six wave functions. Equation 10.53 is a ground-state molecular orbital wave function with an excited-state configuration mixed in. The extension of simple molecular orbital theory to include excited-state configurations is called *configuration interaction*. The CI subscript in Equation 10.53 denotes that ψ_{CI} is a configuration-interaction wave function.

Let's look at ψ_{CI} given by Equation 10.53 in more detail. Using Equations 10.46, Equation 10.53 can be written out as

$$\psi_{CI} = c_1[1s_A(1)1s_A(2) + 1s_A(1)1s_B(2) + 1s_B(1)1s_A(2) + 1s_B(1)1s_B(2)]$$

$$+ c_2[1s_A(1)1s_A(2) - 1s_A(1)1s_B(2) - 1s_B(1)1s_A(2) + 1s_B(1)1s_B(2)]$$

$$= (c_1 - c_2)\psi_{VB} + (c_1 + c_2)\psi_{ionic} \tag{10.54}$$

Thus, we see that ψ_{CI} given by Equation 10.53 is exactly the same as Equation 10.45.

Let's use Equation 10.53 to calculate the electronic energy of H_2. This leads to the 2×2 secular determinantal equation whose smaller root gives $E_{CI}(\zeta, R)$, where, as before, we emphasize that E_{CI} depends upon ζ and R. As with our molecular orbital calculation in Section 10.8, the evaluation of E is lengthy and we leave the details of the calculation to the appendix.

Figure 10.25 shows E_{CI} plotted against R for $\zeta = 1$. Notice that E_{CI} goes to the correct dissociation limit of two isolated ground-state hydrogen atoms ($E = -1 E_h$). The minimum value of E_{CI} is $-1.11865 E_h$ at $R_{eq} = 1.668 a_0$. The energy is improved over the $\zeta = 1$ molecular orbital energy, but the value of R_{eq} is poor in these $\zeta = 1$ calculations. Figure 10.25 also shows the optimized energy, which we obtain from $E_{CI}(\zeta, R)$ by minimizing $E_{CI}(\zeta, R)$ with respect to ζ at various values of R. This results in a significant improvement over the $\zeta = 1$ result, giving a minimum energy

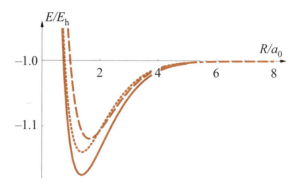

FIGURE 10.25
The configuration-interaction energy E_{CI} of the ground-state energy of H_2 for $\zeta = 1$ (dashed curve) and for an optimized value of ζ (dotted curve) plotted against R. The "exact" results of Kolos and Wolniewicz (solid curve) are shown for comparison.

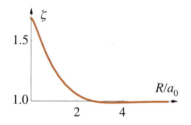

FIGURE 10.26
The optimized value of ζ in Equation 10.53 (the one that minimizes E_{CI}) plotted against R. Note that ζ goes to the correct limits as $R \to 0$ and $R \to \infty$.

of $-1.147\,94\,E_h$ at $R_{eq} = 1.430\,a_0$. Figure 10.26 shows the optimum value of ζ (the one that minimizes E_{CI}) plotted against R. Note that contrary to Figure 10.23, $\zeta \to 1$ as $R \to \infty$, as it should.

The values of c_1 and c_2 in Equation 10.53 also depend upon R. Figure 10.27 shows c_1 and c_2 plotted against R. It turns out that (see the appendix) $c_1 \to 1/\sqrt{2}$ and $c_2 \to -1/\sqrt{2}$ as $R \to \infty$, which as Example 10–9 shows, leads to the correct wave function for two isolated ground-state hydrogen atoms.

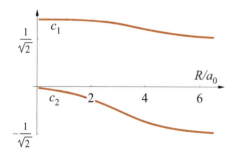

FIGURE 10.27
A plot of c_1 and c_2 for the optimized value of ζ in Equation 10.53 against R. Note that $c_1 \to 1/\sqrt{2}$ and $c_2 \to -1/\sqrt{2}$ as $R \to \infty$.

EXAMPLE 10–9

Show that ψ given by Equation 10.53 goes to the wave function of two isolated ground-state hydrogen atoms as $R \to \infty$. Use the fact that $c_1 \to 1/\sqrt{2}$ and $c_2 \to -1/\sqrt{2}$.

SOLUTION: Using Equation 10.49, we have

$$\psi_{CI} = c_1\psi_1 + c_2\psi_2$$

$$= c_1\sigma_b(1)\sigma_b(2) + c_2\sigma_a(1)\sigma_a(2)$$

$$= \frac{1}{2^{3/2}}[1s_A(1) + 1s_B(1)][1s_A(2) + 1s_B(2)] - \frac{1}{2^{3/2}}[1s_A(1) - 1s_B(1)][1s_A(2) - 1s_B(2)]$$

$$= \frac{1}{2^{1/2}}[1s_A(1)1s_B(2) + 1s_B(1)1s_A(2)]$$

But this result is the wave function of two isolated ground-state hydrogen atoms. Remember that we must use a linear combination of two terms because of the indistinguishability of the two electrons.

Realize that all the calculations that we have carried out so far have been for a minimal basis set, consisting of one atomic orbital situated on each hydrogen nucleus. This minimal basis set leads to two molecular orbitals, a bonding orbital σ_b and an antibonding orbital σ_a. It is important to realize that we obtained two molecular orbitals because we used a linear combination of two atomic orbitals. We used only two atomic orbitals solely for simplicity, and we could just as well have used a linear combination such as

$$\psi = c_1(1s_A + 1s_B) + c_2(2p_{zA} + 2p_{zB}) + c_3(2s_A + 2s_B) \qquad (10.55)$$

and so on. This trial function will lead to a 6×6 secular determinant and consequently six molecular orbitals. Each of these molecular orbitals will have a corresponding energy. Let's order the orbitals, $\psi_1, \psi_2, \ldots, \psi_6$ in order of increasing energy. The ground-state molecular wave function will be given by the Slater determinant

$$\psi_1 = N \begin{vmatrix} \psi_1\alpha(1) & \psi_1\beta(1) \\ \psi_1\alpha(2) & \psi_1\beta(2) \end{vmatrix} \qquad (10.56)$$

where N is just a normalization constant. Equation 10.56 describes two electrons of opposite spin in the molecular orbital ψ_1. The other five orbitals are unoccupied in the ground state, and are called *virtual orbitals* (Figure 10.28). We can use these virtual orbitals to construct excited-state configurations to carry out a configuration-interaction calculation. In this case, we have a total of 6 molecular orbitals and so 12 spin orbitals (an α and a β coupled with each molecular orbital). The total number of configurations

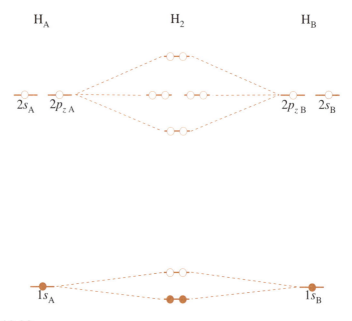

FIGURE 10.28
The six spatial molecular orbitals that are obtained when the LCAO-MO is a linear combination of six atomic orbitals, as in Equation 10.55. Only the molecular orbital of the lowest energy is occupied in the ground electronic state of H_2. The five unoccupied orbitals are called virtual orbitals.

is equal to the number of ways that we can distribute 2 electrons among the 12 spin orbitals, or

$$\frac{12!}{(12-2)!\,2!} = 66$$

Even though the number of configurations used may be large, we'll see that there are readily available computer programs that allow us to do this fairly painlessly. In fact, using more and more terms in Equation 10.55 and more and more configurations leads to the exact result, at least in principle.

EXAMPLE 10–10
Derive a formula for the total number of configurations that you obtain from $2N$ electrons and $2K$ spin orbitals. Apply your result to the minimal basis configuration-interaction calculation that we did in the previous section.

SOLUTION: The total number of configurations is equal to the number of ways of distributing $2N$ identical objects into $2K$ containers, which is given by the binomial coefficient:

$$\text{total number of configurations} = \frac{(2K)!}{(2K - 2N)!(2N)!} \qquad (10.57)$$

For our minimal basis set calculation, $K = 2$ and $N = 1$, and so we have $4!/2!\,2! = 6$ configurations, in agreement with the above calculation.

McLean, Weiss, and Yoshine found a minimum energy of $-1.167\,23\,E_h$ with a five-term CI wave function involving molecular orbitals consisting of $1s$, $2s$, and $2p$ Slater determinants, and Hagstrom and Schull found a minimum energy of $-1.1735\,E_h$ using a 33-term CI function. (See Table 10.4.) We see that we can achieve increasingly better numerical results by using increasingly large basis sets. Our final task in this chapter is to learn how we determine the optimum molecular orbitals for larger basis sets. We're going to do this in a manner similar to that which we used to construct optimum atomic orbitals in the previous chapter. In the next section, we shall develop the Hartree–Fock–Roothaan method, which gives the optimum molecular orbitals in terms of linear combinations of atomic orbitals.

10.10 The Hartree–Fock–Roothaan Equations Give the Optimum Molecular Orbitals

In Chapter 9, we developed the Hartree–Fock theory for atoms and applied it numerically to a helium atom. The Hartree–Fock theory has been extended to molecules in a practical way by Roothaan, and in this section, we shall derive the Hartree–Fock–Roothaan equations and then apply them numerically in a minimal basis set calculation for H_2. The Hartree–Fock–Roothaan equations for H_2 are almost the same as the Hartree–Fock–Roothaan equations for a helium atom. Let's start with the Hamiltonian operator for H_2:

$$\hat{H} = -\frac{1}{2}\sum_{j=1}^{2}\nabla_j^2 + \sum_{j=1}^{2}\left(-\frac{1}{r_{jA}} - \frac{1}{r_{jB}}\right) + \frac{1}{r_{12}} + \frac{1}{R} \qquad (10.58)$$

Because the internuclear repulsion term is a constant for a fixed geometry, we can ignore this term for now and add it later on. Because H_2 is a two-electron molecule, the spatial part and the spin part of the ground-state wave function factor and we can write it as

$$\Psi(\mathbf{r}_1, \mathbf{r}_2) = \psi(\mathbf{r}_1)\psi(\mathbf{r}_2) \qquad (10.59)$$

where the ground-state energy is given by

$$E = \int d\mathbf{r}_1 d\mathbf{r}_2\ \Psi^*(1, 2)\ \hat{H}\ \Psi(1, 2) \qquad (10.60)$$

Substituting $\Psi(\mathbf{r}_1, \mathbf{r}_2)$ from Equation 10.59 into this expression gives (Problem 10–36)

$$E = I_1 + I_2 + J_{11} \tag{10.61}$$

where

$$I_j = \int d\mathbf{r}_j \, \psi^*(\mathbf{r}_j) \left[-\frac{1}{2}\nabla_j^2 - \frac{1}{r_{jA}} - \frac{1}{r_{jB}} \right] \psi(\mathbf{r}_j) \tag{10.62}$$

and

$$J_{11} = \int d\mathbf{r}_1 d\mathbf{r}_2 \, \psi^*(\mathbf{r}_1) \psi^*(\mathbf{r}_2) \frac{1}{r_{12}} \psi(\mathbf{r}_1)\psi(\mathbf{r}_2) \tag{10.63}$$

Notice that $I_1 = I_2$ by symmetry.

Equations 10.61 through 10.63 differ from Equations 9.23 through 9.25 only in the definition of I_j. Equation 9.23 represents the interaction of an electron with the helium nucleus, and Equation 10.62 represents the interaction of an electron with the two hydrogen nuclei. Otherwise the equations are formally identical.

We're now going to carry out a Hartree–Fock calculation, much like we did for a helium atom in the appendix to Chapter 9. First we'll write the LCAO-MO in Equation 10.59 as

$$\psi = c_1\phi_A + c_2\phi_B \tag{10.64}$$

where ϕ_A (ϕ_B) is a normalized Slater orbital with $\zeta = 1.193\,02$ centered on atom A (B). We chose the value $\zeta = 1.193\,02$ because it is the optimum value of ζ at $R = 1.385\,43\,a_0$ for this basis set (see Table 10.4). These are the values of ζ and R_{eq} (to six figures) that we obtained for the optimized molecular orbital theory calculation that we carried out in Section 10.8. Having fixed the value of ζ, we shall determine the values of c_1 and c_2 by a Hartree–Fock self-consistent procedure.

Following what we did in Chapter 9, we substitute Equation 10.64 into Equation 10.60 to obtain

$$E = \frac{2(c_1^2 H_{AA} + 2c_1c_2 H_{AB} + c_2^2 H_{BB}) + c_1^2 G_{AA} + 2c_1c_2 G_{AB} + c_2^2 G_{BB}}{c_1^2 S_{AA} + 2c_1c_2 S_{AB} + c_2^2 S_{BB}} \tag{10.65}$$

where

$$H_{ij} = H_{ji} = \int d\mathbf{r}_1 \, \phi_i(r_1) \left(-\frac{1}{2}\nabla_1^2 - \frac{1}{r_{1A}} - \frac{1}{r_{1B}} \right) \phi_j(r_1)$$

$$= \left\langle \phi_i \left| -\frac{1}{2}\nabla_1^2 - \frac{1}{r_{1A}} - \frac{1}{r_{1B}} \right| \phi_j \right\rangle \qquad i, j = A, B \tag{10.66}$$

TABLE 10.5
The Numerical Values of the Matrix Elements in Equations 10.66 Through 10.68 for $\zeta = 1.193$ at $R = 1.385\,43\,a_0$

$$H_{AA}^{core} = H_{BB}^{core} = -1.132\,95\,E_h$$

$$H_{AB}^{core} = H_{BA}^{core} = -0.974\,75\,E_h$$

$$S_{AA} = S_{BB} = 1$$

$$S_{AB} = S(w) = 0.682\,42$$

$$G_{AA} = 0.745\,64\,c_1^2 + 0.893\,59\,c_1c_2 + 0.813\,35\,c_2^2$$

$$G_{AB} = G_{BA} = 0.446\,75\,c_1^2 + 0.364\,98\,c_1c_2 + 0.446\,75\,c_2^2$$

$$G_{BB} = 0.813\,36\,c_1^2 + 0.893\,49\,c_1c_2 + 0.745\,64\,c_2^2$$

$$G_{ij} = G_{ji} = c_1^2 \left\langle \phi_i(1)\phi_A(2) \left| \frac{1}{r_{12}} \right| \phi_j(1)\phi_A(2) \right\rangle + c_1c_2 \left\langle \phi_i(1)\phi_A(2) \left| \frac{1}{r_{12}} \right| \phi_j(1)\phi_B(2) \right\rangle$$

$$+ c_2c_1 \left\langle \phi_i(1)\phi_B(2) \left| \frac{1}{r_{12}} \right| \phi_j(1)\phi_A(2) \right\rangle + c_2^2 \left\langle \phi_i(1)\phi_B(2) \left| \frac{1}{r_{12}} \right| \phi_j(1)\phi_B(2) \right\rangle$$

$$i, j = A, B \qquad (10.67)$$

and

$$S_{ij} = S_{ji} = \langle \phi_i | \phi_j \rangle \qquad i, j = A, B \qquad (10.68)$$

with $S_{AA} = S_{BB} = 1$. All of these integrals are given in Tables 10.1 and 10.4, and the numerical values of all the terms in Equations 10.65 through 10.68 are given in Table 10.5 for $\zeta = 1.193\,02$ at $R = 1.385\,43\,a_0$ (Problem 10–39).

When we differentiate Equation 10.65 with respect to c_1 and c_2, we obtain the Hartree–Fock–Roothaan equations for this system:

$$(F_{AA} - \varepsilon)c_1 + (F_{AB} - \varepsilon S_{AB})c_2 = 0$$
$$(F_{AB} - \varepsilon S_{AB})c_1 + (F_{BB} - \varepsilon)c_2 = 0 \qquad (10.69)$$

or

$$\begin{pmatrix} F_{AA} & F_{AB} \\ F_{AB} & F_{BB} \end{pmatrix} \begin{pmatrix} c_1 \\ c_2 \end{pmatrix} = \varepsilon \begin{pmatrix} S_{AA} & S_{AB} \\ S_{AB} & S_{BB} \end{pmatrix} \begin{pmatrix} c_1 \\ c_2 \end{pmatrix} \qquad (10.70)$$

with the corresponding secular determinantal equation

$$
\begin{vmatrix}
F_{AA} - \varepsilon & F_{AB} - \varepsilon S_{AB} \\
F_{AB} - \varepsilon S_{AB} & F_{BB} - \varepsilon
\end{vmatrix} = 0
\tag{10.71}
$$

where

$$
F_{ij}(c_1, c_2) = H_{ij} + G_{ij} \qquad i, j = A, B
\tag{10.72}
$$

are the elements of the Fock matrix. As the notation implies, the F_{ij} depend upon c_1 and c_2. Equation 10.71 is a quadratic equation in ε. The matrix elements F_{ij} depend upon c_1 and c_2, whose values we don't know. In the first step, we assume values for c_1 and c_2. Let's assume that $c_1/c_2 = 2$. For our trial function to be normalized, we must have (Problem 10–40)

$$
c_1^2 + 2c_1c_2 S_{AB} + c_2^2 = 1
\tag{10.73}
$$

which, upon using $S_{AB} = 0.682\,42$ from Table 10.5, gives $c_1 = 0.719\,37$ and $c_2 = 0.359\,68$. We can now use these values to calculate the G_{ij} in Table 10.5 and then use Equation 10.72 to calculate the F_{ij}, whose values are given in Table 10.6. This initial guess of c_1 and c_2 gives an energy

$$
\begin{aligned}
E &= 2(c_1^2 H_{AA} + 2c_1c_2 H_{AB} + c_2^2 H_{BB}) + c_1^2 G_{AA} + 2c_1c_2 G_{AB} + c_2^2 G_{BB} \\
&= c_1^2(H_{AA} + F_{AA}) + 2c_1c_2(H_{AB} + F_{AB}) + c_2^2(H_{BB} + F_{BB}) \\
&= -1.805\,53\,E_h
\end{aligned}
\tag{10.74}
$$

If we add the $1/R$ internuclear repulsion term ($R = 1.385\,43\,a_0$), then we get $E_{el} = -1.083\,73\,E_h$.

The next step is to use the F_{ij} to solve Equation 10.71 for ε, which gives $\varepsilon_1 = -0.588\,04$ as the lower eigenvalue. Now that we know ε_1, we can solve Equations 10.70 for c_1 and c_2:

$$
\begin{aligned}
\frac{c_1}{c_2} &= -\frac{F_{AB} - \varepsilon S_{AB}}{F_{AA} - \varepsilon S_{AA}} = -\frac{-0.591\,33 - (-0.588\,04)(0.682\,42)}{-0.410\,68 - (0.588\,04)} \\
&= 1.0715
\end{aligned}
\tag{10.75}
$$

Using this value, Equation 10.73 gives

$$
c_1 = 0.563\,91 \qquad \text{and} \qquad c_2 = 0.526\,27
\tag{10.76}
$$

These values are given in the second line in Table 10.6.

The next step is to use these new coefficients to calculate a new set of G_{ij}. Using $F_{ij} = H_{ij} + G_{ij}$, we calculate a new set of F_{ij}, and a new value of ε from Equation 10.71, then a new set of coefficients from Equation 10.70, and finally a new

TABLE 10.6
The Results of the SCF Calculation of H_2 Using a Minimal Basis Set of Two Slater Determinants with $\zeta = 1.193\,02$ ($R = 1.385\,43\,a_0$) [a]

	c_1	c_2	F_{AA}	F_{AB}	F_{BB}	ε_1	ε_2	E_{el}
Initial guess	0.719 37	0.359 68	−0.410 68	−0.591 33	−0.384 40			−1.083 73
1st iteration	0.563 91	0.526 27	−0.405 41	−0.600 14	−0.402 63	−0.588 04	0.610 49	−1.127 74
2nd iteration	0.547 11	0.543 19	−0.404 23	−0.600 74	−0.403 94	−0.597 15	0.619 13	−1.128 22
3rd iteration	0.545 36	0.544 95	−0.404 11	−0.600 74	−0.404 08	−0.597 25	0.619 22	−1.128 23
4th iteration	0.545 17	0.545 13	−0.404 09	−0.600 74	−0.404 09	−0.597 25	0.619 23	−1.128 23
5th iteration	0.545 15	0.545 15	−0.404 09	−0.600 74	−0.404 09	−0.597 25	0.619 23	−1.128 23

a. ε_1 is the smaller eigenvalue and ε_2 is the larger eigenvalue of Equation 10.71. All quantities are expressed in atomic units.

value of E from Equation 10.74. All these results are given in Table 10.6, along with the results of subsequent iterations. Notice that the energy has converged to the optimized molecular orbital theory result in three iterations.

We've done this calculation to show how to use the Hartree–Fock–Roothaan equations. In fact, the symmetry of H_2 suggests that $c_1 = c_2$ in the ground state, and if we had started with $c_1 = c_2$, the entire calculation would have converged immediately and would be equivalent to the molecular orbital calculation. We chose $c_1/c_2 = 2$ simply to illustrate the convergence numerically.

Although we have derived the Hartree–Fock–Roothaan equations for the special case of a basis set consisting of only two Slater determinants, we shall see in Chapter 12 that the procedure is readily extended to any number of base set functions and that a number of readily available computer packages can handle fairly large basis sets for molecules consisting of tens or even hundreds of atoms. In fact, although in the 1960s and 1970s such calculations belonged to professional quantum chemists, they are now routine and are done as "experiments" in many organic and physical chemistry laboratory courses.

As we use larger and larger basis sets, we achieve lower and lower energies, and eventually reach the Hartree–Fock limit. It is important to remember that if we use a basis set consisting of only a few terms and determine the coefficients self-consistently, then we may not necessarily achieve the Hartree–Fock limit. Thus, an SCF-LCAO-MO is not necessarily the same as a Hartree–Fock orbital. They are the same only if the SCF-LCAO-MO contains enough terms that the Hartree–Fock limit is reached.

Table 10.7 shows the results of a Hartree–Fock–Roothaan calculation on H_2 in which the Hartree–Fock limit is achieved. Note that the basis set in this case consists of 12 Slater orbitals. Both the orbital exponents and the linear coefficients were optimized in this calculation, giving a Hartree–Fock limiting energy of $-1.133\,629\,E_h$, compared to $-1.128\,23\,E_h$ for our minimal basis set calculation (using a fixed value of ζ for simplicity).

TABLE 10.7

The Final Results for a Hartree–Fock–Roothaan Calculation of H_2 in Which the Hartree–Fock Limit Is Achieved [a]

Slater orbital	Value of ζ	Linear coefficient
$1s$	1.188 63	0.903 74
$1s'$	2.500 21	−0.049 78
$2s$	0.794 45	0.011 44
$2s'$	1.730 27	−0.128 32
$3s$	3.436 00	0.001 94
$2p_z$	1.055 29	0.039 84
$2p_z'$	1.985 53	0.024 13
$2p_z''$	4.081 82	−0.000 94
$3p_z$	3.433 59	0.001 29
$3d_{z^2}$	1.266 63	0.006 49
$3d_{z^2}'$	2.680 42	0.003 93
$4f_{z^3}$	2.708 08	0.001 03

a. *Source:* Cade, P.E., Wahl, A. C., Hartree–Fock–Roothaan Wavefunctions for Diatomic Molecules: II. First-Row Homonuclear Systems A_2, A_2^+, and A_2^*. *At. Data Nucl. Data Tables,* **13**, 339 (1974).

A basis set that consists of K atomic orbitals will lead to K molecular orbitals. In the case of H_2, the first of these will be occupied by electrons of opposite spin, the others will be unoccupied, or virtual orbitals. This leads naturally to using these orbitals to form excited-state Slater determinants in a configuration-interaction calculation. This procedure has been well developed, and has constituted one of the principal methods of going beyond the Hartree–Fock approximation, or one of the principal *post-Hartree–Fock methods*.

We should point out here that Hartree–Fock–Roothaan molecular orbitals suffer from the same deficiency as the molecular orbitals that we found earlier in this chapter. At large R, the ground-state molecular orbital does not go to the wave function of two isolated ground-state hydrogen atoms. In other words, it does not give the correct dissociation limit. This may seem disconcerting, but it turns out not to be a problem for several reasons. One reason is that modifications of the Hartree–Fock–Roothaan equations have been developed that do lead to the correct dissociation limit. This method is called *unrestricted Hartree–Fock*. We won't consider any unrestricted Hartree–Fock calculations here, but you should be aware of the method. Another reason is that most post-Hartree–Fock methods, such as configuration interaction, do lead to the correct dissociation limit. We have, in fact, seen that this is so in the previous section. Still another reason, and perhaps most important of all, is that Hartree–Fock calculations usually yield quite acceptable equilibrium molecular geometries, as we shall see in Chapter 12. So, except for the calculation of potential energy curves, or more generally, potential energy surfaces, which are of central importance in chemical kinetics,

Hartree–Fock calculations are quite acceptable. Because of the ready availability of computer packages for Hartree–Fock and post-Hartree–Fock calculations, molecular orbital theory, which is essentially Hartree–Fock theory, has become the "standard" for molecular calculations.

Appendix: Molecular Orbital Theory of H_2

Even a minimal basis set molecular orbital theory calculation for H_2 is a fairly lengthy algebraic exercise. The original calculations were done many years ago in the early days of quantum mechanics and have been greatly improved upon since, but they still reflect the basic features of a molecular calculation. In this appendix, we present the details of a minimal basis set molecular orbital theory and configuration-interaction calculation using Slater orbitals. The definitive reference for this material is the one to Slater in the end-of-chapter references.

We start with the energy given by

$$E_{MO} = \langle \psi_{MO} \mid \hat{H} \mid \psi_{MO} \rangle \tag{1}$$

where $\hat{H}$ is given by Equation 10.35. When Equation 10.39 is substituted into equation 1, we obtain

$$E_{MO} = \zeta^2 \left[\frac{1 - S(w) - 2K(w)}{1 + S(w)} \right] + \zeta \left[\frac{-2 + 2J(w) + 4K(w)}{1 + S(w)} \right]$$
$$+ \zeta \left\{ \frac{5/16 + J'(w)/2 + K'(w) + 2L(w)}{[1 + S(w)]^2} \right\} + \frac{\zeta}{w} \tag{2}$$

after some amount of algebra.

Equation 2 may look complicated, but it's not. First of all, the discussion between equations 4 and 5 below shows that equation 2 is of the form

$$E_{MO} = \zeta^2 T(w) + \zeta V(w) \tag{3}$$

just as is Equation 10.33 for H_2^+. As we mentioned before, this result is a consequence of the form of the Hamiltonian operator. We have encountered the integrals, $S(w)$, $J(w)$, and $K(w)$, in the first two terms of equation 2 in our H_2^+ calculation, but $J'(w)$, $K'(w)$, and $L'(w)$ are new, and are given in Table 10.8. The first three integrals in Table 10.8 are fairly easy to evaluate using an elliptical coordinate system as we did for H_2^+, but the fourth one is famously challenging. It may look fairly ugly, but it makes no difference to a computer. The function $E_1(x)$ that is part of $K'(w)$ is called an exponential integral and is a well-tabulated function (see, for example, the reference to Abramowitz and Stegun at the end of the chapter) and is built into most mathematical computer packages such as *MathCad* and *Mathematica*. The constant γ that appears in $K'(w)$ is called Euler's constant, and is an irrational number that appears naturally in advanced calculus, much like π and e.

TABLE 10.8

The Integrals over $1/r_{12}$ That Occur in the Molecular Orbital Calculation of the Energy of H_2 [a]

$$\left\langle \phi_A(1)\phi_A(2) \left| \frac{1}{r_{12}} \right| \phi_A(1)\phi_A(2) \right\rangle = \frac{5\zeta}{8}$$

$$\left\langle \phi_A(1)\phi_B(2) \left| \frac{1}{r_{12}} \right| \phi_A(1)\phi_B(2) \right\rangle = \zeta J'(w)$$

$$= \zeta \left[\frac{1}{w} - e^{-2w} \left(\frac{1}{w} + \frac{11}{8} + \frac{3w}{4} + \frac{w^2}{6} \right) \right]$$

$$\left\langle \phi_A(1)\phi_A(2) \left| \frac{1}{r_{12}} \right| \phi_A(1)\phi_B(2) \right\rangle = \zeta L(w)$$

$$= \zeta \left[e^{-w} \left(w + \frac{1}{8} + \frac{5}{16w} \right) + e^{-3w} \left(-\frac{1}{8} - \frac{5}{16w} \right) \right]$$

$$\left\langle \phi_A(1)\phi_A(2) \left| \frac{1}{r_{12}} \right| \phi_B(1)\phi_B(2) \right\rangle = \zeta K'(w)$$

$$= \frac{\zeta}{5} \left\{ -e^{-2w} \left(-\frac{25}{8} + \frac{23w}{4} + 3w^2 + \frac{w^3}{3} \right) \right.$$

$$\left. + \frac{6}{w} [S^2(w)(\gamma + \ln w) - S'(w)^2 E_1(4w) + 2S(w)S'(w)E_1(2w)] \right\} \text{ [b]}$$

a. Recall that $w = \zeta R$.

b. $S'(w) = e^w \left(1 - w + \frac{w^2}{3} \right)$, $\gamma = 0.577\,21 \ldots$ (called Euler's constant) and $E_1(x) = \int_x^\infty \frac{e^{-t}}{t} dt$ (called an exponential integral).

Before we dissect equation 2, let's compare the Hamiltonian operators for H_2^+ (Equation 10.8) and H_2 (Equation 10.35). Notice that the Hamiltonian operator for H_2 can be written in the form

$$\hat{H}_{H_2} = \hat{H}_{H_2^+}(1) + \hat{H}_{H_2^+}(2) + \frac{1}{r_{12}} \tag{4}$$

where, as the notation suggests, $H_{H_2^+}(i)$ is the Hamiltonian operator of $H_2^+(i)$ with electron i. Now, let's look at each of the four terms in equation 2 in turn. The first term is two times the kinetic energy of H_2^+ given by Equation 10.31, and the second term is two times the electron–nuclear interaction energy of H_2^+ given by Equation 10.32. The fourth term is simply the internuclear repulsion energy, $1/R$. The only new term in equation 2 is the third term, which is due to the electron–electron interaction, which occurs in the Hamiltonian operator of H_2 but not in that of H_2^+. Thus, we see that the fourth term in equation 2 is due to

$$\left\langle \psi_{MO} \left| \frac{1}{r_{12}} \right| \psi_{MO} \right\rangle = \zeta \left\{ \frac{5/16 + J'(w)/2 + K'(w) + 2L(w)}{[1 + S(w)]^2} \right\} \tag{5}$$

All the functions of w in equation 2 are given in Tables 10.1 and 10.8, and so we know E_{MO} as a function of ζ and $w = \zeta R$. If we set $\zeta = 1$, then minimizing the energy with respect to R gives $E_{MO} = -1.0991 E_H$ at $R_{eq} = 1.603 a_0$. This is best done (and easily done) using a program such as *MathCad* or *Mathematica* (Problem 10–43). We can also treat ζ as a variational parameter and minimize E_{MO} with respect to both ζ and R. (See Problem 10–44.) In this case, we get $E_{MO} = -1.1282 E_h$ and $R_{eq} = 1.385 a_0$ with $\zeta = 1.193$, a considerable improvement over the calculation with $\zeta = 1$ (cf. Table 10.4).

If we plot E_{MO} against R for $\zeta = 1$, we get the black curve in Figure 10.23, which as we discussed in the chapter, goes to the wrong dissociation limit as $R \to \infty$. Nor does optimizing ζ as a function of R improve things. (See Figure 10.23.) We showed in the chapter that this well-known deficiency of molecular orbital theory is obviated by using configuration interaction. In Section 10.9, we showed that we could use Equation 10.53 with ψ_1 and ψ_2 given by the first two of Equations 10.49 as a minimal basis set configuration-interaction wave function for H_2. This wave function leads to the 2×2 determinantal equation

$$\begin{vmatrix} H_{11} - ES_{11} & H_{12} - ES_{12} \\ H_{12} - ES_{12} & H_{22} - ES_{22} \end{vmatrix} = 0 \tag{6}$$

where

$$H_{11} = \langle \psi_1 | \hat{H} | \psi_1 \rangle; \qquad H_{22} = \langle \psi_2 | \hat{H} | \psi_2 \rangle \qquad H_{12} = H_{21} = \langle \psi_1 | \hat{H} | \psi_2 \rangle \tag{7}$$

$$S_{11} = \langle \psi_1 | \psi_1 \rangle; \qquad S_{22} = \langle \psi_2 | \psi_2 \rangle; \qquad S_{12} = \langle \psi_1 | \psi_2 \rangle$$

The integrals S_{ij} are fairly easy to evaluate. We first show that σ_b and σ_a are orthonormal

$$\langle \sigma_b | \sigma_b \rangle = \left\langle \frac{1s_A + 1s_B}{\sqrt{2(1+S)}} \left| \frac{1s_A + 1s_B}{\sqrt{2(1+S)}} \right. \right\rangle$$

$$= \frac{1 + 2S + 1}{2(1+S)} = 1$$

$$\langle \sigma_a | \sigma_a \rangle = \left\langle \frac{1s_A - 1s_B}{\sqrt{2(1-S)}} \left| \frac{1s_A - 1s_B}{\sqrt{2(1-S)}} \right. \right\rangle$$

$$= \frac{1 - 2S + 1}{2(1-S)} = 1$$

$$\langle \sigma_b | \sigma_a \rangle = \left\langle \frac{1s_A + 1s_B}{\sqrt{2(1+S)}} \left| \frac{1s_A - 1s_B}{\sqrt{2(1-S)}} \right. \right\rangle$$

$$= \frac{1 + S - S - 1}{2\sqrt{1 - S^2}} = 0$$

Using these results, we see that

$$
S_{11} = \langle \psi_1 \mid \psi_1 \rangle = \langle \sigma_b(1)\sigma_b(2) \mid \sigma_b(1)\sigma_b(2) \rangle
$$

$$
= \langle \sigma_b(1) \mid \sigma_b(1) \rangle \langle \sigma_b(2) \mid \sigma_b(2) \rangle = 1 \times 1 = 1
$$

$$
S_{22} = \langle \psi_2 \mid \psi_2 \rangle = \langle \sigma_a(1)\sigma_a(2) \mid \sigma_a(1)\sigma_a(2) \rangle
$$

$$
= \langle \sigma_a(1) \mid \sigma_a(1) \rangle \langle \sigma_a(2) \mid \sigma_a(2) \rangle = 1 \times 1 = 1 \tag{8}
$$

$$
S_{12} = \langle \psi_1 \mid \psi_2 \rangle = \langle \sigma_b(1)\sigma_b(2) \mid \sigma_a(1)\sigma_a(2) \rangle
$$

$$
= \langle \sigma_b(1) \mid \sigma_a(1) \rangle \langle \sigma_b(2) \mid \sigma_a(2) \rangle = 0 \times 0 = 0
$$

The matrix elements of $\hat{H}$ are tedious to evaluate, and are given in Table 10.9. All the necessary integrals were given already in Tables 10.1 and 10.8.

TABLE 10.9
The Matrix Elements H_{ij} and S_{ij} That Occur in the Minimal Basis Set Configuration-Interaction Treatment of H_2 [a]

$$
\left.\begin{array}{c} H_{11}(\zeta, w) \\ H_{22}(\zeta, w) \end{array}\right\} = \zeta^2 \left[\frac{1 \mp S(w) \mp 2K(w)}{1 \pm S(w)} \right] + \zeta \left\{ \frac{-2 + 2J(w) \pm 4K(w)}{1 \pm S(w)} \right.
$$

$$
\left. + \frac{\dfrac{5}{16} + \dfrac{J'(w)}{2} + K'(w) \pm 2L(w)}{[1 \pm S(w)]^2} + \frac{1}{w} \right\}
$$

$$
H_{12}(\zeta, \omega) = \zeta \left[\frac{\dfrac{5}{16} - \dfrac{J'(w)}{2}}{1 - S^2(w)} \right]
$$

a. All the functions of w here are given in Tables 10.1 and 10.4.

We can find the energy by solving equation 6 for E,

$$
E_{\pm}(\zeta, w) = \frac{H_{11}(\zeta, w) + H_{22}(\zeta, w)}{2}
$$

$$
\pm \frac{1}{2}\{[H_{11}(\zeta, w) - H_{22}(\zeta, w)]^2 + 4H_{12}^2(\zeta, w)\}^{1/2} \tag{9}
$$

where the $H_{ij}(\zeta, w)$ are given in Table 10.9. Even though equation 9 is very lengthy, all the terms are known functions of ζ and R (through $w = \zeta R$), and can be handled easily by any of a number of easy-to-use computer programs. It turns out in this case that $E_-(R)$ is always the smaller root, so we'll let $E_{CI} = E_-$. Figure 10.25 shows E_{CI}

plotted against R for $\zeta = 1$. Notice that E_{CI} goes to the correct dissociation limit of two isolated ground-state hydrogen atoms ($E = -1 E_h$).

Before we look at the optimized result, let's see why we get the correct dissociation limit in this case ($\zeta = 1$). It turns out that all the integrals (the functions of $w = \zeta R$) in Table 10.9 go to zero as $R \to \infty$ (Problem 10–47), and so

$$H_{11} = H_{22} \longrightarrow \zeta^2 + \zeta(-2 + 5/16) \qquad \text{and} \qquad H_{12} \longrightarrow 5\zeta/16 \qquad (10)$$

Consequently,

$$E_{CI} \longrightarrow H_{11} - H_{12} = \zeta^2 - 2\zeta \qquad (11)$$

When $\zeta = 1$, $E_{CI} = -1 E_h$, which is the exact energy of two isolated ground-state hydrogen atoms.

We can also determine c_1 and c_2 in Equation 10.53 from the two algebraic equations that lead to equation 6. Solving the first of these gives

$$\frac{c_1}{c_2} = -\frac{H_{11} - E}{H_{12}} \qquad (12)$$

We know H_{11} (Table 10.9), E (equation 9), and H_{12} (Table 10.9) as a function of ζ and $w = \zeta R$, and so we know c_1/c_2 as a function of ζ and w. We can then find c_1 and c_2 individually by requiring that ψ_{CI} given by Equation 10.53 be normalized, which amounts to the relation $c_1^2 + c_2^2 = 1$. The result is shown in Figure 10.27. We can now show that $c_1 \to 1/\sqrt{2}$ and $c_2 \to -1/\sqrt{2}$ as $R \to \infty$. Equations 10, 11, and 12 give us

$$\frac{c_1}{c_2} \longrightarrow -\frac{(\zeta^2 - 27\zeta/16) - (\zeta^2 - 2\zeta)}{5\zeta/16} = -1$$

The normalization condition gives $c_1 = -c_2 = 1/\sqrt{2}$. Notice that this result is independent of ζ.

We can use a program such as *MathCad* or *Mathematica* to minimize E_{CI} with respect to ζ at various values of R to generate an optimized E_{CI} as a function of R. Figure 10.25 shows E_{CI} plotted against R as well as the "exact" Kolos–Wolniewicz results.

Problems

10–1. Express the Hamiltonian operator for a hydrogen molecule in atomic units.

10–2. The vibrational energy levels of H_2^+ (in cm^{-1}) are given by Chase, M. W., Jr., et al., *J. Phys. Chem. Ref. Data* 1985, vol. 14, supplement no. 1, also known as the JANAF Thermochemical Tables. These tables were updated in 1998 by M. W. Chase, Jr. as the *NIST–JANAF Thermochemical Tables*, 4th ed., monograph no. 9 (parts 1 and 2), available from the American Institute of Physics. See *www.nist.gov/srd/jpcrd_28.htm#janaf*

$$G(v) = 2323.23 \left(v + \frac{1}{2}\right) - 67.39 \left(v + \frac{1}{2}\right)^2 + 0.93 \left(v + \frac{1}{2}\right)^3 - 0.029 \left(v + \frac{1}{2}\right)^4$$

The molecule dissociates in the limit that $\Delta G(v) = G(v+1) - G(v) \to 0$, and so v_{max}, the maximum vibrational quantum number, is given by $\Delta G(v_{max}) = 0$. Plot $\Delta G(v) = G(v+1) - G(v)$ against v and show that $v_{max} = 18$.

10–3. The JANAF tables (see the previous problem) give a value of $D_0 = 255.76 \text{ kJ·mol}^{-1}$. Use the vibrational data in the previous problem to show that $D_e = 269.45 \text{ kJ·mol}^{-1} = 0.102\,64\,E_h$.

10–4. The value of D_e for H_2^+ is $0.102\,64\,E_h$, yet the minimum energy in Figure 10.2 is $-0.602\,64\,E_h$. Why is there a difference?

10–5. Plot the product $1s_A\,1s_B$ along the internuclear axis for several values of R.

10–6. The overlap integral, Equation 10.13c, and other integrals that arise in two-center systems like H_2^+ are called *two-center integrals*. Two-center integrals are most easily evaluated by using a coordinate system called *elliptical coordinates*. In this coordinate system (Figure 10.4), there are two fixed points separated by a distance R. A point P is given by the three coordinates

$$\lambda = \frac{r_A + r_B}{R}$$

$$\mu = \frac{r_A - r_B}{R}$$

and the angle ϕ, which is the angle that the (r_A, r_B, R) triangle makes about the interfocal axis. The differential volume element in elliptical coordinates is

$$d\mathbf{r} = \frac{R^3}{8}(\lambda^2 - \mu^2)d\lambda\,d\mu\,d\phi$$

Given the above definitions of λ, μ, and ϕ, show that

$$1 \le \lambda < \infty \qquad -1 \le \mu \le 1 \qquad 0 \le \phi \le 2\pi$$

Now use elliptical coordinates to evaluate the overlap integral (Equation 10.13c):

$$S = \int d\mathbf{r}\,1s_A\,1s_B$$

$$= \frac{Z^3}{\pi}\int d\mathbf{r}\,e^{-Zr_A}e^{-Zr_B}$$

10–7. In this problem, we evaluate the overlap integral (Equation 10.13c) using spherical coordinates centered on atom A. The integral to evaluate is (Problem 10–6)

$$S(R) = \frac{1}{\pi} \int d\mathbf{r}_A e^{-r_A} e^{-r_B}$$

$$= \frac{1}{\pi} \int_0^\infty d r_A e^{-r_A} r_A^2 \int_0^{2\pi} d\phi \int_0^\pi d\theta \sin\theta e^{-r_B}$$

where r_A, r_B, and θ are shown in the figure.

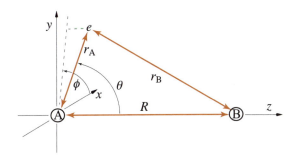

To evaluate the above integral, we must express r_B in terms of r_A, θ, and ϕ. We can do this using the law of cosines:

$$r_B = (r_A^2 + R^2 - 2r_A R \cos\theta)^{1/2}$$

So, the first integral we must consider is

$$I = \int_0^\pi e^{-(r_A^2 + R^2 - 2r_A R \cos\theta)^{1/2}} \sin\theta \, d\theta$$

As usual, let $\cos\theta = x$ to get

$$I = \int_{-1}^1 e^{-(r_A^2 + R^2 - 2r_A R x)^{1/2}} dx$$

Now let $u = (r_A^2 + R^2 - 2r_A R x)^{1/2}$ and show that

$$dx = -\frac{u\,du}{r_A R}$$

Show that the limits of the integration over u are $u = r_A + R$ when $x = -1$ and $u = |R - r_A|$ when $x = 1$. Then show that

$$I = \frac{1}{r_A R} \left[e^{-(R-r_A)}(R + 1 - r_A) - e^{-(R+r_A)}(R + 1 + r_A) \right] \qquad r_A < R$$

$$= \frac{1}{r_A R} \left[e^{-(r_A - R)}(r_A - R + 1) - e^{-(R+r_A)}(R + 1 + r_A) \right] \qquad r_A > R$$

Now substitute this result into $S(R)$ above to get

$$S(R) = e^{-R}\left(1 + R + \frac{R^2}{3}\right)$$

Compare the length of this problem to Problem 10–6.

10–8. Show that the first two integrals in Equation 10.16 come out to be 1/2 and −1, respectively.

10–9. Show that the fourth integral in Equation 10.16 is equal to $1/R$.

10–10. Use the elliptical coordinate system of Problem 10–6 to derive analytic expressions for S, J, and K for the simple molecular orbital treatment of H_2^+.

10–11. Let's use the method that we developed in Problem 10–7 to evaluate the coulomb integral, J, given by Equation 10.18. Let

$$J = -\int \frac{d\mathbf{r}\, 1s_A^* \, 1s_A}{r_B} = -\frac{1}{\pi}\int d\mathbf{r}\, \frac{e^{-2r_A}}{(r_A^2 + R^2 - 2r_A R \cos\theta)^{1/2}}$$

$$= -\frac{1}{\pi}\int_0^\infty dr_A r_A^2 e^{-2r_A}\int_0^{2\pi}d\phi\int_0^\pi \frac{d\theta\,\sin\theta}{(r_A^2 + R^2 - 2r_A R\cos\theta)^{1/2}}$$

Using the approach of Problem 10–7, let $\cos\theta = x$ and $u = (r_A^2 + R^2 - 2r_A R\cos\theta)^{1/2}$ to show that

$$J = \frac{2}{R}\int_0^\infty dr_A r_A e^{-2r_A}\int_{R+r_A}^{|R-r_A|}du = \frac{2}{R}\int_0^\infty dr_A r_A e^{-2r_A}[|R - r_A| - (R + r_A)]$$

$$= e^{-2R}\left(1 + \frac{1}{R}\right) - \frac{1}{R}$$

Hint: You need to use the integrals

$$\int xe^{ax}dx = e^{ax}\left(\frac{x}{a} - \frac{1}{a^2}\right)$$

and

$$\int x^2 e^{ax}dx = e^{ax}\left(\frac{x^2}{a} - \frac{2x}{a^2} + \frac{2}{a^3}\right)$$

10–12. Use the entries in Table 10.1 to plot ΔE_+ against R.

10–13. Use the entries in Table 10.1 to plot ΔE_- against R.

10–14. Show that

$$H_{AA} = H_{BB} = -\frac{1}{2} + J$$

and that

$$H_{AB} = -\frac{S}{2} + K$$

in the simple molecular orbital treatment of H_2^+. The quantities J and K are given by Equations 10.18 and 10.21, respectively.

10–15. Show that

$$\langle \hat{T} \rangle = \left\langle \psi_+ \left| -\frac{1}{2}\nabla^2 \right| \psi_+ \right\rangle = \frac{\dfrac{1}{2} - \dfrac{S}{2} - K}{1 + S}$$

$$\langle \hat{V} \rangle = \left\langle \psi_+ \left| -\frac{1}{r_A} \right| \psi_+ \right\rangle + \left\langle \psi_+ \left| -\frac{1}{r_B} \right| \psi_+ \right\rangle + \left\langle \psi_+ \left| \frac{1}{R} \right| \psi_+ \right\rangle = \frac{-1 + J + 2K}{1 + S} + \frac{1}{R}$$

10–16. Show that $\langle \hat{T} \rangle + \langle \hat{V} \rangle$ from the previous problem agrees with E_+ given by Equation 10.22.

10–17. The change in the total energy when two atoms come together to form a stable bond must be negative, so that $\Delta E < 0$ at R_{eq}. Therefore, $\Delta E = \Delta T + \Delta V < 0$. Show that the virial theorem requires that $\Delta V < 0$. In addition, show that $\Delta T > 0$ upon bond formation. Using the fact that the kinetic energy of a hydrogen atom is $1/2\, E_h$ (see the second entry in Table 10.1), that its potential energy is $-1\, E_h$ (see the fourth entry in Table 10.1), and the result of Example 10–3, show that

$$\Delta T = \langle \hat{T} \rangle_{H_2^+} - \langle \hat{T} \rangle_H = 0.3827\, E_h - 0.5000\, E_h = -0.117\,31\, E_h$$

and

$$\Delta V = \langle \hat{V} \rangle_{H_2^+} - \langle \hat{V} \rangle_H = -0.9475\, E_h - (-1.0000\, E_h) = +0.052\,48\, E_h$$

upon bond formation. The signs of these results are just the opposite from that predicted by the virial theorem, which implies that the simple molecular orbital given by Equation 10.26 does not provide a satisfactory interpretation of bond formation.

10–18. Show that the overlap integral between two Slater $1s$ orbitals of the form $\phi(\zeta, r) = (\zeta^3/\pi)^{1/2} e^{-\zeta r}$ is given by

$$S(\zeta, R) = e^{-\zeta R}\left(1 + \zeta R + \frac{1}{3}\zeta^2 R^2\right)$$

10–19. Verify Equations 10.31 and 10.32.

10–20. In this problem we show that the form of Equations 10.31 and 10.32 are a direct result of the form of the Hamiltonian operator. Start with the Hamiltonian operator of a general molecule consisting of N electrons and n nuclei.

$$\hat{H} = -\frac{1}{2}\sum_{j=0}^{N}\nabla_j^2 + \hat{V}(x_1, y_1, z_1, \ldots, x_N, y_N, z_N, R_1, R_2, \ldots, R_m)$$

10–29. Show that the determinantal wave functions in Equations 10.48 are normalized.

10–30. The six wave functions given by Equations 10.49 are actually linear combinations of the determinantal wave functions given by Equations 10.48. Show that ψ_1 and ψ_2 are the spatial parts of D_1 and D_2, respectively; that two of ψ_3, ψ_4, or ψ_5 are the spatial parts of D_3 and D_6; that the remaining one of ψ_3, ψ_4, or ψ_5 is the spatial part of $(D_4 + D_5)/\sqrt{2}$; and that ψ_6 is the spatial part of $(D_4 - D_5)/\sqrt{2}$. The reason that we take linear combinations $(D_4 \pm D_5)/\sqrt{2}$ is because these combinations are eigenfunctions of $\hat{S}^2$. (We state this without proof.)

10–31. Use the result of the previous problem to show that the complete wave functions (spin functions included) given by Equations 10.49 are

$$\psi_1 = \frac{1}{\sqrt{2}}\sigma_b(1)\sigma_b(2)[\alpha(1)\beta(2) - \alpha(2)\beta(1)]$$

$$\psi_2 = \frac{1}{\sqrt{2}}\sigma_a(1)\sigma_a(2)[\alpha(1)\beta(2) - \alpha(2)\beta(1)]$$

$$\psi_3, \psi_4, \psi_5 = \frac{1}{\sqrt{2}}[\sigma_b(1)\sigma_a(1) - \sigma_a(1)\sigma_b(2)] \begin{cases} \alpha(1)\alpha(2) \\ \alpha(1)\beta(2) + \alpha(2)\beta(1) \\ \beta(1)\beta(2) \end{cases}$$

$$\psi_6 = \frac{1}{\sqrt{2}}[\sigma_b(1)\sigma_a(2) + \sigma_b(2)\sigma_a(1)][\alpha(1)\beta(2) - \alpha(2)\beta(1)]$$

Show that

$$\hat{S}_z\psi_1 = (\hat{S}_{z1} + \hat{S}_{z2})\psi_1 = 0$$

$$\hat{S}_z\psi_2 = (\hat{S}_{z1} + \hat{S}_{z2})\psi_2 = 0$$

$$\hat{S}_z\psi_3 = (\hat{S}_{z1} + \hat{S}_{z2})\psi_3 = 1$$

$$\hat{S}_z\psi_4 = (\hat{S}_{z1} + \hat{S}_{z2})\psi_4 = 0$$

$$\hat{S}_z\psi_5 = (\hat{S}_{z1} + \hat{S}_{z2})\psi_5 = -1$$

$$\hat{S}_z\psi_6 = (\hat{S}_{z1} + \hat{S}_{z2})\psi_6 = 0$$

It turns out that ψ_3, ψ_4, and ψ_5 are wave functions of a triplet state and the others are wave functions of singlet states.

10–32. Show that all six wave functions in the previous problem are antisymmetric.

10–33. Use the symmetry argument developed in Example 10–8 to show that $H_{16} = \langle \psi_1 | \hat{H} | \psi_6 \rangle = 0$ for ψ_1 and ψ_6 given in Equations 10.49.

10–34. In this problem, we shall prove that if an operator $\hat{F}$ commutes with $\hat{H}$, then matrix elements of $\hat{H}$, $H_{ij} = \langle \psi_i | \hat{H} | \psi_j \rangle$, between states with different eigenvalues of $\hat{F}$ vanish. For simplicity, we prove this only for nondegenerate states. Let $\hat{F}$ be an operator that commutes with $\hat{H}$, and let its eigenvalues and eigenfunctions be denoted by λ and ψ_λ, respectively. Show that

$$[\hat{H}, \hat{F}]_{\lambda\lambda'} = \langle \psi_\lambda | [\hat{H}, \hat{F}] | \psi_{\lambda'} \rangle = (\lambda - \lambda') H_{\lambda\lambda'}$$

Now argue that $H_{\lambda\lambda'} = 0$ unless $\lambda = \lambda'$. For degenerate states, it is possible to take linear combinations of the degenerate eigenfunctions and carry out a similar proof.

10–35. Determine the percentage of the correlation energy that the configuration-interaction calculation described in Section 10.9 gives.

10–36. Show that Equations 10.61 through 10.63 result when Equation 10.58 is substituted into Equation 10.60.

10–37. Show that Equation 10.61 is the same as equation 2 of the appendix when we use $\psi_1 = (1s_A + 1s_B)/\sqrt{2(1 + S)}$.

10–38. Substitute Equation 10.64 into Equation 10.60 to derive Equation 10.65.

10–39. Verify the entries in Table 10.5.

10–40. Show that the normalization condition for Equation 10.64 is $c_1^2 + 2c_1c_2S_{AB} + c_2^2 = 1$ where S_{AB} is the overlap integral involving ϕ_A and ϕ_B.

10–41. Use a program such as *MathCad* or *Mathematica* to redo the Hartree–Fock–Roothaan calculation in Section 10.10, starting with $c_1 = c_2$. Explain why the result converges so rapidly.

10–42. Why do the energies in Table 10.6 converge after three or four iterations but the coefficients in the wave function take five iterations?

10–43. Use a program such as *MathCad* or *Mathematica* to plot E_{MO} given by equation 2 of the appendix against R for $\zeta = 1$, and compare your result to Figure 10.23.

10–44. Use a program such as *MathCad* or *Mathematica* and the result of Problem 10–21 to determine ζ as a function of w for E_{MO} given by equation 2 of the appendix. Plot $\zeta(w)$ and $R(w)$ parametrically to give ζ plotted against R. Compare your result to that in Figure 10.24.

10–45. In this problem, we'll use the result of Problem 10–21 to show that $\zeta \to 27/16$ as $w \to 0$ for the molecular orbital calculation for H_2 in the appendix. (See also Figure 10.24.) We'll use the fact that the definition of $K'(w)$ in Table 10.8 has the following expansion:

$$E_1(x) = -\gamma - \ln x + x - x^2/4 + O(x^3)$$

(See the reference to Abramowitz and Stegun at the end of the chapter.) Using this expansion, first show that the second term in the definition of $K'(w)$ (the one multiplied by $6/w$) goes to zero as $w \to 0$. (This is a good example of one in which you should keep track of the order of the terms that you keep or ignore.) Now show that

$$K'(w) = \frac{5}{8} + O(w)$$

$$J'(w) = \frac{5}{8} + O(w)$$

$$L(w) = \frac{5}{8} + O(w)$$

In Problem 10–23, we obtained

$$S(w) = 1 - \frac{1}{6}w^2 + O(w^3)$$

$$J(w) = -1 + \frac{2}{3}w^2 + O(w^3)$$

$$K(w) = -1 = \frac{w^2}{2} + O(w^3)$$

Put this all together to show that

$$T(w) = 1 + O(w)$$

and that

$$V(w) = \frac{1}{w} - \frac{27}{8} + O(w)$$

Finally, use the result of Problem 10–21 to show that

$$\zeta \longrightarrow 27/16 \qquad \text{as } w \longrightarrow 0$$

10–46. This problem complements the previous problem in the sense that we'll show that $\zeta \to 27/32$ as $w \to \infty$ for the molecular orbital calculation for H_2 done in the appendix. (See also Figure 10.24 and Example 10–6.) We'll use the fact that

$$E_1(x) \longrightarrow \frac{e^{-x}}{x}[1 - x + O(x^2)] \qquad \text{as } x \longrightarrow \infty$$

(See the reference to Abramowitz and Stegun at the end of the chapter.) Show that

$$T(w) \longrightarrow 1 + O(w^2 e^{-w})$$

and that

$$V(w) \longrightarrow -\frac{27}{16} - \frac{1}{2w} + O(we^{-w})$$

Now use the result of Problem 10–21 to show that

$$\zeta \longrightarrow 27/32 \qquad \text{as } w \longrightarrow \infty$$

10–47. Show that all the integrals in Table 10.8 go to zero as $R \to \infty$ so that $H_{11} = H_{22} \to \zeta^2 - 27\zeta/16$, $H_{12} \to 5\zeta/16$, and $E_{CI} \to \zeta^2 - \zeta$.

10–48. In the Born–Oppenheimer approximation, we assume that because the nuclei are so much more massive than the electrons, the electrons can adjust essentially instantaneously to any nuclear motion, and hence we have a unique and well-defined energy, $E(R)$, at each internuclear separation R. Under this same approximation, $E(R)$ is the internuclear potential and so is the potential field in which the nuclei vibrate. Argue, then, that under the Born–Oppenheimer approximation, the force constant is independent of isotopic substitution. Using the above ideas, and given that the dissociation energy for H_2 is $D_0 = 430.3 \text{ kJ} \cdot \text{mol}^{-1}$ and that the fundamental vibrational frequency ν is $1.32 \times 10^{14} \text{ s}^{-1}$, calculate D_0 and ν for deuterium, D_2. Realize that the observed dissociation energy is given by

$$D_0 = D_e - \frac{1}{2}h\nu$$

where D_e is the value of $E(R)$ at R_{eq}.

References

Slater, J. C. *Quantum Theory of Molecules and Solids*. Vol. 1: *Electronic Structure of Molecules*. McGraw-Hill: New York, 1963.

Mulliken, R. S., Ermler, W. C. *Diatomic Molecules: Results of ab initio Calculations*. Academic Press: New York, 1977.

Szabo, A., Ostlund, N. S. *Modern Quantum Chemistry*. Macmillan Publishing: New York, 1982.

Carrington, A. The Simplest Molecule. *New Scientist,* **53** (Jan. 14, 1989).

Bates, D. R., Ledsham, K., Stewart, A. L. Wave Functions of the Hydrogen Molecular Ion. *Phil. Trans. Roy. Soc.,* **A 246**, 215 (1953).

Kolos, W., Wolniewicz, L. Improved Theoretical Ground-State Energy of the Hydrogen Molecule. *J. Chem. Phys.,* **49**, 404 (1968).

Herzberg, G. Dissociation Energy and Ionization Energy of Molecular Hydrogen. *Phys. Rev. Letters,* **23**, 1081 (1961).

Kolos, W., Roothaan, C. C. J. Accurate Electronic Wave Functions for the H_2 Molecule. *Rev. Mod. Phys.,* **32**, 219 (1960).

McLean, A. D., Weiss, A., Yoshine, M. Configuration Interaction in the Hydrogen Molecule. *Rev. Mod. Phys.,* **32**, 211 (1960).

Abramowitz, M., Stegun, I. *Handbook of Mathematical Functions*. Dover Publications: Mineola, NY, 1965.

Linus Pauling was born in Condon, near Portland, Oregon, on February 28, 1901, and died in 1994. He received his Ph.D. in chemistry in 1925 from the California Institute of Technology for his dissertation on X-ray crystallography of organic compounds and the structure of crystals. After spending a year studying at the University of Munich, he joined the faculty at the California Institute of Technology, where he remained for almost 40 years. Pauling was a pioneer in the application of quantum mechanics to chemistry. His book *The Nature of the Chemical Bond* (1939) was one of the most influential chemistry texts of the twentieth century. In the 1930s, he became interested in biological molecules and developed a structural theory of protein molecules, work that led to demonstrating that sickle cell anemia is caused by a faulty structure in hemoglobin. In the early 1950s, he proposed the alpha helix as the basic structure of proteins. Pauling was awarded the Nobel Prize in Chemistry in 1954 "for his research into the nature of the chemical bond and its application to the elucidation of the structure of complex structures." During the 1950s, Pauling was in the forefront of the fight against nuclear testing, for which he was awarded the Nobel Peace Prize in 1963. From the early 1980s until his death, he was embroiled in the controversy of advocating the use of vitamin C as protection against the common cold and other serious maladies such as cancer.

CHAPTER 11

Qualitative Theory of Chemical Bonding

In the previous chapter, we discussed the chemical bonding in H_2^+ and H_2 in some quantitative detail. In the next chapter, we shall apply the Hartree–Fock–Roothaan method to polyatomic molecules, but before doing so, we shall discuss chemical bonding in a more qualitative manner in this chapter. In the previous chapter, we built a molecular wave function for the ground electronic state of H_2 by placing both electrons in a bonding molecular orbital that we denoted by σ_b. The corresponding electron configuration is σ_b^2. We shall develop this procedure further and construct a set of molecular orbitals and then place electrons into these orbitals in accord with the Pauli exclusion principle and Hund's rules to build up molecular wave functions for diatomic molecules. This procedure is called molecular orbital theory, and you probably have learned some molecular orbital theory in other classes. We shall call the theory qualitative molecular orbital theory because we don't use it to make quantitative predictions like we do in Hartree–Fock–Roothaan theory, but using only rather general arguments and no detailed numerical calculations, we will be able to predict many interesting properties and trends for homonuclear diatomic molecules. For example, we will predict correctly that oxygen is a paramagnetic molecule, but we won't calculate the degree of paramagnetism numerically.

Just as the electron configurations of atoms lead to atomic term symbols (Section 9.9), the electron configurations of molecules lead to molecular term symbols. In Section 11.4 we shall show how to determine molecular term symbols from electron configurations. Molecular term symbols are used to designate the various electronic states of molecules. They also designate the symmetry properties of molecules, which we shall discuss in Section 11.5.

In the rest of the chapter, we discuss a simple theory of bonding in conjugated and aromatic organic molecules that in spite of its simplicity can be used to make many useful predictions. This theory, called Hückel molecular orbital theory, or usually just Hückel theory, focuses on the π electrons of conjugated systems. Although strictly speaking not a qualitative theory, it is used primarily to predict correlations and trends rather than precise numerical results.

11.1 Molecular Orbitals Can Be Ordered According to Their Energies

In this section, we will construct a set of molecular orbitals and assign electrons to them in accord with the Pauli exclusion principle. This procedure will generate electron configurations for molecules similar to those discussed for atoms in Chapter 9. We will illustrate this procedure in some detail for homonuclear diatomic molecules and then present some results for heteronuclear diatomic molecules.

We will use the LCAO–MO approximation and form molecular orbitals as linear combinations of atomic orbitals. Starting with the $1s$ orbitals on each atom (as we did for a minimal basis set for H_2 in the previous chapter), the first two (unnormalized) molecular orbitals we will discuss are

$$\psi_\pm = 1s_A \pm 1s_B \tag{11.1}$$

These two molecular orbitals, which we called σ_b and σ_a in the previous chapter, are σ orbitals as shown in Figure 11.1. Recall that σ orbitals are symmetric about the internuclear axis, or have no nodal planes containing the internuclear (z) axis.

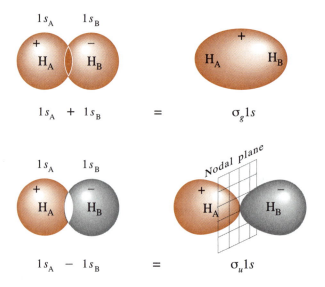

FIGURE 11.1
The linear combination of two $1s$ orbitals to give the bonding ($\sigma 1s$ or $\sigma_g 1s$) and antibonding molecular orbitals ($\sigma^* 1s$ or $\sigma_u 1s$). The bonding orbital, σ_b, concentrates electron density in the region between the two nuclei, whereas the antibonding orbital, σ_a, excludes electron density from that region and even has a nodal plane at the midpoint between the two nuclei.

Because many combinations of atomic orbitals lead to σ orbitals, we must identify which atomic orbitals constitute a particular σ orbital. Molecular orbitals constructed from atomic $1s$ orbitals are denoted by $\sigma 1s$. Because a $\sigma 1s$ orbital may be a bonding orbital or an antibonding orbital, we need to distinguish between the two possibilities.

There are two common ways to make this distinction. One way is to use the notation $\sigma_g 1s$ and $\sigma_u 1s$, which, as in the previous chapter, indicates gerade and ungerade symmetry. We emphasize here, however, that the two molecular orbitals are built entirely from $1s$ orbitals. The other way is to use a superscript asterisk to denote an antibonding orbital, so that the two orbitals in Figure 11.1 are denoted by $\sigma 1s$ (bonding) and $\sigma^* 1s$ (antibonding). Thus, we have two designations of the $\sigma 1s$ orbitals: $\sigma 1s$ and $\sigma^* 1s$, or $\sigma_g 1s$ and $\sigma_u 1s$. Both designations are commonly used, but we will use the g, u notation for molecular orbitals. Note that in the case of $1s$ orbitals, the gerade symmetry leads to a bonding orbital and the ungerade symmetry leads to an antibonding orbital.

Molecular orbitals constructed from other kinds of atomic orbitals are generated in a similar way. We saw in the previous chapter that accurate molecular orbitals are formed from linear combinations of many atomic orbitals, but here we use the simplest approximation, where only atomic orbitals of equal energies are combined to give molecular orbitals. Following the above approach for the $\sigma 1s$ orbitals, the next combinations we consider are $2s_A \pm 2s_B$. These two molecular orbitals look similar to those plotted in Figure 11.1 but are larger in extent because a $2s$ orbital is larger than a $1s$ orbital. In addition, there are spherical nodal planes about each nucleus reflecting the radial nodes of the individual $2s$ wave functions (see Figure 7.2). Following the notation introduced above, the two molecular orbitals $2s_A \pm 2s_B$ are designated $\sigma_g 2s$ and $\sigma_u 2s$. Because an atomic $2s$ orbital is associated with a higher energy than an atomic $1s$ orbital, the energy of the $\sigma_g 2s$ molecular orbital will be higher than that of the $\sigma_g 1s$ molecular orbital. This difference can be demonstrated rigorously by calculating the energies associated with these molecular orbitals, as was done for the $\sigma_g 1s$ and $\sigma_u 1s$ molecular orbitals in Section 10.8. In addition, bonding orbitals are lower in energy than corresponding antibonding orbitals. This then gives an energy ordering $\sigma_g 1s < \sigma_u 1s < \sigma_g 2s < \sigma_u 2s$ for the four molecular orbitals discussed so far.

Now consider linear combinations of the $2p$ orbitals. Although a $2p$ orbital has the same energy as a $2s$ orbital in the case of atomic hydrogen, this is not true for other atoms, in which case $E_{2p} > E_{2s}$. As a result, the molecular orbitals built from $2p$ orbitals will have a higher energy than the $\sigma_g 2s$ and $\sigma_u 2s$ orbitals. Defining the internuclear axis to be the z axis, Figures 11.2 and 11.3 show that the atomic $2p_z$ orbitals combine to give a differently shaped molecular orbital than that made by combining either the atomic $2p_x$ or $2p_y$ orbitals. The two molecular orbitals $2p_{z,A} \pm 2p_{z,B}$ are cylindrically symmetric about the internuclear axis and therefore are σ orbitals. Once again, both a bonding orbital and an antibonding molecular orbital are generated, and the two orbitals are designated by $\sigma_g 2p_z$ and $\sigma_u 2p_z$, respectively.

Unlike the $2p_z$ orbitals, the $2p_x$ and $2p_y$ orbitals combine to give molecular orbitals that are not cylindrically symmetric about the internuclear axis. Figure 11.3 shows that the y–z plane is a nodal plane in both the bonding and antibonding combinations of the $2p_x$ orbitals. As we learned in the previous chapter, molecular orbitals with one nodal plane that contains the internuclear axis are called π orbitals. The bonding and antibonding molecular orbitals that arise from a combination of the $2p_x$ orbitals are denoted $\pi_u 2p_x$ and $\pi_g 2p_x$, respectively. Note that the antibonding orbital $\pi_g 2p_x$ also has a second nodal plane perpendicular to the internuclear axis that is not present in the $\pi_u 2p_x$ bonding orbital. The $2p_y$ orbitals combine in a similar manner, and the resulting

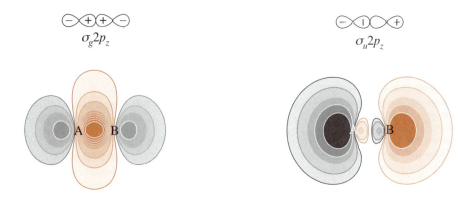

FIGURE 11.2
The $\sigma_g 2p_z$ and $\sigma_u 2p_z$ molecular orbitals formed from linear combinations of the $2p_z$ atomic orbitals. Note that the bonding orbital ($\sigma_g 2p_z$) corresponds to the combination, $2p_{zA} - 2p_{zB}$, and that the antibonding orbital ($\sigma_u 2p_z$) corresponds to the combination, $2p_{zA} + 2p_{zB}$, in contrast to the corresponding combinations of s orbitals. The orange regions correspond to positive values, and the gray and black regions correspond to negative values of the molecular orbitals, respectively.

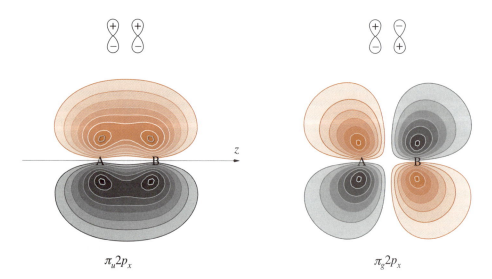

FIGURE 11.3
The bonding $\pi_u 2p_x$ and antibonding $\pi_g 2p_x$ molecular orbitals formed from linear combinations of the $2p_x$ atomic orbitals. The orange regions correspond to positive values, and the gray and black regions correspond to negative values of the molecular orbitals, respectively.

molecular orbitals look like those in Figure 11.3 but are directed along the y axis instead of the x axis. The x–z plane is the nodal plane for the $\pi_u 2p_y$ and $\pi_g 2p_y$ orbitals. Because the $2p_x$ and $2p_y$ orbitals have identical energy and the resulting molecular orbitals differ only in their spatial orientation, the orbitals $\pi_u 2p_x$ and $\pi_u 2p_y$ are degenerate, as are $\pi_g 2p_x$ and $\pi_g 2p_y$. Note that unlike the bonding σ orbitals, the bonding π orbitals have ungerade symmetry and the antibonding π orbitals have gerade symmetry.

The set of molecular orbitals that we have developed so far is enough to discuss the electron configurations of the homonuclear diatomic molecules H_2 through Ne_2. First, however, we need to know the order of these molecular orbitals with respect to energy. The order of the various molecular orbitals depends upon the atomic number (nuclear charge) on the nuclei. As the atomic number increases from three for lithium to nine for fluorine, the energies of the $\sigma_g 2p_z$ and $\pi_u 2p_x$, $\pi_u 2p_y$ orbitals approach each other and actually interchange order in going from N_2 to O_2, as shown in Figure 11.4. The somewhat complicated ordering shown in Figure 11.4, which is consistent with calculations and experimental spectroscopic observations, is reminiscent of the ordering of the energies of atomic orbitals as the atomic number increases.

We will now use Figure 11.4 to deduce electron configurations of the homonuclear diatomic molecules H_2 through Ne_2. We shall do this by placing electrons into these orbitals in accord with the Pauli exclusion principle and Hund's rules, just as we did

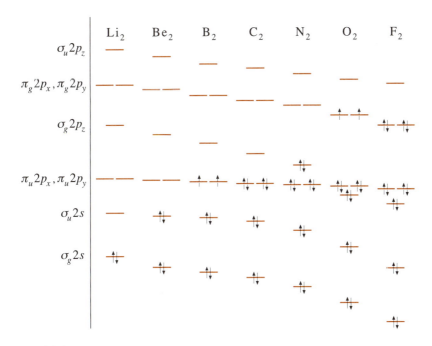

FIGURE 11.4
The relative energies (not to scale) of the molecular orbitals for the homonuclear diatomic molecules Li_2 through F_2. The $\pi_u 2p_x$ and $\pi_u 2p_y$ orbitals are degenerate, as are the $\pi_g 2p_x$ and $\pi_g 2p_y$ orbitals.

for many-electron atoms in Chapter 9. Recall from Section 9.11 that Hund's rules say that if orbitals are degenerate, then we place one electron into each orbital with parallel spins before we place two electrons into any one of the orbitals.

11.2 Electrons Are Placed into Molecular Orbitals in Accord with the Pauli Exclusion Principle

For H_2 through He_2, we need to consider only the $\sigma_g 1s$ and $\sigma_u 1s$ orbitals, the two molecular orbitals of lowest energy. Consider the ground-state electron configuration of H_2. According to the Pauli exclusion principle, two electrons of opposite spin are placed in the $\sigma_g 1s$ orbital. The electron configuration of H_2 is written as $(\sigma_g 1s)^2$. The two electrons in the bonding orbital constitute a bonding pair of electrons and account for the single bond of H_2.

Now consider He_2. This molecule has four electrons, and its ground-state electron configuration is $(\sigma_g 1s)^2(\sigma_u 1s)^2$. This assignment gives He_2 one pair of bonding electrons and one pair of antibonding electrons. Electrons in bonding orbitals tend to draw nuclei together, whereas those in antibonding orbitals tend to push them apart. The result of these opposing forces is that an electron in an antibonding orbital approximately cancels the effect of an electron in a bonding orbital. Thus, in the case of He_2, there is no net bonding. Simple molecular orbital theory predicts that diatomic helium does not exist. (Nevertheless, a very weakly bound molecule, with a bond energy of about $0.01 \, kJ \cdot mol^{-1}$ and a bond length of about 6000 pm, was discovered in 1993.)

The above results are formalized by defining a quantity called *bond order* by

$$\text{bond order} = \frac{1}{2}\left[\left(\begin{array}{c}\text{number of electrons} \\ \text{in bonding orbitals}\end{array}\right) - \left(\begin{array}{c}\text{number of electrons} \\ \text{in antibonding orbitals}\end{array}\right)\right] \quad (11.2)$$

Single bonds have a bond order of one; double bonds have a bond order of two; and so on. The bond order for He_2 is zero. As the following example shows, the bond order does not have to be a whole number; it can be a half-integer.

EXAMPLE 11–1
Determine the bond order of He_2^+.

SOLUTION: The ground-state electron configuration of He_2^+ is $(\sigma_g 1s)^2(\sigma_u 1s)^1$, and so the bond order is

$$\text{bond order} = \frac{1}{2}[(2) - (1)] = \frac{1}{2}$$

Table 11.1 gives the molecular orbital theory results for H_2^+, H_2, He_2^+, and He_2. The qualitative theory that we are developing here does not give us the numerical value

TABLE 11.1
Molecular Properties of H_2^+, H_2, He_2^+, and He_2

Species	Number of electrons	Ground-state electron configuration	Bond order	Bond length/pm	Binding energy/ $kJ \cdot mol^{-1}$
H_2^+	1	$(\sigma_g 1s)^1$	1/2	106	269
H_2	2	$(\sigma_g 1s)^2$	1	75	458
He_2^+	3	$(\sigma_g 1s)^2(\sigma_u 1s)^1$	1/2	108	241
He_2	4	$(\sigma_g 1s)^2(\sigma_u 1s)^2$	0	≈ 6000	≈ 0.01

of the bond lengths and bond energies in Table 11.1, but it does give us the trends. We have to use the Hartree–Fock–Roothaan method or some other numerical method to obtain actual numerical results.

Now let's consider the homonuclear diatomic molecules Li_2 through Ne_2. Each lithium atom has three electrons, so the ground-state electron configuration for Li_2 is $(\sigma_g 1s)^2(\sigma_u 1s)^2(\sigma_g 2s)^2$, and the bond order is one. We predict that a diatomic lithium molecule is stable relative to two separated lithium atoms. Lithium vapor is known to contain diatomic lithium molecules, which have a bond length of 267 pm and a bond energy of 99.8 $kJ \cdot mol^{-1}$. (See Problems 11–1 and 11–2.)

Hartree–Fock–Roothaan contour maps of the electron density in the individual molecular orbitals and the total electron density in Li_2 are shown in Figure 11.5. Each line in the contour maps corresponds to a fixed value of electron density. Contours are generally plotted for fixed increments of electron density. Thus, the distance between contours provides information about how rapidly the electron density is changing. Figure 11.5 shows clearly that there is little difference between the electron densities of the $\sigma_g 1s$ and $\sigma_u 1s$ molecular orbitals of Li_2 and the electron densities of the two $1s$ atomic orbitals of the individual lithium atoms. This observation underlies the common assumption that only electrons in the valence shell need be included in qualitative discussions of chemical bonding. In the case of Li_2, the $1s$ electrons are held tightly about each nucleus and do not participate significantly in the bonding. The ground-state electron configuration of Li_2 can therefore be written as $KK(\sigma_g 2s)^2$, where K represents the filled $n = 1$ shell on a lithium atom.

With increasing nuclear charge across the second row of the periodic table, the $1s$ electrons are held even more tightly than are the $1s$ electrons in lithium. Thus, to a good approximation, only the valence electrons need to be considered in writing electron configurations of diatomic molecules beyond He_2. The $\sigma_g 1s$ and $\sigma_u 1s$ molecular orbitals are equivalent to the filled K shell on each atom.

Diatomic boron is a particularly interesting case. This molecule has a total of six valence electrons (three from each boron atom). According to Figure 11.4, the ground-state electron configuration of B_2 is $KK(\sigma_g 2s)^2(\sigma_u 2s)^2(\pi_u 2p_x)^1(\pi_u 2p_y)^1$. As in the atomic case, Hund's rules apply, so we place one electron into each of the degenerate $\pi_u 2p$ orbitals such that their spins are parallel. Experimental measurements have determined that B_2 does indeed have two unpaired electrons (i.e., is paramagnetic).

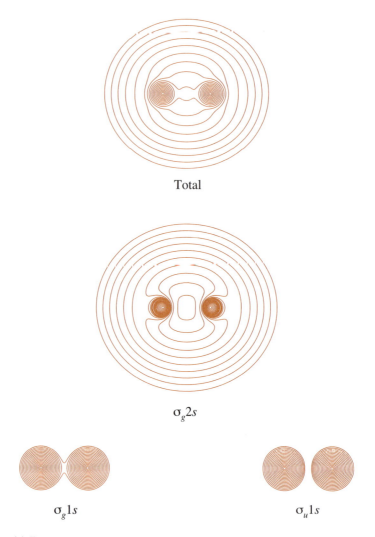

FIGURE 11.5

Electron-density contours for the molecular orbitals of Li_2. Note that the electrons in the $\sigma_g 1s$ and $\sigma_u 1s$ orbitals are tightly held around the nucleus and do not participate to any large extent in the bonding. The electrons in the $\sigma_g 2s$ orbital are the ones responsible for the bonding in Li_2.

EXAMPLE 11–2

Use molecular orbital theory to predict whether or not diatomic carbon exists.

SOLUTION: The ground-state electron configuration of C_2 is $KK(\sigma_g 2s)^2(\sigma_u 2s)^2$ $(\pi_u 2p_x)^2(\pi_u 2p_y)^2$, giving a bond order of two. Thus, we predict that diatomic carbon exists. Experimental measurements have determined that C_2 has no unpaired electrons (i.e., is diamagnetic). The correct prediction of the magnetic properties of B_2 and C_2

corroborates the ordering of the molecular orbital energies given in Figure 11.4 for $Z = 5$ and $Z = 6$. (See Problem 11–4.)

The prediction of the correct electron configuration of an oxygen molecule is one of the most impressive successes of qualitative molecular orbital theory. Oxygen molecules are paramagnetic; experimental measurements indicate that the net spin of the oxygen molecule corresponds to two unpaired electrons of the same spin. The amount of oxygen in the air can be monitored by measuring the paramagnetism of a sample of the air. Because oxygen is the only major component in air that is paramagnetic, the measured paramagnetism of air is directly proportional to the amount of oxygen present. Linus Pauling developed this method, which was used to monitor oxygen levels in submarines and airplanes in World War II, and it is still used by physicians to monitor the oxygen content in blood during anesthesia.

Let's see what molecular orbital theory has to say about this. The predicted ground-state electron configuration of O_2 is $KK(\sigma_g 2s)^2(\sigma_u 2s)^2(\sigma_g 2p_z)^2(\pi_u 2p_x)^2(\pi_u 2p_y)^2$ $(\pi_g 2p_x)^1(\pi_g 2p_y)^1$. Because the $\pi_g 2p_x$ and $\pi_g 2p_y$ orbitals are degenerate, according to Hund's rule, we place one electron in each orbital such that the spins of the electrons are parallel. The occupation of the other molecular orbitals, $KK(\sigma_g 2s)^2(\sigma_u 2s)^2(\sigma_g 2p_z)^2$ $(\pi_u 2p_x)^2(\pi_u 2p_y)^2$, generates no net spin because all these occupied molecular orbitals contain two spin-paired electrons, so we predict that O_2 in its ground state has two unpaired electrons. Thus, the molecular orbital configuration correctly accounts for the paramagnetic behavior of the O_2 molecule.

We can use molecular orbital theory to predict *relative* bond lengths and bond energies, as shown in Example 11–3.

EXAMPLE 11–3
Discuss the relative bond lengths and bond energies of O_2^+, O_2, O_2^-, and O_2^{2-}.

SOLUTION: O_2 has 12 valence electrons. According to Figure 11.4, the ground-state electron configurations and bond orders for these species are as follows:

	Ground-state electron configuration	Bond order
O_2^+	$KK(\sigma_g 2s)^2(\sigma_u 2s)^2(\sigma_g 2p_z)^2(\pi_u 2p_x)^2(\pi_u 2p_y)^2(\pi_g 2p_x)^1$	$2\frac{1}{2}$
O_2	$KK(\sigma_g 2s)^2(\sigma_u 2s)^2(\sigma_g 2p_z)^2(\pi_u 2p_x)^2(\pi_u 2p_y)^2(\pi_g 2p_x)^1(\pi_g 2p_y)^1$	2
O_2^-	$KK(\sigma_g 2s)^2(\sigma_u 2s)^2(\sigma_g 2p_z)^2(\pi_u 2p_x)^2(\pi_u 2p_y)^2(\pi_g 2p_x)^2(\pi_g 2p_y)^1$	$1\frac{1}{2}$
O_2^{2-}	$KK(\sigma_g 2s)^2(\sigma_u 2s)^2(\sigma_g 2p_z)^2(\pi_u 2p_x)^2(\pi_u 2p_y)^2(\pi_g 2p_x)^2(\pi_g 2p_y)^2$	1

We predict that the bond lengths decrease and the bond energies increase with increasing bond order. This prediction is in nice agreement with the experimental values, which are as follows:

	Bond order	Bond length/pm	Bond energy/kJ·mol^{-1}
O_2^+	$2\frac{1}{2}$	112	643
O_2	2	121	494
O_2^-	$1\frac{1}{2}$	135	395
O_2^{2-}	1	149	

Note that removing an electron from O_2 produces a stronger bond, in agreement with the MO prediction.

Figure 11.6 illustrates the correlation between the predicted bond orders and the experimentally measured bond lengths and bond energies of the homonuclear molecules B_2 through Ne_2. The results for the diatomic molecules of the elements in the second row of the periodic table are summarized in Table 11.2.

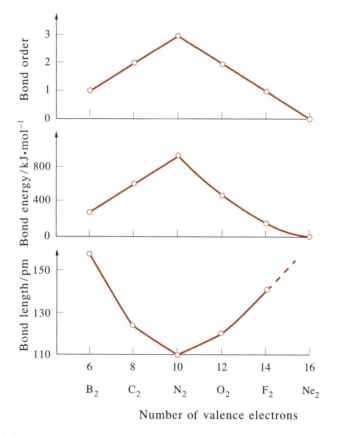

FIGURE 11.6
Plots of various bond properties for the homonuclear diatomic molecules B_2 through Ne_2.

TABLE 11.2
The Ground-State Electron Configurations and Various Physical Properties of Homonuclear Diatomic Molecules of Elements in the Second Row of the Periodic Table

Species	Ground-state electron configuration	Bond order	Bond length/pm	Bond energy/ $kJ \cdot mol^{-1}$
Li_2	$KK(\sigma_g 2s)^2$	1	267	99.8
Be_2	$KK(\sigma_g 2s)^2(\sigma_u 2s)^2$	0	245	≈ 9
B_2	$KK(\sigma_g 2s)^2(\sigma_u 2s)^2(\pi_u 2p_x)^1(\pi_u 2p_y)^1$	1	159	289
C_2	$KK(\sigma_g 2s)^2(\sigma_u 2s)^2(\pi_u 2p_x)^2(\pi_u 2p_y)^2$	2	124	599
N_2	$KK(\sigma_g 2s)^2(\sigma_u 2s)^2(\pi_u 2p_x)^2(\pi_u 2p_y)^2(\sigma_g 2p_z)^2$	3	110	942
O_2	$KK(\sigma_g 2s)^2(\sigma_u 2s)^2(\sigma_g 2p_z)^2(\pi_u 2p_x)^2(\pi_u 2p_y)^2$ $(\pi_g 2p_x)^1(\pi_g 2p_y)^1$	2	121	494
F_2	$KK(\sigma_g 2s)^2(\sigma_u 2s)^2(\sigma_g 2p_z)^2(\pi_u 2p_x)^2(\pi_u 2p_y)^2$ $(\pi_g 2p_x)^2(\pi_g 2p_y)^2$	1	141	154
Ne_2	$KK(\sigma_g 2s)^2(\sigma_u 2s)^2(\sigma_g 2p_z)^2(\pi_u 2p_x)^2(\pi_u 2p_y)^2$ $(\pi_g 2p_x)^2(\pi_g 2p_y)^2(\sigma_u 2p_z)^2$	0		

The idea of atomic orbitals and molecular orbitals is rather abstract and sometimes appears far removed from reality. It so happens, however, that the electron configurations of molecules can be demonstrated experimentally. The approach used is very similar to the photoelectric effect discussed in Chapter 1. If high-energy electromagnetic radiation is directed into a gas, electrons are ejected from the molecules in the gas. The energy required to eject an electron from a molecule, called the *binding energy*, is a direct measure of how strongly bound the electron is within the molecule. The binding energy of an electron within a molecule depends upon the molecular orbital the electron occupies; the lower the energy of the molecular orbital, the more energy needed to remove an electron from that molecular orbital.

The measurement of the energies of the electrons ejected by radiation incident on gaseous molecules is called *photoelectron spectroscopy*. A photoelectron spectrum of N_2 is shown in Figure 11.7. According to Figure 11.4, the ground-state configuration of N_2 is $KK(\sigma_g 2s)^2(\sigma_u 2s)^2(\pi_u 2p_x)^2 (\pi_u 2p_y)^2(\sigma_g 2p_z)^2$. The peaks in the photoelectron spectrum correspond to the energies of occupied molecular orbitals. Photoelectron spectra provide striking experimental support for the molecular orbital picture being developed here.

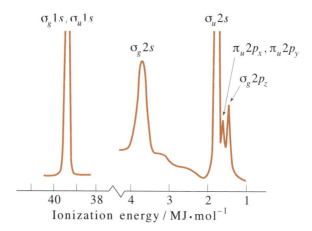

FIGURE 11.7
The photoelectron spectrum of N_2. The peaks in this plot are caused by electrons being ejected from various molecular orbitals.

11.3 Molecular Orbital Theory Also Applies to Heteronuclear Diatomic Molecules

The molecular orbital theory we have developed can be extended to heteronuclear diatomic molecules. It is important to realize that the energies of the atomic orbitals on the two atoms from which the molecular orbitals are constructed will now be different. This difference must be considered in light of the approximation made earlier that only orbitals of equal energy combine to give molecular orbitals. For small changes in atomic number, the energy difference for the same atomic orbital on the two bonded atoms is small (e.g., CO and NO). For many heteronuclear diatomic molecules (e.g., HF and HCl), however, the energies of the respective atomic orbitals can be significantly different, and we will need to rethink which atomic orbitals are involved in constructing the molecular orbitals for such molecules.

Let's consider a cyanide ion (CN^-) first. The atomic numbers of carbon (6) and nitrogen (7) differ by only one unit, so the energy ordering shown in Figure 11.4 may still be valid. The total number of valence electrons is 10 (carbon has four electrons and nitrogen has five electrons in the $n = 2$ shell), and the overall charge on the ion is -1. Accordingly, the ground-state electron configuration of CN^- is predicted to be $KK(\sigma_g 2s)^2(\sigma_u 2s)^2(\pi_u 2p_x)^2(\pi_u 2p_y)^2(\sigma_g 2p_z)^2$, with a bond order of three.

EXAMPLE 11–4
Discuss the bonding in a carbon monoxide molecule, CO.

SOLUTION: A CO molecule has a total of 10 valence electrons. Note that CO is isoelectronic with N_2. The ground-state electron configuration of CO is therefore

$KK(\sigma_g 2s)^2 (\sigma_u 2s)^2 (\pi_u 2p_x)^2 (\pi_u 2p_y)^2 (\sigma_g 2p_z)^2$, so the bond order is three. Because both N_2 and CO have triple bonds and because all three atoms (N, O, C) are approximately the same size, we expect that the bond length and bond energy of CO are comparable with those of N_2. The experimental data are as follows:

	Bond length/pm	Bond energy/kJ·mol^{-1}
N_2	110	942
CO	113	1071

The bond strength of CO is one of the largest known for diatomic molecules.

Figure 11.8 presents the photoelectron spectrum of CO. The energies of the molecular orbitals are revealed nicely by these data. In addition, the photoelectron spectrum exhibits peaks characteristic of the atomic $1s$ orbitals on carbon and oxygen. Notice the high binding energy of the $1s$ atomic orbitals. This energy is a result of their being close to the nuclei, and these data further verify that the $1s$ electrons do not play a significant role in the bonding of these molecules.

Now consider the diatomic molecule HF. This molecule illustrates the case in which the valence electrons on the atoms occupy different electron shells. The energies of the valence electrons in the $2s$ and $2p$ atomic orbitals on the fluorine atom are $-1.572\,E_h$ and $-0.730\,E_h$, respectively (Problem 11–20), and the energy of the valence electron in the $1s$ atomic orbital on the hydrogen atom is $-0.500\,E_h$. Because the $2p$ atomic orbitals

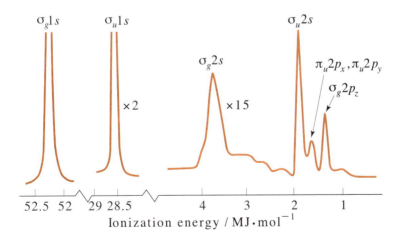

FIGURE 11.8
The photoelectron spectrum of CO. The energies associated with various molecular orbitals are identified. The $\sigma_g 1s$ and $\sigma_u 1s$ orbitals are essentially the $1s$ electrons of the oxygen and carbon atoms, respectively. The relatively large binding energies of these electrons indicate that they are held tightly by the nuclei and play no role in bonding.

take on values of 1, 0, and -1. We now construct a table of all possible combinations of (m_{l1}, m_{s1}) and (m_{l2}, m_{s2}) that correspond to the possible values of M_L and M_S.

		M_S	
	1	0	-1
2	~~$1^+, 1^+$~~	$1^+, 1^-$	~~$1^-, 1^-$~~
M_L 0	$1^+, -1^+$	$1^+, -1^-; 1^-, -1^+$	$1^-, -1^-$
-2	~~$-1^+, -1^+$~~	$-1^+, -1^-$	~~$-1^-, -1^-$~~

In the entries of the above table, the superscripts $+$ and $-$ are used to designate the spin quantum numbers of $m_s = +1/2$ and $m_s = -1/2$, respectively. The numbers in each entry are the corresponding m_l quantum numbers. For example, the entry $1^+, -1^+$ corresponds to $m_{l1} = 1$, $m_{s1} = 1/2$ and $m_{l2} = -1$, $m_{s2} = 1/2$, or $M_L = m_{l1} + m_{l2} = 0$ and $M_S = m_{s1} + m_{s2} = 1$. Not all the entries in the above table are allowed. The Pauli exclusion principle requires that no two electrons in the same orbitals have the same set of quantum numbers; hence the configurations $1^+, 1^+; 1^-, 1^-; -1^+, -1^+;$ and $-1^-, -1^-$ do not correspond to allowed quantum states and are crossed out. This leaves the following combinations of (m_{l1}, m_{s1}) and (m_{l2}, m_{s2}) from which the allowed term symbols are to be derived.

		M_S	
	1	0	-1
2		$1^+, 1^-$	
M_L 0	$1^+, -1^+$	$1^+, -1^-; 1^-, -1^+$	$1^-, -1^-$
-2		$-1^+, -1^-$	

Looking across the middle row, we have three configurations $1^+, -1^+; 1^+, -1^-$ (or $1^-, -1^+$); and $1^-, -1^-$ that correspond to $M_L = 0$ and $M_S = 1, 0, -1$, or a $^3\Sigma$ state. This leaves the following:

		M_S	
	1	0	-1
2		$1^+, 1^-$	
M_L 0		$1^-, -1^+$	
-2		$-1^+, -1^-$	

Two of the remaining terms in the column ($1^+, 1^-$ and $-1^+, -1^-$) correspond to $M_L = 2$ and -2 ($|M_L| = 2$) and $M_S = 0$, or to a $^1\Delta$ state. The remaining term ($1^-, -1^+$) corresponds to $M_L = 0$ and $M_S = 0$ or to a $^1\Sigma$ state. We find that there are three possible molecular states, $^1\Delta$, $^3\Sigma$, and $^1\Sigma$, for B_2. Because Hund's rules apply to molecular electronic states as well as to atomic electronic states, the state with the largest spin multiplicity will be the ground state of B_2. Thus, we predict that the ground state of B_2 is a $^3\Sigma$ state.

EXAMPLE 11–6
Deduce the term symbols for the ground states of O_2 and O_2^+.

SOLUTION: The ground-state electron configuration of O_2 is (Table 11.2) $(1\sigma_g)^2$ $(1\sigma_u)^2(2\sigma_g)^2(2\sigma_u)^2(3\sigma_g)^2(1\pi_{ux})^2(1\pi_{uy})^2(1\pi_{gx})^1(1\pi_{gy})^1$. The only electrons that we need to consider in determining the molecular term symbol are the two that occupy the $1\pi_g$ orbitals. This is identical to what we just discussed for the molecule B_2. Thus, we know that according to Hund's rule, the term symbol for the ground state of O_2 is $^3\Sigma$.

The ground-state electron configuration of O_2^+ is (Example 11–3) $(1\sigma_g)^2(1\sigma_u)^2$ $(2\sigma_g)^2(2\sigma_u)^2(3\sigma_g)^2(1\pi_{ux})^2(1\pi_{uy})^2(1\pi_{gx})^1$. The only electron we need to consider in determining the term symbol is the one electron in the $1\pi_g$ orbital. The allowed values of m_l and m_s for an electron in a $1\pi_g$ orbital are $m_l = \pm 1$ and $m_s = \pm 1/2$. These values correspond to $M_L = 1$ and $M_S = 1/2$, or a term symbol of $^2\Pi$.

11.5 Molecular Term Symbols Designate the Symmetry Properties of Molecular Wave Functions

Term symbols are also used to denote symmetry properties of molecular wave functions. For homonuclear diatomic molecules, inversion through the point midway between the two nuclei leaves the nuclear configuration of the molecule unchanged and so its constituent molecular orbitals have gerade or ungerade symmetry. Because a molecular electronic wave function consists of products of molecular orbitals of g and u symmetry, the overall molecular wave function must be either gerade or ungerade. Consider the simplest case of the product of two molecular orbitals. If both orbitals are gerade, the product is gerade. If both orbitals are ungerade, the product is also gerade because the product of two odd functions is an even function. If the two orbitals have opposite symmetry, the product is ungerade. The resultant symmetry is indicated by either a g or a u right subscript on the molecular term symbol. For example, the ground-state electron configuration of O_2 is $(1\sigma_g)^2(1\sigma_u)^2(2\sigma_g)^2(2\sigma_u)^2(3\sigma_g)^2(1\pi_{ux})^2(1\pi_{uy})^2(1\pi_{gx})^1(1\pi_{gy})^1$. As usual, we can ignore completely filled orbitals and focus on $(1\pi_{gx})^1(1\pi_{gy})^1$. According to Figures 11.2 and 11.3, the symmetry of $(1\pi_g)^1(1\pi_g)^1$ is $g \cdot g = g$, so the molecular term symbol for the ground electronic state of O_2 is $^3\Sigma_g$. Similarly, that for O_2^+ is $^2\Pi_g$.

EXAMPLE 11–7
Determine the symmetry designation (g or u) for the term symbol of the ground-state electron configuration of B_2.

SOLUTION: The ground-state electron configuration is $(1\sigma_g)^2(1\sigma_u)^2(2\sigma_g)^2(2\sigma_u)^2$ $(3\sigma_g)^2(1\pi_{ux})^1(1\pi_{uy})^1$, corresponding to a term symbol of $^3\Sigma$. As usual, we can ignore completely occupied orbitals, so the product of the symmetry of the molecular orbitals occupied by the two unpaired electrons is $u \cdot u = g$, so the term symbol is $^3\Sigma_g$.

The term symbols for heteronuclear diatomic molecules do not have a g or u designation because these molecules do not possess inversion symmetry. In addition to the g/u designation on the term symbols for homonuclear diatomic molecules, Σ electronic states are labeled with a $+$ or $-$ right superscript to indicate the behavior of the molecular wave function when it is reflected through a plane containing the nuclei. Because σ orbitals are symmetric about the internuclear axis, they do not change sign when they are reflected through a plane containing the two nuclei. Figure 11.11 shows that one of the doubly degenerate π_u orbitals changes sign and the other does not. Similarly, one of the doubly degenerate π_g orbitals changes sign and the other one does not (see Figure 11.11). Using these observations, we can determine whether or not a Σ electronic state is labeled with a $+$ or $-$ superscript.

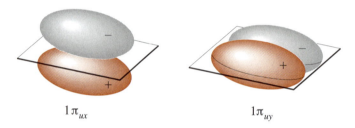

$1\pi_{ux}$ $1\pi_{uy}$

FIGURE 11.11
The behavior of the two $1\pi_u$ orbitals with respect to a plane containing the two nuclei, which we arbitrarily choose as the $y–z$ plane. (See Figure 11.3.)

EXAMPLE 11–8
Determine the complete molecular term symbol of the ground state of O_2.

SOLUTION: According to Example 11–6, the molecular term symbol of O_2 without the $\pm$ designation is $^3\Sigma_g$. The electron configuration is (filled orbitals) $(1\pi_{gx})^1(1\pi_{gy})^1$, so the symmetry with respect to a reflection through the $x–z$ plane is $(+)(-) = (-)$. Therefore, the complete term symbol for the ground state of O_2 is $^3\Sigma_g^-$.

EXAMPLE 11–9
Determine the sign designation $(+)$ or $(-)$ for the ground-state electron configuration of He_2^+.

SOLUTION: The ground-state electron configuration of He_2^+ is $(1\sigma_g)^2(1\sigma_u)^1$, corresponding to a term symbol of $^1\Sigma_u$. Because the $1\sigma_g$ and $1\sigma_u$ orbitals are unchanged upon reflection through a plane containing the two nuclei, the total wave function is unchanged. As a result, the complete term symbol for the ground state of He_2^+ is $^1\Sigma_u^+$.

Table 11.3 lists the term symbols of the ground states of a number of homonuclear diatomic molecules and Problem 11–23 involves the determination of these term symbols.

TABLE 11.3
The Ground-State Electron Configurations and Term Symbols for the First- and Second-Row Homonuclear Diatomic Molecules. (Compare these electron configurations to those given in Table 11.2.)

Molecule	Electron Configuration	Term Symbol
H_2^+	$(1\sigma_g)^1$	$^2\Sigma_g^+$
H_2	$(1\sigma_g)^2$	$^1\Sigma_g^+$
He_2^+	$(1\sigma_g)^2(1\sigma_u)^1$	$^2\Sigma_u^+$
Li_2	$(1\sigma_g)^2(1\sigma_u)^2(2\sigma_g)^2$	$^1\Sigma_g^+$
B_2	$(1\sigma_g)^2(1\sigma_u)^2(2\sigma_g)^2(2\sigma_u)^2(1\pi_u)^1(1\pi_u)^1$	$^3\Sigma_g^-$
C_2	$(1\sigma_g)^2(1\sigma_u)^2(2\sigma_g)^2(2\sigma_u)^2(1\pi_u)^2(1\pi_u)^2$	$^1\Sigma_g^+$
N_2^+	$(1\sigma_g)^2(1\sigma_u)^2(2\sigma_g)^2(2\sigma_u)^2(1\pi_u)^2(1\pi_u)^2(3\sigma_g)^1$	$^2\Sigma_g^+$
N_2	$(1\sigma_g)^2(1\sigma_u)^2(2\sigma_g)^2(2\sigma_u)^2(1\pi_u)^2(1\pi_u)^2(3\sigma_g)^2$	$^1\Sigma_g^+$
O_2^+	$(1\sigma_g)^2(1\sigma_u)^2(2\sigma_g)^2(2\sigma_u)^2(3\sigma_g)^2(1\pi_u)^2(1\pi_u)^2(1\pi_g)^1$	$^2\Pi_g$
O_2	$(1\sigma_g)^2(1\sigma_u)^2(2\sigma_g)^2(2\sigma_u)^2(3\sigma_g)^2(1\pi_u)^2(1\pi_u)^2(1\pi_g)^1(1\pi_g)^1$	$^3\Sigma_g^-$
F_2	$(1\sigma_g)^2(1\sigma_u)^2(2\sigma_g)^2(2\sigma_u)^2(3\sigma_g)^2(1\pi_u)^2(1\pi_u)^2(1\pi_g)^2(1\pi_g)^2$	$^1\Sigma_g^+$

So far we have considered mostly the ground electronic states of diatomic molecules. As we have seen, the electron configuration of the ground electronic state of H_2 is $(1\sigma_g)^2$, whose molecular term symbol is $^1\Sigma_g^+$. The first excited state has the electron configuration $(1\sigma_g)^1(1\sigma_u)^1$, which, as Example 11–10 shows, gives rise to the term symbols $^1\Sigma_u^+$ and $^3\Sigma_u^+$.

EXAMPLE 11–10
Show that the electron configuration $(1\sigma_g)^1(1\sigma_u)^1$ gives rise to the term symbols $^1\Sigma_u^+$ and $^3\Sigma_u^+$.

SOLUTION: The values of m_l are 0 for both electrons, so $M_L = 0$. The possible values of m_{s1} and m_{s2} are $m_{s1} = \pm1/2$ and $m_{s2} = \pm1/2$, respectively, and so $M_S = 1, 0, -1$. We now construct a table of all possible combinations of (m_{l1}, m_{s1}) and (m_{l2}, m_{s2}) that correspond to the possible values of M_L and M_S.

		M_S	
	1	0	-1
M_L 0	$0^+, 0^+$	$0^+, 0^-; 0^-, 0^+$	$0^-, 0^-$

Looking across the table, we see that the entries $0^+, 0^+$; $0^+, 0^-$ (or $0^-, 0^+$); and $0^-, 0^-$ correspond to $M_L = 0$ and $M_S = 1, 0$, and -1, implying that $S = 1$ and giving a $^3\Sigma$ state. The remaining entry $0^-, 0^+$ (or $0^+, 0^-$) corresponds to $M_L = 0$ and $M_S = 0$, or a $^1\Sigma$ state.

The product $1\sigma_g \times 1\sigma_u$ leads to a u state, so we have the states $^3\Sigma_u$ and $^1\Sigma_u$. Furthermore, both the σ_g and σ_u orbitals are symmetric with respect to a reflection through a plane containing the two nuclei, so the complete molecular term symbols are $^3\Sigma_u^+$ and $^1\Sigma_u^+$.

Figure 11.12 shows the internuclear potential energy curves of the ground state and two excited states of H_2. Note that the triplet state corresponding to an electron configuration of $(1\sigma_g)^1(1\sigma_u)^1$ (a $^3\Sigma_u^+$ state) is always repulsive. The second excited state shown in Figure 11.12 corresponds to an electron configuration of $(1\sigma_g)^1(2\sigma_g)^1$, or a

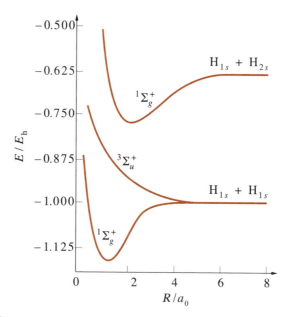

FIGURE 11.12
The internuclear potential energy curves of the ground state and two of the excited electronic states of H_2. Note that the two lowest curves go to $-1\,E_h$ at large distances, indicating two isolated ground-state hydrogen atoms. (The ground state of a hydrogen atom is $-1/2\,E_h$.) The other excited state shown dissociates into one ground-state hydrogen atom and one excited-state hydrogen atom with its electron in the atomic $2s$ orbital (Problem 11–25).

term symbol of $^1\Sigma_g^+$. Like the ground-state H_2 molecule, this excited state has a bond order of one. Because the $2\sigma_g$ orbital is larger than the $1\sigma_g$ orbital, however, we would predict that the bond length of H_2 is longer in this excited state than in the ground state. Experimental measurements confirm this prediction; the bond length is $\approx 35\%$ longer in this $^1\Sigma_g^+$ excited state than it is in the ground state.

11.6 Conjugated Hydrocarbons and Aromatic Hydrocarbons Can Be Treated by a π-Electron Approximation

The molecular orbital energy-level diagram of the valence electrons in ethene, C_2H_4, is shown in Figure 11.13, and the photoelectron spectrum of ethene is shown in Figure 11.14. Ethene has twelve valence electrons, so its ground-state valence electron

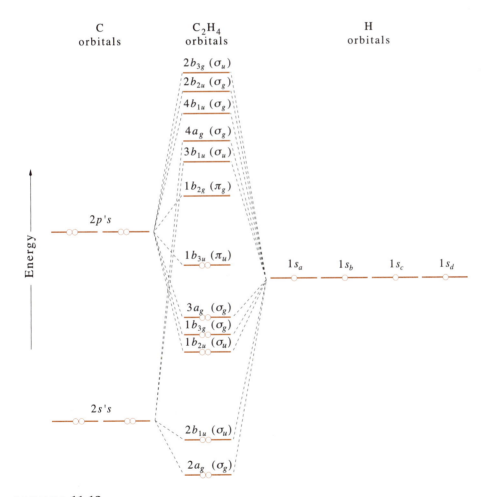

FIGURE 11.13
The molecular orbital energy-level diagram for the valence electrons of C_2H_4. The first five orbitals are σ orbitals, the sixth (HOMO) is a π_u orbital, and the seventh (LUMO) is a π_g orbital.

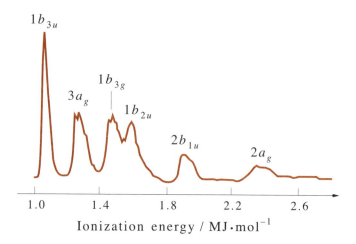

FIGURE 11.14
The photoelectron spectrum of ethene. The energies of the peaks in the photoelectron spectrum can be used to determine the energy spacings between the lowest six molecular orbitals in Figure 11.13.

configuration is $(2a_g)^2(2b_{1u})^2(1b_{2u})^2(1b_{3g})^2(3a_g)^2(1b_{3u})^2$. We simply consider $1a_g$, $2b_{1u}$, etc., to be a shorthand notation for the molecular orbitals. Although not evident from the notation, the lowest five states in Figure 11.13 are σ orbitals; the sixth state, the highest occupied molecular orbital (HOMO), is a π_u orbital, and the seventh state, the lowest unoccupied molecular orbital (LUMO), is a π_g orbital, as indicated in the figure. Figure 11.13 shows that the transition from the highest occupied molecular orbital to the lowest unoccupied molecular orbital is a $\pi_u \rightarrow \pi_g$ transition. It turns out that this is the case for all unsaturated hydrocarbons; the chemically active electrons are the π electrons. This observation suggests that we can develop a simplified molecular orbital treatment of unsaturated hydrocarbons that includes only the π orbitals. In this approximation, the relatively complicated energy-level diagram in Figure 11.13 consists of simply two molecular orbitals, a π_u orbital and a π_g orbital (Figure 11.16).

Ethene is a planar molecule, all of whose bond angles are approximately 120°. You learned in organic chemistry that the carbon atoms in ethene form sp^2 hybrid orbitals and that each C–H bond results from an overlap of the $1s$ hydrogen orbital with an sp^2 hybrid orbital on each carbon atom. Part of the C–C bond in ethene results from the overlap of an sp^2 hybrid orbital from each carbon atom. All five bonds are σ bonds and collectively are called the σ-*bond framework* of the ethene molecule (Figure 11.15). If this σ-bond framework lies in the x–y plane, thus implying that the $2p_x$ and $2p_y$ orbitals are used to construct the hybrid orbitals, then the overlap of the $2p_z$ orbitals, which are perpendicular to the plane of the σ-bond framework, produces a π bond between the carbon atoms. We shall see that in large systems, such as conjugated polyenes and benzene, the π orbitals can be delocalized over the entire molecule. In such cases, we could view the π electrons moving in some fixed, effective, electrostatic potential due to the electrons in the σ framework. This approximation is called the π-*electron*

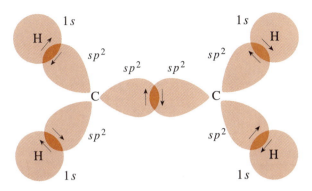

FIGURE 11.15
The (planar) σ-bond framework of an ethene molecule.

approximation. The π-electron approximation can be developed formally by starting with the Schrödinger equation, but we will simply accept it here as a physically intuitive approach to the bonding in unsaturated hydrocarbons.

We now turn our attention to describing the delocalized molecular orbitals occupied by these π electrons. Realize that the Hamiltonian operator we are considering contains an effective potential due to the electrons in the σ framework and that the explicit form of this effective Hamiltonian operator has not been specified in our treatment so far, nor will we need to. With this in mind, let's return to ethene. Here, each carbon atom contributes a $2p_z$ orbital to the delocalized π orbital, and using the same approach as we used to describe the σ bond of the wave function of H_2, we would write the wave function of the π orbital of ethene, ψ_π, as

$$\psi_\pi = c_1 2p_{z1} + c_2 2p_{z2} \tag{11.6}$$

The secular determinant associated with this wave function is

$$\begin{vmatrix} H_{11} - E S_{11} & H_{12} - E S_{12} \\ H_{12} - E S_{12} & H_{22} - E S_{22} \end{vmatrix} = 0 \tag{11.7}$$

where the H_{ij} are integrals involving the effective Hamiltonian operator and the S_{ij} are overlap integrals involving $2p_z$ atomic orbitals. Because the carbon atoms in ethene are equivalent, $H_{11} = H_{22}$. The diagonal matrix elements of the effective Hamiltonian operator in the secular determinant are called *Coulomb integrals,* and the off-diagonal matrix elements of the effective Hamiltonian operator are called *resonance integrals* or *exchange integrals.* Note that the resonance integral involves two atomic centers because it has contributions of atomic orbitals from two different carbon atoms.

To determine the energies and the associated molecular orbitals, we need to either specify the effective Hamiltonian operator or propose approximations for evaluating the various entries in the secular determinant. Here, we examine an approximation proposed by Erich Hückel in 1930, which along with various extensions and modifications has

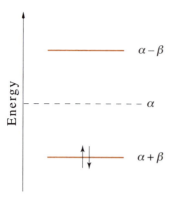

FIGURE 11.16
The ground-state electron configuration of the π electrons in ethene.

found wide use in organic chemistry. There are three simple assertions of *Hückel molecular orbital theory*, at least in its simplest form. First, the overlap integrals, S_{ij}, are set to zero unless $i = j$, where $S_{ii} = 1$. Second, all of the Coulomb integrals are assumed to be the same for all equivalent carbon atoms and are commonly denoted by α. Third, the resonance integrals involving nearest-neighbor carbon atoms are assumed to be the same and are denoted by β; the remaining resonance integrals are set equal to zero. Thus, the Hückel secular determinant for ethene (Equation 11.7) is given by

$$\begin{vmatrix} \alpha - E & \beta \\ \beta & \alpha - E \end{vmatrix} = 0 \tag{11.8}$$

where the two roots are $E = \alpha \pm \beta$.

There are two π electrons in ethene. In the ground state, both electrons occupy the orbital of lowest energy. Because β is intrinsically negative, the lowest energy is $E = \alpha + \beta$, and the π-electronic energy of ethene is $E_\pi = 2\alpha + 2\beta$. Figure 11.16 shows an energy-level diagram for the π electrons of ethene (cf. Figure 11.13). Because α is used to specify the zero of energy, the two energies found from the secular determinant, $E = \alpha \pm \beta$, must correspond to bonding and antibonding orbitals.

EXAMPLE 11–11
Find the bonding and antibonding Hückel molecular orbitals for ethene.

SOLUTION: The equations for c_1 and c_2 associated with Equation 11.7 are

$$c_1(\alpha - E) + c_2\beta = 0 \quad \text{and} \quad c_1\beta + c_2(\alpha - E) = 0$$

For $E = \alpha + \beta$, either equation yields $c_1 = c_2$. Thus,

$$\psi_b = c_1(2p_{z1} + 2p_{z2})$$

The value of c_1 can be found by requiring that the wave function be normalized. The normalization condition on ψ_π gives $c_1^2(1 + 2S + 1) = 1$. Using the Hückel assumption that $S = 0$, we find that $c_1 = 1/\sqrt{2}$.

Substituting $E = \alpha - \beta$ into either of the equations for c_1 and c_2 yields $c_1 = -c_2$, or

$$\psi_a = c_1(2p_{z1} - 2p_{z2})$$

The normalization condition gives $c^2(1 - 2S + 1) = 1$, or $c_1 = 1/\sqrt{2}$.

The case of butadiene is more interesting than that of ethene. Although butadiene exists in both the cis and trans configurations, we will ignore that and picture the butadiene molecule as simply a linear sequence of four carbon atoms, each of which contributes a $2p_z$ orbital to the π-electron orbital (Figure 11.17).

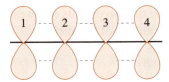

FIGURE 11.17
A schematic representation of the $2p_z$ orbitals of each of the carbon atoms in the butadiene molecule.

Because we are considering the linear combination of four atomic orbitals, the dimension of the secular determinant will be 4×4 and will give rise to four different energies and four different π molecular orbitals. We can write all of the molecular orbitals, ψ_i, by the single expression

$$\psi_i = \sum_{j=1}^{4} c_{ji} 2p_{zj} \qquad i = 1, 2, 3, 4 \tag{11.9}$$

where the c_{ji} are the coefficients of the $2p_z$ atomic orbital on the jth carbon atom ($2p_{zj}$) in the ith molecular orbital. The secular determinantal equation for the butadiene molecule is a 4×4 equation of the form $|H_{ij} - ES_{ij}| = 0$. Using the Hückel approximations, $H_{jj} = \alpha$, $S_{jj} = 1$, $S_{ij} = 0$ if $i \neq j$ and the $H_{ij} = \beta$ for neighboring carbon atoms, and $H_{ij} = 0$ for distant carbon atoms, the secular determinant becomes (Problem 11–29)

$$\begin{vmatrix} \alpha - E & \beta & 0 & 0 \\ \beta & \alpha - E & \beta & 0 \\ 0 & \beta & \alpha - E & \beta \\ 0 & 0 & \beta & \alpha - E \end{vmatrix} = 0 \tag{11.10}$$

These molecular orbitals are shown schematically in Figure 11.19. Notice that as the energy of the molecular orbital increases, so do the number of nodes. This is a general result for molecular orbitals.

EXAMPLE 11–12
Show that ψ_1 in Equation 11.15 is normalized and that it is orthogonal to ψ_2.

SOLUTION: We want to show first that

$$\int d\mathbf{r}\, \psi_1^* \psi_1 = \langle \psi_1 \mid \psi_1 \rangle = 1$$

Using the fact that Hückel theory (as we have discussed it) sets all the overlap integrals to zero, we have

$$\langle \psi_1 \mid \psi_1 \rangle = (0.3717)^2 + (0.6015)^2 + (0.6015)^2 + (0.3717)^2 = 1.000$$

To show that ψ_1 is orthogonal to ψ_2, we must show that

$$\langle \psi_1 \mid \psi_2 \rangle = 0$$

Once again, because all the overlap integrals equal zero, we have

$$\langle \psi_1 \mid \psi_2 \rangle = \int d\mathbf{r}\, \psi_1^* \psi_2 = (0.3717)(0.6015) + (0.6015)(0.3717)$$

$$- (0.6015)(0.3717) - (0.3717)(0.6015) = 0$$

It is straightforward to show that all four molecular orbitals in Equation 11.15 are normalized and that they are mutually orthogonal.

11.7 Hückel Molecular Orbital Theory Can Be Used to Calculate Bond Orders

We can calculate the π-electron energy in terms of the coefficients c_{ji} in Equation 11.9. The π-electron energy is given by

$$E_\pi = \sum n_j E_j \tag{11.16}$$

where n_j is the number of electrons in the ith molecular orbital and E_j is its energy. The energy of the ith molecular orbital is given by

$$E_i = \langle \psi_i \mid \hat{H}_{\text{eff}} \mid \psi_i \rangle \tag{11.17}$$

where

$$\psi_i = \sum_r c_{ri} 2p_{zr} \tag{11.18}$$

If we substitute Equation 11.18 into Equation 11.17, we obtain (Problem 11–33)

$$E_i = \sum_r c_{ri}^2 \alpha_r + \sum_{\substack{r \\ r \neq s}} \sum_s c_{ri} c_{si} \beta_{rs} \tag{11.19}$$

Finally, we use Equation 11.16 to write

$$E_\pi = \sum_r \alpha_r \left(\sum_i n_i c_{ri}^2 \right) + \sum_{\substack{r \\ r \neq s}} \sum_s \beta_{rs} \left(\sum_i n_i c_{ri} c_{ci} \right) \tag{11.20}$$

The summation over i in the first term in Equation 11.20 has the following interpretation. The probability density of an electron in the molecular orbital ψ_i is given by

$$\psi_i^2 = \left(\sum_{j=1}^4 c_{ji} 2p_{zj} \right)^2$$

$$= \sum_{j=1}^4 \sum_{k=1}^4 c_{ji} c_{ki} 2p_{zj} 2p_{zk}$$

$$= \sum_{j=1}^4 c_{ji}^2 2p_{zj}^2 + \sum_{\substack{j=1 \\ j \neq k}}^4 \sum_{k=1}^4 c_{ji} c_{ki} 2p_{zj} 2p_{zk} \tag{11.21}$$

According to Example 11–12,

$$\sum_{j=1}^4 c_{ji}^2 = 1 \qquad i = 1, 2, 3, 4 \tag{11.22}$$

Equation 11.22 suggests that we interpret c_{ji}^2 as the fractional π-electronic charge on the jth carbon atom due to an electron in the ith molecular orbital. Thus, the total π-electronic charge on the jth carbon atom is

$$q_j = \sum_i n_i c_{ji}^2 \tag{11.23}$$

where n_i is the number of electrons in the ith molecular orbital. For butadiene, for example,

$$q_1 = 2\,c_{11}^2 + 2\,c_{12}^2 + 0\,c_{13}^2 + 0\,c_{14}^2$$

$$= 2(0.3717)^2 + 2(0.6015)^2 = 1.000$$

The other q's are also equal to unity, indicating that the π electrons in butadiene are uniformly distributed over the molecule.

The second term in Equation 11.20 suggests that the product $c_{ri}c_{si}$ is the π-electronic charge in the ith molecular orbital between the adjacent carbon atoms r and s. We define the π-*bond order* between adjacent carbon atoms r and s by

$$p_{rs}^\pi = \sum_i n_i c_{ri} c_{si} \tag{11.24}$$

In terms of p_{rs}^π, we can write Equation 11.20 as (Problem 11–34)

$$E_\pi = \sum_r p_{rr}^\pi \alpha_r + \sum_r \sum_{\substack{s \\ r \neq s}} p_{rs}^\pi \beta_{rs} \tag{11.25}$$

As Equation 11.25 implies, a bond order has physical meaning only between r and s for which there is a nonzero value of β.

For butadiene, we have

$$p_{12}^\pi = 2\,c_{11}c_{21} + 2\,c_{12}c_{22} + 0\,c_{13}c_{23} + 0\,c_{14}c_{24}$$

$$= 2(0.3717)(0.6015) + 2(0.6015)(0.3717)$$

$$= 0.8943$$

$$p_{23}^\pi = 2\,c_{21}c_{31} + 2\,c_{22}c_{32} + 0\,c_{23}c_{33} + 0\,c_{24}c_{34}$$

$$= 2(0.6015)(0.6015) + 2(0.3717)(-0.3717)$$

$$= 0.4473$$

Clearly, $p_{12}^\pi = p_{34}^\pi$ by symmetry. If we recall that there is a σ bond between each carbon atom above, then we can define a total bond order

$$p_{rs}^{\text{total}} = 1 + p_{rs}^\pi \tag{11.26}$$

where the first term on the right side is due to the σ bond between atoms r and s. For butadiene, we find that

$$p_{12}^{\text{total}} = p_{34}^{\text{total}} = 1.894$$

$$p_{23}^{\text{total}} = 1.447 \tag{11.27}$$

Equations 11.27 are in excellent agreement with the experimental observations involving the reactivity of these bonds in butadiene. Figure 11.20 shows the correlation of

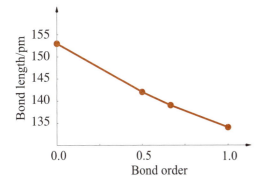

FIGURE 11.20
The correlation of experimental carbon–carbon bond lengths with π-electron bond order.

experimental carbon–carbon bond lengths with the total bond order calculated from Hückel theory.

As our final example, we consider benzene. Benzene has six carbon atoms, each contributing a $2p_z$ orbital from which the π molecular orbitals are to be constructed. Because we are considering linear combinations of six atomic orbitals, the dimension of the secular determinant will be 6×6 and will give rise to six different energies and six different π molecular orbitals. The Hückel secular determinantal equation for benzene is given by (Problem 11–35)

$$\begin{vmatrix} \alpha - E & \beta & 0 & 0 & 0 & \beta \\ \beta & \alpha - E & \beta & 0 & 0 & 0 \\ 0 & \beta & \alpha - E & \beta & 0 & 0 \\ 0 & 0 & \beta & \alpha - E & \beta & 0 \\ 0 & 0 & 0 & \beta & \alpha - E & \beta \\ \beta & 0 & 0 & 0 & \beta & \alpha - E \end{vmatrix} = 0 \qquad (11.28)$$

This 6×6 secular determinant leads to a sixth-degree polynomial for E. Using the same approach as for butadiene, we let $x = (\alpha - E)/\beta$. The resulting determinant can be expanded to give

$$x^6 - 6x^4 + 9x^2 - 4 = 0 \qquad (11.29)$$

The six roots to this equation are $x = \pm 1$, ± 1, and ± 2 (Problem 11–36), giving the following energies for the six molecular orbitals:

$$E_1 = \alpha + 2\beta \qquad E_2 = E_3 = \alpha + \beta \qquad E_4 = E_5 = \alpha - \beta \qquad E_6 = \alpha - 2\beta \quad (11.30)$$

The Hückel energy-level diagram for benzene is given in Figure 11.21. The six π electrons are placed into the three lowest-energy molecular orbitals. The total π-electron energy in benzene is given by

$$E_\pi = 2(\alpha + 2\beta) + 4(\alpha + \beta) = 6\alpha + 8\beta \qquad (11.31)$$

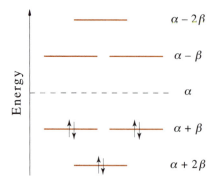

FIGURE 11.21
The ground-state electron configuration of the π electrons in benzene.

Compared with the π-electron energy of three ethene molecules, the delocalization (or resonance) energy in benzene is 2β. Thus, if we use the value $\beta = -75 \, \text{kJ} \cdot \text{mol}^{-1}$, we see that Hückel molecular orbital theory predicts that benzene is stabilized by about $150 \, \text{kJ} \cdot \text{mol}^{-1}$.

The resulting six π molecular orbitals of benzene are given by

$$\psi_1 = \frac{1}{\sqrt{6}}(2p_{z1} + 2p_{z2} + 2p_{z3} + 2p_{z4} + 2p_{z5} + 2p_{z6}) \qquad E_1 = \alpha + 2\beta$$

$$\psi_2 = \frac{1}{\sqrt{4}}(2p_{z2} + 2p_{z3} - 2p_{z5} - 2p_{z6}) \qquad E_2 = \alpha + \beta$$

$$\psi_3 = \frac{1}{\sqrt{3}}(2p_{z1} + \frac{1}{2}2p_{z2} - \frac{1}{2}2p_{z3} - 2p_{z4} - \frac{1}{2}2p_{z5} + \frac{1}{2}2p_{z6}) \qquad E_3 = \alpha + \beta$$

$$\psi_4 = \frac{1}{\sqrt{4}}(2p_{z2} - 2p_{z3} + 2p_{z5} - 2p_{z6}) \qquad E_4 = \alpha - \beta$$

$$\psi_5 = \frac{1}{\sqrt{3}}(2p_{z1} - \frac{1}{2}2p_{z2} - \frac{1}{2}2p_{z3} + 2p_{z4} - \frac{1}{2}2p_{z5} - \frac{1}{2}2p_{z6}) \qquad E_5 = \alpha - \beta$$

$$\psi_6 = \frac{1}{\sqrt{6}}(2p_{z1} - 2p_{z2} + 2p_{z3} - 2p_{z4} + 2p_{z5} - 2p_{z6}) \qquad E_6 = \alpha - 2\beta$$

$$(11.32)$$

EXAMPLE 11–13
Draw the π molecular orbitals for benzene and indicate the nodal planes.

SOLUTION: The solution is illustrated in Figure 11.22. Note that, as we found for ethene and butadiene, the energy increases with the number of nodal planes.

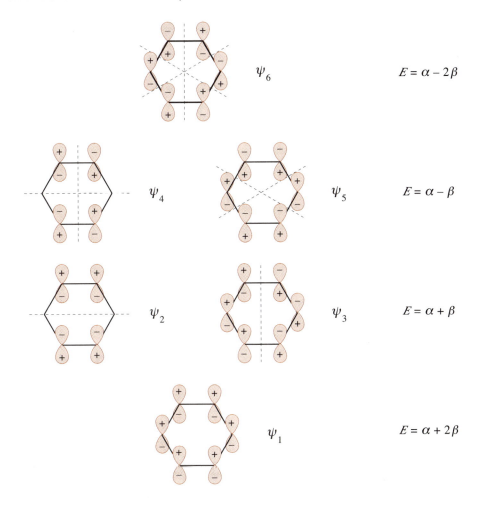

ψ_6 $E = \alpha - 2\beta$

ψ_4 ψ_5 $E = \alpha - \beta$

ψ_2 ψ_3 $E = \alpha + \beta$

ψ_1 $E = \alpha + 2\beta$

FIGURE 11.22
The solution for Example 11–13.

EXAMPLE 11–14
Use the Hückel molecular orbitals of benzene given by Equations 11.32 to calculate
the π-electronic charge on each carbon atom and the total bond orders in benzene.

SOLUTION: Using Equation 11.23, we find the total π-electronic charge on the nth
carbon atom to be

$$q_n = 2(c_{n1}^2 + c_{n2}^2 + c_{n3}^2)$$

Therefore,

$$q_1 - 2\left(\frac{1}{6} + \frac{1}{3}\right) = 1$$

$$q_2 = 2\left(\frac{1}{6} + \frac{1}{4} + \frac{1}{12}\right) = 1$$

$$q_3 = 2\left(\frac{1}{6} + \frac{1}{4} + \frac{1}{12}\right) = 1$$

$$q_4 = 2\left(\frac{1}{6} + \frac{1}{3}\right) = 1$$

$$q_5 = 2\left(\frac{1}{6} + \frac{1}{4} + \frac{1}{12}\right) = 1$$

$$q_6 = 2\left(\frac{1}{6} + \frac{1}{4} + \frac{1}{12}\right) = 1$$

Thus, we see that the π electrons are distributed uniformly around the benzene ring. Using Equation 11.24 for the π-bond orders, we have

$$p_{rs}^\pi = 2(c_{r1}c_{s1} + c_{r2}c_{s2} + c_{r3}c_{s3})$$

Therefore,

$$p_{12}^\pi = 2\left(\frac{1}{6} + \frac{1}{6}\right) = \frac{2}{3}$$

$$p_{23}^\pi = 2\left(\frac{1}{6} + \frac{1}{4} - \frac{1}{12}\right) = \frac{2}{3}$$

$$p_{34}^\pi = 2\left(\frac{1}{6} + \frac{1}{6}\right) = \frac{2}{3}$$

$$p_{45}^\pi = 2\left(\frac{1}{6} + \frac{1}{6}\right) = \frac{2}{3}$$

$$p_{56}^\pi = 2\left(\frac{1}{6} + \frac{1}{4} - \frac{1}{12}\right) = \frac{2}{3}$$

$$p_{61}^\pi = 2\left(\frac{1}{6} + \frac{1}{6}\right) = \frac{2}{3}$$

Thus, we find that all the bonds in benzene are equivalent, in nice agreement with the chemical properties of benzene.

11.8 Hückel Molecular Orbital Theory Can Be Formulated in Matrix Notation

Before we leave this chapter, let's cast all our Hückel theory results in matrix notation. Not only is this a more economical notation, but it also lends itself to standard computational algorithms. A secular determinant such as Equation 11.10 arises from a matrix eigenvalue problem of the form

$$\mathsf{H} \mathsf{c} = E \, \mathsf{S} \mathsf{c} \tag{11.33}$$

Because of our assumption that $S_{ij} = \delta_{ij}$, the matrix S is the unit matrix I. Therefore, Equation 11.33 becomes

$$\mathsf{H} \mathsf{c} = E \, \mathsf{c} \tag{11.34}$$

If H is an $N \times N$ matrix, it will have N eigenvectors and N corresponding eigenvalues. If we let the eigenvectors be c^a and the corresponding eigenvalues be E_a, then Equation 11.34 becomes

$$\mathsf{H} \mathsf{c}^a = E_a \, \mathsf{c}^a \tag{11.35}$$

for $a = 1, 2, \ldots, N$. Because H is symmetric, we can choose the eigenvectors to be orthonormal, which says that

$$(\mathsf{c}^a)^{\mathsf{T}} \mathsf{c}^b = \sum_{j=1}^{N} (c_i^a)^* c_i^b = \delta_{ab} \tag{11.36}$$

where $(\mathsf{c}^a)^{\mathsf{T}}$ represents the complex conjugate of the transpose of c^b. (If c^a is a column vector, then $(\mathsf{c}^a)^{\mathsf{T}}$ is a row vector, and $(\mathsf{c}^a)^{\mathsf{T}} \mathsf{c}^b$ is a number.)

We now define a matrix C whose columns are the (normalized) eigenvectors; in particular, the ath column of C is the normalized c^a. The matrix elements of C are given by

$$C_{ia} = c_i^a \tag{11.37}$$

Thus, C_{ia} is the ith element of the eigenvector c^a. The ath molecular orbital can be written in terms of the C_{ia} by

$$\psi_a = \sum_{j=1}^{N} C_{ia} 2p_{zi} \tag{11.38}$$

Equation 11.38 is the generalization of Equation 11.19.

11–2. The vibrational energy (in cm^{-1}) of Li_2 in its ground electronic state is given in the JANAF Thermochemical Tables (see References) as

$$G(v) = 351.39 \left(v + \frac{1}{2}\right) - 2.578 \left(v + \frac{1}{2}\right)^2 - 0.00647 \left(v + \frac{1}{2}\right)^3$$

$$- 9.712 \times 10^{-5} \left(v + \frac{1}{2}\right)^4$$

Calculate the number of vibrational energy levels. (See Problem 10–2.)

11–3. The rotational constant of B_2 in its ground vibrational state is given in the JANAF Thermochemical Tables (see References) as 1.228 cm^{-1}. Calculate the bond length of B_2.

11–4. The rotational constant of C_2 in its ground vibrational state is given in the JANAF Thermochemical Tables (see References) as 1.811 cm^{-1}. Calculate the bond length of C_2.

11–5. Use molecular orbital theory to explain why the dissociation energy of N_2 is greater than that of N_2^+, but the dissociation energy of O_2^+ is greater than that of O_2.

11–6. Discuss the bond properties of F_2 and F_2^+ using molecular orbital theory.

11–7. Predict the relative stabilities of the species N_2, N_2^+, and N_2^-.

11–8. Predict the relative bond strengths and bond lengths of diatomic carbon, C_2, and its negative ion, C_2^-.

11–9. The force constants for the diatomic molecules B_2 through F_2 are given in the table below. Is the order what you expect? Explain.

Diatomic molecule	$k/N \cdot m^{-1}$
B_2	350
C_2	930
N_2	2260
O_2	1140
F_2	450

11–10. Write out the ground-state molecular orbital electron configurations for Na_2 through Ar_2. Would you predict a stable Mg_2 molecule?

11–11. In Section 11.2, we constructed molecular orbitals for homonuclear diatomic molecules using the $n = 2$ atomic orbitals on each of the bonded atoms. In this problem, we will consider the molecular orbitals that can be constructed from the $n = 3$ atomic orbitals. These orbitals are important in describing diatomic molecules of the first row of the transition metals. Once again we choose the z axis to lie along the molecular bond. What are the designations for the $3s_A \pm 3s_B$ and $3p_A \pm 3p_B$ molecular orbitals? The $n = 3$ shell also contains a set of five $3d$ orbitals. (The shapes of the $3d$ atomic orbitals are shown in Figure 7.8.) Given that molecular orbitals with two nodal planes that contain the internuclear axis are called δ orbitals, show that ten $3d_A \pm 3d_B$ molecular orbitals consist of a bonding

σ orbital, a pair of bonding π orbitals, a pair of bonding δ orbitals, and their corresponding antibonding orbitals.

11–12. Determine the largest bond order for a first-row transition-metal homonuclear diatomic molecule. (See the previous problem.)

11–13. Determine the ground-state molecular orbital electron configuration of NO^+ and NO. Compare the bond order of these two species.

11–14. Figure 11.10 plots a schematic representation of the energies of the molecular orbitals of HF. How will the energy-level diagram for the diatomic OH radical differ from that of HF? What is the highest occupied molecular orbital of OH?

11–15. Using Figure 11.10, you found that the highest occupied molecular orbital for HF is a fluorine $2p$ atomic orbital. The measured ionization energies for an electron from this nonbonding molecular orbital of HF is $1550 \ kJ \cdot mol^{-1}$. However, the measured ionization energy of a $2p$ electron from a fluorine atom is $1795 \ kJ \cdot mol^{-1}$. Why is the ionization energy of an electron from the $2p$ atomic orbital on fluorine greater for the fluorine atom than for an HF molecule?

11–16. In this problem, we consider the heteronuclear diatomic molecule CO. The ionization energies of an electron from the valence atomic orbitals on the carbon atom and the oxygen atom are listed below.

Atom	Valence orbital	Ionization energy / $MJ \cdot mol^{-1}$
O	$2s$	3.116
	$2p$	1.524
C	$2s$	1.872
	$2p$	1.023

Use these data to construct a molecular orbital energy-level diagram for CO. What are the symmetry designations of the molecular orbitals of CO? What is the electron configuration of the ground state of CO? What is the bond order of CO? Is CO paramagnetic or diamagnetic?

11–17. The molecule BF is isoelectronic with CO. However, the molecular orbitals for BF are different from those for CO. Unlike CO, the energy difference between the $2s$ orbitals of boron and fluorine is so large that the $2s$ orbital of boron combines with a $2p$ orbital on fluorine to make a molecular orbital. The remaining $2p$ orbitals on fluorine combine with two of the $2p$ orbitals on B to form π orbitals. The third $2p$ orbital on B is nonbonding. The energy ordering of the molecular orbitals is $\psi(2s_B + 2p_F) < \psi(2p_B - 2p_F) < \psi(2s_B - 2p_F) < \psi(2p_B + 2p_F) < \psi(2p_B)$. What are the symmetry designations of the molecular orbitals of BF? What is the electron configuration of the ground state of BF? What is the bond order of BF? Is BF diamagnetic or paramagnetic? How do the answers to these last two questions compare with those obtained for CO (Problem 11–16)?

11–18. The photoelectron spectrum of O_2 exhibits two bands of $52.398 \ MJ \cdot mol^{-1}$ and $52.311 \ MJ \cdot mol^{-1}$ that correspond to the ionization of an oxygen $1s$ electron. Explain this observation.

11–19. The experimental ionization energies for a fluorine $1s$ electron from HF and F_2 are 66.981 and 67.217 $MJ \cdot mol^{-1}$, respectively. Explain why these ionization energies are different even though the $1s$ electrons of the fluorine are not involved in the chemical bond.

11–20. Go to the website *www.ccl.net/cca/data/atomic-RHF-wavefunctions/tables* and verify that the $2s$ and $2p$ orbital energies of a fluorine atom are $-1.572\ E_h$ and $-0.730\ E_h$, respectively.

11–21. When we built up the molecular orbitals for diatomic molecules, we combined only those orbitals with the same energy because we said that only those with similar energies mix well. This problem is meant to illustrate this idea. Consider two atomic orbitals χ_A and χ_B. Show that a linear combination of these orbitals leads to the secular determinant

$$\begin{vmatrix} \alpha_A - E & \beta - ES \\ \beta - ES & \alpha_A - E \end{vmatrix} = 0$$

where

$$\alpha_A = \int \chi_A h^{\text{eff}} \chi_A d\tau \qquad \alpha_B = \int \chi_B h^{\text{eff}} \chi_B d\tau$$

$$\beta = \int \chi_B h^{\text{eff}} \chi_A d\tau = \int \chi_A h^{\text{eff}} \chi_B d\tau \qquad S = \int \chi_A \chi_B d\tau$$

where h^{eff} is some effective one-electron Hamiltonian operator for the electron that occupies the molecular orbital ϕ. Show that

$$(1 - S^2)E^2 + (2\beta S - \alpha_A - \alpha_B)E + \alpha_A \alpha_B - \beta^2 = 0$$

It is usually a satisfactory first approximation to neglect S. Doing this, show that

$$E_\pm = \frac{\alpha_A + \alpha_B \pm [(\alpha_A - \alpha_B)^2 + 4\beta^2]^{1/2}}{2}$$

Now if χ_A and χ_B have the same energy, show that $\alpha_A = \alpha_B = \alpha$ and that

$$E_\pm = \alpha \pm \beta$$

giving one level of β units below α and one level of β units above α—that is, one level of β units more stable than the isolated orbital energy and one level of β units less stable. Now investigate the case in which $\alpha_A \neq \alpha_B$, say, $\alpha_A < \alpha_B$. Show that

$$E_\pm = \frac{\alpha_A + \alpha_B}{2} \pm \frac{\alpha_A - \alpha_B}{2} \left[1 + \frac{4\beta^2}{(\alpha_A - \alpha_B)^2} \right]^{1/2}$$

$$= \frac{\alpha_A + \alpha_B}{2} \pm \frac{\alpha_A - \alpha_B}{2} \left[1 + \frac{2\beta^2}{(\alpha_A - \alpha_B)^2} - \frac{2\beta^4}{(\alpha_A - \alpha_B)^4} + \cdots \right]$$

$$= \frac{\alpha_A + \alpha_B}{2} \pm \frac{\alpha_A - \alpha_B}{2} \pm \frac{\beta^2}{\alpha_A - \alpha_B} + \cdots$$

where we have assumed that $\beta^2 < (\alpha_A - \alpha_B)^2$ and have used the expansion

$$(1+x)^{1/2} = 1 + \frac{x}{2} - \frac{x^2}{8} + \cdots$$

Show that

$$E_\pm = \alpha_A \pm \frac{\beta^2}{\alpha_A - \alpha_B} + \cdots$$

Using this result, discuss the stabilization–destabilization of α_A and α_B versus the case above in which $\alpha_A = \alpha_B$. For simplicity, assume initially that $\alpha_A - \alpha_B$ is large.

11–22. Show that filled orbitals can be ignored in the determination of molecular term symbols.

11–23. Deduce the ground-state term symbols of all the diatomic molecules given in Table 11.3.

11–24. Determine the ground-state molecular term symbols of O_2, N_2, N_2^+, and O_2^+.

11–25. Calculate the sum of the energies of one hydrogen atom in the $1s$ state and one in the $2s$ state and show that your result is consistent with Figure 11.12.

11–26. The highest occupied molecular orbitals for an excited electronic configuration of the oxygen molecule are $(1\pi_g)^1(3\sigma_u)^1$. What are the molecular term symbols for oxygen with this electronic configuration?

11–27. Show that the π molecular orbital corresponding to the energy $E = \alpha - \beta$ for ethene is $\psi_\pi = \frac{1}{\sqrt{2}}(2p_{z1} - 2p_{z2})$.

11–28. Generalize our Hückel molecular orbital treatment of ethene to include overlap of $2p_{z1}$ and $2p_{z2}$. Determine the energies and the orbitals in terms of the overlap integral, S.

11–29. Verify Equation 11.10.

11–30. Show that

$$\begin{vmatrix} x & 1 & 0 & 0 \\ 1 & x & 1 & 0 \\ 0 & 1 & x & 1 \\ 0 & 0 & 1 & x \end{vmatrix} = 0$$

gives the algebraic equation $x^4 - 3x^2 + 1 = 0$.

11–31. Show that the four π molecular orbitals for butadiene are given by Equations 11.15.

11–32. Show that the four molecular orbitals for butadiene (Equations 11.15) satisfy Equation 11.25.

11–33. Derive Equation 11.19.

11–34. Show that

$$E_\pi = \sum_r p_{rr}^\pi \alpha_r + 2 \sum_r \sum_{s>r} p_{rs}^\pi \beta_{rs}$$

where α_r is the coulomb integral associated with the rth atom and β_{rs} is the exchange integral between atoms r and s. The relation serves as a good check on the calculated energy levels.

11–35. Derive the Hückel theory secular determinant for benzene (see Equation 11.28).

11–36. Show that the six roots of Equation 11.25 are $E_1 = \alpha + 2\beta$, $E_2 = E_3 = \alpha + \beta$, $E_4 = E_5 = \alpha - \beta$, and $E_6 = \alpha - 2\beta$.

11–37. Calculate the Hückel π-electronic energies of cyclobutadiene. What do Hund's rules say about the ground state of cyclobutadiene? Compare the stability of cyclobutadiene with that of two isolated ethylene molecules.

11–38. Calculate the Hückel π-electronic energy of trimethylenemethane:

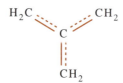

11–39. Calculate the π-electronic energy levels and the total π-electronic energy of bicyclobutadiene:

11–40. Show that the Hückel molecular orbitals of benzene given in Equation 11.29 are orthonormal.

11–41. Set up the Hückel molecular orbital theory determinantal equation for naphthalene.

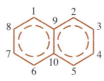

11–42. Use a program such as *MathCad* or *Mathematica* to show that a Hückel calculation for naphthalene, $C_{10}H_8$, gives the molecular orbital energy levels $E_i = \alpha + m_i \beta$, where the 10 values of m_i are 2.3028, 1.6180, 1.3029, 1.0000, 0.6180, -0.6180, -1.0000, -1.3029, -1.6180, and -2.3028. Calculate the ground-state π-electron energy and the delocalization energy of naphthalene.

11–43. Use a program such as *MathCad* or *Mathematica* to determine the 10 π orbitals of napthalene. Calculate the π-electronic charge on each carbon atom and the various bond orders.

11–44. Using Hückel molecular orbital theory, determine whether the linear state $(H\!-\!H\!-\!H)^+$ or the triangular state

of H_3^+ is the more stable state. Repeat the calculation for H_3 and H_3^-.

11–45. Set up a Hückel theory secular determinant for pyridine.

11–46. Calculate the delocalization energy, the charge on each carbon atom, and the bond orders for the allyl radical, cation, and anion. Sketch the molecular orbitals for the allyl system.

11–47. Because of the symmetry inherent in the Hückel theory secular determinant of linear and cyclic conjugated polyenes, we can write mathematical formulas for the energy levels for an arbitrary number of carbon atoms in the system (for present purposes, we consider cyclic polyenes with only an even number of carbon atoms). These formulas are

$$E_n = \alpha + 2\beta \cos \frac{\pi n}{N+1} \qquad n = 1, 2, \ldots, N \qquad \text{linear chains}$$

and

$$E_n = \alpha + 2\beta \cos \frac{2\pi n}{N} \qquad n = 0, \pm 1, \ldots, \pm\left(\frac{N}{2}-1\right), \frac{N}{2} \qquad \text{cyclic chains (N even)}$$

where N is the number of carbon atoms in the conjugated π system.

(a) Use these formulas to verify the results given in the chapter for butadiene and benzene.

(b) Now use these formulas to predict energy levels for linear hexatriene (C_6H_8) and octatetraene (C_8H_{10}). How does the stabilization energy of these molecules per carbon atom vary as the chains grow in length?

(c) Compare the results for hexatriene and benzene. Which molecule has a greater stabilization energy? Why?

11–48. The problem of a linear conjugated polyene of N carbon atoms can be solved in general. The energies E_j and the coefficients of the atomic orbitals in the jth molecular orbital are given by

$$E_j = \alpha + 2\beta \cos \frac{j\pi}{N+1} \qquad j = 1, 2, 3, \ldots, N$$

and

$$c_{jk} = \left(\frac{2}{N+1}\right)^{1/2} \sin \frac{jk\pi}{N+1} \qquad k = 1, 2, 3, \ldots, N$$

Determine the energy levels and the wave functions for butadiene using these formulas.

11–49. We can calculate the electronic states of a hypothetical one-dimensional solid by modeling the solid as a one-dimensional array of atoms with one orbital per atom, and using Hückel theory to calculate the allowed energies. Use the formula for E_j in the previous problem to show that the energies will form essentially a continuous band of width 4β. *Hint:* Calculate $E_N - E_1$ and let N be very large so that you can use $\cos x \approx 1 - x^2/2 + \cdots$.

11–50. The band of electronic energies that we calculated in the previous problem can accommodate N pairs of electrons of opposite spins, or a total of $2N$ electrons. If each atom contributes one electron (as in the case of a polyene), the band is occupied by a total of N electrons. Using some ideas you may have learned in general chemistry, would you expect such a system to be a conductor or an insulator?

11–51. Verify Equation 11.39.

11–52. Write out the C matrix for butadiene and show that it is orthogonal. Recall (MathChapter G) that $C^T = C^{-1}$, or that $CC^T = C^TC = I$ for an orthogonal matrix.

11–53. Show that Equations 11.42 and 11.43 are equivalent to Equations 11.23 and 11.25.

11–54. Use a program such as *MathCad* or *Mathematica* to show that $HC = CD$ for butadiene.

11–55. Use a program such as *MathCad* or *Mathematica* to show that the R matrix (Equation 11.41) for butadiene is

$$
\begin{pmatrix}
1.0000 & 0.8943 & 0 & -0.4473 \\
0.8943 & 1.0000 & 0.4473 & 0 \\
0 & 0.4473 & 1.0000 & 0.8943 \\
-0.4473 & 0 & 0.8943 & 1.0000
\end{pmatrix}
$$

Interpret this result.

11–56. Use a program such as *MathCad* or *Mathematica* to show that the R matrix (Equation 11.41) for cyclobutadiene is

$$
\begin{pmatrix}
1.0 & 0.5 & 0 & 0.5 \\
0.5 & 1.0 & 0.5 & 0 \\
0 & 0.5 & 1.0 & 0.5 \\
0.5 & 0 & 0.5 & 1.0
\end{pmatrix}
$$

Interpret this result.

11–57. Use a program such as *MathCad* or *Mathematica* to show that the R matrix (Equation 11.41) for bicyclobutadiene is

$$
\begin{pmatrix}
0.621268 & 0.485071 & 0.485071 & 0.621268 \\
0.485071 & 1.37873 & -0.621268 & 0.485071 \\
0.485071 & -0.621268 & 1.37873 & 0.485071 \\
0.621268 & 0.485071 & 0.485071 & 0.621268
\end{pmatrix}
$$

Interpret this result.

References

DeKock, R. L., Gray, H. B. *Chemical Structure and Bonding*. University Science Books: Sausalito, CA, 1989.

JANAF Thermochemical Tables: Chase, M. W. Jr., et al., *J. Phys. Chem. Ref. Data* 1985, vol. 14, Supplement No. 1.

JANAF Thermochemical Tables (update): *http://www.nist.gov/srd/jpcrd_28.htm#janaf*

Levine, I. N. *Quantum Chemistry,* 5th ed. Prentice Hall: Upper Saddle River, NJ, 2000.

Pilar, F. L. *Elementary Quantum Chemistry,* 2nd ed. Dover Publications: Mineola, NY, 1990.

Streitwieser, A. *Molecular Orbital Theory for Organic Chemists*. Wiley & Sons: New York, 1961.

John Pople was born in Somerset, England, on October 31, 1925, and died in 2004 in Chicago. He received his Ph.D. in mathematics from Cambridge University in 1951. He emigrated to the United States in 1964 to the Carnegie Institute of Technology (now Carnegie Mellon University), where he remained until 1974. Over a period of years, Pople developed computational algorithms for the ab initio calculation of molecular properties based upon Gaussian orbitals. The computer programs developed by Pople and his many collaborators were initially freely distributed, but later were packaged as a commercially available program called Gaussian, one of the most widely used computational quantum-chemical programs. The availability of such programs has made it possible for chemistry students at all levels to calculate molecular properties. Pople shared the 1998 Nobel Prize in Chemistry with Walter Kohn, who developed density functional theory.

Clemens Roothaan was born in 1918 in Nijmegen, the Netherlands. In 1936, he entered the Technical University of Delft and after graduating in 1940 continued there as a graduate student in physics. In 1943 Roothaan returned to his family's home in response to increasing Nazi oppression. Due to his younger brother's involvement with the underground resistance, he was arrested and spent the remainder of World War II in police lockup and eventually in concentration camps in the Netherlands and Germany. After the war, he received his master's degree from the University of Delft, and then went to the University of Chicago to work for Robert Mulliken, while teaching at the Catholic University of America in Washington, DC. In 1950, he joined the physics department at the University of Chicago, where he remained until his retirement in 1988. Roothaan was a pioneer in the application of computers to quantum chemistry.

CHAPTER 12

The Hartree–Fock–Roothaan Method

We developed the Hartree–Fock approximation for atoms in Chapter 9 and then the Hartree–Fock–Roothaan approximation using a minimal basis set for H_2 in Chapter 10. In this chapter, we shall develop the Hartree–Fock–Roothaan approximation for any molecule. This approximation gives us the optimal molecular orbital representation of the electronic structure of a molecule. It is an approximation because the very concept of molecular orbitals, or atomic orbitals in the case of atoms, assumes that the electrons interact in some average, or self-consistent, potential. The Hartree–Fock–Roothaan approximation is the standard starting point for ab initio molecular (and atomic) calculations and is the workhorse of molecular quantum chemistry. By an ab initio calculation, we mean one in which no empirical parameters are introduced; the calculation is done "from the beginning."

In the course of performing Hartree–Fock–Roothaan calculations, or Hartree–Fock calculations for short, we are led naturally to a discussion of basis sets. The contribution of Roothaan to molecular Hartree–Fock calculations was to introduce a basis set, which converts the Hartree–Fock coupled differential equations into a set of matrix equations for the coefficients of the basis functions in the basis set. Modern quantum-chemical calculations use basis sets consisting of linear combinations of Gaussian functions, and a notation for these basis sets has evolved that has become pervasive in the quantum-chemical literature. We'll learn what terms such as STO-3G and 6-31G* mean in describing basis sets.

There are a number of commercially available and on-line computer programs that make it possible for even beginning students to carry out reliable molecular calculations. Three of the most popular programs are Gaussian 03, GAMESS, and SPARTAN, with Gaussian being the most widely used. The 03 with the name Gaussian indicates the year (2003) that the current version was released. All these programs run on PC, Mac, and Unix computers. In Section 12.5, we shall illustrate how easy and user-friendly it is to set up a Hartree–Fock calculation using Gaussian 03. In the next section, we shall assess the results of Hartree–Fock calculations for various properties for a variety of molecules, and in the last section we shall briefly describe some commonly used post-Hartree–Fock methods.

12.1 The Hartree–Fock–Roothaan Equations Give the Optimum Molecular Orbitals as Linear Combinations of Atomic Orbitals

The Hartree–Fock approximation is the standard first approximation for all atomic and molecular calculations in modern quantum chemistry. Because there are a number of commercial and even free computer programs available, Hartree–Fock calculations are now routine, and as we said earlier, are even used in many organic chemistry and physical chemistry laboratory courses. We introduced Hartree–Fock theory for multielectron atomic systems in Chapter 9 and then for a minimal basis set calculation for H_2 in Chapter 10.

In this section we shall restate the principal Hartree–Fock equations for polyatomic molecules and then introduce a basis set to express the molecular orbitals as linear combinations of atomic orbitals, thus leading to the Hartree–Fock–Roothaan equations, which serve as the starting point for almost all molecular calculations. You should be aware that although the equations have summations and look complicated in the general case, we are going to let computers manipulate them and solve them, and they don't mind lots of summations and integrals because that's exactly what they're good at. Let's now discuss Hartree–Fock theory once again.

In Section 9.6, we developed the Hartree–Fock method for atoms. For simplicity, we considered only closed-shell systems in which we have $2N$ electrons occupying N doubly occupied spatial orbitals. In such cases, the wave function is given by one Slater determinant. We shall consider only closed-shell systems here, which fortunately describe most molecules in their ground state.

The Hamiltonian operator for a $2N$-electron molecule with M nuclei in the Born–Oppenheimer approximation is given by

$$\hat{H} = -\frac{1}{2} \sum_{i=1}^{2N} \nabla_i^2 - \sum_{i=1}^{2N} \sum_{A=1}^{M} \frac{Z_A}{r_{iA}} + \sum_{i=1}^{2N} \sum_{j>i} \frac{1}{r_{ij}} + \sum_{A} \sum_{B<A}^{M} \frac{Z_A Z_B}{R_{AB}} \quad (12.1)$$

The first term here represents the kinetic energy of the electrons, the second term represents the interaction of each electron with each nucleus, the third term represents the electron–electron interactions, and the fourth term represents the internuclear interactions. Because the internuclear repulsion terms are constants for a given molecular geometry, we can ignore them for now and simply include them later.

The wave function is given by the (normalized) Slater determinant

$$\Psi(1, 2, \ldots, 2N) = \frac{1}{\sqrt{(2N)!}} \begin{vmatrix} \psi_1\alpha(1) & \psi_1\beta(1) & \cdots & \psi_N\alpha(1) & \psi_N\beta(1) \\ \psi_1\alpha(2) & \psi_1\beta(2) & \cdots & \psi_N\alpha(2) & \psi_N\beta(2) \\ \vdots & \vdots & \ddots & \vdots & \vdots \\ \psi_1\alpha(2N) & \psi_1\beta(2N) & \cdots & \psi_N\alpha(2N) & \psi_N\beta(2N) \end{vmatrix}$$

and the energy is given by

$$E = \langle \Psi^*(1, 2, \ldots, 2N) \mid \hat{H} \mid \Psi(1, 2, \ldots, 2N) \rangle \quad (12.2)$$

As in the atomic case, it is a straightforward but worthwhile exercise (Problem 12–1) to show that Equation 12.2 can be written as

$$E = 2 \sum_{j=1}^{N} I_j + \sum_{i=1}^{N} \sum_{j=1}^{N} (2J_{ij} - K_{ij}) \tag{12.3}$$

where

$$I_j = \int d\mathbf{r}_j \, \psi_j^*(\mathbf{r}_j) \left(-\frac{1}{2} \nabla_j^2 - \sum_{A}^{M} \frac{Z_A}{r_{jA}} \right) \psi_j(\mathbf{r}_j) \tag{12.4}$$

$$J_{ij} = \iint d\mathbf{r}_1 d\mathbf{r}_2 \, \psi_i^*(\mathbf{r}_1) \psi_j^*(\mathbf{r}_2) \frac{1}{r_{12}} \psi_i(\mathbf{r}_1) \psi_j(\mathbf{r}_2) \tag{12.5}$$

$$K_{ij} = \iint d\mathbf{r}_1 d\mathbf{r}_2 \, \psi_i^*(\mathbf{r}_1) \psi_j^*(\mathbf{r}_2) \frac{1}{r_{12}} \psi_i(\mathbf{r}_2) \psi_j(\mathbf{r}_1) \tag{12.6}$$

The factors of 2 in Equation 12.3 occur because we are considering a closed-shell system of $2N$ electrons, N of which have spin function α and N of which have spin function β. The J_{ij} integrals are called *coulomb integrals* and the K_{ij} integrals are called *exchange integrals* if $i \neq j$. Note that $K_{ii} = J_{ii}$ (Problem 12–2). Note also that the I_j in Equation 12.4 differs from the I_j in the atomic case (Equation 9.67) because we sum over electron–nuclear interactions in Equation 12.4. In the atomic case, there is only one term (one nucleus) in the summation.

As in the atomic case, the spatial orbitals $\psi_i(\mathbf{r}_i)$ are determined by applying the variational principle to Equation 12.3. When we do this, we find that the spatial orbitals that minimize the energy E satisfy the equations

$$\hat{F}(\mathbf{r}_1) \psi_i(\mathbf{r}_1) = \varepsilon_i \psi_i(\mathbf{r}_1) \qquad i = 1, 2, \ldots, N \tag{12.7}$$

where $\hat{F}(\mathbf{r}_1)$, the Fock operator, is given by

$$\hat{F}(\mathbf{r}_1) = \hat{f}(\mathbf{r}_1) + \sum_{j=1}^{N} [2\hat{J}_j(\mathbf{r}_1) - \hat{K}_j(\mathbf{r}_1)] \tag{12.8}$$

where

$$\hat{f}(\mathbf{r}_1) = -\frac{1}{2} \nabla_1^2 - \sum_{A} \frac{Z_A}{r_{1A}} \tag{12.9}$$

$\hat{J}_j(\mathbf{r}_1)$, called the *coulomb operator*, is given by

$$\hat{J}_j(\mathbf{r}_1) \psi_i(\mathbf{r}_1) = \psi_i(\mathbf{r}_1) \int d\mathbf{r}_2 \, \psi_j^*(\mathbf{r}_2) \frac{1}{r_{12}} \psi_j(\mathbf{r}_2) \tag{12.10}$$

and $\hat{K}_j(\mathbf{r}_1)$, called the *exchange operator*, is given by

$$\hat{K}_j(\mathbf{r}_1)\psi_i(\mathbf{r}_1) = \psi_j(\mathbf{r}_1) \int d\mathbf{r}_2 \, \psi_j^*(\mathbf{r}_2)\frac{1}{r_{12}}\psi_i(\mathbf{r}_2) \tag{12.11}$$

The eigenvalue in Equation 12.7 is called the Hartree–Fock orbital energy. Note that the summation in the definition of the Fock operator $\hat{F}(\mathbf{r})$ goes from 1 to N because there are N spatial orbitals. Note also that the integrals in Equations 12.10 and 12.11 are functions of $\mathbf{r}_1$ because $1/r_{12} = 1/|\mathbf{r}_1 - \mathbf{r}_2|$ and we integrate over $\mathbf{r}_2$, leaving a function of $\mathbf{r}_1$ behind. Furthermore, notice that the coulomb operator operating on $\psi_i(\mathbf{r}_1)$ gives $\psi_i(\mathbf{r}_1)$ times a function of $\mathbf{r}_1$. The exchange operator acts differently, however; the function that it acts upon ends up under the integral sign.

We can obtain an expression for the energy of the ith molecular orbital by multiplying Equation 12.7 from the left by $\psi_i^*(\mathbf{r}_1)$ and integrating over $\mathbf{r}_1$ to obtain

$$\varepsilon_i = \int d\mathbf{r}_1 \, \psi_i^*(\mathbf{r}_1)\hat{F}(\mathbf{r}_1)\psi_i(\mathbf{r}_1) \tag{12.12}$$

Using the above definition of the Fock operator, Equation 12.12 becomes

$$\varepsilon_i = I_i + \sum_{j=1}^{N}(2J_{ij} - K_{ij}) \tag{12.13}$$

where I_i, J_{ij}, and K_{ij} are given by Equations 12.4 through 12.6, respectively. If we compare this result to Equation 12.3, we see that

$$E = \sum_{i=1}^{N}(I_i + \varepsilon_i) \tag{12.14}$$

Note that E is *not* simply the sum of the Hartree–Fock orbital energies.

Equations 12.7 through 12.14 are almost the same as Equations 9.58 through 9.65. The only difference is that in the atomic case there is only one term in the summation over A in Equation 12.9 because there is only one nucleus. As in the atomic case, the Fock operator in Equation 12.7 depends upon all the orbitals, and so cannot be evaluated from Equations 12.10 and 12.11 until all the orbitals are known. Thus, Equation 12.7 represents a set of N *coupled* equations, which must be solved by a self-consistent procedure in which one assumes an initial set of orbitals $\psi_i(\mathbf{r}_i)$ and then calculates an initial set of Fock operators. Using these Fock operators, we can now solve Equation 12.7 to find a new set of orbitals. These new orbitals are used to calculate a new set of Fock operators, which in turn are used to calculate a still new set of orbitals. This cyclic procedure is continued until the orbitals of one cycle are essentially the same as those in the previous cycle, or, in other words, until they are self-consistent.

We have already carried out two fairly simple Hartree–Fock calculations explicitly. In Chapter 9, we carried out a Hartree–Fock SCF calculation for a helium atom, and in Chapter 10, we carried out a similar calculation for a hydrogen molecule. In each case, we expressed the orbital (an atomic orbital in the case of a helium atom and a molecular orbital in the case of a hydrogen molecule) as a linear combination of two Slater orbitals. In doing so, the Hartree–Fock equations are converted into matrix equations, which can be solved routinely using matrix methods. This procedure was developed in the 1950s, just as computers were becoming generally available, by Clemens Roothaan of the University of Chicago. He expressed the molecular orbitals, ψ, as linear combinations of basis functions, $\phi_v(\mathbf{r})$, $v = 1, \ldots, K$,

$$\psi = \sum_{v=1}^{K} c_v \phi_v \tag{12.15}$$

which were usually, but not necessarily, taken to be Slater orbitals. As the basis set becomes larger and larger, Equation 12.15 leads to more and more accurate representations of the molecular orbitals. The limit is called the *Hartree–Fock limit*, and gives the best molecular orbitals. Don't forget, however, that even these optimum molecular orbitals constitute an approximation to the true molecular wave function because the Hartree–Fock approximation assumes that each electron experiences an average potential of all the other electrons. In other words, the motion of the electrons is uncorrelated.

If we substitute Equation 12.15 into Equation 12.7 and then multiply both sides from the left by $\phi_\mu^*(\mathbf{r}_1)$ and integrate over $\mathbf{r}_1$, then we obtain

$$\sum_v F_{\mu v} c_v = \varepsilon \sum_v S_{\mu v} c_v \qquad \mu = 1, 2, \ldots, K \tag{12.16}$$

where the Fock matrix elements are given by

$$F_{\mu v} = \int d\mathbf{r}_1 \, \phi_\mu^*(\mathbf{r}_1) \hat{F}(\mathbf{r}_1) \phi_v(\mathbf{r}_1) \tag{12.17}$$

and the overlap matrix elements are given by

$$S_{\mu v} = \int d\mathbf{r}_1 \, \phi_\mu^*(\mathbf{r}_1) \phi_v(\mathbf{r}_1) \tag{12.18}$$

Both of these matrices are $K \times K$ Hermitian matrices; they are real and symmetric if the basis set is chosen to be a set of real functions, which is usually the case. The set of algebraic equations for the c_v (Equation 12.16) are the *Hartree–Fock–Roothaan equations*. They can be written more compactly by writing them in matrix notation:

$$\mathbf{F}\mathbf{c} = \varepsilon \mathbf{S}\mathbf{c} \tag{12.19}$$

In Equation 12.19, $\mathbf{F}$ and $\mathbf{S}$ are $K \times K$ matrices and $\mathbf{c}$ is a $K \times 1$ column vector. The equation is very similar to our standard matrix eigenvalues problem (MathChapter H) and will lead to a $K \times K$ secular determinant, which will give us K orbital energies ε and K eigenvectors $\mathbf{c}$, which gives us K molecular orbitals. In the ground electronic state, the N spatial orbitals of lowest energy will be occupied by $2N$ electrons.

Realize that the elements of F given by Equation 12.17 depend upon the c_j's in Equations 12.8 through 12.11. We can emphasize this by writing Equation 12.19 as

$$F(c)c = \varepsilon S c \tag{12.20}$$

As usual, Equation 12.20 must be solved by an iterative, self-consistent procedure.

The Hartree–Fock–Roothaan method was very well developed in the 1960s by Roothaan and numerous coworkers. A typical result for the ground electronic state of N_2 is given in Table 12.1. The first column in the table gives the basis sets for the resultant occupied molecular orbitals in the ground electronic state of N_2. The notation $\sigma_g 1s$ designates $1s_A + 1s_B$, and so forth. Similarly, $\sigma_u 1s$ designates $1s_A - 1s_B$, and so on. Because N_2 is a homonuclear diatomic molecule, the coefficients of similar atomic orbitals on the two nuclei are such that they always occur as sums or differences. For example, $\sigma_g 1s(5.682\,98)$ is the sum of two Slater $1s$ orbitals (one centered on each nucleus) with an orbital exponent $\zeta = 5.682\,98$; $\sigma_u 1s(5.955\,34)$ is the difference of Slater $1s$ orbitals with $\zeta = 5.955\,34$. The second column gives the coefficients of the members of the basis set in column 1 for the $1\sigma_g$ molecular orbital, the one of lowest energy. The third column gives the coefficients of the basis set for the $1\sigma_u$ molecular orbital of second lowest energy. The other columns give the coefficients for the indicated molecular orbitals.

Recall that σ orbitals are symmetric about the internuclear axis, so they consist of sums and differences of p_z, d_{z^2}, and f_{z^3} atomic orbitals. Pi orbitals, on the other hand, have a node containing the internuclear axis, and so consist of sums and differences of p_x, p_y, and the appropriate d and f orbitals. (See Figure 11.3 for a pictorial representation of both a π_u and a π_g orbital.)

EXAMPLE 12–1
Write out the basis set orbital $\sigma_g 2s(2.438\,75)$ in the first column in Table 12.1.

SOLUTION: The $\sigma_g 2s(2.438\,75)$ signifies the sum of two Slater $2s$ orbitals, each with orbital exponent $\zeta = 2.438\,75$. Using the fact that

$$S_{2s}(r) = \left(\frac{\zeta^5}{3\pi}\right)^{1/2} re^{-\zeta r}$$

we see that the $\sigma_g 2s(2.438\,75)$ orbital is given by

$$\sigma_g 2s(2.438\,75) = N[S_{2s}(r_A, 2.438\,75) + S_{2s}(r_B, 2.438\,75)]$$

$$= N\left(\frac{\zeta^5}{3\pi}\right)^{1/2} (r_A e^{-\zeta r_A} + r_B e^{-\zeta r_B})$$

where N is a normalization constant and $\zeta = 2.438\,75$.

TABLE 12.1

The Basis Set for the Ground State of N_2 at Its Equilibrium Internuclear Separation, 1.094 Å, and Values of the Expansion Coefficients for the Occupied MOs [a,b]

Basis set functions	MO					
	$1\sigma_g$	$1\sigma_u$	$2\sigma_g$	$2\sigma_u$	$3\sigma_g$	$1\pi_u$
$\sigma_g 1s$ (5.682 98)	0.923 19		−0.279 31		0.074 84	
$\sigma_g 1s$ (10.342 40)	0.152 04		−0.006 15		0.002 62	
$\sigma_g 2s$ (1.453 49)	0.000 90		0.141 06		−0.458 59	
$\sigma_g 2s$ (2.438 75)	−0.000 03		0.599 48		−0.176 62	
$\sigma_g 3s$ (7.040 41)	−0.085 01		−0.023 33		−0.006 78	
$\sigma_g 2p$ (1.282 61)	0.000 41		0.116 02		0.429 14	
$\sigma_g 2p$ (2.569 88)	0.001 04		0.259 07		0.484 53	
$\sigma_g 2p$ (6.216 98)	0.001 09		0.010 92		0.024 78	
$\sigma_g 3d$ (1.341 42)	0.000 17		0.036 26		0.046 96	
$\sigma_g 3d$ (2.916 81)	0.000 98		0.039 48		0.030 65	
$\sigma_g 3d$ (5.520 63)	−0.000 18		−0.002 75		−0.001 58	
$\sigma_g 4f$ (2.594 49)	0.000 32		0.013 34		0.010 86	
$\sigma_u 1s$ (5.955 34)		0.934 06		−0.243 70		
$\sigma_u 1s$ (10.658 79)		0.114 83		−0.000 00		
$\sigma_u 2s$ (1.570 44)		−0.011 56		0.364 37		
$\sigma_u 2s$ (2.489 65)		0.004 79		0.547 02		
$\sigma_u 3s$ (7.291 69)		−0.053 43		−0.030 54		
$\sigma_u 2p$ (1.485 49)		−0.006 79		−0.413 55		
$\sigma_u 2p$ (3.499 90)		0.002 94		−0.109 45		
$\sigma_u 3d$ (1.690 03)		−0.001 21		−0.035 53		
$\pi_u 2p$ (1.384 36)						0.469 21
$\pi_u 2p$ (2.532 88)						0.398 69
$\pi_u 2p$ (5.691 76)						0.031 41
$\pi_u 3d$ (2.057 07)						0.059 38
$\pi_u 3d$ (2.706 50)						0.017 38
$\pi_u 4f$ (3.068 96)						0.012 33

a. The orbital energies (in hartrees) are $\varepsilon_{1\sigma_g} = -15.681\,95$, $\varepsilon_{1\sigma_u} = -15.678\,33$, $\varepsilon_{2\sigma_g} = -1.473\,60$, $\varepsilon_{2\sigma_u} = -0.777\,96$, $\varepsilon_{3\sigma_g} = -0.634\,95$, and $\varepsilon_{1\pi_u} = -0.615\,44$.

b. *Source:* Cade, P. E., Wahl, A. C., Hartree–Fock–Roothaan Wavefunctions for Diatomic Molecules: II. First-Row Homonuclear Systems. *At. Data Nucl. Data Tables,* **13**, 339 (1974).

The difficulty with calculations on polyatomic molecules is that there are basis functions centered on all the nuclei, and so many (actually, most) of the above integrals will involve basis functions centered on more than two nuclei. These three- and four-centered integrals (multicentered integrals) are very difficult to evaluate, to say nothing of the fact that the number of basis functions, and so the number of integrals, increases significantly for polyatomic molecules. The resolution to this difficulty was proposed as early as 1950 by the British chemist, Frank Boys, who recognized that multicentered integrals can be evaluated analytically in simple form if we use Gaussian-type functions instead of Slater orbitals as the basis functions. Gaussian functions are of the form

$$g_s(r, \alpha) = \left(\frac{2\alpha}{\pi}\right)^{3/4} e^{-\alpha r^2}$$

$$g_x(r, x, \alpha) = \left(\frac{128\alpha^5}{\pi^3}\right)^{1/4} x e^{-\alpha r^2}$$

$$g_y(r, y, \alpha) = \left(\frac{128\alpha^5}{\pi^3}\right)^{1/4} y e^{-\alpha r^2} \tag{12.21}$$

$$g_z(r, z, \alpha) = \left(\frac{128\alpha^5}{\pi^3}\right)^{1/4} z e^{-\alpha r^2}$$

and so on. (See also Problem 12–28.) The factors in front of these expressions are normalization constants (Problem 12–9). Note that these Gaussian functions have the same angular symmetry as Slater orbitals. The primary difference is the Gaussian factor, $e^{-\alpha r^2}$, instead of an exponential factor, $e^{-\zeta r}$. We shall see below that g_s Gaussian functions are used to represent all s orbitals, be they $1s$, $2s$, $3s$, and so forth. There is no such thing as a g_{1s} Gaussian function, for example. Similarly, g_x Gaussian functions are used to represent all p_x orbitals, be they $2p_x$, $3p_x$, and so forth.

The advantage of using Gaussian functions is that the product of two Gaussian functions centered at two different positions can be written as a single Gaussian function centered at one position. This property reduces all three- and four-centered integrals to two-centered integrals, which in turn can be evaluated analytically. Problems 12–10 through 12–17 review the properties of Gaussian functions and show how multicentered integrals are easy to evaluate using Gaussian functions.

The idea of using Gaussian functions lay fairly dormant until John Pople and numerous coworkers developed their use into an extremely successful methodology that dominates essentially all modern molecular calculations. The basic idea is to replace Slater orbitals with Gaussian functions. The problem with this idea is that Slater orbitals and Gaussian functions have very different behavior for small values of r (and for large values of r as well). Figure 12.2 compares a normalized Slater $1s$ orbital with $\zeta = 1$,

$$S_{1s} = \left(\frac{1}{\pi}\right)^{1/2} e^{-r} \tag{12.22}$$

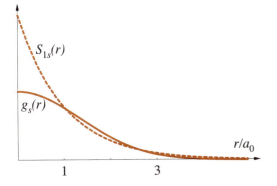

FIGURE 12.2
A comparison of a Slater $1s$ orbital with $\zeta = 1$ (dotted curve) (Equation 12.22) with a Gaussian function with $\alpha = 0.2710$ (solid curve) (Equation 12.23). This value of α is chosen to maximize the overlap between the two functions.

with a Gaussian s function,

$$g_s = \left(\frac{2\alpha}{\pi}\right)^{3/4} e^{-\alpha r^2} \tag{12.23}$$

with $\alpha = 0.2710$. This value of α has been chosen to maximize the overlap between $S_{1s}(r)$ and $g_s(r)$ (Problem 12–19). Note that the Slater orbital has a cusp at $r = 0$ and is larger than the Gaussian function at large values of r.

The discrepancy between a Slater orbital and a Gaussian function turns out to be significant in molecular calculations. To overcome this difficulty, we use a linear combination of Gaussian functions to curve-fit one Slater orbital, the fit improving with N, the number of Gaussian functions used. Figure 12.3 shows this fit as a function

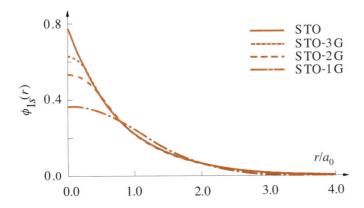

FIGURE 12.3
The Slater orbital, STO (solid curve), compared with various contracted Gaussian functions consisting of one, two, and three primitive Gaussian functions (various dashed lines).

TABLE 12.2
Orbital Exponents, ζ, for the Slater Orbitals
of the Atoms of the First Two Rows of the
Periodic Table

Atom	ζ_{1s}	$\zeta_{2s} = \zeta_{2p}$
H	1.24	
He	1.69	
Li	2.69	0.80
Be	3.68	1.15
B	4.68	1.50
C	5.67	1.72
N	6.67	1.95
O	7.66	2.25
F	8.56	2.55
Ne	9.64	2.88

Because of this convenient scaling relation, Gaussian fits are usually made to Slater orbitals with $\zeta = 1.00$ and then the various Gaussian exponents are scaled according to Equation 12.25. The linear coefficients are the same, so all we have to do is to scale the exponents.

EXAMPLE 12–5
Given the fit given by Equation 12.24,

$$S_{1s}(r, 1.0000) = 0.4446\, g_s(r, 0.109\,82) + 0.5353\, g_s(r, 0.405\,77)$$
$$+ 0.1543\, g_s(r, 2.2277)$$

write out the corresponding fit for $S_{1s}(r, \zeta = 1.24)$.

SOLUTION: Using Equation 12.25, we find that

$$\alpha_1 = (1.24)^2(0.109\,82) = 0.1688$$

$$\alpha_2 = (1.24)^2(0.405\,77) = 0.6239$$

$$\alpha_3 = (1.24)^2(2.2277) = 3.425$$

Therefore,

$$S_{1s}(r, \zeta = 1.24) = 0.4446\, g_s(r, 0.1688) + 0.5353\, g_s(r, 0.6239) + 0.1543\, g_s(r, 3.425)$$

12.3 Extended Basis Sets Must Be Used to Obtain Accurate Results

The STO-NG orbitals that we have discussed so far constitute a minimal basis set. Hartree–Fock–Roothaan calculations with minimal basis sets usually give adequate, but not great, results. Bond lengths typically are in error by several hundredths of angstroms and bond angles are in error by several degrees. One inadequacy of a minimal basis set is due to the fact that the orbital exponents are fixed, and so an orbital is unable to contract or expand in different molecular environments. The fixed orbital exponents simply make the orbitals too rigid. Although it's impractical to vary orbital exponents because of their nonlinearity, we can achieve almost the same result by using a linear combination of orbitals of the same type (same symmetry) but with different orbital exponents.

For example, a $2s$ orbital be written as

$$\phi_{2s}(r) = S_{2s}(r, \zeta_1) + d S_{2s}(r, \zeta_2) \tag{12.26}$$

where d is a parameter to be determined from the Hartree–Fock–Roothaan equations. The advantage of doing this can be seen as follows. The Slater orbitals $S_{2s}(r, \zeta_1)$ and $S_{2s}(r, \zeta_2)$ represent different-sized $2s$ orbitals for $\zeta_1 \neq \zeta_2$ (see Figure 12.4).

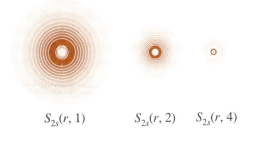

$S_{2s}(r, 1)$ $S_{2s}(r, 2)$ $S_{2s}(r, 4)$

FIGURE 12.4
An illustration of the dependence of the size of a $2s$ orbital on the value of the orbital exponent. The values of ζ are 1.00, 2.00, and 4.00.

Using a linear combination of $S_{2s}(r, \zeta_1)$ and $S_{2s}(r, \zeta_2)$, we can construct an atomic orbital whose size can range between that specified by $S_{2s}(r, \zeta_1)$ and $S_{2s}(r, \zeta_2)$ by varying the constant d in Equation 12.26, as shown in Figure 12.5. Because both functions are of the same type (S_{2s} in this case), the linear combination retains the desired symmetry of the atomic orbital. Basis sets generated from a sum of two orbitals are called *double-zeta basis sets* because each orbital in the basis set is the sum of two orbitals that differ only in their value of the orbital exponent, ζ (zeta).

Usually only the valence orbitals are expressed by a double-zeta representation. The inner-shell electrons are still described by a single orbital. For example, the electrons in the $1s$ atomic orbital on a carbon atom would be described by a single $1s$ orbital with ζ given in Table 12.2, whereas the electrons in the $2s$ atomic orbital would be described by a linear combination of two $2s$ orbitals with different values of the orbital exponent, ζ. Basis sets that describe the inner-shell electrons by a single orbital and the valence shell electrons by a sum of orbitals are called *split-valence basis sets*. In the simplest split-valence basis set, hydrogen and helium atoms are represented by two

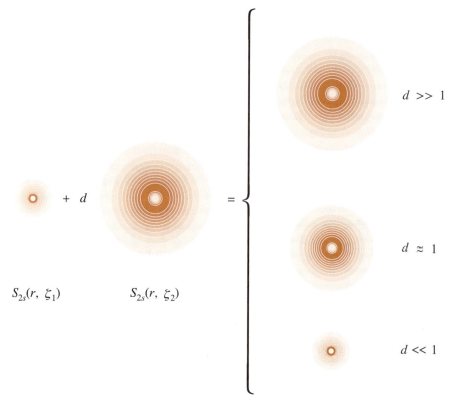

FIGURE 12.5
A linear combination of two orbitals of the same type (S_{2s} in the case shown) but with different orbital exponents ζ_1 and ζ_2 can generate an atomic orbital of adjustable size by varying the constant d in Equation 12.26.

s-type orbitals and the next two rows of atoms in the periodic table are represented by two complete sets of valence s and p orbitals. For the atoms hydrogen through argon, we have

H, He:	$1s$
	$1s'$
Li through Ne:	$1s$
	$2s'$, $2p'_x$, $2p'_y$, $2p'_z$
	$2s''$, $2p''_x$, $2''_y$, $2p''_z$
Na through Ar:	$1s$
	$2s$, $2p_x$, $2p_y$, $2p_z$
	$3s'$, $3p'_x$, $3p'_y$, $3p'_z$
	$3s''$, $3p''_x$, $3p''_y$, $3p''_z$

EXAMPLE 12-6

How many orbitals are used in a split-valence basis set calculation for methane, CH_4?

SOLUTION: Each hydrogen atom is assigned two $1s$ orbitals; the carbon atom is assigned one $1s$ orbital, two $2s$ orbitals, and two of each of the $2p$ orbitals. Thus, we have a total of $4 \times 2 + 1 + 2 + 2 \times 3 = 17$ orbitals.

Up to this point, we have based our discussion on Slater orbitals. The term "double-zeta" basis set comes from the use of the sum of Slater orbitals with different orbital exponents (typically denoted by ζ for Slater orbitals). Nevertheless, the Gaussian basis sets other than the STO-NG minimal basis sets that are in common use nowadays are no longer based on Slater orbitals. Consider the fairly simple basis set designated by 3-21G. The 3 indicates that each nonvalence orbital is represented by a sum of three Gaussian functions, which as a whole is called a *contracted Gaussian function*; the dash indicates that the basis set is a split-valence basis set; the 21 indicates that each valence orbital is represented by two contracted Gaussian functions, one consisting of two primitive Gaussian functions and the other consisting of one (hardly a contraction); and the G emphasizes that all the orbitals are expressed in terms of Gaussian functions.

For example, for a hydrogen atom, which has no nonvalence orbitals, we write

$$\phi'_{1s}(\mathbf{r}) = \sum_{i=1}^{2} d'_{i,1s} g_s(\mathbf{r}, \alpha'_{i,1s})$$

$$\phi''_{1s}(\mathbf{r}) = g_s(\mathbf{r}, \alpha''_{i,1s})$$

(12.27)

By convention, $\phi'_{1s}(\mathbf{r})$ is such that it is more compact than $\phi''_{1s}(\mathbf{r})$, meaning that $\phi''_{1s}(\mathbf{r})$ is of greater extent than $\phi'_{1s}(\mathbf{r})$. This will result when the orbital exponent in $\phi''_{1s}(\mathbf{r})$ is smaller than those in $\phi'_{1s}(\mathbf{r})$. We shall refer to these orbitals as the compact and extended orbitals.

For the atoms lithium through fluorine, we have

$$\phi_{1s}(\mathbf{r}) = \sum_{i=1}^{3} d_{i,1s} g_s(\mathbf{r}, \alpha_{i,1s})$$

$$\phi'_{2s}(\mathbf{r}) = \sum_{i=1}^{2} d'_{i,2s} g_s(\mathbf{r}, \alpha'_{i,2sp})$$

$$\phi''_{2s}(\mathbf{r}) = g_s(\mathbf{r}, \alpha''_{2sp})$$

(12.28)

$$\phi'_{2p}(\mathbf{r}) = \sum_{i=1}^{2} d'_{i,2p} g_{2p}(\mathbf{r}, \alpha'_{i,2sp})$$

$$\phi''_{2p}(\mathbf{r}) = g_{2p}(\mathbf{r}, \alpha''_{2sp})$$

Let's be sure to understand everything in Equations 12.28. First note that the $1s$ orbital, a nonvalence orbital, is represented by a contraction of three Gaussian functions, and hence the 3 in the 3-21G designation. The $\phi'_{2s}(\mathbf{r})$ orbital, the compact $2s$ valence orbital, is represented by a contraction of two Gaussian functions, and hence the 2 in the 3-21G designation. The primitive Gaussian functions in ϕ'_{2s} are of the same form as those used in the $1s$ contraction; only the values of the parameters are different. The $\phi''_{2s}(\mathbf{r})$ orbital, the extended valence orbital, is represented by just one Gaussian function, and hence the 1 in the 3-21G designation. Finally, and importantly, note that the compact $2s$ and $2p$ orbitals, $\phi'_{2s}(\mathbf{r})$ and $\phi'_{2p}(\mathbf{r})$, have the same orbital exponents, and that the same is true for the extended valence orbitals $\phi''_{2s}(\mathbf{r})$ and $\phi''_{2p}(\mathbf{r})$. This is generally the case for Gaussian basis sets.

The contribution of a carbon atom, for example, to a molecular orbital is given by a linear combination of the atomic orbitals in Equations 12.28:

$$\psi_C = c_1\phi_{1s} + c_2\phi'_{2s} + c_3\psi''_{2s} + c_4\phi'_{2p} + c_5\phi''_{2p}$$

The molecular Hartree–Fock calculation treats the ϕ's as known quantities and determines the linear coefficients, c_j. The question that arises here is where do these "known" quantities come from. In other words, how do we know the values of the d's and α's in Equations 12.28? A number of ways have been used by quantum chemists. We saw in the case of STO-NG basis sets that the contracted Gaussian functions were determined by curve-fitting linear combinations of Gaussian functions to Slater orbitals. For split-valence basis sets and other extended basis sets, the parameters of contracted Gaussian functions like those in Equations 12.28 are determined directly from optimization of atomic energies, or perhaps the energies and bond lengths of a few small test molecules. This procedure and variations of this procedure have been used to produce sets of atomic orbitals consisting of contracted Gaussian functions to be used in molecular calculations. We shall see in the next section that these basis sets are available on-line.

EXAMPLE 12–7
How many primitive Gaussian functions and how many contracted Gaussian functions are used in a 3-21G calculation of a methane molecule, CH_4?

SOLUTION: Each hydrogen atom is assigned two $1s$ orbitals; one of them is represented by a contraction of two Gaussian functions and the other by one Gaussian function. Thus, the four hydrogen atoms are assigned $4 \times 3 = 12$ primitive Gaussian functions and $4 \times 2 = 8$ contracted Gaussian functions. The carbon atom is assigned one $1s$ orbital, which is represented by a contraction of three Gaussian functions; two valence $2s$ orbitals, one of which is represented by a contraction of two Gaussian functions and the other by just one Gaussian function; each of one set of the three $2p$ valence orbitals is represented by a contraction of two Gaussian functions, and each of the $2p$ orbitals in the other set is represented by just one Gaussian function. Thus, we have total of $1 \times 3 + 1 \times 2 + 1 \times 1 + 3 \times 2 + 3 \times 1 = 15$ primitive Gaussian functions and nine contracted Gaussian functions for the carbon atom. The total number of primitive Gaussian functions for the molecule is $12 + 15 = 27$ and the total number of contracted Gaussian functions is 17.

We have used the 3-21G basis set to illustrate a split-valence basis set, but there is a whole family of split-valence basis sets. To specify how many primitive Gaussian functions are used in the contractions that are used for the various orbitals, we will use the notation N-MPG, where N, M, and P are integers. In this notation, N is the number of primitive Gaussian functions that are contracted to represent the inner-shell (nonvalence) orbitals, the hyphen indicates that we have a split-valence basis set, and the integers M and P designate the number of primitive Gaussian functions that are contracted to represent the two sets of split-valence-shell orbitals; M is the number of primitive Gaussian functions that are contracted to represent the more compact orbital (the one with the larger orbital exponents), and P is the number of primitive Gaussian functions that are contracted to represent the more extended orbital [the one with the smaller orbital exponent(s)]. The G simply tells us that we are using Gaussian functions.

For example, one popular split-valence basis set is the 6-31G basis set. Consider a carbon atom in the 6-31G basis. The 6 tells us that the $1s$ orbital on the carbon atom (the nonvalence orbital) is represented by a contraction of six Gaussian functions. The hyphen indicates a split-valence basis set, telling us that the valence orbitals are split. One of the $2s$ orbitals, the compact one, is represented by a contraction of three Gaussian functions (hence the 3), with relatively large values of the α's. The other $2s$ orbital, the extended one, is represented by one Gaussian function (hence the 1), with a relatively small value of α. The $2p$ orbitals are represented by contractions of three Gaussian functions (the compact $2p$ orbitals) and one Gaussian function (the extended $2p$ orbitals). (See Problem 12–30.)

EXAMPLE 12–8
Describe the orbitals of a carbon atom in a 6-311G basis set. This basis set describes a triple split-valence basis set.

SOLUTION: In this case, the $1s$ orbital (the nonvalence orbital) is represented by a contraction of six Gaussian functions. The valence orbitals are split into three orbitals each. One of the $2s$ orbitals, the compact one, is represented by a contraction of three Gaussian functions, with relatively large values of the α's. Another $2s$ orbital is represented by one Gaussian function, with a value of α smaller than those in the compact orbital, and another $2s$ orbital is represented by one Gaussian function with the smallest value of α. The $2p$ orbitals are also split into three, and are represented by contractions of three, one, and one Gaussian functions, with decreasing values of the α's. (See Problem 12–31.)

12.4 Asterisks in the Designation of a Basis Set Denote Orbital Polarization Terms

Consider the formation of the minimal basis set σ 1s molecular orbital in H_2, formed from a 1s orbital on each hydrogen atom. As we discussed in Section 10.6, the electron distribution about each hydrogen atom does not remain spherically symmetric as the two atoms approach each other. We took this effect into account by constructing the molecular orbital from a linear combination of a 1s orbital and a $2p_z$ orbital on each hydrogen atom instead of from just a 1s orbital. In this manner, we accommodate the fact that atomic orbitals distort as atoms are brought together, which is called *polarization*. We can account for polarization by including orbitals of a higher orbital angular-momentum quantum number, l, to a given orbital, just as we added a $2p_z$ orbital to a 1s hydrogen orbital in Section 10.6. In addition, d character can be added to the description of the valence electrons in $2p$ orbitals, thereby providing a representation of the asymmetric shape of the electron density along the chemical bonds involving $2p$ orbitals (Figure 12.6). The use of such functions for the second-row elements in the periodic table is denoted by an asterisk *; for example, the notation 6-31G* [also designated by 6-31G(d)] designates a 6-31G basis set to which we add d orbitals to the p orbitals. The presence of a double asterisk, **, denotes that polarization is also being taken into account for the orbital descriptions on hydrogen atoms by adding a p-type orbital to the hydrogen 1s orbital. In particular, the basis set 6-31G** [also denoted by 6-31G(d,p)] is useful in systems in which hydrogen bonding occurs.

$$2s(\mathbf{r}) + (30\%)2p_x(\mathbf{r}) =$$

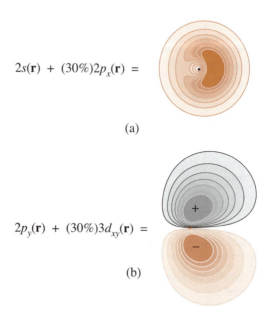

(a)

$$2p_y(\mathbf{r}) + (30\%)3d_{xy}(\mathbf{r}) =$$

(b)

FIGURE 12.6
An illustration of how polarization can be achieved by adding (a) a p orbital to an s orbital and (b) a d orbital to a p orbital.

EXAMPLE 12–9
Write the general expressions for the 1s orbital on a carbon atom in the 6-31G** basis sets.

SOLUTION: This is a split-valence basis set. The first number of the designation indicates the total number of Gaussian functions that are contracted to represent the 1s orbital. Thus, we have

$$\phi_{1s}(r) = \sum_{i=1}^{6} d_{1si} g_s(r, \alpha_{1si})$$

The ** on the 6-31G** tells us that a set of Gaussian p orbitals is added to the representation of the 1s orbital. Thus, we have

$$\phi_{1s}(r) = \sum_{i=1}^{6} d_{1si} g_s(r, \alpha_{1si}) + d_{2p_x} g_x(r, \alpha_{2p}) + d_{2p_y} g_y(r, \alpha_{2p}) + d_{2p_z} g_z(r, \alpha_{2p})$$

The values of the d_{1si} and the α_{1si} are fixed by some chosen method as described above, and a standard Gaussian exponent of $\alpha_{2p} = 1.1$ is commonly used for the added p functions. The coefficients d_{2p_z}, d_{2p_y}, and d_{2p_x} are optimized as part of the Hartree–Fock variational calculation.

In addition to adding polarization to 3-21G and 6-31G basis sets, for example, it is sometimes desirable to add what are called *diffuse functions*. Diffuse functions are Gaussian functions with fairly small orbital exponents, which cause the Gaussian functions to be large in extent. Diffuse functions are particularly useful for describing electrons that are relatively far from the nucleus, such as lone-pair electrons or electrons in anions. We usually add one diffuse function, composed of a single Gaussian function, for each valence orbital of the atoms beyond H and He. A basis set with diffuse functions is designated by inserting a + sign before the G of the basis set designation to which the diffuse functions are added. For example, the 6-31+G* [6-31+G(d)] basis set for a carbon atom consists of the orbitals

1s
$2s'$, $2p'_x$, $2p'_y$, $2p'_z$
$2s''$, $2p''_x$, $2p''_y$, $2p''_z$
five 3d orbitals
$2s+$, $2p_x+$, $2p_y+$, $2p_z+$

for a total of 18 basis functions, where the orbitals $2s+$ and so on are the diffuse orbitals.

Of the basis sets that we have discussed, 3-21G is considered to be a small basis set and the 6-31+G* and 6-31+G** basis sets are considered to be moderate-sized. To give an idea about the relative results from these basis sets, we give the bond lengths

TABLE 12.3

Calculated and Experimental Bond Lengths (Angstroms) and Bond Angles (Degrees) for Small Molecules for Different Basis Sets

Molecule	Geometrical parameter	STO-3G	3-21G	3-21G*	6-31G*	6-31G**	6-31+G**	Expt.
H_2	r(HH)	0.712	0.735	0.735	0.730	0.732	0.733	0.742
CH_4	r(CH)	1.083	1.083	1.083	1.084	1.084	1.084	1.092
NH_3	r(NH)	1.003	1.003	1.003	1.002	1.001	1.000	1.012
	∠(HNH)	104.2	112.4	112.4	107.2	107.6	108.9	106.7
H_2O	r(OH)	0.990	0.967	0.967	0.947	0.943	0.943	0.958
	∠(HOH)	100.0	107.6	107.6	105.5	105.9	107.1	104.5
HF	r(FH)	0.956	0.937	0.937	0.911	0.901	0.902	0.917
SiH_4	r(SiH)	1.422	1.487	1.475	1.475	1.476	1.476	1.481
PH_3	r(PH)	1.378	1.423	1.402	1.403	1.405	1.405	1.420
	∠(HPH)	95.0	96.1	95.2	95.4	95.6	95.7	93.3
H_2S	r(SH)	1.329	1.350	1.327	1.326	1.327	1.328	1.336
	∠(HSH)	92.5	95.8	94.4	94.4	94.4	94.4	92.1
HCl	r(ClH)	1.313	1.293	1.267	1.267	1.266	1.261	1.275

and bond angles obtained from Hartree–Fock–Roothaan calculations for a number of small molecules in Table 12.3.

All the results in Table 12.3 are easily calculated using any number of commercially available quantum chemistry programs. We're going to learn a little about using these programs in the next section. These programs have all the above basis sets built into them, so you never have to input them. Nevertheless, it's nice to have some knowledge about these basis sets rather than to just use them blindly. You can see the basis set used by any of the programs at *http://gnode2.pnl.gov/bse/portal* (formerly *http://www.emsl.pnl.gov/forms/basisforms.html*), a website maintained by the government-supported Pacific Northwest Laboratory in Richland, Washington. For example, if you ask for the 6-31G** basis set that is used by Gaussian 94 for a carbon atom, you obtain the result shown in Figure 12.7. The first column of numbers lists the orbital exponents and the other columns list the linear coefficients. For the two sets of rows headed by SP, the second column gives the coefficients for the s orbitals and the third column gives the coefficients for the p orbitals. Notice that the 2s and 2p orbitals have the same orbital exponents. As we've said before, this is generally the case and leads to great computational efficiency.

Figure 12.8 compares the 1s orbital of a carbon atom from the 6-31** basis set shown in Figure 12.7 to the Hartree–Fock 1s orbital given in the website *http://www.ccl.net/cca/data/atomic-RHR-wavefunctions/tables* given in Chapter 9.

```
C     0
S     6        1.00
            3047.5249000                    0.0018347
             457.3695100                    0.0140373
             103.9486900                    0.0688426
              29.2101550                    0.2321844
               9.2866630                    0.4679413
               3.1639270                    0.3623120
SP    3        1.00
               7.8682724                   -0.1193324              0.0689991
               1.8812885                   -0.1608542              0.3164240
               0.5442493                    1.1434564              0.7443083
SP    1        1.00
               0.1687144                    1.0000000              1.0000000
D     1        1.00
               0.8000000                    1.0000000
****
```

FIGURE 12.7
A screen shot of the 6-31G** basis set for a carbon atom taken from *http://gnode2.pnl.gov/bse/portal*

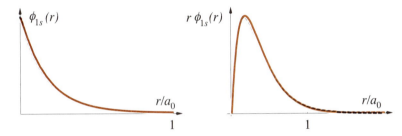

FIGURE 12.8
A comparison of the 1s orbital given by Figure 12.7 (solid orange curve) and the Hartree–Fock 1s orbital (dashed, black curve) given in the website *http://www.ccl.net/cca/data/atomic-RHR-wavefunctions/tables*

EXAMPLE 12–10
Use the entries in Figure 12.7 to write out the contracted Gaussian functions for the $2s$ and $2p$ orbitals.

SOLUTION: The compact $2s$ orbital is given by

$$\phi'_{2s} = -0.119\,3324\,g_s(r,\,7.868\,2724) - 0.160\,8542\,g_s(r,\,1.881\,2885)$$

$$+ 1.143\,4564\,g_s(r,\,0.544\,2493)$$

and the extended $2s$ orbital is given by

$$\phi''_{2s} = g_s(r,\,0.168\,7144)$$

where $g_s(r, \alpha)$ is given by Equation 12.21. Note that the orbital exponent of ϕ_{2s}'' is much smaller than the orbital exponents of ϕ_{2s}'. This causes the ϕ_{2s}'' orbital to be more extended than the ϕ_{2s}' orbital. Figure 12.9 compares the ϕ_{2s}' and ϕ_{2s}'' orbitals.

The compact $2p_x$ orbital is given by

$$\phi_{2p_x}' = 0.068\,9991\,g_x(r, x, 7.868\,2724) + 0.316\,4240\,g_x(r, x, 1.881\,2885)$$

$$+ 0.744\,3083 g_x(r, x, 0.544\,2493)$$

and the extended $2p_x$ orbital is given by

$$\phi_{2p_x}'' = g_x(r, x, 0.168\,7144)$$

where $g_x(r, x, \alpha)$ is given by Equation 12.21. There are similar results for $2p_y$ and $2p_z$ orbitals. Note that the $2s$ and $2p$ orbitals share the same orbital exponents.

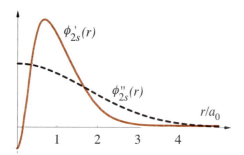 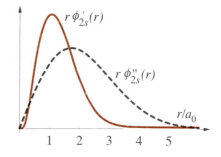

FIGURE 12.9
A comparison of the compact (solid curve) and extended (dashed curve) $2s$ orbitals for the 6-31G** basis set given in Figure 12.7.

When you go to the Pacific Northwest Laboratory website, you'll see that there are over 100 basis sets to choose from. The ones that we have discussed in this chapter are due to Pople and his coworkers and are sometimes called Pople-type basis sets. Another type of basis set has a designation such as cc-pVTZ, which stands for *correlation consistent, polarized valence triple zeta*. These cc-pxxx basis sets were developed by Thom Dunning for calculations that include electron correlation (see Section 12.7) and are commonly used in post-Hartree–Fock calculations.

12.5 Gaussian 03 and WebMO Are the Quantum Chemistry Programs Most Commonly Used by Undergraduate Chemistry Students

We have said that there are a number of commercially available computer programs that can be used by nonexperts to calculate molecular properties. Some of the most widely used programs are Gaussian 03, a commercially available program developed by Gaussian, Inc. (*www.gaussian.com*), GAMESS (General Atomic and Molecular Electronic Structure System, *www.msg.ameslab.gov/GAMESS/*), maintained by the Mark Gordon Research Group at the University of Iowa, and SPARTAN, another commercially available program developed by WaveFunction, Inc. (*www.wavefun.com*). When the first edition of this book was published in 1983, molecular calculations could be done only by professional quantum chemists, but they can now be carried out by any undergraduate chemistry student, thanks to the availability of programs such as the ones listed above. Not only do many college chemistry departments have a license for one or more of these programs, but the Shoder Education Foundation (*http://chemistry. ncsm.edu*, formerly *www.shoder.org/chemistry*) has developed a North Carolina High School Computational Chemistry Server, where high school students carry out molecular calculations. Table 12.4 lists some of the most popular quantum chemistry programs along with their websites.

In this section, we shall use an acetaldehyde molecule (CH_3CHO) to illustrate what is involved in running a Gaussian calculation, but the steps involved are very similar for the other programs. To carry out a calculation, you must specify the starting geometry or the coordinates of each atom in the molecule of interest and the *model chemistry* to use. By the term "model chemistry," we mean the combination of level

TABLE 12.4
Some Popular Quantum Chemistry Programs and Their Websites

Name	Website	Cost
ACES II	*www.aces2.de/*	free
ADF	*www.scm.com/*	commercial
Dalton	*www.kjemi.uio.no/software/dalton/*	free
GAMESS	*www.msg.ameslab.gov/GAMESS*	free
Gaussian	*www.gaussian.com*	commercial
Jaquar	*www.schrodinger.com/*	commercial
Molpro	*www.molpro.net/*	commercial
MOPAC	*www.openmopac.net/*	free
NWChem	*www.emsl.pnl.gov/docs/nwchem/nwchem.html*	free
Q-Chem	*www.q-chem.com/*	commercial
SPARTAN	*www.wavefun.com/*	commercial

of theory (such as Hartree–Fock, configuration interaction, and so on) and the basis set. In our case, we shall do a Hartree–Fock calculation using a 6-311G(d,p) basis set. We designate this model chemistry by HF/6-311G(d,p). Early versions of Gaussian required that you create an input file containing the initial coordinates of all the atoms in the molecule, which could be quite demanding for large molecules. Fortunately, there are now graphical user interfaces that make setting up, running, and viewing the results of a Gaussian calculation much easier. We will illustrate the use of these interfaces for GaussView, developed by Gaussian, Inc., and packaged with Gaussian 03, and WebMO, a free web-based interface used by many educational institutions. WebMO also supports the use of GAMESS and several other quantum chemistry programs in addition to Gaussian.

The first step in setting up a Gaussian calculation using GaussView is to build the molecule pictorially from atoms and various bond types. Once the molecule is built, an initial set of atomic coordinates is automatically generated from the pictorial representation. This set of atomic coordinates is used as an initial geometry in the Gaussian calculation. We're using an acetaldehyde molecule as a simple illustrative example, but the pictorial molecule builder in GaussView and WebMO includes segments such as saturated rings, aromatic rings, and segments of biochemical molecules to build more complicated and larger molecules easily.

The next step in the process is shown in Figure 12.10. You select Job Type [such as a single-point (fixed geometry) energy calculation, an optimization of the geometry, or an IR frequency calculation], and then Method, which specifies the model chemistry, which in our case is HF/6-311G(d,p). Upon submitting the job, you get a multipage output containing a plethora of information.

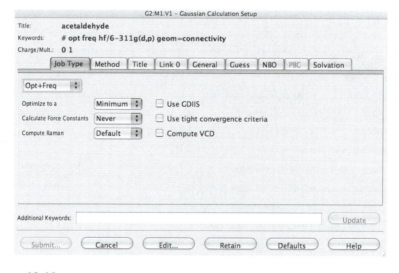

FIGURE 12.10
A screen shot of the Gaussian Calculation Setup page in GaussView. This window specifies the model chemistry used in the calculation.

```
Orbital symmetries:
     Occupied   (A') (A') (A') (A') (A') (A') (A') (A') (A") (A')
                (A") (A')
     Virtual    (A') (A") (A') (A") (A') (A') (A') (A") (A') (A')
                (A") (A') (A') (A") (A') (A') (A') (A') (A') (A')
                (A") (A') (A") (A') (A") (A') (A") (A") (A') (A')
                (A") (A') (A') (A') (A") (A') (A") (A') (A') (A")
                (A') (A') (A') (A") (A') (A') (A") (A') (A") (A')
                (A') (A") (A') (A") (A') (A') (A') (A") (A') (A')
                (A") (A') (A') (A') (A') (A')
The electronic state is 1-A'.
Alpha  occ. eigenvalues --   -20.55164 -11.32875 -11.23673  -1.41113  -1.01990
Alpha  occ. eigenvalues --    -0.80098  -0.67954  -0.62903  -0.60721  -0.56246
Alpha  occ. eigenvalues --    -0.51064  -0.42043
Alpha  virt. eigenvalues --    0.15524   0.15741   0.18606   0.21232   0.23115
Alpha  virt. eigenvalues --    0.35797   0.38132   0.45501   0.48483   0.55389
Alpha  virt. eigenvalues --    0.60494   0.62239   0.74508   0.77882   0.78158
Alpha  virt. eigenvalues --    0.79350   0.82958   0.89886   0.99840   1.00528
Alpha  virt. eigenvalues --    1.11277   1.19368   1.19466   1.42478   1.45121
Alpha  virt. eigenvalues --    1.63127   1.68850   1.70047   1.73056   1.73860
Alpha  virt. eigenvalues --    1.75130   1.76317   1.87138   1.98182   2.01892
Alpha  virt. eigenvalues --    2.16117   2.16405   2.29676   2.39840   2.51390
Alpha  virt. eigenvalues --    2.57097   2.70200   2.71707   2.72122   2.76078
Alpha  virt. eigenvalues --    2.86622   2.87903   3.03472   3.04821   3.11440
Alpha  virt. eigenvalues --    3.28956   3.29205   3.46515   3.74522   3.85183
Alpha  virt. eigenvalues --    3.91373   4.17384   4.18437   4.27206   4.52726
Alpha  virt. eigenvalues --    5.43760   5.56623   6.15357  25.01426  25.08726
Alpha  virt. eigenvalues --   51.66346
```

FIGURE 12.11
A screen shot of part of the output of Gaussian 03, showing the HF/6-311G(d,p) orbital energies of acetaldehyde. The 24 electrons in acetaldehyde occupy the 12 molecular orbitals of lowest energies. The remaining 66 unoccupied orbitals are virtual orbitals.

Figure 12.11 shows a screen shot of part of this output, showing the orbital energies of the 78 molecular orbitals involved. Let's make sure that we see why there are 78 basis functions, consequently yielding 78 orbitals, in this 6-311G(d,p) basis set. The basic 6-311G basis set (a triple split-valence basis set) has three $1s$ orbitals on each of the four hydrogen atoms and a $1s$ orbital and three sets of $2s$, $2p_x$, $2p_y$, and $2p_z$ orbitals on the oxygen atom and each of the two carbon atoms, for a total of 51 basis functions (so far). The (d,p) augmentation adds three p orbitals to each hydrogen atom and five d orbitals to the oxygen atom and the two carbon atoms, adding 27 more basis functions to the 51 to give a total of 78. This information is given in the Gaussian 03 output.

EXAMPLE 12–11
How many primitive Gaussian functions are there in the 6-311G(d,p) basis set for acetaldehyde?

SOLUTION: The $1s$, $1s'$, and $1s''$ hydrogen orbitals in this triple split-valence basis set are represented by three, one, and one Gaussian functions, respectively, for a total of five for each hydrogen atom. Each of the three p orbitals added to each hydrogen atom is represented by one Gaussian function, for a total of eight Gaussian functions for each hydrogen atom, for a total of 32 primitive Gaussian functions on the four hydrogen atoms. The $1s$ orbitals of the oxygen atom and the carbon atoms are represented by six Gaussian functions; each of the $2s$, $2p_x$, $2p_y$, and $2p_z$ orbitals is represented by three Gaussian functions, and each of the $2s'$, $2p'_x$, $2p'_y$, $2p'_z$, and so forth, orbitals is represented by one Gaussian function; and each of the d orbitals added to the oxygen atom and the two carbon atoms accounts for six Gaussian functions (see Problem 12–28), for a total of $6 + 12 + 4 + 4 + 6 = 32$ Gaussian functions for the oxygen atom and each carbon atom. The total number of Gaussian functions involved in the 6-311G(d,p) basis set for acetaldehyde is $32 + 3 \times 32 = 128$. This information is given in the Gaussian 03 output.

Table 12.5 compares the calculated bond lengths and bond angles of acetaldehyde using the HF/6-311G(d,p) model chemistry with the experimental values. You can see from this table that Hartree–Fock calculations yield fairly good geometries. We have excised only the orbital energies and the optimized geometry from the Gaussian 03 output, but there is a great deal more information listed, such as how the electron charge is distributed over the atoms in the molecule, IR and Raman vibrational spectral properties such as frequencies and intensities, electrical properties such as dipole moments and other multipole moments, NMR properties, and much more.

Another commonly used interface is WebMO. As we said above, WebMO (*www. webmo.net*) is a free web-based interface for a number of computational chemistry programs. WebMO installs on any Unix computer (the server). It and any computational

TABLE 12.5

A Comparison of the Bond Lengths and Bond Angles for Acetaldehyde Calculated with an HF/6-311G(d,p) Model Chemistry with Experimental Values [a]

	Calculated value	Experimental value
R(C,O)	1.1819	1.216
R(C,H) (aldehyde)	1.0986	1.114
R(C,C)	1.5034	1.501
R(C,H) (methyl)	1.0819	1.086
R(C,H) (methyl)	1.0870	1.086
R(C,H) (methyl)	1.0870	1.086
A(COH)	124.5673	123.9
A(CCH) (aldehyde)	115.1790	117.5
A(HCH) (methyl)	110.2088	108.3
A(HCH) (methyl)	110.0354	108.3
A(HCH) (methyl)	107.3411	108.3

a. The bond lengths (the R's) are in angstroms and the bond angles (the A's) are in degrees.

programs also installed on the server can then be accessed from a standard web browser on any other computer, even a PC running Windows or a Mac in a dormitory or at home. This type of setup is used by almost one thousand chemistry departments around the world.

To log into WebMO, you go to the internet address of the server and enter your user name and password, which you obtain from your system administrator. (Even if you do not have access to WebMO, you can log in as a guest and receive one minute of CPU time to run a calculation (which is enough time to run a number of Hartree–Fock calculations on the molecules that we have been considering) by going to *www.webmo.net/demo/*, then log onto Demo Server, and then follow the instructions.) After logging in, you will see a **WebMO Job Manager** page, which displays a history of your past jobs. After clicking on **Create New Job**, you come to a pictorial molecule builder, which is comparable to that in GaussView. Once the molecule is built and the initial geometry is specified, you select **Choose Computational Engine**, where you can choose Gaussian from a list of computational programs that are available. Figure 12.12 shows the **Configure Gaussian Job Options** page, which is comparable to the **Gaussian Calculation Setup** page in Figure 12.10. Once you click on **Submit Job**, your job is run by Gaussian and you are returned to the **WebMO Job Manager** page. Clicking on a completed job displays a summary of the calculation, visualization of calculated quantities (dipole moment, partial charges, vibrational normal modes (Figure 12.13), spectra, and molecular orbitals), and a link to the output text of the program.

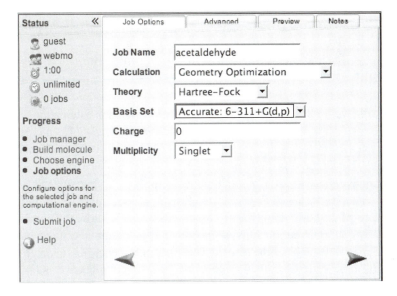

FIGURE 12.12
A screen shot of the Configure Gaussian Job Options page in WebMO.

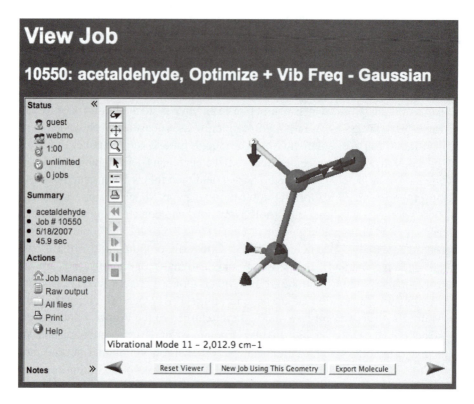

FIGURE 12.13
A screen shot of the View Job page in WebMO displaying one of the vibrational normal modes of acetaldehyde.

12.6 Hartree–Fock–Roothaan Calculations Have Been Carried Out for Many Molecules

In this section, we'll review results of Hartree–Fock–Roothaan calculations for a number of molecules. There is a huge literature on molecular Hartree–Fock calculations, and the primary goal of this section is to provide you with the ability to peruse this literature for yourself. The National Institute of Standards and Technology has a website (*http:// srdata.nist.gov/cccbdb*) called Computational Chemistry Comparison and Benchmark Data Base (cccbdb). Its stated goals are "(1) Provide a benchmark set of molecules for the evaluation of ab initio computational methods and (2) Allow the comparison between different ab initio computational methods for the prediction of thermochemical properties." The website gives the results of calculations of energies, geometries, vibrational frequencies, entropies, dipole moments, quadrupole moments, polarizabilities, barriers to internal rotation, and other properties for hundreds of molecules. For each molecule, it lists the results of a given property using different model chemistries. It also lists experimental values for all these properties. Furthermore, this website has a link called "bad calculations," which illustrates calculations that produce poor results for certain model chemistries and molecules and also a tutorial, which includes a glossary of computational terms.

Figure 12.14 is a screen shot from the NIST cccbdb website of the result for the standard molar entropy (in units of $J \cdot K^{-1} \cdot mol^{-1}$) of acetaldehyde at 298 K calculated using various model chemistries. We chose to display the results for the standard molar entropy because its calculation (through statistical thermodynamic formulas) involves both the geometry of the molecule and its vibrational frequencies. The first row shows the results of Hartree–Fock calculations using increasingly sophisticated basis sets. The rest of the rows show results for various post-Hartree–Fock methods, which we shall describe briefly in the next section. The NIST cccbdb website gives an experimental result of 263.95 $J \cdot K^{-1} \cdot mol^{-1}$.

		STO-3G	3-21G	3-21G*	6-31G	6-31G*	6-31G**	6-31+G**	6-311G*	6-311G**	6-31G(2df,p)	cc-pVDZ	cc-pVTZ	aug-cc-pVDZ	6-311+G(3df,2p)	6-311+G(3df,2pd)
hartree fock	HF	265.2	263.2	263.3	263.6	263.6	263.4	263.5	263.2	262.9	262.5	262.3	262.5	262.4	262.8	
density functional	BLYP	266.2	263.0	263.0	263.4	263.5	263.5	263.9	263.2	263.2	262.9	262.8	263.1	263.4		
	B3LYP	265.9	262.8	262.9	263.4	263.4	263.5	263.8	263.2	263.1	262.8	262.6	263.0	263.1	263.2	
	B3LYPultrafine					263.5										
	B3PW91	265.8	263.0	263.1	263.6	263.7	263.7	264.0	263.3	263.2	263.0	262.8	263.2	263.2		
	mPW1PW91	265.7	262.9	263.2	263.7	263.6	263.6	263.9	263.3	263.3	263.0	262.7	263.0	263.2		
	PBEPBE	265.9	262.9	263.7	263.4	263.4	263.5	263.8	263.2	263.2	262.9	262.9	263.1	263.4		
Moller Plesset perturbation	MP2FC	266.2	263.0	263.1	263.9	264.1	264.1	264.6	263.6	263.4	263.2	263.0	263.4	263.4		
	MP2FU		263.0			263.9	264.0	264.5	263.6	263.4		263.0				
	MP4					264.2										
Configuration interaction	CID					263.9										
	CISD					263.9										
Quadratic configuration interaction	QCISD			263.2		264.0	264.3	264.5	263.6	263.4						
	QCISD(T)					264.1										
Coupled Cluster	CCD					263.9										
	CCSD					266.4										
	CCSD(T)					263.8	264.1									

FIGURE 12.14
A screen shot from the NIST website (*http://srdata.nist.gov/cccbdb*) for the standard molar entropy (in units of $J \cdot K^{-1} \cdot mol^{-1}$) of acetaldehyde at 298 K calculated using various model chemistries.

Another excellent source of the results of ab initio molecular calculations is the book by Hehre, Radom, Schleyer, and Pople referred to at the end of the chapter. This book presents a thorough discussion and comparison of calculations of many different molecular properties for many molecules. A more recent and equally valuable source is the book *A Guide to Molecular Mechanics and Quantum Chemical Calculations* by Hehre. There is also a Quantum Chemistry Literature Database website (*http://qcldb2.ims.ac.jp/*), maintained by a consortium of Japanese quantum chemists, which is a database of those papers published after 1978 that treat ab initio calculations of atomic and molecular structures.

The first thing that we shall investigate is how dependent the results of Hartree–Fock calculations are on the basis set. To keep things manageable, we'll follow Szabo and Ostlund (see the end-of-chapter references) and focus on the ten-electron series CH_4, NH_3, H_2O, and HF. The Hartree–Fock method is variational with respect to the energy, and so we expect the energy to decrease with an increase in the basis set. Table 12.6 shows the Hartree–Fock energies for a series of increasingly large basis sets. Notice that

TABLE 12.6

The Hartree–Fock Energies (in Hartrees) of the Ten-Electron Series CH_4, NH_3, H_2O, and HF as a Function of the Basis Set Used [a]

	STO-3G	3-21G	6-31G*	6-31G**	6-311G*	6-311G**	Hartree–Fock limit
H_2O	−74.9659	−75.5860	−76.0107	−76.0236	−76.0324	−76.0470	−76.065
NH_3	−55.4554	−55.8722	−56.1844	−56.1955	−56.2010	−56.2104	−56.255
CH_4	−39.7269	−39.9769	−40.1952	−40.2021	−40.2027	−40.2090	−40.225
HF	−98.5728	−99.4602	−100.0029	−100.0117	−100.0345	−100.0469	−100.071

a. *Source: http://srdata.nist.gov/cccbdb*

the energy does decrease from left to right in the table. Note also that even the 6 311G** basis set does not provide a Hartree–Fock limit and that its energies are about 0.02 to 0.03 hartrees (about 50 to 80 $kJ \cdot mol^{-1}$) higher than the Hartree–Fock limit in each case.

Certainly, geometries are more interesting than energies. Table 12.7 lists the calculated bond lengths (in angstroms) and bond angles for CH_4 through HF for various basis sets. Note that the various calculated values do not necessarily vary monotonically because bond lengths and bond angles are not variational like the energy. The agreement with the experimental values is pretty good, being about 0.02 Å in error for the bond lengths and one or two degrees for the bond angles. Even the 3-21G basis set yields acceptable geometries. This relatively small basis set is often used to determine an initial geometry to be used in more sophisticated calculations involving large basis sets.

TABLE 12.7

The Hartree–Fock Bond Lengths (in angstroms) and Bond Angles of the Ten-Electron Series CH_4, NH_3, H_2O, and HF as a Function of the Basis Set Used [a]

	3-21G	6-31G*	6-31G**	6-311G*	6-311G**	Experimental
CH_4	1.083	1.084	1.084	1.083	1.084	1.094
	109.5	109.5	109.5	109.5	109.5	109.5
NH_3	1.003	1.003	1.001	0.9990	1.001	1.012
	112.4	107.2	107.6	107.4	107.4	106.7
H_2O	0.9665	0.9473	0.9430	0.9394	0.9410	0.958
	107.7	105.5	106.0	107.5	105.4	104.5
HF	0.9376	0.9109	0.9005	0.8973	0.8960	0.917

a. *Source: http://srdata.nist.gov/cccbdb*

TABLE 12.8
The Hartree–Fock Dipole Moments (in Debyes) of the Ten-Electron Series NH_3, H_2O, and HF as a Function of the Basis Set Used

	STO-3G	3-21G	6-31G*	6-31G**	6-311G*	6-311G**	Experimental
H_2O	1.709	2.388	2.199	2.148	2.317	2.318	1.850
NH_3	1.876	1.752	1.920	1.839	1.845	1.725	1.470
HF	1.252	2.174	1.972	1.944	1.980	1.980	1.820

Dipole moments (see Problems 12–54 and 12–55) are more sensitive to the model chemistry used than is geometry. Table 12.8 gives the calculated and experimental dipole moments of NH_3 through HF (the dipole moment of CH_4 is zero by symmetry). The STO-3G basis set in this case doesn't even give the correct order $H_2O > HF > NH_3$, and the 3-21G basis set and larger give results that are too large. You've got to use post-Hartree–Fock methods to obtain dipole moments that agree well with experimental values.

Figures 12.15 through 12.17 show results for molecules other than CH_4 through HF. Figure 12.15 shows HF/6-31G** bond lengths of a number of hydrides of the general formula AH_n plotted against the experimental values. The solid line has a slope of one, so points on or near this line are in good agreement with experiment. Figure 12.16 shows a similar plot for HF/6-31G** bond lengths for molecules containing two second- and third-row atoms. Both Figures 12.15 and 12.16 show that Hartree–Fock geometries are quite good. Figure 12.17 shows the HF/6-31G** dipole moments plotted against the experimental values for a large number of molecules. Notice that many of the points lie below the line of unit slope, indicating that Hartree–Fock dipole moments are often too large. Finally, Tables 12.9 and 12.10 give bond lengths and bond angles for a variety of organic compounds.

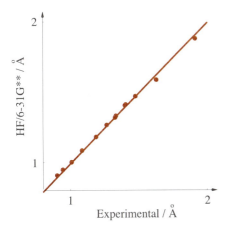

FIGURE 12.15
The bond lengths of a number of hydrides of the general formula AH_n calculated with the HF/6-31G** model chemistry plotted against the experimental values. The solid line has a unit slope.

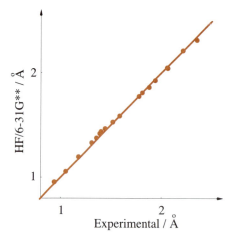

FIGURE 12.16
The A–B bond length of molecules of the general formula H_nABH_m, where A and B are second- and third-row atoms calculated with the HF/6-31G** model chemistry plotted against the experimental values. The solid line has a unit slope.

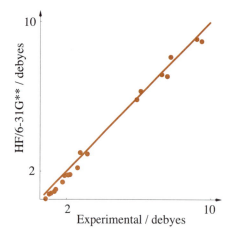

FIGURE 12.17
The dipole moments of selected molecules calculated with the HF/6-31G** model chemistry plotted against the experimental values. The solid line has a unit slope.

Remember that all the results that we have presented here are Hartree–Fock results, and in most cases, we haven't yet reached the Hartree–Fock limit. In carrying out any ab initio calculation, there are two sources of error. One of them is due to the fact that we cannot employ an infinite basis set, and so the results from any finite basis set are subject to a *basis set truncation error*. The other source of error is due to the level of theory that we use, which varies from Hartree–Fock to full configuration interaction. Figure 12.18 is a schematic illustration of the various computational pathways from a Hartree–Fock calculation with a minimal basis set to an essentially exact calculation. Of course, using increasingly better model chemistries comes at a cost.

Hartree–Fock gives reasonably good geometries, but its neglect of electron correlation is a serious deficiency when it comes to calculating bond energies and reaction energies. Even though Hartree–Fock calculations give over 95% of the total energy,

TABLE 12.9
Hartree–Fock Bond Lengths (in Å) of a Number of Organic Molecules as a Function of the Basis Set Used

		Bond length/Å					
Bond type	Molecule	STO-3G	3-21G	3-21G*	6-31G*	6-31G**	Experimental
C–C	Ethane		1.542	1.542	1.547	1.527	1.536
	Propane		1.541	1.541	1.528	1.528	1.526
	Acetonitrile	1.488	1.457	1.457		1.467	1.458
	Acetaldehyde	1.537	1.507	1.507	1.504	1.503	1.501
	1,3-Cyclopentadiene	1.522	1.519	1.519	1.507	1.506	1.506
	Cyclobutane	1.554	1.571	1.571	1.548	1.545	1.548
C=C	Cyclopropene	1.277	1.282	1.282	1.276	1.276	1.300
	Allene	1.288	1.292	1.292	1.296	1.296	1.308
	Cyclobutene	1.314	1.326	1.326	1.322	1.322	1.332
	Ethene	1.306	1.315	1.315	1.317	1.316	1.339
C≡C	Acetylene	1.180	1.188	1.188	1.185	1.186	1.203
	Propyne	1.170	1.188	1.188	1.184	1.187	1.206
C–C	Benzene		1.385	1.385	1.386	1.386	1.397
C≡N	Hydrogen cyanide	1.153	1.137	1.137	1.133	1.135	1.153
	Acetonitrile	1.154	1.139	1.139	1.133	1.135	1.157
	Hydrogen isocyanide	1.170	1.160	1.160	1.154	1.155	1.169
C=O	Carbon dioxide	1.188	1.156	1.156	1.143	1.143	1.162
	Formaldehyde	1.216	1.212	1.212		1.184	1.193
	Acetone	1.219	1.211	1.210		1.192	1.222
C–Cl	Tetrachloromethane		1.832	1.772	1.766	1.766	1.767
	Chloromethane	1.802	1.802	1.806	1.785	1.784	1.781

the correlation energy—the difference between the Hartree–Fock energy and the exact energy—can be quite significant for chemical purposes. For example, the electron correlation energy of H_2, which is the difference between the Hartree–Fock and the exact energy, is 0.0404 hartrees, which looks pretty good until you realize that one hartree is equivalent to 2625 kJ·mol^{-1}, in which case the correlation energy is 106 kJ·mol^{-1}, a sizable fraction of a typical chemical bond. In the next section, we shall discuss several methods that improve upon the Hartree–Fock results. Collectively, these results are called post-Hartree–Fock methods.

TABLE 12.10
Hartree–Fock Bond Angles of a Number of Organic Molecules as a Function of the Basis Set Used

	3-21G	3-21G*	6-31G*	6-31G**	6-311G**	Experimental
Formaldehyde						
O–C–H	122.5	122.5	122.2	122.1	122.1	121.9
H–C–H	114.9	114.9	115.7	115.7	115.8	116.1
1-Butene						
C=C–C	124.9	124.9	125.4	125.4	125.4	125.4
C–C–C	111.2	111.2	112.5	112.5	112.6	112.1
Dimethyl ether						
C–O–C	114.0	114.0	113.8	113.9	113.9	111.2
Formamide						
H–C–O		122.4	122.3	122.3	122.3	122.5
H–N–C		121.9	121.8	121.6	121.5	120.0
Propane						
C–C–C	111.6	111.6	112.8	112.9	112.9	112.4
Chloromethane						
Cl–C–H	106.2	108.1	108.5	108.4	108.2	108.4
H–C–H	112.5	110.8	110.4	110.5	110.7	110.5
Dichloromethane						
Cl–C–Cl	110.8	112.1	112.8	112.8	112.8	111.8
H–C–H	114.5	111.5	111.2	111.2	111.7	112.0
Methylene (CH_2)						
H–C–H	131.3	131.3	130.7	131.1	131.7	135.5
Propene						
C–C–C	124.7	124.7	125.2	125.2	125.2	124.8

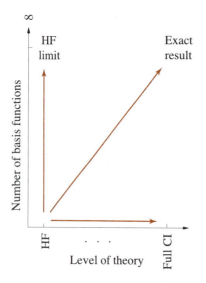

FIGURE 12.18
A schematic illustration of the various pathways from a Hartree–Fock calculation with a minimal basis set to an essentially exact calculation.

12.7 Post-Hartree–Fock Methods Can Yield Almost Exact Results for Molecular Properties

Figure 12.19 shows a screen shot from the NIST cccbdb website (*http://srdata.nist.gov/cccbdb*) for the energy of a water molecule calculated by various model chemistries. This figure and Figure 12.14 show the results of a number of post-Hartree–Fock calculations. We'll discuss each of these methods briefly in turn, starting with Møller–Plesset perturbation theory, followed by configuration interaction, quadratic configuration interaction, coupled-cluster theory, and finally density functional theory. Recall that we defined electron correlation as the difference between the exact nonrelativistic ground-state electronic energy of a system within the Born–Oppenheimer approximation and the Hartree–Fock limiting energy, E_{HF}. In an equation, we have

$$E_{corr} = E_{exact} - E_{HF}$$

Because post-Hartree–Fock methods give some amount of the correlation energy, they are also called *electron correlation methods*.

		Methods with standard basis sets											
		STO-3G	3-21G	3-21G*	6-31G	6-31G*	6-31G**	6-31+G**	6-311G*	6-311G**	6-31G(2df,p)	cc-pVDZ	cc-pVTZ
hartree fock	HF	-74.962123	-75.582180	-75.582180	-75.981579	-76.006967	-76.019835	-76.027450	-76.028621	-76.043232	-76.025542	-76.023274	-76.053989
density functional	BLYP	-75.288175	-75.947470	-75.947470	-76.362896	-76.384765	-76.395105	-76.412358	-76.411656	-76.424689	-76.401663	-76.395087	-76.437682
	B3LYP	-75.318996	-75.970186	-75.970186	-76.382336	-76.405174	-76.415957	-76.430267	-76.430153	-76.443668	-76.422297	-76.416847	-76.456060
	B3LYPultrafine					-76.405174							
	B3PW91	-75.302244	-75.945691	-75.945691	-76.354645	-76.377833	-76.388768	-76.400817	-76.400952	-76.415103	-76.395036	-76.390795	-76.426904
	mPW1PW91	-75.311801	-75.951441	-75.952149	-76.361072	-76.383844	-76.395030	-76.407058	-76.406640	-76.421567	-76.401895	-76.397119	-76.432831
	PBEPBE	-75.234755	-75.882699	-75.882698	-76.296016	-76.318702	-76.329437	-76.344832	-76.342604	-76.356589	-76.337247	-76.330388	-76.369303
Moller Plesset perturbation	MP2FC	-75.002262	-75.704022	-75.704022	-76.109395	-76.193068	-76.216005	-76.229327	-76.231172	-76.260192	-76.263562	-76.224886	-76.314877
	MP2FU		-75.705840			-76.195465	-76.218668	-76.232105	-76.249874	-76.279116		-76.227209	
	MP4		-75.712387			-76.203470			-76.242809				
Configuration interaction	CID					-76.193807			-76.230314				-76.309010
	CISD					-76.194427							-76.310094
Quadratic configuration interaction	QCISD		-75.711927			-76.202283	-76.225166	-76.237629	-76.239537	-76.267919		-76.234571	
	QCISD(T)					-76.204123	-76.227878	-76.240776	-76.243066			-76.237591	-76.328602
Coupled Cluster	CCD		-75.710840			-76.201465	-76.224252	-76.235967	-76.238258			-76.233666	-76.319660
	CCSD					-76.202061	-76.224989	-76.237161	-76.239290			-76.234426	-76.320776
	CCSD(T)					-76.005821	-76.227804	-76.240613				-76.237526	-76.328437

FIGURE 12.19

A screen shot from the NIST cccbdb website (*http://srdata.nist.gov/cccbdb*) for the energy of a water molecule calculated by various model chemistries.

Møller–Plesset Perturbation Theory

The basic idea of Møller–Plesset perturbation theory is to start with the Hartree–Fock wave function as the unperturbed wave function and then to improve it systematically by perturbation theory. It turns out that the Møller–Plesset (MP) perturbation theory energy through first order is equal to the Hartree–Fock energy. Second-order MP perturbation theory is the first correction to Hartree–Fock results and is designated by MP2 in Figure 12.19. The modifiers FC and FU denote *frozen core*, in which excitations from the nonvalence orbitals are not included, and *full*, in which excitations from all orbitals are included. The designation MP4 denotes fourth-order MP perturbation theory. All the programs that we discussed in Section 12.5 have built-in MP perturbation theory routines. Figure 12.20 shows a plot of the relative times required for a calculation of the optimized geometry of a methanol molecule using various computational methods plotted against the size of the basis set. Note that the plot is a logarithmic plot; an MP2 calculation requires about a factor of two more time and that an MP4 calculation requires about a factor of 20 more time than a similar Hartree–Fock calculation. Some comparisons of MP perturbation theory results with the results of other post-Hartree–Fock methods are shown in Figure 12.19 and Table 12.11 (see end of section).

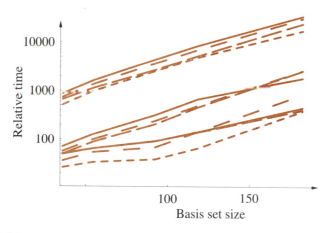

FIGURE 12.20
A logarithmic plot of the relative times required for a geometry optimization calculation of a methanol molecule using various computational methods as a function of the size of the basis set. The lower set of four curves are HF, MP2, BLYP, and B3LYP; the middle set of three curves are CISD, CCD, and QCISD; and the upper set of four curves are CCSD, CCSDT, QCISDT, and MP4. (Data taken from *http://srdata.nist.gov/cccbdb.*)

Configuration Interaction

The two lines in Figure 12.19 labeled CID and CISD are configuration-interaction results. We presented a simple configuration-interaction calculation for H_2 using a minimal basis set in Section 10.9. If we use a basis set consisting of K basis functions, then a Hartree–Fock calculation yields K molecular orbitals, and so $2K$ spin orbitals. For the ground state of a closed-shell molecule consisting of $2N$ electrons, the $2N$ spin orbitals of lowest energy will be occupied and the remaining $2K - 2N$ spin orbitals will be unoccupied (virtual spin orbitals). The starting point of a configuration-interaction calculation for a $2N$-electron closed-shell system is Ψ_0, the Slater determinant formed from the $2N$ lowest-energy spin orbitals. Using Ψ_0 as our reference state, we describe all the excited configurations in terms of Ψ_0 by designating how they differ from Ψ_0. First, we have all those determinants that differ from Ψ_0 by replacing an occupied spin orbital by a virtual spin orbital. It is standard notation to denote occupied spin orbitals by χ_i, χ_j, ... and virtual spin orbitals by χ_a, χ_b, ... and to let Ψ_i^a be the determinant that we obtain by promoting an electron from the occupied spin orbital χ_i to the virtual spin orbital χ_a. The resulting determinant Ψ_i^a is called a *singly excited determinant*. Now let's define a single excitation operator $\hat{C}_1$ by

$$\hat{C}_1 \Psi_0 = \sum_i \sum_a c_i^a \Psi_i^a \tag{12.29}$$

where the c_i^a are constants that are determined variationally. Equation 12.29 represents the sum of all possible single excitations from the ground-state determinant: excitations from each and every occupied spin orbital to each and every virtual spin orbital (Problem 12–42).

We can continue this procedure and consider all possible doubly excited configurations

$$\hat{C}_2 \Psi_0 = \sum_{i<j} \sum_{a<b} c_{ij}^{ab} \Psi_{ij}^{ab} \tag{12.30}$$

where the summations go over all values of i and j (with $i < j$) and all values of a and b (with $a < b$). The restriction on the summation indices ($i < j$ and $a < b$) ensures that a given doubly excited configuration occurs only once in the summation. The full configuration-interaction wave function Ψ_{CI} is given by

$$\begin{aligned}
\Psi_{CI} &= c_0 \Psi_0 + \hat{C}_1 \Psi_0 + \hat{C}_2 \Psi_0 + \cdots + \hat{C}_{2N} \Psi_0 \\
&= c_0 \Psi_0 + \sum_i \sum_j c_i^a \Psi_i^a + \sum_{i<j} \sum_{a<b} c_{ij}^{ab} \Psi_{ij}^{ab} + \cdots + \sum_{i<j<\cdots} \sum_{a<b<\cdots} c_{ij\cdots}^{ab\cdots} \Psi_{ij\cdots}^{ab\cdots}
\end{aligned} \tag{12.31}$$

where the last term on the second line involves the promotion of all $2N$ electrons from the occupied spin orbitals to all possible virtual spin orbitals.

The full configuration-interaction (full CI) wave function given by Equation 12.31 is an exact solution of the $2N$-electron problem for that basis set. In the limit of an infinite basis set, Equation 12.31 provides an *exact* solution to the Schrödinger equation. The number of terms in Equation 12.31 can be in the millions. For example, a full CI calculation on N_2 using a 6-31G* basis set involves over a million configurations. Consequently, a full CI calculation on any but the smallest molecules with modest basis sets is out of the question. Fortunately, early extensive CI calculations in the 1970s showed that doubly excited configurations make a major contribution to the energy. Because of this, a practical approach to CI calculations has been to truncate Ψ_{CI} in Equation 12.31.

If we set all the $\hat{C}_n$ except for $\hat{C}_2$ equal to zero, then we have what is called the *configuration-interaction doubles* (CID) approximation. It turns out computationally that it involves little extra computational effort to include single excitations, and so a frequently used CI approximation is CISD, *configuration-interaction singles and doubles*.

There are a number of examples where Hartree–Fock gives qualitatively incorrect results that are remedied by CI. We saw in Chapter 10 that Hartree–Fock gives the incorrect dissociation limit of a diatomic molecule, but that CI gives the correct dissociation limit. Configuration interaction also gives the correct order of the ionization energies of N_2, which, as we pointed out at the end of Section 12.1, Hartree–Fock does not. Another example involves the calculation of the dipole moment of a carbon monoxide molecule, which long offered a challenge for quantum chemists. (See Problems 12–54 through 12–57 for a review of dipole moments.) A simple electronegativity argument predicts that the oxygen atom should be the negative end and that the carbon atom should be the positive end of the dipole moment, whereas the opposite is true experimentally. Hartree–Fock, even with a very large basis set, gives the wrong direction of the dipole moment; in other words, it predicts that the oxygen atom is the negative end. Configuration interaction, however, as well as the other post-Hartree–Fock methods that we discuss in this section, gives the correct direction of the dipole moment.

Even though configuration interaction leads to huge computational demands, it is exact in principle and is also variational. Being variational is a desirable feature because you know that the energy is always an upper bound to the true energy. At one time, it was believed by most computational chemists that configuration interaction provided a practical method for molecular calculations. In fact, almost half of all ab initio molecular calculations in the 1960s were CI calculations. However, as more and more CI calculations were done, it became clear that the rate of convergence of CI calculations was distressingly slow. Including more and more terms in Equation 12.31 did indeed give better results, but at a disappointingly slow rate. In addition to this, configuration interaction suffers from another deficiency that is particularly important for calculations on large molecules.

Consider two N-electron molecules. The energy of these two molecules when they are widely separated is twice the energy of a single molecule. Suppose, however, that we treat the two molecules as a $2N$-electron system, but in a geometric setup where the two molecules are widely separated. Surely, the energy that we calculate should be twice the energy of a single molecule, provided they are sufficiently far from each other. The fact is, however, that truncated versions of CI such as CISD and CID do *not* give the correct energy; it does not come out to be the energy of two isolated molecules. Surely, a sensible method should give the energy of n widely separated molecules treated as one "supermolecule" as being n times the energy of a single molecule. This requirement is called *size consistency*. It turns out that a full CI calculation is size-consistent because it is exact within the basis set, but truncated versions of CI are not necessarily size-consistent. There have been a number of remedies proposed for this deficiency; the entries in Figures 12.14 and 12.19 labeled QCISD and QCISD(T) are called *quadratic configuration interaction* and are versions of CI that are size-consistent. Figure 12.20 shows, however, that these are quite time-demanding.

Coupled-Cluster Theory

The next computational method that we shall discuss (the one labeled CCD in Figures 12.14 and 12.19) is not only size-consistent, but is more accurate than CI for a comparable amount of computer resources. Nevertheless, it is not variational, meaning that a larger basis set does not necessarily yield a more accurate energy. This method, called *coupled-cluster theory*, is one of the most successful and widely used post-Hartree–Fock methods, particularly for small molecules. Coupled-cluster theory was originally developed for the study of atomic nuclei in the late 1950s and then later formulated for molecular electronic structure calculations in the late 1970s. The derivation of the fundamental equation of coupled-cluster theory is fairly involved, but nowadays it is simply presented as an ansatz. The fundamental equation of coupled-cluster theory for an N-electron molecule is

$$\Psi_{\text{exact}} = e^{\hat{T}} \Psi_0 \tag{12.32}$$

where Ψ_0 is usually the reference Hartree–Fock determinantal wave function and $\hat{T}$ is an operator,

$$\hat{T} = \hat{T}_1 + \hat{T}_2 + \hat{T}_3 + \cdots + \hat{T}_N \tag{12.33}$$

where

$$\hat{T}_1 \Psi_0 = \sum_{i,a} t_i^a \Psi_i^a \tag{12.34}$$

$$\hat{T}_2 \Psi_0 = \sum_{\substack{i<j \\ a<b}} t_{ij}^{ab} \Psi_{ij}^{ab} \tag{12.35}$$

and so on, where the $t_i^a, t_{ij}^{ab}, \ldots$ are constants that are determined from a set of nonlinear simultaneous algebraic equations (that we shall not present here).

Equation 12.32 is formally exact, but numerically intractable as it stands. If we ignore all the terms in Equation 12.33 except $\hat{T}_2$, then we have what is called the *coupled-cluster doubles* (CCD) approximation,

$$\Psi_{\text{CCD}} = e^{\hat{T}_2} \Psi_0 \tag{12.36}$$

The coupled-cluster doubles approximation is *much* different than the configuration-interaction doubles approximation. To see why, expand the exponential in Equation 12.36 according to $e^x = 1 + x + x^2/2! + x^3/3! + \cdots$ and write

$$\Psi_{\text{CCD}} = \Psi_0 + \hat{T}_2 \Psi_0 + \frac{1}{2}\hat{T}_2^2 \Psi_0 + \cdots \tag{12.37}$$

The first two terms of Equation 12.37 constitute the CID approximation, but there are many additional terms in Equation 12.37. Using the definition of $\hat{T}_2$ in Equation 12.35, we have

$$\hat{T}_2^2 \Psi_0 = \hat{T}_2(\hat{T}_2\Psi_0) = \hat{T}_2 \sum_{\substack{i<j \\ a<b}} t_{ij}^{ab} \Psi_{ij}^{ab}$$

$$= \sum_{\substack{i<j \\ a<b}} t_{ij}^{ab} \hat{T}_2 \Psi_{ij}^{ab} = \sum_{\substack{i<j \\ a<b}} \sum_{\substack{k<l \\ c<d}} t_{ij}^{ab} t_{kl}^{cd} \Psi_{ijkl}^{abcd} \qquad (12.38)$$

This term represents quadruple excitations, albeit only those that are in a sense products of double excitations. It turns out, however, that such product terms often make dominant numerical contributions to quadruple excitations. Similarly, further terms in Equation 12.37 account for many sextuple, octuple, ... and further excitations. Thus, although CCD is obtained by replacing $\hat{T}$ by $\hat{T}_2$ in Equation 12.32, its exponential form introduces many excitations beyond double excitations. Furthermore, this is done at little added time and expense because the number of coefficients to be determined is roughly the same as for CID. If you let $\hat{T}$ in Equation 12.32 be $\hat{T}_1 + \hat{T}_2$ instead of just $\hat{T}_2$, then you have CCSD, *coupled-cluster singles and doubles*. All the standard computational chemistry programs include coupled-cluster routines. The common versions are CCSD and CCD.

As in the case of configuration interaction, coupled-cluster calculations are currently limited to fairly small molecules with fairly small basis sets. Let's compare the results of a Hartree–Fock, a CISD, and a CCSD calculation using a cc-pVTZ basis set for the enthalpy of reaction at 298 K, ΔH_r° for the dissociation of HCl(g):

$$HCl(g) \longrightarrow H(g) + Cl(g) \qquad (12.39)$$

If you go to the NIST cccbdb website and click on Reaction Data and the User Specified Reaction at 298 K, you see a form asking you to enter the reactants and products of the reaction. After doing this, you're taken to a page that lists experimental results ($431.6 \pm 0.1 \, \text{kJ} \cdot \text{mol}^{-1}$ in this case) and the errors associated with calculations done using various model chemistries. Click on HF/cc-pVTZ to find the result $302.6 \, \text{kJ} \cdot \text{mol}^{-1}$. Similarly, the result for CISD/cc-pVTZ is $412.6 \, \text{kJ} \cdot \text{mol}^{-1}$ and that for CCSD/cc-pVTZ is $402.6 \, \text{kJ} \cdot \text{mol}^{-1}$. Notice that the Hartree–Fock result is in error by over $100 \, \text{kJ} \cdot \text{mol}^{-1}$, but that both the CISD and CCSD results are within 10% of the experimental value.

It's interesting to calculate these values for ΔH_r° yourself using the individual values of ΔH_f°. The NIST cccbdb website gives the following values of ΔH_f°, the standard heat of formation (without the zero-point energy), at 298 K.

Method	$\Delta H_f^\circ / E_h$		
	HCl(g)	H(g)	Cl(g)
HF/cc-pVTZ	−460.103 547	−0.497 449	−459.483 044
CISD/cc-pVTZ	−460.315 791	−0.497 449	−459.657 404
CCSD/cc-pVTZ	−460.326 255	−0.497 449	−459.663 893

The HF/cc-pVTZ value of the standard enthalpy of the reaction described by Equation 12.39 is

$$\Delta H_r^\circ = \Delta H_f^\circ[H(g)] - \Delta H_f^\circ[Cl(g)] - \Delta H_f^\circ[HCl(g)]$$

$$= -0.497\,449E_h - 459.483\,044E_h + 460.103\,547E_h$$

$$= 0.123\,054E_h = 323.1\,kJ\cdot mol^{-1}$$

Notice that ΔH_r° is obtained from the small difference between two fairly large numbers. This is why the accuracy of molecular energies is so important. The final value of ΔH_r° here is of the same order of magnitude as the correlation energy. The value that we obtain here differs from the one above because the zero-point energy is not included in the value of ΔH_f°. If you include this, you get the same value as above (Problem 12–48). The corresponding values for CISD/cc-pVTZ and CCSD/cc-pVTZ are

$$\Delta H_r^\circ(CISD/cc\text{-}pVTZ) = -(460.315\,791E_h) - (-0.497\,449E_h) - (-459.657\,404E_h)$$

$$= -0.160\,938E_h = -422.5\,kJ\cdot mol^{-1}$$

$$\Delta H_r^\circ(CCSD/cc\text{-}pVTZ) = (-460.326\,255E_h) - (-0.497\,449E_h) - (-459.663\,893E_h)$$

$$= -0.164\,913E_h = -433.0\,kJ\cdot mol^{-1}$$

Density Functional Theory

All the molecular theories that we have discussed so far are based upon the wave function of the molecule. The wave function of an N-electron molecule is a function of $3N$ spatial coordinates and N spin coordinates, $\psi(x_1, y_1, z_1, \sigma_1, \ldots, x_N, y_N, z_N, \sigma_N)$, and physical properties, such as energies, geometries, and dipole moments, are obtained from integrals over all these coordinates. For example, let's look at the calculation of a property corresponding to a one-electron operator of the form

$$\hat{A} = \sum_{i=1}^{N} \hat{A}(x_i, y_i, z_i) \tag{12.40}$$

The expectation value of this operator is given by (Problem 12–51)

$$A = \langle \psi | \hat{A} | \psi \rangle = \sum_{i=1}^{N} \int \cdots \int \psi^*(x_1, \ldots, \sigma_N)\hat{A}(x_i, y_i, z_i)$$

$$\times \psi(x_1, \ldots, \sigma_N)d\tau_1 d\tau_2 \cdots d\tau_N$$

$$= \iiint dxdydz\,\hat{A}(x, y, z)\rho(x, y, z) \tag{12.41}$$

where

$$\rho(x, y, z) = N \int \cdots \int d\sigma_1 d\tau_2 \cdots d\tau_N \psi^*(x, y, z, \sigma_1, x_2, \ldots, \sigma_N)$$

$$\times \psi(x, y, z, \sigma_1, x_2, \ldots, \sigma_N)$$

is the probability density of finding an electron in the volume element $dx\,dy\,dz$ surrounding the point (x, y, z), and where $d\tau_i = dx_i\,dy_i\,dz_i\,d\sigma_i$ symbolically signifies an integration over the spatial coordinates x_i, y_i, and z_i and a summation over the spin coordinate σ_i.

The important point of Equation 12.41 is that you can calculate A from the electron density; you don't need the entire wave function. The wave function contains much more information than we actually need. Not only is the electron density conceptually simpler than the wave function, it can be determined experimentally by X-ray diffraction or electron diffraction techniques. Thus, it would appear to be desirable if we could formulate quantum chemistry in terms of the electron density $\rho(x, y, z)$ instead of the much more detailed many-electron wave function $\psi(x_1, y_1, \ldots, \sigma_N)$.

The reason that the calculation of A depends only upon the electron density is because $\hat{A}(x, y, z)$ is a one-electron operator. The calculation of the energy involves the integral

$$E = \int \cdots \int d\tau_1 \cdots d\tau_N \psi^*(x_1, \ldots, \sigma_N)\hat{H}\psi(x_1, \ldots, \sigma_N)$$

The Hamiltonian operator consists of one- and two-electron operators, so it wouldn't seem likely that you could express E in terms of the electron density. Ever since the early work in quantum mechanics in the 1920s, people tried to express atomic and molecular properties in terms of electron density, but little progress was made until 1964 when Pierre Hohenberg and Walter Kohn published two remarkable results. First, they showed that it was indeed possible to express the ground-state energy, as well as all other ground-state molecular properties, as an integral involving the electron density. We shall express this result by the notation

$$E = E[\rho] \tag{12.42}$$

where the square bracket here is standard notation and denotes an integral involving $\rho(x, y, z)$ in this case. Then, they showed that Equation 12.42 is variational in the sense that if E_0 and ρ_0 are the exact quantities, then

$$E_0 = E[\rho_0] \leq E[\rho] \tag{12.43}$$

where ρ is any trial electron density. The equality holds if $\rho = \rho_0$.

These two Hohenberg–Kohn theorems, as they are now called, opened up an entirely different approach to quantum chemistry, bypassing wave functions in favor of electron densities, which has seen an explosive growth since the 1990s (Figure 12.21). As we're going to see, one drawback is that the first Hohenberg–Kohn theorem is simply an *existence theorem*. In other words, although they proved rigorously that it is *possible* to express the ground-state energy as an integral involving $\rho(x, y, z)$, nobody knows (yet?) just what this relation is. The search for this relation, or at least good approximations to it, has generated an enormous amount of quantum-chemical research since the 1980s. This theory is called *density functional theory* (DFT), and all molecular computational programs have routines for various forms of DFT built into them. There

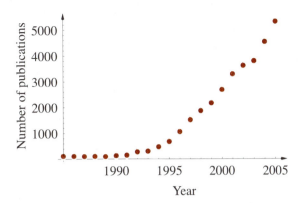

The number of publications where the phrases "density functional theory" or "DFT" appear in the title or abstract from a *Chemical Abstracts* search covering the years from 1990 to 2005.

are even dedicated DFT programs. Much of the appeal of density functional theory is that its results are comparable to those of the post–Hartree–Fock methods that we have discussed so far, but that the time required is comparable to that of Hartree–Fock calculations as Figure 12.20 shows.

Before we go on to discuss density functional theory, let's see where the name comes from. Recall that a function is a rule that assigns one number to another number. For example, $y = f(x) = x^3$ is a rule to generate the number y from the number x. We say that $y = f(x)$ is a *mapping* from one number into another. A *functional*, on the other hand, is the mapping of a function into a number. The classic example of a functional is a definite integral, such as

$$I[f] = \int_1^2 dx f(x)$$

Given $f(x)$, the integral gives us a number. The notation $I[f]$ means that I, a number, is a functional of f. Sometimes it is written as $I[f(x)]$. The first Hohenberg–Kohn theorem, Equation 12.41, simply says that the energy is a functional of the electron density.

As we said above, although Hohenberg and Kohn proved that the energy can be expressed as a functional of the electron density, their proof was an existence theorem and does not provide a prescription for it. Much of the research in density functional theory has involved the development of various approximations for $E[\rho]$. These approximations are designated by the initials of the quantum chemists who have proposed them. For example, one successful, commonly used approximation is due to Becke, Lee, Yang, and Parr and is called the BLYP functional. (See Figures 12.14 and 12.19.) The other density functional theory entries in those figures denote energy functionals due to other quantum chemists.

A density functional calculation is formulated by combining an energy functional with a basis set. The combination called BLYP/6-31G*, which uses a BLYP functional

TABLE 12.11

A Comparison of Atomization Energies (in $kJ \cdot mol^{-1}$) Calculated by Various Post-Hartree–Fock Methods [a,b]

Molecule	HF	BLYP	B3LYP	BVWN	MP2	CISD	CCSD	QCISD	Experimental
CH_4	1257	1631	1637	1657	1482	1470	1488	1481	1642
NH_3	712	1130	1128	1144	972	951	968	965	1158
H_2O	551	867	869	875	790	755	770	769	918
HF	344	520	514	524	495	477	477	477	566
C_2H_2	1138	1604	1598	1590	1530	1415	1464	1469	1627
C_2H_4	1649	2210	2217	2211	2048	1960	2015	2015	2225
C_2H_6	2117	2765	2780	2782	2546	2468	2527	2523	2788
HCN	774	1281	1279	1260	1202	1070	1123	1128	1263
CO	704	1077	1052	1050	1064	940	983	993	1072
H_2CO	995	1514	1485	1500	1404	1287	1341	1347	1494
CH_3OH	1387	1989	1987	1987	1819	1719	1780	1779	2012
O_2	121	572	522	534	492	359	406	414	494
HOOH	458	1058	1014	1038	919	795	865	865	1056
CO_2	982	1644	1572	1587	1594	1330	1437	1455	1598

a. All the calculations were done using a 6-31G* basis set.
b. *Source:* Johnson, B. J., Gil, P. M. W., Pople, J. A., The Performance of a Family of Density Functional Methods. *J. Chem. Phys.,* **98**, 5612 (1993), and the NIST website, *http://srdata.nist.gov/cccbdb*

with a 6-31G* basis set, yields very good molecular properties. Table 12.11 shows molecular properties using several different DFT schemes (BLYP, B3LYP, and BVWN). We can summarize these results by quoting the last sentence in the abstract to the 1993 paper by Johnson, Gil, and Pople (see Table 12.11): "The density functional vibrational frequencies compare favorably with the ab initio results, while for atomization energies, two of the DFT methods give excellent agreement with experiment and are clearly superior to all other methods considered." A look at Table 12.11 shows that DFT methods give molecular results that are usually at least as good as any of the post-Hartree–Fock methods that we have discussed earlier.

Because of the relatively light computational demands of DFT calculations, density functional theory can be applied to much larger systems than the other post-Hartree–Fock methods. (See Figure 12.20.) Several excellent reviews of density functional theory that discuss calculations on larger molecules are given at the end of the chapter. We'll discuss just one such calculation here. Table 12.12 shows transition metal–CO bond lengths and bond dissociation energies for the octahedral complexes $Cr(CO)_6$, $Mo(CO)_6$, and $W(CO)_6$. These systems, which contain well over 100 electrons, are very expensive using other post-Hartree–Fock methods. You can see from Table 12.12 that the DFT results are in quite good agreement with the experimental results.

TABLE 12.12

The Bond Lengths (in Å) and the First Bond Dissociation Energies (in $kJ \cdot mol^{-1}$) for the Octahedral Complexes $Cr(CO)_6$, $Mo(CO)_6$, and $W(CO)_6$ [a]

| | Bond length/Å | | | | | | Bond dissociation/$kJ \cdot mol^{-1}$ | | |
| | $Cr(CO)_6$ | | $Mo(CO)_6$ | | $W(CO)_6$ | | | | |
Method	Cr–C	C–O	Mo–C	C–O	W–C	C–O	$Cr(CO)_6$	$Mo(CO)_6$	$W(CO)_6$
MP2	1.883	1.168	2.066	1.164	2.054	1.116	243	193	230
CCSD(T)	1.938	1.172					192	169	201
B88P91	1.910	1.153	2.076	1.153	2.049	1.155	193	166	183
B3LYP	1.921	1.155	2.068	1.155	2.078	1.156	170	168	187
Experimental	1.918	1.141	2.063	1.145	2.058	1.148	154	169	192

a. *Source:* Koch, W., Hertwig, R. H., Density Functional Theory: Applications to Transition Metal Problems. *Encyclopedia of Computational Chemistry*, von Ragué Schleyer, P. R., Allinger, N. L., Clarke, T., Gasteiger, J., Kollman, P., Schaefer, H. F. III, Eds. Wiley & Sons: New York, 1998.

Is density functional theory an ab initio theory? If the exact energy functional were known, then it would be an ab initio theory, but unfortunately, it is not known. Although some energy functionals that have been proposed are based upon first principles and do not contain any empirical parameters, the more successful ones, in the sense of yielding the most accurate molecular properties, do contain parameters that must be determined by some means. To be sure, this is usually done by fitting a functional to a few ab initio atomic properties and then using the resulting functional to predict molecular properties, but this procedure destroys the ab initio "purity" of the method to some quantum chemists.

Perhaps the most serious deficiency of density functional theory is that the method cannot be improved in a systematic manner. The other electron correlation methods that we have discussed can achieve increasingly better results by including more terms in their expansions. Their implementation is strictly a matter of computer resources, which certainly have been expanding at a prodigious rate. Improvements in density functional theory, however, depend upon the construction of better functionals. Just what improvements lie ahead depends upon the insight and ingenuity of researchers in the field.

Problems

12–1. Derive Equations 12.3 through 12.6.

12–2. Show that the coulomb integral J_{ii} is equal to the exchange integral K_{ii}.

12–3. Explain why the subscript i in Equation 12.7 runs from 1 to N even though it is describing a $2N$-electron molecule.

12–4. Show that the Hartree–Fock energy is not equal to the sum of the Hartree–Fock orbital energies.

12–5. Write out the basis set orbital $\sigma_u 3s\,(7.291\,69)$ in the first column of Table 12.1.

12–6. Write out the basis set orbital $\sigma_u 3d\,(1.690\,03)$ in the first column of Table 12.1.

12–7. Use the information in Table 12.1 to write out the $2\sigma_u$ molecular orbital in terms of the atomic orbitals.

12–8. Show that the matrix elements of the Fock operator consist of integrals of the form

$$(\mu\nu\,|\,\sigma\lambda) = \int d\mathbf{r}_1 d\mathbf{r}_2\,\phi_\mu(\mathbf{r}_1)\phi_\nu(\mathbf{r}_1)\frac{1}{r_{12}}\phi_\sigma(\mathbf{r}_2)\phi_\lambda(\mathbf{r}_2)$$

12–9. Show that the Gaussian functions given by Equation 12.21 are normalized.

12–10. Show that a three-dimensional Gaussian function centered at $\mathbf{r}_0 = x_0\mathbf{i} + y_0\mathbf{j} + z_0\mathbf{k}$ is a product of three one-dimensional Gaussian functions centered on x_0, y_0, and z_0.

12–11. Show that

$$\int_{-\infty}^{\infty} e^{-(x-x_0)^2} dx = \int_{-\infty}^{\infty} e^{-x^2} dx = 2\int_0^{\infty} e^{-x^2} dx = \pi^{1/2}$$

12–12. The Gaussian integral

$$I_0 = \int_0^{\infty} e^{-ax^2} dx$$

can be evaluated by a trick. First write

$$I_0^2 = \int_0^{\infty} dx e^{-ax^2} \int_0^{\infty} dy e^{-ay^2} = \int_0^{\infty}\int_0^{\infty} dx dy e^{-a(x^2+y^2)}$$

Now convert the integration variables from cartesian coordinates to polar coordinates and show that

$$I_0 = \frac{1}{2}\left(\frac{\pi}{a}\right)^{1/2}$$

12–13. Show that the integral

$$I_{2n} = \int_0^{\infty} x^{2n} e^{-ax^2} dx$$

can be obtained from I_0 in the previous problem by differentiating n times with respect to a. Show that

$$I_{2n} = \frac{1\cdot 3\cdot 5\cdots(2n-1)}{2(2a)^n}\left(\frac{\pi}{a}\right)^{1/2}$$

12–14. Show that the product of a (not normalized) Gaussian function centered at $\mathbf{R}_A$ and one centered at $\mathbf{R}_B$—that is,

$$\phi_1 = e^{-\alpha|\mathbf{r} - \mathbf{R}_A|^2} \qquad \text{and} \qquad \phi_2 = e^{-\beta|\mathbf{r} - \mathbf{R}_B|^2}$$

is a Gaussian function centered at

$$\mathbf{R}_p = \frac{\alpha \mathbf{R}_A + \beta \mathbf{R}_B}{\alpha + \beta}$$

For simplicity, work in one dimension and appeal to Problem 12–10 for the three-dimensional proof.

12–15. Show explicitly that if

$$\phi_s(\alpha, \mathbf{r} - \mathbf{R}_A) = \left(\frac{2\alpha}{\pi}\right)^{3/4} e^{-\alpha|\mathbf{r} - \mathbf{R}_A|^2}$$

and

$$\phi_s(\beta, \mathbf{r} - \mathbf{R}_B) = \left(\frac{2\beta}{\pi}\right)^{3/4} e^{-\beta|\mathbf{r} - \mathbf{R}_B|^2}$$

are normalized Gaussian s functions, then

$$\phi_s(\alpha, \mathbf{r} - \mathbf{R}_A)\phi_s(\beta, \mathbf{r} - \mathbf{R}_B) = K_{AB}\phi_{1s}(p, \mathbf{r} - \mathbf{R}_p)$$

where $p = \alpha + \beta$, $\mathbf{R}_p = (\alpha \mathbf{R}_A + \beta \mathbf{R}_B)/(\alpha + \beta)$ (see the previous problem), and

$$K_{AB} = \left[\frac{2\alpha\beta}{(\alpha + \beta)\pi}\right]^{3/4} e^{-\frac{\alpha\beta}{\alpha+\beta}|\mathbf{R}_A - \mathbf{R}_B|^2}$$

12–16. Plot the product of the two Gaussian functions $\phi_1 = e^{-2(x-1)^2}$ and $\phi_2 = e^{-3(x-2)^2}$. Interpret the result.

12–17. Using the result of Problem 12–15, show that the overlap integral of the two normalized Gaussian functions

$$\phi_s = \left(\frac{2\alpha}{\pi}\right)^{3/4} e^{-\alpha|\mathbf{r} - \mathbf{R}_A|^2} \qquad \text{and} \qquad \phi_s = \left(\frac{2\beta}{\pi}\right)^{3/4} e^{-\beta|\mathbf{r} - \mathbf{R}_B|^2}$$

is

$$S(|\mathbf{R}_A - \mathbf{R}_B|) = \left(\frac{4\alpha\beta}{\pi}\right)^{3/4} \left(\frac{\pi}{\alpha + \beta}\right)^{3/2} e^{-\frac{\alpha\beta|\mathbf{R}_A - \mathbf{R}_B|^2}{\alpha+\beta}}$$

Plot this result as a function of $|\mathbf{R}_A - \mathbf{R}_B|$.

12–18. One criterion for determining the best possible "fit" of a Gaussian function to a Slater orbital is to minimize the integral of the square of their difference. For example, we can find the optimal value of α in $g_s(r, \alpha)$ by minimizing

$$I = \int d\mathbf{r} \, [S_{1s}(r, 1.00) - g_s(r, \alpha)]^2$$

with respect to α. If the two functions $S_{1s}(r, 1.00)$ and $g_s(r, \alpha)$ are normalized, show that minimizing I is equivalent to maximizing the overlap integral of $S_{1s}(r, 1.00)$ and $g_s(r, \alpha)$:

$$S(\alpha) = \int d\mathbf{r} \, S_{1s}(r, 1.00) g_s(r, \alpha)$$

12–19. Show that $S(\alpha)$ in the previous problem is given by

$$S(\alpha) = 4\pi^{1/2} \left(\frac{2\alpha}{\pi} \right)^{3/4} \int_0^\infty r^2 e^{-r} e^{-\alpha r^2} dr$$

Using a program such as *Mathematica* or *MathCad*, show that the maximum occurs at $\alpha = 0.270\,95$.

12–20. Compare $S_{1s}(r, 1.00)$ and $g_s(r, 0.270\,95)$ graphically by plotting them on the same graph.

12–21. To show how to determine a fit of Gaussian functions to a Slater orbital $S_{1s}(\zeta \neq 1)$ when we know the fit to $S_{1s}(\zeta = 1)$, start with the overlap integral of $S_{1s}(r, \zeta)$ and $g_s(r, \beta)$:

$$S = 4\pi^{1/2} \left(\frac{2\beta}{\pi} \right)^{3/4} \zeta^{3/2} \int_0^\infty r^2 e^{-\zeta r} e^{-\beta r^2} dr$$

Now let $u = \zeta r$ to get

$$S = 4\pi^{1/2} \left(\frac{2\beta/\zeta^2}{\pi} \right)^{3/4} \int_0^\infty u^2 e^{-u} e^{-(\beta/\zeta^2)u^2} du$$

Compare this result for S with that in Problem 12–19 to show that $\beta = \alpha\zeta^2$ or, in more detailed notation,

$$\alpha(\zeta) = \alpha(\zeta = 1.00) \times \zeta^2$$

12–22. What is the difference between a double-zeta (DZ) basis set and a double split-valence basis set?

12–23. What is meant by a triple-zeta basis set?

12–24. How many basis functions are used in a 6-31G basis set calculation of an ethane molecule? How many for a propane molecule?

12–25. How many basis functions are used in a 6-31G basis set calculation of a benzene molecule? How many for a toluene molecule?

12–26. How many primitive Gaussian functions and how many contracted Gaussian functions are used in a 6-31G basis set calculation of an ethane molecule? How many for a propane molecule?

12–27. How many primitive Gaussian functions and how many contracted Gaussian functions are used in a 6-31G basis set calculation of a benzene molecule? How many for a toluene molecule?

12–28. Even though there are only five d orbitals, it is customary in Gaussian basis sets to use six "d" orbitals:

$$g_{xx} = \left(\frac{2048\alpha^7}{9\pi^3}\right)^{1/4} x^2 e^{-\alpha r^2} \qquad g_{yy} = \left(\frac{2048\alpha^7}{9\pi^3}\right)^{1/4} y^2 e^{-\alpha r^2}$$

$$g_{zz} = \left(\frac{2048\alpha^7}{9\pi^3}\right)^{1/4} z^2 e^{-\alpha r^2} \qquad g_{xy} = \left(\frac{2048\alpha^7}{9\pi^3}\right)^{1/4} xy e^{-\alpha r^2}$$

$$g_{xz} = \left(\frac{2048\alpha^7}{9\pi^3}\right)^{1/4} xz e^{-\alpha r^2} \qquad g_{yz} = \left(\frac{2048\alpha^7}{9\pi^3}\right)^{1/4} yz e^{-\alpha r^2}$$

Not all those functions have the same angular symmetry as d orbitals. However, show that $g_{xy}, g_{xz}, g_{yz}, (2g_{zz} - g_{xx} - g_{yy})/2$, and $(3/4)^{1/2}(g_{xx} - g_{yy})$ do have the correct symmetry. Show also that a sixth linear combination, $(g_{xx} + g_{yy} + g_{zz})/5^{1/2}$, has the symmetry of an s orbital.

12–29. Problem 12–28 describes the Gaussian functions that correspond to d orbitals. Rather than use the five combinations that have the correct symmetry for d orbitals, it is easier computationally to use all six Gaussian "d" orbitals in forming basis sets. Given that, show that a 6-31G* basis set calculation of an ethane molecule uses 42 basis functions. How many are used in a similar calculation for a benzene molecule? How many in a 6-31G** basis set calculation for a benzene molecule?

12–30. Go to the website *http://gnode2.pnl.gov/bse/portal* to obtain the information to write out the contracted Gaussian functions for the 6-31G basis set of a nitrogen atom. Show that the two 2s orbitals are normalized. Plot them and show that the one represented by a single Gaussian function is larger in extent than the other.

12–31. Go to the website *http://gnode2.pnl.gov/bse/portal* to obtain the information to write out the contracted Gaussian functions for the 6-311G basis set of a carbon atom. Show that the three 2s orbitals are normalized. Plot them and show that they are progressively larger in extent.

12–32. Go to the website *http://gnode2.pnl.gov/bse/portal* to obtain the information to write out the contracted Gaussian functions for the 6-31G* basis set of an oxygen atom. Show that the two 2s orbitals are normalized. Plot them and show that the one represented by a single Gaussian function is larger in extent than the other.

12–33. Go to the website *http://gnode2.pnl.gov/bse/portal* to obtain the STO-3G 1s and 2s orbitals of a carbon atom and show that they are normalized.

12–34. Show that the $1s$, $2s'$, $2p'_x$, $2p'_y$, and $2p'_z$ orbitals in the 6-31G** basis set given in Figure 12.7 are normalized.

12–35. If you have access to Gaussian, verify that a 6-311G(d,p) calculation on acetaldehyde uses 78 basis functions and 128 primitive Gaussian functions.

For Problems 12–36 to 12–40: Even if you don't have unlimited access to GaussView or WebMO, you can log into WebMO as a guest and receive one minute of CPU time to run calculations. One minute is quite enough time to run a number of Hartree–Fock calculations on the molecules that we have been considering. Just go to the website, www.webmo.net/demo/, then log onto the Demo Server, and follow the instructions.

12–36. Use either GaussView or WebMO to confirm the HF/6-31G* entry for a methane molecule in Table 12.6.

12–37. Use either GaussView or WebMO to confirm the HF/6-311G* entry for NH_3 in Table 12.7.

12–38. Use either GaussView or WebMO to confirm the HF/6-311G* entry for CH_4 in Table 12.7.

12–39. Use either GaussView or WebMO to confirm the HF/6-31G* entry for HF in Table 12.8.

12–40. Use either GaussView or WebMO to confirm the HF/6-31G* entry for a formaldehyde molecule in Table 12.10.

12–41. Use the NIST website *http://srdata.nist.gov/cccbdb* to verify the entries in Table 12.6. (The Hartree–Fock limits are from other sources.)

12–42. Show that the number of singly excited determinants that occur in a configuration-interaction calculation is given by $2N(2K - 2N)$ for a molecule with $2N$ electrons and $2K$ spin orbitals. Show that the number of n-tuple excitations is given by

$$\binom{2N}{n}\binom{2K - 2N}{n}$$

12–43. Use the result of the previous problem to show that the number of single and double excitations for a CI calculation on N_2 using a basis set consisting of 100 basis functions is over a million.

12–44. Why does Ψ_{CI} in Equation 12.31 truncate at the $2N$th term?

12–45. Why are many more excited terms used in a CCD calculation than a CID calculation?

12–46. What do you think a CISDQ calculation means?

12–47. In this problem, we'll give a simple proof that CCD is size-consistent but that CID is not. Consider n widely separated molecules, A. Argue that the wave function of these n molecules is given by $\psi(nA) = \psi(A)^n$ and the energy is $E(nA) = nE(A)$. For one molecule, CCD gives $\psi(A) = \exp(\hat{T}_2)\psi_0(A)$, where $\psi_0(A)$ is the reference wave function of an A molecule. If we let the reference wave function of the n widely separated A molecules be $\Psi_0(nA) = \psi_0(1)\psi_0(2)\cdots\psi_0(n)$, then show that

$$\Psi_{CCD} = \exp[\hat{T}_2(1) + \hat{T}_2(2) + \cdots + \hat{T}_2(n)]\,\Psi_0(nA)$$
$$= [\exp(\hat{T}_2)\psi_0(A)]^n = \psi^n(A)$$

and that the corresponding energy is $nE(A)$. Now show that in contrast, if $\psi(A) = (1 + \hat{C}_2)\psi_0(A)$ as in CID, then

$$\Psi_{CID} = [1 + \hat{C}_2(1) + \cdots + \hat{C}_2(n)]\Psi_0(nA)$$

$$\neq [(1 + \hat{C}_2)\psi_0(A)]^n$$

12–48. Correct the values of ΔH_r° at 298 K that we obtained for the dissociation reaction $HCl(g) \rightarrow H(g) + Cl(g)$ in Section 12.7 for the effect of the zero-point vibrational energy of $HCl(g)$ and the spin-orbit interaction energy of $Cl(g)$. (Go to the NIST ccbdb website for the values of these quantities.)

12–49. Go to the NIST cccbdb website and determine the value of ΔH_r° at 298 K for the dissociation reaction $HF(g) \rightarrow H(g) + F(g)$ from the Reaction Data section. Now calculate ΔH_r° from the values of ΔH_f° of the individual species and compare your results. Why do they differ?

12–50. Use the NIST cccbdb website to calculate the standard molar enthalpy of combustion of propane at 289 K with a CCSD/cc-pVTZ model chemistry and compare your result to the experimental value.

12–51. Derive Equation 12.41.

12–52. The classic example of a functional is a definite integral. Give two other examples of functionals.

12–53. Why might you not expect that the energy can be expressed as a functional of the electron density?

12–54. The units of dipole moment given by Gaussian are called debyes (D, after the Dutch-American chemist, Peter Debye, who was awarded the Nobel Prize in Chemistry in 1936 for his work on dipole moments). One debye is equal to 10^{-18} esu·cm, where esu (electrostatic units) is the non-SI unit for electric charge. Given the protonic charge is 4.803×10^{-10} esu, show that the conversion factor between debyes and C·m (coulomb·meters) is $1 D = 3.33 \times 10^{-30}$ C·m.

12–55. Show that a dipole of one debye (1 D) is equivalent to 0.39345 au. (See the previous problem for the definition of a debye.)

12–56. The Gaussian output for a water molecule calculated with an HF/6-31G* model chemistry gives $R(OH) = 0.947$ Å, $A(HOH) = 105.5°$, and the partial charges on the hydrogen atoms as $+0.41$ au and on the oxygen atom as -0.82 au. Use these values to calculate the dipole moment of a water molecule and compare your result to the experimental value.

12–57. The Gaussian output for an ammonia molecule calculated with an HF/6-31G* model chemistry gives $R(NH) = 1.000$ Å, $A(HNH) = 107.1°$, and the partial charges on the hydrogen atoms as $+0.37$ au and that on the nitrogen atom as -1.11 au. Use these values to calculate the dipole moment of an ammonia molecule and compare your result to the experimental value.

12–58. Use either GaussView or WebMO to confirm the BLYP/6-31G* entry for H_2O in Table 12.11.

12–59. Use either GaussView or WebMO to confirm the BLYP/6-31G* entry for NII_3 in Table 12.11.

12–60. Use either GaussView or WebMO to confirm the BLYP/6-31G* entry for CH_4 in Table 12.11.

References

Basis sets: *http://www.emsl.pnl.gov/forms/basisform.html* or *https://bse.pnl.gov/bse/portal*

NIST Computation Chemistry Comparison and Benchmark Data Base: NIST Standard Reference Database Number 101, Release 12, Aug. 2005, Russell D. Johnson III (Editor); *http://srdata.nist .gov/cccbdb*

Quantum Chemistry Literature Database: *http://qcldb2.ims.ac.jp/*

Gaussian home page: *http://www.gaussian.com*

GAMESS homepage: *www.msg.ameslab.gov/GAMESS/*

WebMO homepage: *www.webmo.net*

SPARTAN home page: *http://www.wavefun.com*

North Carolina High School computational chemistry server: *http://chemistry.ncssm.edu*

Hehre, W., Radom, L., Schleyer, P. V., Pople, J. *Ab Initio Molecular Orbital Theory*. Wiley & Sons: New York, 1986.

Hehre, W. *A Guide to Molecular Mechanics and Quantum Chemical Calculations*. Wavefunction, Inc.: Irvine, CA, 2005.

Levine, I. N. *Quantum Chemistry,* 5th ed. Prentice Hall: Upper Saddle River, NJ, 2000.

Szabo, A., Ostlund, N. *Modern Quantum Chemistry: Introduction to Advanced Electronic Structure Theory*. Dover Publications: Mineola, NY, 1982.

Gotwals, R. R., Jr., Sendlinger, S. A. *A Chemistry Educator's Guide to Molecular Modeling*. Available in pre-publication form at *http://www.chemistry.ncssm.edu/book/index.html*

References for Post-Hartree–Fock Methods

General

Head-Gordon, M. Quantum Chemistry and Molecular Processes. *J. Phys. Chem.,* **100**, 13213 (1996).

Raghavachari, K., Anderson, J. B. Electron Correlation Effects in Molecules. *J. Phys. Chem.,* **100**, 12960 (1996).

Jensen, F. *Introduction to Computational Chemistry,* 2nd ed. Wiley & Sons: New York, 2006.

Cramer, C. *Essentials of Computational Chemistry: Theories and Models,* 2nd ed. Wiley & Sons: New York, 2004.

Configuration Interaction

Szabo, A., Ostlund, N. *Modern Quantum Chemistry: Introduction to Advanced Electronic Structure Theory*. Dover Publications: Mineola, NY, 1982.

Lee, T. J., Remington, R. B., Yamaguchi, Y., Schaefer, H. F., III. The Effects of Triple and Quaduple Excitations in Configuration Interaction Procedures for the Quantum Mechanical Prediction of Molecular Properties. *J. Chem. Phys.,* **89**, 408 (1988).

Christoffersen, R. E. *Basic Principles and Techniques of Molecular Quantum Mechanics*. Springer-Verlag: New York, 1989.

Coupled-Cluster Theory

Levine, I. N. *Quantum Chemistry,* 5th ed. Prentice Hall: Upper Saddle River, NJ, 2000.
Bartlett, R. Coupled-Cluster Approach to Molecular Structure and Spectra: A Step Toward Predictive Quantum Chemistry. *J. Phys. Chem.,* **93**, 1697 (1989).

Density Functional Theory

Parr, R. G., Yang, W. *Density Functional Theory of Atoms and Molecules*. Oxford University Press: New York, 1989.
Koch, W., Holthausen, M. C. *A Chemist's Guide to Density Functional Theory,* 2nd ed. Wiley-VCH: New York, 2001.
Levine, I. N. *Quantum Chemistry,* 5th ed. Prentice Hall: Upper Saddle River, NJ, 2000.
Ziegler, T. Approximate Density Functional Theory as a Practical Tool in Molecular Energetics and Dynamics. *Chem. Rev.,* **91**, 651 (1991).
Kohn, W., Becke, A. D., Parr, R. G. Density Functional Theory of Electronic Structure. *J. Phys. Chem.,* **100**, 12974 (1996).

Answers to the Numerical Problems

Chapter 1

1–1. $\nu = \dfrac{c}{\lambda} = \dfrac{3.00 \times 10^8 \text{ m·s}^{-1}}{200 \times 10^{-9} \text{ m}} = 1.50 \times 10^{15}$ Hz

$\tilde{\nu} = \dfrac{1}{\lambda} = 5.00 \times 10^6 \text{ m}^{-1} = 5.00 \times 10^4 \text{ cm}^{-1}$

$E = h\nu = \dfrac{hc}{\lambda} = 9.93 \times 10^{-19} \text{ J} = 0.993 \text{ aJ}$

1–2. $\nu = c\tilde{\nu} = (3.00 \times 10^8 \text{ m·s}^{-1})(10^3 \text{ cm}^{-1}) = 3 \times 10^{13}$ Hz

$\lambda = \dfrac{1}{\tilde{\nu}} = 10^{-3} \text{ cm} = 10^{-5} \text{ m}$

$E = h\nu = 2 \times 10^{-20}$ J

1–3. $\tilde{\nu} = 0.666 \text{ cm}^{-1}$

$\lambda = \dfrac{c}{\nu} = 0.015 \text{ m}$

$E = h\nu = 1.3 \times 10^{-23}$ J

1–6. (a) 9.67×10^{-6} m (b) 9.67×10^{-7} m (c) 2.90×10^{-7} m

1–7. 11 000 K

1–8. 3×10^{-10} m; X ray

1–11. $E = \dfrac{hc}{\lambda} = 2 \times 10^{-15}$ J

1–13. (a) 1.07×10^{16} photons (b) 5.41×10^{15} photons (c) 2.68×10^{15} photons

1–14. $\lambda_{max} = 1.01 \times 10^{-5}$ m; infrared

1–15. $\nu = 4.738 \times 10^{14}$ Hz; $E = 3.319 \times 10^{-19} \text{ J} = 0.3139 \text{ aJ}$

1–16. 1.70×10^{15} photon·s^{-1}

1–17. 5300 K

1–18. $\phi = 3.52 \times 10^{-19}$ J $= 2.20$ eV;   KE $= 1.32 \times 10^{-19}$ J

1–19. $\phi = 4.40$ eV $= 7.05 \times 10^{-19}$ J; $E = 9.93 \times 10^{-19}$ J;   KE $= 2.88 \times 10^{-19}$ J

1–20. KE $= 0.805$ eV $= 1.29 \times 10^{-19}$ J; $E = 8.64 \times 10^{-19}$ J; $\phi = 7.35 \times 10^{-19}$ J $=$
4.59 eV; $\nu_0 = 1.11 \times 10^{15}$ Hz

1–21. $h = 6.60 \times 10^{-34}$ Hz; $\phi = 3.59 \times 10^{-19}$ J $= 2.24$ eV

1–23. $\nu = 3.286 \times 10^{15}$ Hz; $\lambda = 9.117 \times 10^{-8}$ m
1 Rydberg $= 13.60$ eV $= 1312$ kJ·mol^{-1}

1–24. $\lambda = 121.57$ nm; 102.57 nm; 97.253 nm

1–25. $n = 3$

1–26. $n = 2$

1–28. $\lambda = 91.17$ nm; $E = -2.179 \times 10^{-18}$ J

1–29. $(109\ 737\ \text{cm}^{-1})(0.999\ 45) = 109\ 678\ \text{cm}^{-1}$

1–30. $\mu = 7.00$; 0.9798

1–31. $I = 2.68 \times 10^{-47}$ kg·m^2; $r = 1.28 \times 10^{-10}$ m $= 1.28$ Å

1–33. Infrared

1–34. 54.394 eV $= 5248.2$ kJ·mol^{-1}

1–35. $v_1 = 2.188 \times 10^6$ m·s^{-1}, $v_2 = 1.094 \times 10^6$ m·s^{-1}, $v_3 = 7.292 \times 10^5$ m·s^{-1}

1–36. The hydrogen is present as H(g); 31 800 K

1–37. $\mu = 1.69 \times 10^{-28}$ kg; $r_0 = 0.284$ pm, $\nu = 4.59 \times 10^{17}$ Hz; $E = -4.06 \times 10^{-16}$ J

1–38. (a) 0.123 nm   (b) 2.86×10^{-3} nm   (c) 0.332 nm

1–39. (a) 1.602×10^{-17} J·electron^{-1}, 1.23×10^{-10} m $= 1.23$ Å   (b) 6.02×10^{-18} J

1–40. 0.082 V

1–41. 1.3×10^{-18} J/α-particle, 5.1 pm $= 0.051$ Å

1–42. 2500 K

1–43. 2.188×10^7 m·s^{-1}; 1.094×10^8 m·s^{-1}

1–44. $\sin\theta = \lambda/d = 0.006$, $\theta = 0.006$ radians, $\tan\theta = x/2.00$ m, or $x = 0.012$ m $=$
12 mm

1–45. $\tan \theta = 1.50 \times 10^{-2}$ m$/3.00$ m $= 5.00 \times 10^{-3}$, or $\theta = 5.00 \times 10^{-3}$ radians,
$\sin \theta = 5.00 \times 10^{-3}$, $d = \lambda/\sin \theta = 0.139$ mm

1–46. 3.6×10^7 m·s^{-1}

1–47. 6.6×10^{-23} kg·m·s^{-1}, compared to 1.993×10^{-24} kg·m·s^{-1}

1–49. 2.9×10^{-23} s

1–50. 7×10^{-25} J

1–51. 7×10^{-22} J

MathChapter A

A–1. (a) $(2 - i)^3 = 8 - (3)(4)i + 3(2)i^2 - i^3 = 8 - 12i - 6 + i = 2 - 11i$

(b) $e^{\pi i/2} = \cos \dfrac{\pi}{2} + i \sin \dfrac{\pi}{2} = i$

(c) $e^{-2+i\pi/2} = e^{-2}\left(\cos \dfrac{\pi}{2} + i \sin \dfrac{\pi}{2}\right) = e^{-2}i$

(d) $(\sqrt{2} + 2i)\left(\cos \dfrac{\pi}{2} - i \sin \dfrac{\pi}{2}\right) = (\sqrt{2} + 2i)(-i) = 2 - i\sqrt{2}$

A–2. (a) x

(b) $z^2 = (x + 2iy)^2 = x^2 + 4ixy - 4y^2$, Re $z^2 = x^2 - 4y^2$

(c) Im $z^2 = 4xy$

(d) $zz^* = (x + 2iy)(x - 2iy) = x^2 + 4y^2$, Re $zz^* = x^2 + 4y^2$

(e) Im $zz^* = 0$

A–3. (a) $z = 6e^{i\pi/2}$

(b) $r = (16 + 2)^{1/2} = 3\sqrt{2}$; $\theta = \tan^{-1} \dfrac{-\sqrt{2}}{4} = -0.340$ (note quadrant);
$z = 3\sqrt{2}e^{-0.340i}$

(c) $r = \sqrt{5}$; $\theta = \tan^{-1} 2 = 1.107 + \pi$ (note quadrant); $z = \sqrt{5}e^{(\pi+1.107)i}$

(d) $z = (\pi^2 + e^2)^{1/2}e^{i \tan^{-1}(e/\pi)} = (\pi^2 + e^2)^{1/2}e^{0.713i}$

A–4. (a) $z = \cos \dfrac{\pi}{4} + i \sin \dfrac{\pi}{4} = \dfrac{\sqrt{2}}{2} + i\dfrac{\sqrt{2}}{2}$

(b) $z = 6\left(\cos \dfrac{2\pi}{3} + i \sin \dfrac{2\pi}{3}\right) = -3 + i3\sqrt{3}$

(c) $z = e^{\ln 2}\left(\cos \dfrac{\pi}{4} - i \sin \dfrac{\pi}{4}\right) = 2\left(\dfrac{\sqrt{2}}{2} - i\dfrac{\sqrt{2}}{2}\right) = \sqrt{2} - i\sqrt{2}$

(d) $z = (\cos 2\pi - i \sin 2\pi) + (\cos 4\pi + i \sin 4\pi) = 1 + 1 = 2$

A–11. They do not oscillate.

A–13. $i^i = (e^{i\pi/2})^i = e^{-\pi/2}$

A–15. $x = 2; -1 \pm i\sqrt{3}$

Chapter 2

2–1. (a) $y(x) = c_1 e^{3x} + c_2 e^x$

(b) $y(x) = c_1 + c_2 e^{-6x}$

(c) $y(x) = c_1 e^{-3x}$

(d) $y(x) = c_1 e^{(-1+\sqrt{2})x} + c_2 e^{(-1-\sqrt{2})x}$

(e) $y(x) = c_1 e^{2x} + c_2 e^x$

2–2. (a) $y(x) = 2e^{2x}$ (b) $y(x) = -3e^{2x} + 2e^{3x}$ (c) $y(x) = 2e^{2x}$

2–4. (a) $x(t) = \dfrac{v_0}{\omega} \sin \omega t$ (b) $x(t) = A \cos \omega t + \dfrac{v_0}{\omega} \sin \omega t$

2–5. $c_1 = A \sin \phi = B \cos \psi$, $c_2 = A \cos \phi = -B \sin \psi$

2–6. (a) $y(x) = e^{-x}(c_1 \cos x + c_2 \sin x)$

(b) $y(x) = e^{3x}(c_1 \cos 4x + c_2 \sin 4x)$

(c) $y(x) = e^{-\beta x}(c_1 \cos \omega x + c_2 \sin \omega x)$

(d) $y(x) = e^{-2x}(\cos x - \sin x)$

2–7. The motion is oscillatory with frequency $(1/2\pi)(k/m)^{1/2}$ and amplitude $v_0(m/k)^{1/2}$.

2–8. $\xi(t) = e^{-\gamma t/2m} \cos(\omega' t + \phi)$ with $\omega' = (4km - \gamma^2)^{1/2}/2m$

2–10. $\psi(x) = A \sin \dfrac{n\pi x}{a}$, $n = 1, 2, \ldots$

2–14. $\psi(x, y) = A \sin \dfrac{n_x \pi x}{a} \sin \dfrac{n_y \pi y}{b}$, $n_x, n_y = 1, 2, \ldots$

2–15. $\psi(x, y, z) = A \sin \dfrac{n_x \pi x}{a} \sin \dfrac{n_y \pi y}{b} \sin \dfrac{n_z \pi z}{c}$, $n_x, n_y, n_z = 1, 2, \ldots$

2–25. Height $= v_0^2/2g$; time to return $= 2v_0/g$

2–26. Natural frequency $= (g/l)^{1/2}/2\pi$

MathChapter B

B–1. $\langle x \rangle = 0.30$, $\langle x^2 \rangle = 5.80$, $\sigma^2 = 5.71 > 0$

B–2. $\langle n \rangle = \lambda$, $\langle n^2 \rangle = \lambda^2 + \lambda$, $\sigma^2 = \lambda > 0$

B–3. $c = \lambda$, $\langle x \rangle = 1/\lambda$, $\langle x^2 \rangle = 2/\lambda^2$, $\sigma^2 = 1/\lambda^2$, Prob $[x \geq a] = e^{-\lambda a}$

B–12. $1/2$ for all n

Chapter 3

3–1. (a) $\pm x^2$ (b) $(x^3 - a^3)e^{-ax}$ (c) $\dfrac{1}{4} - 1 + 3 = \dfrac{9}{4}$ (d) $6xy^2z^4 + 2x^3z^4 + 12x^3y^2z^2$

3–2. (a) nonlinear (b) nonlinear (c) linear (d) nonlinear (e) linear (f) nonlinear

3–3. (a) $-\omega^2$ (b) $i\omega$ (c) $\alpha^2 + 2\alpha + 3$ (d) 6

3–5. (a) $\hat{A}^2 = \dfrac{d^4}{dx^4}$

(b) $\hat{A}^2 = \dfrac{d^2}{dx^2} + 2x\dfrac{d}{dx} + 1 + x^2$

(c) $\hat{A}^2 = \dfrac{d^4}{dx^4} - 4x\dfrac{d^3}{dx^3} + (4x^2 - 2)\dfrac{d^2}{dx^2} + 1$

3–6. (a) commute

(b) do not commute

(c) do not commute (watch $\pm$ signs)

(d) commute

3–7. Same result as ordinary algebra only if $\hat{P}$ and $\hat{Q}$ commute.

3–13. The square root of reciprocal length

3–25. No, because $\hat{H}\psi(x)$ is just a function.

3–26. $\langle x \rangle = a/2$, $\langle x^2 \rangle = \dfrac{a^2}{3} - \dfrac{a^2}{8\pi^2}$

3–27. $\langle p \rangle = 0$, $\langle p^2 \rangle = \dfrac{n^2\pi^2\hbar^2}{a^2} = \sigma_p^2$, $\sigma_p = h/a$ for $n = 2$

3–32. 1, 2, 1, 1

MathChapter C

C–1. $\sqrt{14}$

C–2. $(x^2 + y^2)^{1/2}$, $(x^2 + y^2 + z^2)^{1/2}$

C–3. $\cos\dfrac{\pi}{2} = 0$

C–6. $\cos\theta = \dfrac{-3}{(6)^{1/2}(14)^{1/2}} = -0.327$, $\theta = 109°$

C–8. $5\,\mathbf{i} + 5\,\mathbf{j} - 5\,\mathbf{k}$

C–15. $3\,\mathbf{i} - \mathbf{j} + \mathbf{k}$

Chapter 4

4–1. (a) normalizable, $1/\pi^{1/4}$

(b) not normalizable

(c) normalizable, $1/(2\pi)^{1/2}$

(d) not normalizable

(e) normalizable, 2

4–2. (a) $2/\sqrt{\pi}$ (b) normalized   (c) normalized

4–5. (a) not normalizable

(b) acceptable

(c) acceptable

(d) not normalizable

(e) not normalizable

4–7. $\langle E \rangle = 6\hbar^2/ma^2$, $\langle E^2 \rangle = 126\hbar^4/m^2a^4$, $\sigma_E^2 = 90\hbar^4/m^2a^4$

4–9. $\langle p \rangle = 0$, $\sigma_p^2 = \dfrac{h^2}{4}\left(\dfrac{n_x^2}{a^2} + \dfrac{n_y^2}{b^2}\right)$

4–10. $\langle E \rangle = \dfrac{5\hbar^2}{m}\left(\dfrac{1}{a^2} + \dfrac{1}{b^2}\right)$

4–11. (a) $[\hat{A}, \hat{B}] = 2\dfrac{d}{dx}$ (b) $[\hat{A}, \hat{B}] = 2$ (c) $[\hat{A}, \hat{B}] = -f(0)$ (d) $[\hat{A}, \hat{B}] = 4x\dfrac{d}{dx} + 3$

4–12. The subscripts occur as x, y, z and as cyclic permutations of x, y, z.

4–15. $[\hat{X}, \hat{P}_y] = [\hat{Y}, \hat{P}_z] = 0$; $[\hat{X}, \hat{P}_x] = [\hat{Y}, \hat{P}_y] = i\hbar$

4–16. No

4–17. $\Delta x \Delta p_y = \Delta y \Delta p_x = 0$

4–18. d/dx is not Hermitian; id/dx is Hermitian; d^2/dx^2 is Hermitian; id^2/dx^2 is not Hermitian; xd/dx is not Hermitian; x is Hermitian

4–25. $\phi_0(x) = 1$, $\phi_1(x) = x$, $\phi_2(x) = \frac{1}{2}(3x^2 - 1)$

4–27. $\langle n \mid m \rangle = \delta_{nm}$; $\mid \phi \rangle = \displaystyle\sum_n c_n \mid n \rangle$; $c_n = \langle n \mid \phi \rangle$

4–29. $f_0(x) = 1$, $f_1(x) = \sqrt{3}(1 - 2x)$, $f_2(x) = \sqrt{5}(1 - 6x + 6x^2)$

4–31. Only if $\hat{A}$ and $\hat{B}$ commute.

4–39. $\langle x \rangle = \dfrac{a}{2} - \dfrac{48a}{25\pi^2}\cos\omega_{23}t$; $\omega_{23} = (E_3 - E_2)/\hbar$

4–40. $\langle x \rangle = \dfrac{a}{2} - \dfrac{32a}{225\pi^2} \cos \omega_{14}t$; $\omega_{14} = (E_4 - E_1)/\hbar$; amplitude $= \dfrac{64a}{225\pi^2}$ compared to $\dfrac{96a}{25\pi^2}$ in the previous problem

4–52. 0.52 ($v_0 = 1.966$)

4–53. $4/(4 + v_0)$

MathChapter D

D–1. $1.25 \times 10^{-3}\%$; $4.97 \times 10^{-3}\%$; $11.1 \times 10^{-3}\%$; $\ldots$; 0.468%

D–2. 0.249%; 0.499%; 0.748%; $\ldots$; 4.92%

D–3. $1 + \dfrac{x}{2} - \dfrac{x^2}{8} + O(x^3)$

D–4. $1 - \dfrac{x}{2} + \dfrac{3x^2}{8} + O(x^3)$

D–6. 1

D–7. $3/2$

D–8. $1/3$

D–17. 1

D–18. $\dfrac{a^3}{3} - \dfrac{a^4}{4} + O(a^5)$

D–21. $-1/2$

D–22. The first expansion is valid for $x < 1$ and the second expansion is valid for $x > 1$. The sum of the two expansions is valid for no value of x.

Chapter 5

5–4. The period, τ, which is the time it takes to undergo one cycle, is $2\pi/\omega = 1/v$.

5–9. $479 \, \text{N} \cdot \text{m}^{-1}$

5–10. $1.81 \times 10^{10} \, \text{m}^{-1}$

5–11. $\gamma_3 = -6D\beta^3$

5–12. $\tilde{x}_e = 0.01961$; $\tilde{x}_e \tilde{\omega}_e = 56.59 \, \text{cm}^{-1}$

5–13. $313 \, \text{N} \cdot \text{m}^{-1}$

5–14. $\tilde{\omega}_e = 556 \, \text{cm}^{-1}$; $E_0 = 5.52 \times 10^{-21} \, \text{J}$

5–15. $\tilde{\omega}_e = 2169.0 \text{ cm}^{-1}$; $\tilde{x}_e\tilde{\omega}_e = 13.0 \text{ cm}^{-1}$

5–16. 2558.549 cm^{-1}, 5026.64 cm^{-1}, 7404.31 cm^{-1}, 9691.54 cm^{-1}

5–17. $\tilde{\omega}_e = 2989 \text{ cm}^{-1}$; $\tilde{x}_e\tilde{\omega}_e = 51.6 \text{ cm}^{-1}$

5–18. $\tilde{\omega}_e = 3841 \text{ cm}^{-1}$; $\tilde{x}_e\tilde{\omega}_e = 1.45 \text{ cm}^{-1}$

5–19. $\tilde{D}_e = 37\,400 \text{ cm}^{-1}$

5–20. Around 68

5–39. $8.0 \text{ pm} = 0.080 \text{ Å}$; less than 6%

5–41. ≈ 1; $1 - e^{-1.906} = 0.851$

5–42. (a) 3, 3, 9 (b) 3, 2, 4 (c) 3, 3, 30 (d) 3, 3, 6

MathChapter E

E–2. $\left(1, \dfrac{\pi}{2}, 0\right)$; $\left(1, \dfrac{\pi}{2}, \dfrac{\pi}{2}\right)$; $(1, 0, \phi)$; $(1, \pi, \phi)$

E–3. (a) a sphere of radius 5 centered at the origin

(b) a cone about the z axis

(c) the y–z axis

E–4. $2\pi a^3/3$

E–5. $2\pi a^2$

E–6. 4/15

E–10. 0, 1/3

E–11. $8\pi/3$

Chapter 6

6–4. $\mu \approx 10^{-25} \text{ kg}$, $r \approx 10^{-10} \text{ m}$ and so $I \approx 10^{-45} \text{ kg·m}^2$ and $B \approx 10^{10} \text{ Hz}$

6–5. $3.35 \times 10^{-47} \text{ kg·m}^2$, $142 \text{ pm} = 1.42 \text{ Å}$

6–6. $113 \text{ pm} = 1.13 \text{ Å}$

6–7. $127.5 \text{ pm} = 1.275 \text{ Å}$

6–8. $305.5 \text{ pm} = 3.055 \text{ Å}$

6–9. $1.964 \times 10^{11} \text{ Hz}$, $1.964 \times 10^5 \text{ MHz}$, 6.552 cm^{-1}

6–10. 1.36×10^{11} revolution$\cdot$s^{-1}

6–11. $2\tilde{B} = 2.96$ cm^{-1}

6–12. 1.894×10^{-46} kg$\cdot$m^2

6–13. $\tilde{B}_{HI} = 6.428$ cm^{-1}, $I_{HI} = 4.355 \times 10^{-47}$ kg$\cdot$m^2, $\bar{l}_{HI} = 161.9$ pm $= 1.619$ Å,
$\tilde{B}_{DI} = 3.254$ cm^{-1}, $I_{DI} = 8.604 \times 10^{-47}$ kg$\cdot$m^2, $\bar{l}_{DI} = 161.7$ pm $= 1.617$ Å

6–14. $\tilde{\nu}_R = 2143.0$ cm$^{-1} + (3.74$ cm$^{-1})(J + 1)$; $\tilde{\nu}_P = 2143.0$ cm$^{-1} - (3.74$ cm$^{-1})J$

6–15. $\tilde{\nu}_R = 963.7$ cm$^{-1} + (1.52$ cm$^{-1})(J + 1)$; $\tilde{\nu}_P = 963.7$ cm$^{-1} - (1.52$ cm$^{-1})(J)$

6–16. $\tilde{\nu}_R(0 \rightarrow 1) = 2905.6$ cm^{-1}, $\tilde{\nu}_R(1 \rightarrow 2) = 2925.2$ cm^{-1}, $\tilde{\nu}_P(0 \rightarrow 1) = 2864.4$ cm^{-1},
$\tilde{\nu}_P(2 \rightarrow 1) = 2842.9$ cm^{-1}

6–17. $\tilde{B}_0 = 6.47$ cm^{-1}, $\tilde{B}_1 = 6.28$ cm^{-1}, $\tilde{\alpha}_e = 0.19$ cm^{-1}, $\tilde{B}_e = 6.56$ cm^{-1}

6–18. Yes

Molecule	$^{74}Ge^{32}S$	$^{74}Ge^{32}S$
$(\tilde{\nu}, J = 0)/$cm^{-1}	0.372 381	0.375 505
$(\tilde{\nu}, J = 1)/$cm^{-1}	0.744 761	0.750 009
$\Delta\tilde{\nu}/$cm^{-1}	0.372 379	0.375 504

6–19. $\tilde{\alpha}_e = 0.005\,92$ cm^{-1}, $\tilde{B}_e = 0.820\,04$ cm^{-1}

6–20. 6; 4; 9; 18

6–21. $\tilde{B}_0 = 8.35$ cm^{-1}, $\tilde{B}_1 = 8.12$ cm^{-1}, $\tilde{B}_e = 8.47$ cm^{-1}, $\tilde{\alpha}_e = 0.23$ cm^{-1}

6–22. $\tilde{B} = 10.40$ cm^{-1}, $\tilde{D} = 4.43 \times 10^{-4}$ cm^{-1}

6–23. $\tilde{B} = 1.9227$ cm^{-1}, $\tilde{D} = 6.387 \times 10^{-6}$ cm^{-1}

6–24. $0 \rightarrow 1$, 21.1847 cm^{-1}; $1 \rightarrow 2$, 42.3566 cm^{-1}; $2 \rightarrow 3$, 63.5030 cm^{-1}; $3 \rightarrow 4$,
84.6110 cm^{-1}

6–25. ratio $= \dfrac{\tilde{D}J^2(J + 1)^2}{\tilde{B}J(J + 1) - \tilde{D}J^2(J + 1)^2}$

For H^{35}Cl, ratio $= 5.64 \times 10^{-3}$; for ^{35}Cl^{35}Cl, ratio $= 8.38 \times 10^{-5}$

6–39. $\hat{L}_x = i\hbar \left(\sin\phi \dfrac{\partial}{\partial\theta} + \cot\theta \cos\phi \dfrac{\partial}{\partial\phi} \right)$, $\hat{L}_y = -i\hbar \left(\cos\phi \dfrac{\partial}{\partial\theta} - \cot\theta \sin\phi \dfrac{\partial}{\partial\phi} \right)$

6–40. (a) 0 (b) $2\hbar^2$ (c) $2\hbar^2$ (d) $2\hbar^2$

6–43. $I_{0 \rightarrow 1}$, $1/\sqrt{3}$; $I_{0 \rightarrow 2}$, $2/\sqrt{15}$; $I_{0 \rightarrow 1}/I_{0 \rightarrow 2} = \sqrt{5}/2$

6–44. 2 to 3 for H^{35}Cl and ≈ 30 for ^{127}I^{35}Cl at 300 K

MathChapter F

F–1. 5, 5, 5

F–2. $-5, -5$

F–3. 0, 0

F–4. $x^4 - 3x^2 = 0$; $x = 0, 0, \pm\sqrt{3}$

F–5. $x^4 - 4x^2 = 0$; $x = 0, 0, \pm 2$

F–6. $\cos^2\theta + \sin^2\theta = 1$

F–7. $\lambda = 1, \ 1 \pm \sqrt{2}$

F–8. (9/5, 1/5)

F–9. (1, 3, -4)

F–10. $x^4 - 4x^2 = 0$; $x = 0, 0, \pm 2$

Chapter 7

7–9. 0.762

7–10. 1.3 a_0; 2.7 a_0

7–18. $\langle r \rangle_{20} = 6\, a_0$; $\langle r \rangle_{21} = 5\, a_0$

7–28. n^2

7–29. 1312 kJ·mol^{-1}; 5248 kJ·mol^{-1}

7–30. 0.999 728

7–35. $\Delta E = (1.391 \times 10^{-22}\,\text{J})m_l$, $m_l = 0, \pm 1$; $E_{2p} - E_{1s} = 1.635 \times 10^{-18}$ J

7–38. $2p\ ^2P_{1/2}$ and $2p\ ^2P_{3/2}$ states: 0.3652 cm^{-1} (calc) vs. 0.3659 cm^{-1} (expt)
$3p\ ^2P_{1/2}$ and $3p\ ^2P_{3/2}$ states: 0.1082 cm^{-1} (calc) vs. 0.1084 cm^{-1} (expt)
$4p\ ^2P_{1/2}$ and $4p\ ^2P_{3/2}$ states: 0.0457 cm^{-1} (calc) vs. 0.00457 cm^{-1} (expt)

7–39. $3d\ ^2D_{3/2}$ and $3d\ ^2D_{5/2}$ states: 0.0361 cm^{-1} (calc) vs. 0.0362 cm^{-1} (expt)
$4d\ ^2D_{3/2}$ and $4d\ ^2D_{5/2}$ states: 0.0152 cm^{-1} (calc) vs. 0.0153 cm^{-1} (expt)
$4f\ ^2F_{5/2}$ and $4f\ ^2F_{7/2}$ states: 0.0076 cm^{-1} (calc) vs. 0.0077 cm^{-1} (expt)

7–40. $\alpha = 7.297\,353\,01 \times 10^{-3}$; $\alpha^{-1} = 137.035\,991$

7–41. $2p\ ^2P_{1/2}$ and $2p\ ^2P_{3/2}$ states: 5.844 cm^{-1} (calc) vs. 5.857 cm^{-1} (expt)
$3p\ ^2P_{1/2}$ and $3p\ ^2P_{3/2}$ states: 1.731 cm^{-1} (calc) vs. 1.736 cm^{-1} (expt)
$4p\ ^2P_{1/2}$ and $4p\ ^2P_{3/2}$ states: 0.7305 cm^{-1} (calc) vs. 0.7321 cm^{-1} (expt)

7–42. $3d\ ^2D_{3/2}$ and $3d\ ^2D_{5/2}$ states: 0.5772 cm^{-1} (calc) vs. 0.5784 cm^{-1} (expt)

$4d\ ^2D_{3/2}$ and $4d\ ^2D_{5/2}$ states: 0.2435 cm^{-1} (calc) vs. 0.2440 cm^{-1} (expt)

$4f\ ^2F_{5/2}$ and $4f\ ^2F_{7/2}$ states: 0.1217 cm^{-1} (calc) vs. 0.1220 cm^{-1} (expt)

7–44. 18 states; $m_j = \dfrac{9}{2}, \dfrac{7}{2}, \dfrac{5}{2}, \dfrac{3}{2}, \dfrac{1}{2}, -\dfrac{1}{2}, -\dfrac{3}{2}, -\dfrac{5}{2}, -\dfrac{7}{2}, -\dfrac{9}{2}$

$m_j = \dfrac{7}{2}, \dfrac{5}{2}, \dfrac{3}{2}, \dfrac{1}{2}, -\dfrac{1}{2}, -\dfrac{3}{2}, -\dfrac{5}{2}, -\dfrac{7}{2}$

7–45. $2(2l + 1)$

7–47. $E_{\text{upper}} = (82\ 258.9206 - 0.1556)$ cm^{-1}; $E_{\text{lower}} = (0 + 0.4669)$ cm^{-1};

$\Delta E = 82\ 258.2981$ cm^{-1}

7–48. $E_{\text{upper}} = (82\ 259.2865 - 0.3112)$ cm^{-1}; $E_{\text{lower}} = (0 - 0.4669)$ cm^{-1};

$\Delta E = 82\ 259.4422$ cm^{-1}

MathChapter G

G–1. $C = \begin{pmatrix} 5 & -3 & -2 \\ -11 & 4 & -6 \\ -3 & -1 & -1 \end{pmatrix}$ $\qquad D = \begin{pmatrix} -7 & 6 & 1 \\ 19 & -2 & 12 \\ 6 & 5 & 5 \end{pmatrix}$

G–4. If A, B, and C correspond to $\hat{L}_x$, $\hat{L}_y$, and $\hat{L}_z$, respectively, then the results are similar to the commutation relations of $\hat{L}_x$, $\hat{L}_y$, and $\hat{L}_z$.

G–7. $\det C_3 = 1$, $\operatorname{Tr} C_3 = -1$, $\det \sigma_v = -1$, $\operatorname{Tr} \sigma_v = 0$, $\det \sigma_v' = -1$, $\operatorname{Tr} \sigma_v' = 0$, $\det \sigma_v'' = -1$, $\operatorname{Tr} \sigma_v'' = 0$

G–8. They all are.

G–9. $A^{-1} = \begin{pmatrix} 0 & \sqrt{2} \\ \sqrt{2} & -1 \end{pmatrix}$; $\qquad B^{-1} = -\dfrac{1}{4}\begin{pmatrix} 1 & -2 & -1 \\ 1 & -6 & 3 \\ -2 & 4 & -2 \end{pmatrix}$

G–10. The prescription for finding the inverse of a matrix has you divide by its determinant.

G–11. $D = \begin{pmatrix} 2 & 0 & 0 \\ 0 & 1 & 0 \\ 0 & 0 & 0 \end{pmatrix}$ $\qquad$ D is diagonal, and its elements are the eigenvalues of A.

G–12. $x^T = \dfrac{1}{13}(24, -19, -8)$

Chapter 8

8–3. You get the exact result for the energy because the trial function is the exact ground-state wave function.

8–4. $E_{\min} = \dfrac{\hbar}{\sqrt{2}}\left(\dfrac{k}{\mu}\right)^{1/2} = 0.7071\,\hbar\left(\dfrac{k}{\mu}\right)^{1/2}$

8–5. $E_{min} = \dfrac{\sqrt{7}}{5}\hbar\left(\dfrac{k}{\mu}\right)^{1/2} = 0.5292\,\hbar\left(\dfrac{k}{\mu}\right)^{1/2}$

8–6. The value of c will come out to be equal to zero because $e^{-\alpha x^2/2}$ is the exact ground-state wave function of a harmonic oscillator.

8–7. $E_{min} = -\dfrac{3}{8}\dfrac{e^2}{4\pi\epsilon_0 a_0} = -\dfrac{3}{8}\left(\dfrac{m_e e^4}{16\pi^2\epsilon_0^2\hbar^2}\right)$

8–8. The value of c_2 will come out to be equal to zero because $e^{-\alpha r}$ is the exact ground-state wave function of a hydrogen atom. The value of α will be equal to $1/a_0$ and c_1 is just a normalization constant.

8–9. The variational energy is the exact energy because $e^{-\beta x^2}$ is the form of the exact ground state wave function of a harmonic oscillator.

8–10. $E_{min} = \dfrac{3}{2}\hbar\left(\dfrac{k}{\mu}\right)^{1/2}$ for $\phi(r) = e^{-\alpha r^2}$ (this is the exact energy);

$E_{min} = 3^{1/2}\hbar\left(\dfrac{k}{\mu}\right)^{1/2}$ for $\phi(r) = e^{-\alpha r}$

8–11. $E_{min} = \dfrac{(3)(6)^{1/3}}{8}c^{1/3}\dfrac{\hbar^{4/3}}{\mu^{2/3}}$

8–12. E_{min} is the lesser of $\dfrac{\hbar^2}{8ma^2} + V_0 a\left(\dfrac{1}{4} + \dfrac{1}{\pi^2}\right)$ or $\dfrac{\hbar^2}{2ma^2} + \dfrac{V_0 a}{4}$

8–13. $E_{min} = 0.6816\dfrac{\hbar^2}{ma^2}$ for $\alpha = 4$ and $E_{min} = 0.6219\dfrac{\hbar^2}{ma^2}$ for $\alpha = 12$

8–14. $E_{min} = 0.6381\dfrac{\hbar^2}{ma^2}$ for $\alpha = 4$ and $E_{min} = 0.8432\dfrac{\hbar^2}{ma^2}$ for $\alpha = 12$

8–15. $E_{min} = 5\hbar^2/ma^2$

8–16. $E_{min} = 7\hbar^2/ma^2$

8–19. The ground-state energy of a helium atom is the negative of the sum of the ionization energy of a helium atom $(198\,310.6672\text{ cm}^{-1})$ and that of a helium ion $(438\,908.8863\text{ cm}^{-1})$. The sum is equal to $637\,219.5535\text{ cm}^{-1}$ or $E = -2.903\,386\,E_h$.

8–21. $E_{min} = \dfrac{3}{2}h\nu + \dfrac{7\gamma_4}{32\alpha^2} - \dfrac{1}{2}\left(4h^2\nu^2 + \dfrac{3h\nu\gamma_4}{2\alpha^2} + \dfrac{11\,\gamma_4^2}{64\,\alpha^4}\right)^{1/2}$

8–25. $\phi(x) = \left(1.280\,59\sin\dfrac{\pi x}{a} + 0.600\,07\sin\dfrac{2\pi x}{a}\right)/a^{1/2}$

8–26. The off-diagonal matrix element $\langle 1 \mid x \mid 3 \rangle = 0$. We say that $\sin 3\pi x/a$ does not mix with $\sin \pi x/a$ in this case. The same is true for $\sin 5\pi x/a$, or the sine of any odd multiple of $\pi x/a$.

8–27. The off-diagonal terms are equal to $-32V_0/225\pi^2$ and $\varepsilon_{min} = 5.998\,62$ compared to $5.155\,95$ in Example 8–4. The off-diagonal terms are equal to $-48/1225\pi^2$, and $\varepsilon_{min} = 5.999\,95$ for $\sin 6\pi x/a$.

8–29. $\varepsilon_{min} = 5.15403$

8–35. $\psi_1(x) = \left(\dfrac{4\alpha^3}{\pi} \right)^{1/4} x e^{-\alpha x^2/2}$; $E^{(1)} = \dfrac{\gamma_3}{6} \langle 1 \mid x^3 \mid 1 \rangle + \dfrac{\gamma_4}{24} \langle 1 \mid x^4 \mid 1 \rangle = \dfrac{15\gamma_4}{96\alpha^2}$

8–36. $E^{(1)} = \dfrac{V_0 a}{2} \left(\dfrac{1}{2} + \dfrac{1 - \cos n\pi}{n^2\pi^2} \right)$

8–37. $\hat{H}^{(1)} = cx^4 - \dfrac{kx^2}{2}$; $E^{(1)} = \dfrac{3c}{4\alpha^2} - \dfrac{k}{4\alpha}$

8–38. $a = D\beta^2$; $b = -D\beta^3$; $c = 7D\beta^4/12$; $E^{(1)}(v = 0) = \dfrac{3c}{4\alpha^2}$; $E^{(1)}(v = 1) = \dfrac{15c}{4\alpha^2}$; $E^{(1)}(v = 2) = \dfrac{39c}{4\alpha^2}$

MathChapter H

H–1. $\lambda = 2, 0$; $\begin{pmatrix} 1 \\ 1 \end{pmatrix}$, $\begin{pmatrix} 1 \\ -1 \end{pmatrix}$

H–2. $\lambda = 3, -1$; $\begin{pmatrix} 1 \\ -1 \end{pmatrix}$, $\begin{pmatrix} 1 \\ 1 \end{pmatrix}$

H–3. $\lambda = \dfrac{1 + \sqrt{5}}{2}, 1, \dfrac{1 - \sqrt{5}}{2}$; $\begin{pmatrix} (1 + \sqrt{5})/2 \\ 0 \\ 1 \end{pmatrix}$, $\begin{pmatrix} 0 \\ 1 \\ 0 \end{pmatrix}$, $\begin{pmatrix} (1 - \sqrt{5})/2 \\ 0 \\ 1 \end{pmatrix}$

H–4. $\lambda = 2, 1, 0$; $\begin{pmatrix} 1 \\ 0 \\ -1 \end{pmatrix}$, $\begin{pmatrix} 0 \\ 1 \\ 0 \end{pmatrix}$, $\begin{pmatrix} 1 \\ 0 \\ 1 \end{pmatrix}$

H–7. $S = \begin{pmatrix} -\dfrac{1}{\sqrt{2}} & 0 & \dfrac{1}{\sqrt{2}} \\ 0 & 1 & 0 \\ \dfrac{1}{\sqrt{2}} & 0 & \dfrac{1}{\sqrt{2}} \end{pmatrix}$

H–8. $S = \begin{pmatrix} \dfrac{1}{\sqrt{2}} & \dfrac{1}{\sqrt{2}} \\ \dfrac{1}{\sqrt{2}} & -\dfrac{1}{\sqrt{2}} \end{pmatrix}$; $D = \begin{pmatrix} 2 & 0 \\ 0 & 0 \end{pmatrix}$

H–9. $S = \begin{pmatrix} \dfrac{1}{\sqrt{2}} & \dfrac{1}{\sqrt{2}} \\ \dfrac{1}{\sqrt{2}} & -\dfrac{1}{\sqrt{2}} \end{pmatrix}$; $\quad D = \begin{pmatrix} 3 & 0 \\ 0 & -1 \end{pmatrix}$

H–10. $S = \begin{pmatrix} \dfrac{1+\sqrt{5}}{(10+2\sqrt{5})^{1/2}} & 0 & \dfrac{1-\sqrt{5}}{(10-2\sqrt{5})^{1/2}} \\ 0 & \dfrac{1}{\sqrt{2}} & 0 \\ \dfrac{2^{1/2}}{(5+\sqrt{5})^{1/2}} & 0 & \dfrac{2^{1/2}}{(5-\sqrt{5})^{1/2}} \end{pmatrix}$;

$D = \begin{pmatrix} \dfrac{1+\sqrt{5}}{2} & 0 & 0 \\ 0 & 1 & 0 \\ 0 & 0 & \dfrac{1-\sqrt{5}}{2} \end{pmatrix}$

H–11. $S = \begin{pmatrix} \dfrac{1}{\sqrt{2}} & 0 & \dfrac{1}{\sqrt{2}} \\ 0 & 1 & 0 \\ -\dfrac{1}{\sqrt{2}} & 0 & \dfrac{1}{\sqrt{2}} \end{pmatrix}$; $\quad D = \begin{pmatrix} 2 & 0 & 0 \\ 0 & 1 & 0 \\ 0 & 0 & 0 \end{pmatrix}$

H–13. $\lambda = -2+a, \quad -1+a, \quad -1+a, \ 1+a, \ 1+a, \ 2+a$;

$\begin{pmatrix} -1 \\ 1 \\ -1 \\ 1 \\ -1 \\ 1 \end{pmatrix}, \begin{pmatrix} -1 \\ 0 \\ 1 \\ -1 \\ 0 \\ 1 \end{pmatrix}, \begin{pmatrix} -1 \\ 1 \\ 0 \\ -1 \\ 1 \\ 0 \end{pmatrix}, \begin{pmatrix} 1 \\ 0 \\ -1 \\ -1 \\ 0 \\ 1 \end{pmatrix}, \begin{pmatrix} -1 \\ -1 \\ 0 \\ 1 \\ 1 \\ 0 \end{pmatrix}, \begin{pmatrix} 1 \\ 1 \\ 1 \\ 1 \\ 1 \\ 1 \end{pmatrix}$

Chapter 9

9–18. It has the same form as the classical expression for the coulombic interaction of two charge distributions.

9–22. The two hydrogen atoms are isolated from each other.

9–23. The angular equation associated with the Schrödinger equation is the same in both cases. The radial equation, however, is different.

9–24. The radial equation associated with the Schrödinger equation is different in the two cases; the potential energy is purely coulombic in the case of a hydrogen atom, but is more complicated (see Figure 9.2 and Problem 9–20) for the Hartree–Fock approximation.

9–30. 0 and $\hbar/2$

9–31. 0 and 0

9–35.
$$\psi_{1s}(r) = 0.285\,107\,S_{1s}(r,\,5.7531) + 0.474\,813\,S_{1s}(r,\,3.7156)$$
$$-\,0.001\,620\,S_{3s}(r,\,9.9670) + 0.052\,852\,S_{3s}(r,\,3.7128)$$
$$+\,0.243\,499\,S_{2s}(4.4661) + 0.000\,106\,S_{2s}(r,\,1.2919)$$
$$-\,0.000\,032\,S_{2s}(0.8555)$$

where the S's are normalized Slater orbitals.

9–36. The orbital energy of the $2p$ orbital of a neon atom is given as $\varepsilon_{2p} = -0.850\,410\,E_h$, or $I_{2p} = 2.23\,MJ \cdot mol^{-1}$. Similarly, we have $\varepsilon_{2s} = -1.930\,391\,E_h$, $(I_{2s} = 5.07\,MJ \cdot mol^{-1})$, and $\varepsilon_{1s} = -32.772\,442\,E_h$, $(I_{1s} = 86.0\,MJ \cdot mol^{-1})$. For an argon atom, $\varepsilon_{3p} = -0.591\,016\,E_h$, $(I_{3p} = 1.55\,MJ \cdot mol^{-1})$, $\varepsilon_{3s} = -1.277\,352\,E_h$, $(I_{3s} = 3.35\,MJ \cdot mol^{-1})$, $\varepsilon_{2p} = -9.571\,464\,E_h$, $(I_{2p} = 25.13\,MJ \cdot mol^{-1})$, $\varepsilon_{2s} = -12.322\,152\,E_h$, $(I_{2s} = 32.35\,MJ \cdot mol^{-1})$, and $\varepsilon_{1s} = -118.610\,349\,E_h$, $(I_{1s} = 311.4\,MJ \cdot mol^{-1})$.

9–38. $E_{corr} = E_{exact} - E_{HF} = -2.903\,724\,375\,E_h + 2.861\,68\,E_h = -0.042\,04\,E_h$.
For Eckart, $\% = 100 \times (2.8757 - 2.861\,68)/0.042\,04 = 33.3\%$
For Hylleraas, $\% = 100 \times (2.903\,63 - 2.861\,68)/0.042\,04 = 99.8\%$

9–42. $^1S_0,\,^3S_1$

9–43. $^2P_{3/2},\,^2P_{1/2}$

9–44. $(1 \times 1)(^1S) + (3 \times 3)(^3P) + (1 \times 5)(^1D) = 15$

9–45. $45,\,(1 \times 1)(^1S) + (1 \times 5)(^1D) + (3 \times 3)(^3P) + (3 \times 7)(^3F) + (1 \times 9)(^1G) = 45$

9–46. $^1P_1,\,^3P_2,\,^3P_1,\,^3P_0;\,^3P_0$

9–47. $20;\,^1D_2,\,^3D_3,\,^3D_2,\,^3D_1;\,^3D_1$

9–48. $^1S_0,\,^1D_2,\,^1G_4,\,^3P_2,\,^3P_1,\,^3P_0,\,^3F_4,\,^3F_3,\,^3F_2;\,^3F_2$

9–49. $^2P_{3/2},\,^2P_{1/2},\,^2D_{5/2},\,^2D_{3/2},\,^4S_{3/2};\,^4S_{3/2}$

9–50. $[Ne]3s^2;\,^1S_0$

9–51. $^3F_2$ (see Problem 9–48)

9–52. $^1S_0$

9–53. $^3P_2$: five-fold degenerate, $^3P_1$: three-fold degenerate, $^3P_0$: singly degenerate, $^1P_1$: three-fold degenerate

9–54. $\Delta E = (34\,548.766 - 16\,956.172)\,cm^{-1} = 17\,592.594\,cm^{-1} \rightarrow 5684.210\,\text{Å}$; $5684.210\,\text{Å}/1.000\,29 = 5682.6\,\text{Å}$

9–55. The results are given in the table that follows:

Sharp series	Principal series	Diffuse series	Fundamental series
11 404 Å	5895.9 Å	8194.8 Å	18 459 Å
11 382 Å	5889.9 Å	8183.3 Å	
6160.7 Å	3302.9 Å	5688.2 Å	12 678 Å
6154.2 Å	3302.3 Å	5682.7 Å	
5153.6 Å	2853.0 Å	4982.9 Å	
5149.1 Å	2852.8 Å	4978.6 Å	

9–56. $E = -(43\,487.150 + 610\,079.0 + 987\,661.027)\,\text{cm}^{-1} = -1\,641\,227.177\,\text{cm}^{-1} = -7.4780\,E_h$ compared to the Hartree–Fock value, $-7.432\,727\,E_h$.

9–60. $4(\zeta_i^3\zeta_j^3)^{1/2}/(\zeta_i + \zeta_j)^2$

Chapter 10

10–4. The value of D_e represents the energy difference between the minimum of the energy curve and the dissociation limit. The dissociation limit is that of a ground-state hydrogen atom, whose electronic energy is $-1/2\,E_h$. Thus, $-1/2\,E_h - (-0.602\,64\,E_h) = 0.102\,64\,E_h$.

10–35. $E_{corr} = E_{exact} - E_{HF} = (-1.1744 + 1.1336)\,E_h = -0.0408\,E_h$;
$\% = 100 \times (-1.1479 + 1.1336)/0.0408 = 35.1\%$

10–41. If you set $c_1 = c_2$ right at the beginning of the calculation, then the calculation converges immediately because $c_1 = c_2$ by the symmetry of H_2.

10–42. This observation is due to the fact that if you know a wave function through first order in a perturbation, then you know the energy through third order. The values of the energy are more stable than those of the wave function.

10–48. $\nu_{D_2} = 9.332 \times 10^{13}\,\text{Hz}$; $D_0^{D_2} = 439.8\,\text{kJ·mol}^{-1}$. The experimental value of $D_0^{D_2}$ is $439.6\,\text{kJ·mol}^{-1}$.

Chapter 11

11–1. $268\,\text{pm} = 2.68\,\text{Å}$

11–2. ≈ 48

11–3. $159\,\text{pm} = 1.59\,\text{Å}$

11–4. $124\,\text{pm} = 1.24\,\text{Å}$

11–5. N_2 has a bond order of 3 and N_2^+ has a bond order of 5/2; O_2^+ has a bond order of 5/2 and O_2 has a bond order of 2.

11–6. F_2 has a bond order of 1; F_2^+ has a bond order of 3/2.

11–7. The relative bond orders are 3, 5/2, and 5/2.

11–8. The bond order of C_2 is 2; that of C_2^- is 5/2.

11–9. The magnitude of the force constants is in accord with the bond orders.

11–10. (L represents the filled $n = 2$ shell.)

$$\begin{aligned}
&\text{Na}_2 && KKLL(\sigma_g 3s)^2 \\
&\text{Mg}_2 && KKLL(\sigma_g 3s)^2(\sigma_u 3s)^2 \\
&\text{Al}_2 && KKLL(\sigma_g 3s)^2(\sigma_u 3s)^2(\pi_u 3p)^2 \\
&\text{Si}_2 && KKLL(\sigma_g 3s)^2(\sigma_u 3s)^2(\pi_u 3p)^4 \\
&\text{P}_2 && KKLL(\sigma_g 3s)^2(\sigma_u 3s)^2(\pi_u 3p)^4(\sigma_g 3p_z)^2 \\
&\text{S}_2 && KKLL(\sigma_g 3s)^2(\sigma_u 3s)^2(\pi_u 3p)^4(\sigma_g 3p_z)^2(\pi_g 3p)^2 \\
&\text{Cl}_2 && KKLL(\sigma_g 3s)^2(\sigma_u 3s)^2(\pi_u 3p)^4(\sigma_g 3p_z)^2(\pi_g 3p)^4 \\
&\text{Ar}_2 && KKLL(\sigma_g 3s)^2(\sigma_u 3s)^2(\pi_u 3p)^4(\sigma_g 3p_z)^2(\pi_g 3p)^4(\sigma_u 3p_z)^2
\end{aligned}$$

11–12. 6, Cr_2

11–13. The bond order of NO^+ is 3 and that of NO is 5/2.

11–14. The nonbonding orbitals $2p_{xO}$ and $2p_{yO}$.

11–16. CO has a bond order of 3 and is diamagnetic.

11–17. BF has a bond order of 1 and is paramagnetic.

11–18. The energies of the $1s$ electrons of different spin differ because of spin-orbit coupling.

11–24. See Table 11.3.

11–25. $E_{1s} = -1/2\, E_h$ and $E_{2s} = -1/8\, E_h$ and their sum is $-5/8\, E_h = -0.625\, E_h$.

11–26. $^3\Pi$ and $^1\Pi$

11–28. $E_\pm = (\alpha \pm \beta)/(1 \pm S)$; $\psi_\pm = (2p_{zA} + 2p_{zB})/\sqrt{2(1 \pm S)}$

11–37. $x^4 - 4x^2 = 0$; $x = 2, 0, 0, -2$; $E = \alpha + 2\beta, \alpha, \alpha, \alpha - 2\beta$. We predict that the ground state is a triplet. The two molecules have the same stability. Cyclobutadiene has no stabilization energy.

11–38. $x^4 - 3x^2 = 0$; $x = \sqrt{3}, 0, 0, -\sqrt{3}$; $E = \alpha + \sqrt{3}\beta, \alpha, \alpha, \alpha - \sqrt{3}\beta$; $E_\pi = 2(\alpha + \sqrt{3}\beta) + 2\alpha = 4\alpha + 2\sqrt{3}\beta$

11–39. $x^4 - 5x^2 + 4x = 0$; $x = 1, 0, -\frac{1}{2} \pm \frac{1}{2}\sqrt{17}$; $E = \alpha + 2.5616\beta, \alpha, \alpha - \beta, \alpha - 1.5616\beta$; $E_\pi = 2(\alpha + 2.5616\beta) + 2\alpha = 4\alpha + 5.1231\beta$

11–42. $E_\pi = 2(\alpha + 2.3028\beta) + 2(\alpha + 1.6180\beta) + 2(\alpha + 1.3028\beta) + 2(\alpha + \beta) + 2(\alpha + 0.6180\beta) = 10\,\alpha + 13.6832\beta$; $E_{deloc} = 3.6832\beta$

11–43. All the partial charges come out to be one and the various bond orders are $p_{23}^\pi = p_{18}^\pi = p_{67}^\pi = p_{45}^\pi = 0.724\,56$, $p_{78}^\pi = p_{34}^\pi = 0.603\,17$, $p_{19}^\pi = p_{29}^\pi = p_{6,10}^\pi = p_{5,10}^\pi = 0.554\,70$, and $p_{9,10}^\pi = 0.518\,23$. (See Problem 10–41 for the numbering convention used.)

11–44.

	$E_{H_3^+}$	E_{H_3}	$E_{H_3^-}$
trangular	$2\alpha + 4\beta$	$3\alpha + 3\beta$	$4\alpha + 2\beta$
linear	$2\alpha + 2\sqrt{2}\beta$	$3\alpha + 2\sqrt{2}\beta$	$4\alpha + 2\sqrt{2}\beta$

Therefore, we predict that H_3^+ is triangular; H_3 is triangular; and H_3^- is linear.

11–46.

	E_{deloc}	q_1	q_2	q_3	p_{12}^π	p_{23}^π
radical	$2\sqrt{2} - 2$	1	1	1	$\sqrt{2}$	$\sqrt{2}$
carbonium	$2\sqrt{2} - 2$	1/2	1	1/2	$\sqrt{2}$	$\sqrt{2}$
carbanion	$2\sqrt{2} - 2$	3/2	1	3/2	$\sqrt{2}$	$\sqrt{2}$

11–47. For hexatriene: $E_1 = \alpha + 1.802\beta$, $E_2 = \alpha + 1.247\beta$, $E_3 = \alpha + 0.4450\beta$, $E_4 = \alpha - 0.4450\beta$, $E_5 = \alpha - 1.247\beta$, $E_6 = \alpha - 1.802\beta$; $E_{deloc} = 0.9880\beta$; $E_{deloc} = 0.1647$ per carbon atom
For octatetraene: $E_1 = \alpha + 1.879\beta$, $E_2 = \alpha + 1.532\beta$, $E_3 = \alpha + \beta$, $E_4 = \alpha + 0.3473\beta$, $E_5 = \alpha - 0.3473\beta$, $E_6 = \alpha - \beta$, $E_7 = \alpha - 1.532\beta$, $E_8 = \alpha - 1.879\beta$; $E_{deloc} = 1.517\beta$; $E_{deloc} = 0.1896$ per carbon atom

11–49. $E_1 - E_N = 2\beta\left(\cos\dfrac{\pi}{N+1} - \cos\dfrac{N\pi}{N+1}\right) \to 2\beta(\cos 0 - \cos \pi) = 4\beta$

11–50. A conductor

11–52. $C = \begin{pmatrix} 0.3717 & 0.6015 & 0.6015 & 0.3717 \\ 0.6015 & 0.3717 & -0.3717 & -0.6015 \\ 0.6015 & -0.3717 & -0.3717 & 0.6015 \\ 0.3717 & -0.6015 & 0.6015 & -0.3717 \end{pmatrix}$

Chapter 12

12–3. Equation 12.7 gives the N spatial orbitals, each of which is doubly occupied by electrons of opposite spin.

12–5. $\sigma_u 3s(7.291\,69) = N[S_{3s}(r_A, 7.291\,69) - S_{3s}(r_B, 7.291\,69)]$, where $S_{3s}(r, 7.291\,69)$ is a Slater $3s$ orbital (Equation 9.15) with $\zeta = 7.291\,69$ and N is a normalization constant.

12–6. $\sigma_u 3d(1.690\ 03) = N[S_{3d_{z^2}}(r_A, 1.690\ 03) - S_{3d_{z^2}}(r_B, 1.690\ 03)]$, where $S_{3d_{z^2}}(r, 1.690\ 03)$ is a Slater $3d_{z^2}$ orbital (Equation 9.15) with $\zeta = 1.690\ 03$ and N is a normalization constant.

12–7. The equation is

$$2\sigma_u = -0.243\ 70\ [S_{1s}(r_A, 5.955\ 34) - S_{1s}(r_B, 5.955\ 34)]$$
$$- 0.000\ 0\ [S_{1s}(r_A, 10.658\ 79) - S_{1s}(r_B, 10.658\ 79)]$$
$$+ 0.364\ 37\ [S_{2s}(r_A, 1.570\ 44) - S_{2s}(r_B, 1.570\ 44)]$$
$$+ 0.547\ 02\ [S_{2s}(r_A, 2.489\ 65) - S_{2s}(r_B, 2.489\ 65)]$$
$$- 0.030\ 54\ [S_{3s}(r_A, 7.291\ 69) - S_{3s}(r_B, 7.291\ 69)]$$
$$- 0.413\ 55\ [S_{2p_z}(\mathbf{r}_A, 1.485\ 49) + S_{2p_z}(\mathbf{r}_B, 1.485\ 49)]$$
$$- 0.109\ 45\ [S_{2p_z}(\mathbf{r}_A, 3.499\ 90) + S_{2p_z}(\mathbf{r}_B, 3.499\ 90)]$$
$$- 0.035\ 53\ [S_{3d_{z^2}}(\mathbf{r}_A, 1.690\ 03) - S_{3d_{z^2}}(\mathbf{r}_B, 1.690\ 03)]$$

12–22. In a double-zeta (DZ) basis set, all atomic orbitals are represented by sums of two Slater orbitals with different values of ζ; in a double split-valence basis set, only the valence atomic orbitals are represented by sums of two Slater orbitals with different values of ζ.

12–23. In a triple-zeta (TZ) basis set, all atomic orbitals are represented by sums of three Slater orbitals with different values of ζ.

12–24. C_2H_6: $2 \times 9 + 6 \times 2 = 30$, two for each hydrogen atom and 9 for each carbon atom. C_3H_8: $3 \times 9 + 8 \times 2 = 43$

12–25. C_6H_6: $6 \times 9 + 6 \times 2 = 66$ (see previous problem). C_7H_8: $7 \times 9 + 8 \times 2 = 79$

12–26. C_2H_6: 30 contracted Gaussian functions (see Problem 12–24); $2 \times (6 + 4 \times 3 + 4 \times 1) + 6 \times (3 + 1) = 44 + 24 = 68$ primitive Gaussian functions
C_3H_8: 43 contracted Gaussian functions; $3 \times (6 + 4 \times 3 + 4 \times 1) + 8 \times (3 + 1) = 66 + 32 = 98$ primitive Gaussian functions

12–27. C_6H_6: $6 \times 9 + 6 \times 2 = 66$ contracted Gaussian functions (see Problem 11–25); $6 \times (6 + 4 \times 3 + 4 \times 1) + 6 \times (3 + 1) = 132 + 24 = 156$ primitive Gaussian functions
C_7H_8: $7 \times 9 + 8 \times 2 = 79$ contracted Gaussian functions; $7 \times (6 + 4 \times 3 + 4 \times 1) + 8 \times (3 + 1) = 154 + 24 = 178$ primitive Gaussian functions

12–29. $6 \times (1 + 4 + 4 + 6) + 6 \times 2 = 102$; $6 \times (1 + 4 + 4 + 6) + 6 \times (2 + 3) = 120$

12–30. For the $2s$ orbitals, $2s'(r) = -0.114\ 9610\ g_s(r, 11.626\ 3580) - 0.169\ 1180\ g_s(r, 2.716\ 2800) + 1.145\ 8520\ g_s(r, 0.772\ 2180)$; $2s''(r) = g_s(r, 0.212\ 0313)$

12–31. For the $2s$ orbital, $2s'(r) = 0.114\,660\,g_s(r,\,20.964\,200) +$ $0.919\,9990\,g_s(r,\,4.803\,3100) - 0.003\,030\,68\,g_s(r,\,1.459\,3300)$; $2s''(r) =$ $g_s(r,\,0.483\,4560)$; $2s'''(r) = g_s(r,\,0.145\,5850)$

12–32. $1s(r) = 0.001\,8311\,g_s(r,\,5484.671\,700) + 0.013\,9501\,g_s(r,\,825.234\,9500) +$ $0.068\,4451\,g_s(r,\,188.046\,9600) + 0.232\,7143\,g_s(r,\,52.964\,500) +$ $0.0470\,1930\,g_s(r,\,16.897\,5700) + 0.358\,5209\,g_s(r,\,5.799\,6353)$; $2s'(r) =$ $-0.110\,7775\,g_s(r,\,15.539\,6160) - 0.148\,0263\,g_s(r,\,3.599\,9336) +$ $1.130\,7670\,g_s(r,\,1.013\,7618)$; $2p'(r) = 0.070\,8743\,g_p(\mathbf{r},\,15.539\,6160) +$ $0.339\,7528\,g_p(\mathbf{r},\,3.599\,9336) + 0.727\,1586\,g_p(\mathbf{r},\,1.013\,7618)$; $2s''(r) =$ $g_s(r,\,0.270\,0058)$; $2p''(\mathbf{r}) = g_p(\mathbf{r},\,0.270\,0058)$; $d(\mathbf{r}) = g_d(\mathbf{r},\,0.800\,0000)$

12–33. $1s(r) = 0.154\,328\,97\,g_s(r,\,71.616\,8370) + 0.535\,328\,14\,g_s(r,\,13.045\,0960) +$ $0.444\,634\,54\,g_s(r,\,3.530\,5122)$; $2s(r) = -0.099\,967\,23\,g_s(r,\,2.941\,2494) +$ $0.399\,512\,83\,g_2(r,\,0.683\,4831) + 0.700\,115\,47\,g_s(r,\,0.222\,2899)$

12–43. $K = 100$, $N = 7$, and so the number of single and double excitations $=$
$$\frac{14!}{13!\,1!}\frac{186!}{185!\,1!} + \frac{14!}{12!\,2!}\frac{186!}{184!\,2!} = 2604 + 1\,565\,655 = 1\,568\,259$$

12–44. The $2N$th term in Equation 12.31 represents terms in which all the electrons occupy virtual orbitals.

12–45. See Equation 12.37.

12–46. CISDQ means a configuration interaction in which single, double, and quadruple excitations are used.

12–52. The number of maxima or the number of zeroes of $f(x)$ in an interval

12–53. Because part of the Hamiltonian operator (the interelectronic interaction terms) consists of two-electron operators (the $1/r_{ij}$).

12–56. 2.3 D compared to an experimental value of 1.85 D

12–57. 2.0 D compared to an experimental value of 1.47 D

Index

Illustration Credi

Chapter-Opening Photos

Chapter 1: Max Planck, reprinted with permission from AIP Emilio Segre Visual Archives, W. F. Meggers Collection.

Chapter 2: Louis de Broglie, reprinted with permission from AIP Emilio Segre Visual Archives, Brittle Books Collection.

Chapter 3: Erwin Schrödinger, reprinted with permission from AIP Emilio Segre Visual Archives.

Chapter 4: Niels Bohr and Werner Heisenberg, photograph by Paul Ehrenfest, Jr., reprinted with permission from AIP Emilio Segre Visual Archives, Weisskopf collection.

Chapter 5: E. Bright Wilson, Harvard University News Office, courtesy of the Caltech Archives.

Chapter 6: Pieter Zeeman, reprinted with permission from AIP Emilio Segre Visual Archives, W. F. Meggers Gallery of Nobel Laureates.

Chapter 7: George Uhlenbeck and Samuel Goudsmit, reprinted with permission from AIP Emilio Segre Visual Archives.

Chapter 8: Douglas Hartree, reprinted with permission from AIP Emilio Segre Visual Archives, Hartree Collection.

Vladimir Fock, reprinted with permission from AIP Emilio Segre Visual Archives, gift of Tatiana Yudovina.

Chapter 9: Charlotte Moore, courtesy of National Institute of Standards and Technology, Gaithersburg, MD.

Chapter 10: Robert Mulliken, reprinted with permission from AIP Emilio Segre Visual Archives.

Chapter 11: Linus Pauling, courtesy of the Caltech Archives.

Chapter 12: John Pople, courtesy of John Pople.

Clemens Roothaan, courtesy of Clemens Roothaan.

690

$$\sin(x \pm y) = \sin x \cos y \pm \cos x \sin y$$

$$\cos(x \pm y) = \cos x \cos y \mp \sin x \sin y$$

$$\sin x \sin y = \frac{1}{2}\cos(x - y) - \frac{1}{2}\cos(x + y)$$

$$\cos x \cos y = \frac{1}{2}\cos(x - y) + \frac{1}{2}\cos(x + y)$$

$$\sin x \cos y = \frac{1}{2}\sin(x + y) + \frac{1}{2}\sin(x - y)$$

$$e^{\pm ix} = \cos x \pm i \sin x$$

$$\cos x = \frac{e^{ix} + e^{-ix}}{2} \qquad \sin x = \frac{e^{ix} - e^{-ix}}{2i}$$

$$\cosh x = \frac{e^{x} + e^{-x}}{2} \qquad \sinh x = \frac{e^{x} - e^{-x}}{2}$$

$$f(x) = f(a) + f'(a)(x - a) + \frac{1}{2!}f''(a)(x - a)^2 + \frac{1}{3!}f'''(a)(x - a)^3 + \cdots$$

$$e^{x} = 1 + x + \frac{x^2}{2!} + \frac{x^3}{3!} + \frac{x^4}{4!} + \cdots$$

$$\cos x = 1 - \frac{x^2}{2!} + \frac{x^4}{4!} - \frac{x^6}{6!} + \cdots$$

$$\sin x = x - \frac{x^3}{3!} + \frac{x^5}{5!} - \frac{x^7}{7!} + \cdots$$

$$\ln(1 + x) = x - \frac{x^2}{2} + \frac{x^3}{3} - \frac{x^4}{4} + \cdots \qquad -1 < x \le 1$$

$$\frac{1}{1 - x} = 1 + x + x^2 + x^3 + x^4 + \cdots \qquad x^2 < 1$$

$$(1 \pm x)^n = 1 \pm nx + \frac{n(n - 1)}{2!}x^2 \pm \frac{n(n - 1)(n - 2)}{3!}x^3 + \cdots \qquad x^2 < 1$$

$$\int_0^\infty x^n e^{-ax}\, dx = \frac{n!}{a^{n+1}} \qquad (n \text{ positive integer})$$

$$\int_0^\infty e^{-ax^2}\, dx = \left(\frac{\pi}{4a}\right)^{1/2}$$

$$\int_0^\infty x^{2n} e^{-ax^2}\, dx = \frac{1 \cdot 3 \cdot 5 \cdots (2n - 1)}{2^{n+1}a^n}\left(\frac{\pi}{a}\right)^{1/2} \qquad (n \text{ positive integer})$$

$$\int_0^\infty x^{2n+1} e^{-ax^2}\, dx = \frac{n!}{2a^{n+1}} \qquad (n \text{ positive integer})$$